Long-term Limnological Research and Monitoring at Crater Lake, Oregon

AF181540

Developments in Hydrobiology 191

Series editor
K. Martens

Long-term Limnological Research and Monitoring at Crater Lake, Oregon

A benchmark study of a deep and exceptionally clear montane caldera lake

Edited by

G.L. Larson[1], R. Collier[2] and M.W. Buktenica[3]

[1]*USGS Forest and Rangeland Ecosystem Science Center,*
3200 SW Jefferson Way, Corvallis, OR 97331, USA
[2]*College of Oceanic and Atmospheric Sciences, Oregon State University,*
Ocean Admin Bldg 104, Corvallis, Ore 97331, USA
[3]*U. S. National Park Service, Crater Lake National Park,*
PO Box 7, Crater Lake, OR 97604, USA

Reprinted from Hydrobiologia, Volume 574 (2007)

Springer

A C.I.P. Catalogue record for this book is available from the Library of Congress.

ISBN-13: 978-94-007-9867-0

Published by Springer,
P.O. Box 17, 3300 AA Dordrecht, The Netherlands

Cite this publication as Hydrobiologia vol. 574 (2006).

Cover illustration: Crater Lake and Wizard Island, Crater Lake National Park, Oregon, USA.
© Sean Bagshaw | OutdoorExposurePhoto.com

Printed on acid-free paper

All Rights reserved
© 2007 Springer
Softcover re-print of the Hardcover 1st edition 2007

No part of this material protected by this copyright notice may be reproduced or utilized in any form or by any means, electronic or mechanical, including photocopying, recording or by any information storage and retrieval system, without written permission from the copyright owner.

TABLE OF CONTENTS

Long-term limnological research and monitoring at Crater Lake, Oregon
G.L. Larson, R. Collier, M.W. Buktenica 1–11

Subaqueous geology and a filling model for Crater Lake, Oregon
M. Nathenson, C.R. Bacon, D.W. Ramsey 13–27

Evaporation and the hydrologic budget of Crater Lake, Oregon
K.T. Redmond 29–46

Long-term observations of deepwater renewal in Crater Lake, Oregon
G.B. Crawford, R.W. Collier 47–68

Thermal, chemical, and optical properties of Crater Lake, Oregon
G.L. Larson, R.L. Hoffman, D.C. McIntire, M.W. Buktenica, S.F. Girdner 69–84

The extent and significance of petroleum hydrocarbon contamination in Crater Lake, Oregon
D.R. Oros, R.W. Collier, B.R.T. Simoneit 85–105

Ultraviolet radiation and bio-optics in Crater Lake, Oregon
B.R. Hargreaves, S.F. Girdner, M.W. Buktenica, R.W. Collier, E. Urbach, G.L. Larson 107–140

Predicting Secchi disk depth from average beam attenuation in a deep, ultra-clear lake
G.L. Larson, R.L. Hoffman, B.R. Hargreaves, R.W. Collier 141–148

Measurements of spectral optical properties and their relation to biogeochemical variables and processes in Crater Lake, Crater Lake National Park, OR
E.S. Boss, R. Collier, G. Larson, K. Fennel, W.S. Pegau 149–159

Bacterioplankton communities of Crater Lake, OR: dynamic changes with euphotic zone food web structure and stable deep water populations
E. Urbach, K.L. Vergin, G.L. Larson, S.J. Giovannoni 161–177

Seasonal and interannual variability in the taxonomic composition and production dynamics of phytoplankton assemblages in Crater Lake, Oregon
C.D. McIntire, G.L. Larson, R.E. Truitt 179–204

Nutrient limitation in Crater Lake, Oregon
A.W. Groeger 205–216

Distribution and abundance of zooplankton populations in Crater Lake, Oregon
G.L. Larson, C.D. McIntire, M.W. Buktenica, S.F. Girdner, R.E. Truitt 217–233

Variability of kokanee and rainbow trout food habits, distribution, and population dynamics, in an ultraoligotrophic lake with no manipulative management
M.W. Buktenica, S.F. Girdner, G.L. Larson, C.D. McIntire 235–264

Seasonal nutrient and plankton dynamics in a physical-biological model of Crater Lake
K. Fennel, R. Collier, G. Larson, G. Crawford, E. Boss 265–280

Hydrobiologia (2007) 574:1–11
DOI 10.1007/s10750-006-0342-6

CRATER LAKE, OREGON

Long-term limnological research and monitoring at Crater Lake, Oregon

Gary L. Larson · Robert Collier · Mark W. Buktenica

© Springer Science+Business Media B.V. 2007

Abstract Crater Lake is located in the caldera of Mount Mazama in Crater Lake National Park, Oregon. The lake has a surface area of about 53 km^2 at an elevation of 1882 m and a maximum depth of 594 m. Limited studies of this ultraoligotrophic lake conducted between 1896 and 1981, lead to a 10-year limnological study to evaluate any potential degradation of water quality. No long-term variations in water quality were observed that could be attributed to anthropogenic activity. Building on the success of this study, a permanent limnological program has been established with a long-term monitoring program to insure a reliable data base for use in the future. Of equal importance, this program serves as a research platform to develop and communicate to the public a better understanding of the coupled biological, physical, and geochemical processes in the lake and its surrounding environment. This special volume represents our current state of knowledge of the status of this pristine ecosystem including its special optical properties, algal nutrient limitations, pelagic bacteria, and models of the inter-relationships of thermal properties, nutrients, phytoplankton, deep-water mixing, and water budgets.

Keywords Crater Lake · Limnology · Lake monitoring · Water quality · Secchi disk

Guest Editors: Gary L. Larson, Robert Collier, and Mark W. Buktenica
Long-term Limnological Research and Monitoring at Crater Lake, Oregon

G. L. Larson (✉)
USGS Forest and Rangeland Ecosystem Science Center, 3200 SW Jefferson Way, Corvallis, OR 97331, USA
e-mail: gary_l._larson@usgs.gov

R. Collier
College of Oceanography and Atmospheric Sciences, Oregon State University, Corvallis, OR 97331, USA

M. W. Buktenica
Crater Lake National Park, Crater Lake, OR 97604, USA

Introduction

It is clearly accepted by the academic community and natural resource managers that many important questions in ecosystem science, environmental biology, and earth science can only be addressed with long-term data. For example, fundamental research questions focus on: populations or predator–prey systems that oscillate over decades; pools of materials such as nutrients in deep reservoirs that turn over very slowly; and responses to external forcing such as climatic cycles or global warming that manifest over long timescales. The observation of a system's response to these dynamics is a robust tool to develop and test hypotheses on earth and ecosystem function. Such research also informs complex environmental management decisions that may respond to, or effect long-term change. This special volume highlights and integrates research results built on the foundation of more than two decades of interdisciplinary research and monitoring of the physical, biological, and geochemical systems of Crater Lake, Oregon. This issue connects with past research, showcases the results of ongoing studies, and suggests future directions for research and monitoring. These research reports demonstrate numerous examples of how long-term data are essential to our basic understanding of the ecosystem and also highlight the integration of focused process studies with long-term monitoring programs that support the understanding of the system functions over all timescales.

Long-term monitoring of lakes typically begins in response to perceived or documented deterioration of water quality. The literature is replete with examples of lake systems that have been impacted by human activities, e.g., National Research Council, 1992; Sakamoto, 1997. These activities range from on site disturbances such as siltation, sewage, and introductions of invasive species, to allochthonous disturbances such as nutrient loading from atmospheric deposition and global climate change. However, it is often difficult to separate anthropogenic impacts from significant natural variations that occur over relatively short time scales. Magnuson (1990) illustrates the problem of detecting long-term change in his example of the annual duration of ice on Lake Mendota from 1855 to 1995, a period of 140 years. Although there is a statistical decrease of the annual duration of ice cover on Lake Mendota, the interannual variation is large and changes in ice duration can not be easily observed for any period of 10–20 consecutive years. Magnuson (1990) refers to these long-term changes as the invisible present. Conversely, more rapid and detectable changes in lakes due to human activities, e.g., Lake Tahoe (Goldman, 1988; Jassby et al. 2003) and Lake Washington (Edmondson, 1991), could be viewed as the visible present. But even in these systems, only long-term datasets, coupled with sustained research efforts, have identified the complex network of natural and anthropogenic processes controlling variation of the lake. Owing to the difficulties of separating normal variation from human related changes in most lakes, lake monitoring requires a long-term commitment and the use of carefully standardized procedures to insure that the data bases contain internally compatible information for evaluating long-term change.

Program History

Limnological studies of Crater Lake, located in Crater Lake National Park, Oregon, between 1896 and 1970, were sparse, fragmented, and sampling methods were often not comparable. Enough was known about the lake from these studies, however, to indicate that this very deep lake was transparent (Secchi disk clarity measurements in the high 30-m range), low in nitrogen, slightly alkaline, thermally stratified in summer, well oxygenated throughout the water column, low in primary production, and contained low densities of phytoplankton and zooplankton (Larson, 1996). However, limnological studies conducted from 1978 to 1981 suggested that the water quality of the lake might have deteriorated because the clarity of the lake had decreased as compared to observations made years earlier (D. Larson, 1972; 1983). A panel of limnologists reviewed the available limnological information in 1982 and concluded that the database was insufficient to determine if the lake had actually changed (Larson, 1996). One of the recommendations of the review panel was to monitor and document the basic characteristics of the lake. Congress passed Public Law 97–250 in the fall of 1982 that directed the Secretary of the Interior to conduct a 10-year study to examine the lake for possible deterioration of water quality.

The main goals of the limnological program at Crater Lake as developed by the National Park Service were to: (1) develop a reliable data base for use in the future; (2) develop a better understanding of physical, chemical, and biological characteristics, and processes of the lake; (3) establish a long-term monitoring program; and (4) investigate the possibility of long-term changes in the lake; and (5) if changes were found and related to human activities, identify the cause(s) and recommend ways of mitigating the change(s). The first field season was conducted during the summer of 1983, although the National Park Service sampled the lake in the summer of 1982.

The final report from the mandated 10-year program (Larson et al., 1993) concluded that the lake had not declined in water quality, including clarity, within the limits of the methods used and the period of time studied. Recognizing the complexity identified during the 10-year program, Superintendent Robert Benton emphasized the need for a long-term study to understand the dynamics of the lake. His vision was realized by 1994 when the National Park Service provided base funding to continue the lake monitoring program indefinitely.

Facilities and long-term sampling protocols of the monitoring program

Lake sampling from 1983 through 1985 was confined to summer and early fall because the boats had to be transported to the lake by helicopter or winched down the side of the caldera. Access to the lake in winter was not feasible owing to the extensive snow pack. In 1985, with the construction of a boathouse on Wizard Island that could be used to store the boats within the caldera, winter sampling of the lake became possible. The lake crew was transported to the island by helicopter for the first winter sampling in January 1986. Winter sampling since then, however, has been infrequent owing to budgetary and logistical constraints.

Two research boats were used during the early years of the study—a 5.5 m Boston Whaler and an 8-m pontoon vessel with a flat deck and an open center well. In 1995, a new research vessel, the *Neuston*, was helicoptered onto the lake. This 33-foot V-hulled aluminum vessel with twin-V8 engines was equipped with hydraulic winches, radar, depth finder, and GPS. The workboat, *Ouzel*, replaced the Whaler in 1997. In a typical year, the lake is accessible in summer and early fall using a trail down the northeast edge of the caldera to Cleetwood Cove. At the end of the field season (mid-September), the boats are returned to the boathouse and winterized.

Based on earlier studies and the information collected from studies of Crater Lake conducted from 1982 through 1984, a study plan and standard sampling procedures were finalized between 1985 and 1988. A set of working hypotheses and objectives were developed to relate the general physical, chemical, and biological lake characteristics, lake organization and structure, biological features, optical properties, paleolimnology, and evaluation of long-term system dynamics (summarized in Larson, 1996). Baseline sampling of the lake typically occurs in late June or early July, mid-July, mid-August, and early September. Sampling routinely includes temperature and conductivity profiles, optical properties (e.g., Secchi disk, photometer, spectraradiometer, *in situ* chlorophyll fluorometer and beam transmissometer), water quality and nutrients of the water column to a depth of 550 m, profiles of extracted chlorophyll (to 300 m), primary production (to 180 m), phytoplankton (to 200 m), zooplankton (to 200 m), and analyses of selected intra-caldera springs/streams. Fish sampling typically occurs between June and September and includes species, size, age, growth, and food habits. Annual population estimates of fish abundance and distribution using an echo sounder began in 1996. Recently, a data management program was designed and implemented by 2005. The database is actively maintained and web accessible as an essential resource in support of the ongoing monitoring and research programs.

Hydrology, physical limnology, and optical properties

Hydrology

Nathenson et al. (2007) reviewed the results of a 2000 bathymetric survey of Crater Lake in relation to the question of lake filling rate after the formation of the Mt. Mazama caldera about 7,800 years ago. Drowned beaches were apparent in the high resolution bathymetric data and resulted from lower equilibrium surface water levels during dry periods after the lake filled. One of the most significant observations was a large wave-cut platform 4 m below the current range of lake elevations, which suggested that the lake has been at its current level for "a very long time." They presented the hypothesis that significant seepage of lake water occurs through glacial debris sandwiched between two lava flows that extends down to a depth that is currently ~39 meters below the lake surface along the N.E. caldera wall. Various models for filling the lake were tested, starting from the initial collapse of the caldera. Based on the models and historic precipitation estimates relative to modern deposition, the lake reached its current level between 420 to 740 years after the formation of the caldera. The models used the drowned beaches and historical periods of low precipitation to characterize seepage of water out of the caldera. The preferred model parameterized seepage as proportional to water depth with about 45% of the total seepage responding to the total lake depth and 55% proportional to the lake elevation above the permeable debris layer. This model suggested that ~95% of modern precipitation equilibrates lake level at the observed wave-cut platform elevation. The filling model placed constraints on the post-caldera geologic history and suggested that most of the andesitic volcanism ceased within a few hundred years after the formation of the caldera. The geologic constraints refined in the new bathymetric survey allowed for an improved hydrologic model where a long record of lake level variation can be tied to climate variation for net precipitation (see also Redmond, 2007). Future refinements could focus on all terms in the hydrologic budget and further test the hypothesis that significant seepage occurs through the northeast wall of the caldera.

Redmond (2007) integrated the long-time series of direct lake level observations of Crater Lake with new data on lake physics, climate forcing (weather buoy) and hourly resolution lake-level and precipitation observations. Evaporation was directly estimated based on thermodynamic models and heat flux observations, and was lower than previously estimated as a residual of other terms. Seasonally, the greatest evaporation occurred in autumn and the least in spring. Long-term evaporation estimates were built on buoy proxy relationships. The standard deviation of the water-year precipitation was 4.6 times larger than the deviation in evaporation. Thus, lake level was controlled more by changes in precipitation than by changes of evaporation. There should be sufficient data over a range of lake elevations to begin testing the seepage model of Nathenson et al. (2007) with the evaporation-constrained water-loss model proposed by Redmond (2007).

Physical limnology

Based on a 14-year data set, Crawford and Collier (2007) discussed how the vertical distribution of heat, salt, and oxygen in Crater Lake were controlled by vertical mixing and convective renewal / ventilation of the deep hypolimnion. There were no statistically significant trends in the heat or salt content over that period, however. Deep convection events, which have the potential to transport significant amounts of nitrogen out of the hypolimnion into the euphotic zone, were observed in some years while no ventilation occurred in other years. Controls on this process were shown to be complex and may be related to the evolution of a cold thermocline during the first few weeks after reverse stratification is first established. Thermobaric instabilities and horizontal (edge) effects were evident. Co-variation of oxygen and nitrate with renewal events were observed, but significant complexity in this relationship emphasized that the impact of biogeochemical cycling of organic matter on the oxygen and especially, nitrogen fluxes, controlled these distributions as much as direct physical exchange. No simple, short term relationships were recorded between these observations of ventilation and stocks of phytoplankton (biomass or chl-*a*) or the export flux of organic matter (sediment trap fluxes).

Optical Properties

One of the most cited and treasured features of Crater Lake is its remarkable clarity and color. The lake's water clarity inspires both the public's appreciation of this resource and significant attention of the scientific community. In fact, many of the studies reported in this volume focus directly or indirectly on water clarity, from the time-series observations of Secchi disk depths to basic scientific research on the inherent optical properties of the lake's water, particles, and dissolved substances.

Optical measurements taken during the course of this study have documented the exceptional clarity of the lake. Secchi disk clarity readings varied from 18.1 m (July 1995) to 41.5 m (June 1997), with an average reading of about 30 m. Light transmission readings (25 cm beam transmissometer at 660 nm) throughout the water column between June and September typically range between 88.8 and 90.7%, relative to air (pure water at 91.3%). The depth of 1% of incident solar radiation during thermal stratification varies annually between 80 an 100 m. No long-term changes in the optical measurements were observed (see Larson et al., 2007b Hargreaves et al., 2007b and Boss et al., 2007).

As part of their discussion of the time series of water quality properties, Larson et al. (2007b) summarized the substantial body of Secchi disk clarity measurements. This included the large number of measurements taken since 1983 and all the related data collected over the previous century. No long-term trend was apparent in the time series; variations within a single year often encompassed much of the variation seen in the full dataset. Since the Secchi disk depth was always considerably shallower than the suspended particle and biomass maxima, its large variability in this deep, oligotrophic environment was sensitive to many secondary processes such as mixed layer depth, atmospheric deposition, and runoff from erosive precipitation events. In order to develop a better understanding of the short-term controls on clarity, as indexed by the Secchi disk depth, Larson et al. (2007a) discussed a strategy that could substantially increase the amount of clarity data that can be collected in lakes, especially high-mountain systems where conditions and access are often suboptimal. By cross-calibrating Secchi disk readings with in situ data collected from a beam transmissometer in Crater Lake, they discussed a protocol whereby a Secchi disk "proxy" can be developed based on the depth-integrated light transmission. This proxy can be applied to transmissometer data collected under any conditions (weather, sun, time of day) and should vastly improve the ability to study processes that control clarity. It is particularly exciting that the new instrumental data can be tied back to the long-term records of Secchi disk readings that were more commonly available and broadly recognized.

Boss et al. (2007) took advantage of state-of-the-art instrumentation to characterize the spectral inherent optical properties (IOP), in situ. The IOPs were controlled by water properties and were measured in situ over a relatively short optical path length (≤ 25 cm). Comparing profiles of spectral absorption, attenuation, and backscattering from June and September 2001, they discussed a variety of physical properties and biogeochemical processes that control the IOP. As noted by the authors, the rich dataset reflecting the chemical and physical properties of suspended particles was remarkable, especially when compared to the limited data that can be produced by collecting discrete samples for analysis in the laboratory. The beam attenuation profile, similar to that used by Larson et al. (2007a) to derive the Secchi disc proxy, was highly correlated with suspended particulate organic carbon (POC) and this offered a powerful new technique to study the change in suspended organic carbon concentration over time.

Boss et al. (2007) also found that chlorophyll absorption, relative to total particulate matter attenuation, increased rapidly between the lake surface and the depth of the chlorophyll maximum, demonstrating the extreme effect of photoacclimation by phytoplankton. Nearly 30% of the pigment at the chlorophyll maximum was contributed by submicron cells of unknown taxonomy. At pelagic stations, the particulate matter concentration maximum was dominated by particles with a low index of refraction, suggesting the dominance of organic matter biomass (phytoplankton and detritus). However, vertical variations in the derived particulate size distributions and indices of refraction demonstrated the considerable vertical structure of particle assemblages in the lake. Colored dissolved organic material was quite low in concentration and shows increases in concentrations at depths below the POC maxima by September, suggesting a seasonal accumulation of dissolved organic matter that was released by biological cycling within the ecosystem. Data collected near a intra-caldera wall stream after a rain storm emphasized the common observation of the dramatic increases in lithogenous particulate matter carried into the lake by these events. Hargreaves et al. (2007) discussed the impact of these events on the pelagic water column as the lithogenous particles settle through the water column over a time scale of days to weeks.

Continuing with an extensive review of the optical properties of Crater Lake, Hargreaves et al. (2007) focused on the long-wavelength ultraviolet radiation (UVR) by making use of downwelling spectral irradiance profiles and upwelling radiance collected by a variety of radiometers and photometers. These apparent optical properties (AOP) were dependent on the IOP of the water and the properties of the incident downwelling and upwelling light fields. Hargreaves et al. (2007) derived a series of "proxies" for the attenuation of UV radiation based on other bio-optical measurements, including the particle attenuation coefficient measurements by the beam transmissometer, the chlorophyll fluorescence, attenuation measurements by a photometer in the "blue" spectral range, and Secchi disk depths – which offered a long time series of UVR attenuation estimates. The authors observed a correlation between the atmospheric column inventory of ozone and the standing stock phytoplankton in the lake and discussed the hypothesis that UV-B irradiance, as controlled by stratospheric ozone variations, had an inhibitory effect on the plankton. They demonstrated that the upper waters of Crater Lake were among the clearest in the world, especially to UV radiation. Compared to surface irradiance, the intensity of UV-B averaged 27% at 20 meters depth and 5% at 40 meters depth—the deepest known penetration of UV into any aquatic system. There was no evidence of significant long-term changes in UV transparency or other spectral properties of the water since spectral irradiance data were first collected in the 1960s. Based on the developed proxies, no secular changes could be resolved since the first optical measurements were made in 1896.

General water quality, nutrients, and petroleum-derived contaminants

Water quality and nutrients

The public and scientific communities were concerned that the water quality of the lake had decreased in the late 1970s and early 1980s. The present work suggests that the lake remains in pristine conditions and, within the limits of the methods used, we are not able to identify any changes in general water quality responding to human activities over the period of observation, although potential impacts on the ecosystem from introduced fishes remain uncertain (Larson et al., 2007b). The lake is usually isothermal in winter except for a slight increase in temperature in the lower depths of the lake from ongoing geothermal inputs (see also Crawford and Collier, 2007). Mixing of the upper water column in winter by wind energy and convection extends to depths of 200–250 m. Although occurring infrequently with varying intensities each year in winter and spring, cold (< 4.0°C) down-welling surface waters sink to the lake bottom that result in the compensating exchange of deep lake nutrients (e.g., nitrate) into the upper strata of the lake and re-oxygenation of the deep lake waters (see Crawford and Collier, 2007).

Thermal stratification is established in summer and early fall. The seasonal epilimnion extends to a maximum depth of about 20 m and the maximum depth of the metalimnion is about 100 m; thus, most of the lake retains a cold hypolimnion during the period of thermal stratification. The year-round temperatures below 300 m remain near 3.5°C. Maximum near-surface water temperatures are typically in the high teens in August.

The concentration of dissolved oxygen is dominated by equilibration with the atmosphere in the winter, with a slight subsurface supersaturation (+5%, 70 m) within the productivity maximum.

Below the productivity maximum, there is a steady decrease (–10%) to the bottom due to respiration and microbial degradation of settling organic matter. The steady input of low-temperature hydrothermal fluids establish the relatively high alkalinity and conductivity of the lake, accounting for the bulk of the dissolved ions entering the lake. Phosphorus and particularly nitrogen are low in concentration. Total phosphorus and orthophosphate-P concentrations were fairly uniform throughout the water column. Nitrate-N was below detection limits in the upper 200 m of the lake and increased in concentration with depth. Reduced nitrogen (Kjeldahl-N and ammonia-N) were highest in concentration in the upper 200 m of the lake, but these are also near detection limits. No long-term changes were resolved for any water quality variable.

In comparison with other lakes, most of the physical and chemical conditions of Crater Lake were consistent with those that typify high-mountain ultraoligotrophic lakes. Weathering of the volcanic terrain provides slightly elevated phosphorus concentrations that are not utilized due to the extreme limitation by nitrogen. The alkalinity and conductivity of the lake are significantly higher than other high-elevation mountain lakes in the region with inflow of hydrothermal fluids. Thus, some of the physical and chemical properties of the lake are unique and do not conform to the range of characteristics typically associated with oligotrophic lakes (see Larson et al., 2007b).

Much has been written about the apparent decline of lake clarity during the period from the late 1960s and 1978–1982 (e.g., D. Larson, 1983; Dahm et al., 1990). One hypothesis for the decline was increased lake productivity from sewage contamination, principally nitrate-N, from intra-caldera Spring 42 located below a sewage system (septic tank and drain field) on the outer flank of the caldera rim to the west of the area called Rim Village. A panel of experts reviewed the water quality from the spring in 1986 and concluded that a tracer dye study was not feasible owing to the inaccessibility of the lake for 8–9 months each year. The panel recommended removal of the sewage septic/drain-field from the caldera rim. Superintendent Benton accepted this recommendation and included it in the overall plan to reconstruct Rim Village and Crater Lake Lodge. Several years were required to develop alternative plans to remove the sewage system from the Rim Village area and receive public comment about the various alternatives. In 1991 the sewage system servicing Rim Village was disconnected and sewage was piped to a park sewage facility located on the flanks of the mountain at an elevation of about 1921 m. The septic drain field was removed in 1992. No long-term changes have been observed in the water quality and nitrate-N concentrations in Spring 42 (or nearby springs) since the septic facilities were removed from the caldera rim. The Nitrate-N concentrations in Spring 42 vary inversely with the depth of the snow-pack on the caldera-rim in May. That is, in years with low snow-pack and the discharge of the spring is low (visual observation), the concentrations of nitrate-N are relatively high (see Larson et al., 2007b).

Petroleum-derived contaminants

Visitors to the park have the opportunity to explore the lake and its surroundings from many vantage points – including the classic rim drive and boat tour on the lake. The contamination of the lake by hydrocarbons and combustion products from the operation of vehicles within the park is a concern of the NPS and the public. Oros et al. (2007) conducted the first detailed investigation of the quantity and quality both natural and anthropogenic hydrocarbons in the lake. Petroleum-derived hydrocarbons can be detected at very low concentrations in slicks on the lake surface and in some lake sediments, particularly near the Cleetwood Cove area that is the focus of highest boat use and fuel transfers. Combustion products, including PAH, are found at very low to undetectable concentrations in bulk lake sediments. These concentrations are similar to those reported in some of the most remote environments on earth. Although these concentrations are orders of magnitude below threshold effect levels for sensitive aquatic organisms, these refractory compounds are present and show evidence of both direct input and atmospheric transport to the lake surface.

Biological Components

Bacterioplankton

Urbach et al. (2007) applied modern molecular techniques to investigate bacterioplankton populations in the lake over a 3-year period. Extending their initial survey (Urbach et al., 2001), this was the first study of bacterial phylogeny in the lake and correlates these distributions with the phytoplankton, zooplankton, and water column parameters regularly monitored. Two groups of bacteria and archaea consistently dominated in the deep waters, reflecting a stable community, whereas the euphotic zone populations varied in response to other elements of the annual food web structure. A statistical relationship was observed between the microbial community and *Daphnia* density that could suggest top-down control by consumers including kokanee, *Daphnia*, and bacterivorous protists. These results highlighted many poorly known linkages in the food web and identified numerous targets for future research.

Phytoplankton, primary production, and chlorophyll

Primary production and phytoplankton dynamics in the ecosystem was examined by combining detailed analyses of the phytoplankton community with the distribution of chlorophyll-a pigments, ^{14}C primary productivity (McIntire et al., 2007), and nutrient enrichment experiments (Groeger, 2007). The phytoplankton assemblage of Crater Lake to a depth of 200 m is low in density and diverse in species composition. During the period from 1984 to 2000, 134 taxa were identified to genus and 102 were identified to the species or variety level of classification. Overall, the dominant taxa by density or biovolume included *Nitzschia gracilis*, *Stephanodiscus hantzschii*, *Ankisrodesmus spiralis*, *Mougeotia parvula*, *Dinobryon sertularia*, *Tribonema affine*, *Asphanocapsa delicatissima*, *Synechocyctis* sp., *Gymnodinium inversum*, and *Peridinium inconspicuum*. The flora was uniformly distributed to the depth of mixing, about 200 m, when the lake was not thermally stratified. During thermal stratification, typically from late July to October, the epilimnion typically was dominated by *Nitzschia gracilis*. However, the assemblage in the epilimnion also included small-bodied species of cyanobacteria, e.g., *Aphanocapsa delicatissima* and *Synechocystis* sp. The hypolimnion often was dominated in density by *Aphanocapsa delicatissima*. The small cell size of *A. delicatissima*, however, contributed less to the total cell biovolume in the hypolimnion than did some of the larger bodied species, e.g., *Tribonema affine*, *Mougeotia parvula*, *Stephanodiscus hantzschii*, *Asterionella*, and *Synechocytis* sp. Although phytoplankton cell density and biovolume integrated from the lake surface to a depth of 200 m exhibited seasonal and inter-annual variations, there were no long-term changes observed.

The depth of maximum carbon–14 primary productivity during periods of maximum thermal stratification (August and September) usually occurred between 60 and 80 m below the lake surface, and chlorophyll maxima occurred between 120 and 140 m. During periods of strong thermal stratification, a near-surface productivity maximum also develops. During periods when the lake was not thermally stratified, maximum primary production occurred in the upper 40 m of the water column and chlorophyll concentration was essentially uniform to the depth of mixing, e.g., about 200 m below the lake surface. Temporal patterns of primary production and chlorophyll concentration when integrated to a depth of 200 m below the lake surface were difficult to interpret. Both variables were at relatively high levels in the late 1980s and early 1990s and then declined through 2000. Bioassays suggest that primary production is limited by nitrogen and iron (see Groeger, 2007). Hargreaves et al. (2007) also hypothesized that increasing UV-B radiation over this same period might inhibit the phytoplankton, as correlated with a drop in Chl-*a*, which may also impact primary productivity. Temporal changes in species composition, density, and biomass of the phytoplankton assemblages were not clearly linked with those of the zooplankton assemblage (see McIntire et al., 2007).

Zooplankton

The zooplankton assemblages in Crater Lake from 1988 to 2000 included 10 species of rotifer and 2 species of crustacean (excluding rare taxa). During thermal stratification in summer and fall vertical partitioning of the water column to a depth of 200 m was observed for the maximum densities of species with similar food habits and/or feeding mechanisms. The assemblages exhibited consistency in species richness, but there was considerable interannual variation in total density and biomass. Co-generic replacement was not observed. Between 1985 and 1990 the assemblages switched from periods of dominance of *Keratella cochlearis-Polyarthra* and *Daphnia*. *Asplanchna* was the dominant species in 1991 and 1992 and the zooplankton biomass collapsed in 1993 when *Asplanchna* decreased greatly in biomass. *Kellicottia* dominated the assemblages between 1994 and 1997, but then decreased in biomass as *Daphnia* dramatically increased in biomass and dominated the assemblages by 1998. When *Daphnia* biomass decreased by 2000, the biomass of *Keratella* increased and dominated the assemblage. During an initial study of the Crater Lake zooplankton assemblages, Karnaugh (1988) observed a switch from a dominance of *Keratella* in 1985 and 1986 to a dominance of Daphnia in 1987. Thus, the zooplankton assemblage typically shifts between the dominance of *Keratella cochlearis-Polyarthra* and *Daphnia*. The unexpected occurrence and dominance of *Asplanchna*, a predatory species, in 1991 and 1992, the collapse of the zooplankton biomass in 1993, and the high biomass of *Kellicottia* from 1994–1997 coincided with a change in the typical switch in dominance between *Keratella cochlearis-Polyarthra* and *Daphnia* (see Larson et al., 2007c). The value of long-term monitoring to document changes in the species composition and abundances of the zooplankton assemblages was clearly evident. Additional studies are needed, however, to: understand the population dynamics of the taxa; understand the reason(s) for the infrequent presence of *Daphnia* and *Asplanchna*; explain the absence of diaptomid and cylopoid copepods that are common in many Cascade Mountain Range lakes and ponds; and define the roles and significance of the zooplankton taxa in the pelagic food web.

Fish

This young, isolated caldera lake was naturally fishless, but during the period from 1888 and 1941 five salmonid species were introduced into the lake. During this study only two fish species were found – rainbow trout (*Oncorhynchus mykiss*) and kokanee (*O. nerka*). Rainbow trout inhabited the near-shore area of the lake and kokanee were primarily pelagic, although distribution and diel migration of kokanee varied over the duration of the study and appeared to be most closely associated with prey availability, maximization of bioenergetic efficiency, and fish density. Kokanee displayed temporal variation in population demographics that appeared to be related to density dependent growth and associated changes in reproduction and abundance. Rainbow trout were not as abundant as kokanee, and their population demographics were more consistent relative to kokanee. There was some evidence that the population dynamics of rainbow trout were influenced by the availability of kokanee as prey. Kokanee fed primarily on small-bodied prey from the mid-water column and the lake surface, where as rainbow trout fed primarily on large-bodied prey from the benthos and lake surface. Cladoceran zooplankton abundance may be regulated by kokanee predation, and conversely kokanee growth and reproductive success may be influenced by the availability of *Daphnia* as prey (see Buktenica et al., 2007).

Coupled biological-physical model

Fennel et al. (2007) developed a biological model of the ecosystem coupled to the lake physics that captured the processes thought to determine nutrient cycling and phytoplankton production at first-order with a simplified description of the food web. The vertical model (1D) simulated the seasonal evolution of two functional phytoplankton groups, total chlorophyll, and zooplankton, and was in good agreement with the observations of the monitoring program. The model did not explicitly include fish. Nitrogen was the biological "currency" of the model and is balanced across all of the dissolved and particulate pools, including organisms, detritus, and total dissolved nitrogen. The physical model, with 150 vertical layers, was forced at the surface by an annual climatology of wind stress, heat, and water fluxes to predict a surface boundary layer depth with a turbulent mixing profile below. The annual development of two different phytoplankton maxima and a separate, deep chlorophyll maximum followed from the differential ability of the two phytoplankton groups to increase their chlorophyll-to-biomass ratio in response to different light levels (photoacclimation). The model was used to investigate the simple hypothesis developed from a two-box vertical model where the mixing of "new" nitrogen upwards from the hypolimnion controlled, and was balanced by, the export flux of particulate organic nitrogen (PON) downwards from the epilimnion (Dymond et al., 1996). It was recognized that if a significant pool of active dissolved organic nitrogen was present and correctly accounted for in the total dissolved nitrogen pool, the modeled and observed vertical flux of PON agreed. This was a significant hypothesis resulting from the modeling experiment and will be tested in future field programs. Overall, this was a critical step in understanding how physical processes interact with biological cycling through the nitrogen cycle.

Summary

The primary goals of the limnological investigations of Crater Lake initiated in 1983 were to develop a reliable database for use in the future, improve our understanding of physical, chemical and biological characteristics and processes, and establish a long-term monitoring program to examine the limnological characteristics of the lake through time. The establishment of a long-term monitoring program in 1994 and the promotion of lake research and education under the 2005 Crater Lake National Park general management plan will help guarantee this vision. The research projects presented in this special volume of Hydrobiologia demonstrate significant progress in characterizing complex lake systems and identify many remaining issues and mysteries. In future years, park management will have a baseline of information from which to assess the status of the lake relative to natural processes and anthropogenic disturbance. The lake also represents a valuable natural laboratory for the study of aquatic ecosystem processes, particularly in light of its exceptional clarity, depth, nearly pristine condition, and level of environmental protection.

Acknowledgements The long-term limnological studies of Crater Lake have been an exceptional example of cooperation among individuals representing science and park management. We are grateful to the National Park Service (NPS) for the vision and support for the program. Since the retirement of Robert Benton, we are very grateful and appreciative for the support from Superintendents David Morris, Al Hendricks, and Charles Lundy. Dr. Douglas Larson made many important contributions to our understanding of the limnology of Crater Lake, starting with his Ph.D. research in 1968–69. He conducted additional studies of the lake between 1978 and 1981 and first focused public attention on the potential decline in lake clarity. Dr. Larson was hired part-time in 1982 by the NPS to initiate a 10 year limnological study of Crater Lake and continued his work at Crater Lake through the 1984 field season.

Dr. Jack Dymond initiated the collaboration of research scientists at the OSU College of Oceanic and Atmospheric Sciences with the NPS and USGS limnologists and earth scientists studying the lake in 1983. Dymond started what has become one of the longest records of settling particle fluxes in the world and initiated the first submersible dives to the lake floor in search of active hydrothermal systems. Although Jack died in 2003, the synergy between basic research and resource management that he inspired is well represented in the work presented in this special volume.

There have been many other significant contributors to the program. Resource management specialist Jon Jarvis, who is now the Pacific West Regional Director of NPS, provided exceptional leadership during the early years of the 10 year study. We extend our sincere thanks to park biologist Scott Girdner for his dedication and many significant contributions to the program. We also want to thank the following for their assistance, contributions, and dedication to the program: James Larson, Shirley Clark, Raymond Herrmann, Mac Brock, James Milestone, Mark Forbs, Dennis Fenn, Jerry McCrea, Edward Starkey, Manuel Nathensen, Charles Bacon, Greg Crawford, James McManus, Chris Moser, Kelly Redmond, John Salinas, Robert Hoffman, Emanuel Boss, Katia Fennel, Bruce Hargreaves, Allan Groeger, Scott Rumsey, Stanford Loeb, Stanley Dodson, W.T. Edmondson, Charles Goldman, John Reuter, C. David McIntire, Robert Truitt, Ruth Jacobs, Michael Hurley, John (Jack) Walstad, Peter Nelson, Stanley Gregory, Rainier Farmer, Dan Harland, Mike Conrady, Cameron Jones, Norman Anderson, and numerous interpreters, rangers, maintenance, and administrative staffs at the park.

References

Boss, E., R. Collier, G. Larson & W. Pegau, 2007. Measurements of spectral optical properties and their relation to biogeochemical variable and processes in Crater Lake, Crater Lake National Park, OR. Hydrobiologia 574: 149–159.

Buktenica, M., S. Girdner, G. Larson & C. D. McIntire, 2007. Variability of kokanee and rainbow trout food habits, distribution, and population dynamics, in an Ultraoligotrophic Lake with no manipulative management. Hydrobiologia 574: 235–264.

Crawford, G. B. & R. W. Collier, 2007. Long-term observations of deepwater renewal in Crater Lake, Oregon. Hydrobiologia 574: 47–68.

Dalm, C. N., D. W. Larson, N. S. Geiger & L. K. Herrera, 1990. Secchi disk, photometry, and phytoplankton data from Crater Lake: Long-term trends and relationships. In Drake E. T., G. L. Larson, J. Dymond & R. Collier (eds), Crater Lake: An Ecosystem Study. Pacific Division, American Association of the Advancement of Science, San Francisco, California:143–152.

Dymond, J., R. Collier, J. McManus & G. Larson, 1996. Unbalanced particle flux budgets in Crater Lake, Oregon: Implications for edge effects and sediment focusing in lakes. Limnology and Oceanography 41: 732–743.

Edmondson, W. T., 1991. The uses of ecology: Lake Washington and beyond. University of Washington Press, Seattle.

Fennel, K., R. Collier, G. Larson, G. Crawford & E. Boss, 2007. Seasonal nutrient and plankton dynamics in a physical-biological model of Crater Lake. Hydrobiologia 574: 265–280.

Goldman, C. R., 1988. Primary productivity, nutrients, and transparency during the early onset of eutrophication in ultra-oligotrophic Lake Tahoe, California-Nevada. Limnology and Oceanography 33: 1321–1333.

Groeger, A., 2007. Nutrient limitation in Crater Lake, Oregon. Hydrobiologia 574: 205–216.

Hargreaves, B., S. Girdner, M. Buktenica, R. Collier, E. Urbach & G. L. Larson, 2007. Ultraviolet radiation and bio-optics in Crater Lake, Oregon. Hydrobiologia 574: 107–140.

Jassby, A.D., J. E. Reuter & C. R. Goldman, 2003. Determining long-term water quality change in the presence of climate variability: Lake Tahoe (U.S.A.). Can. J. Fish. Aquat. Sci. 60: 1452–1461.

Karnaugh, E. N., 1988. Structure, abundance, and distribution of pelagic zooplankton in a deep, oligotrophic caldera lake. MS thesis, Department of Fisheries and Wildlife, Oregon State University, Corvallis, Oregon.

Larson, D. W., 1972. Temperature, transparency, and phytoplankton productivity in Crater Lake, Oregon. Limnology and Oceanography 17: 410–417.

Larson, D. W., 1983. The Crater Lake study: detection of possible optical deterioration of a rare, unusually deep caldera lake in Oregon, USA. Verhandlungen International Vereingung fur Theoretische und Angewandte Limnologie 22: 513–517.

Larson, G. L., 1996. Development of a 10-year limnological study of Crater Lake, Crater Lake National Park, Oregon, USA. Lake and Reservoir Management 12: 221–229.

Larson, G. L., R. Hoffman, B. Hargreaves & R. Collier, 2007a. Predicting Secchi disk depth from average beam attenuation in a deep, ultra-clear lake. Hydrobiologia 574: 141–148.

Larson, G. L., R. L. Hoffman, C. D. McIntire, M. W. Buktenica & S. F. Girdner, 2007b. Thermal, chemical, and optical properties of Crater Lake, Oregon, Hydrobiologia 574: 69–84.

Larson, G. L., C. D. McIntire & R. W. Jacobs (eds), 1993. Crater Lake Limnological Studies, Final Report. National Park Service Technical Report NPS/PNR/OSU/NRTR-93/03.

Larson, G. L., C. D. McIntire, M. Buktenica, S. Girdner & R. Truitt, 2007c. Distribution and abundance of zooplankton populations in Crater Lake, Oregon. Hydrobiologia 574: 217–233.

Magnuson, J. J., 1990. Long-term ecological research and the invisible present. BioScience 40:495–501.

McIntire, C. D., G. Larson & R. Truitt, 2007. Seasonal and interannual variability in the taxonomic composition and production dynamics of phytoplankton assemblages in Crater Lake, Oregon. Hydrobiologia 574: 179–204.

Nathenson, M., C. Bacon & D. Ramsey, 2007. Subaqueous geology and a filling model for Crater Lake, Oregon. Hydrobiologia 574: 13–27.

National Research Council, 1992. Restoration of aquatic ecosystems: science, technology, and public policy. National Academy of Sciences.

Oros, D, R. Collier & B. Simonet, 2007. The extent and significance of petroleum hydrocarbon contamination in Crater Lake, Oregon. Hydrobiologia 574: 85–105.

Redmond, K., 2007. Evaporation and the Hydrologic Budget of Crater Lake, Oregon. Hydrobiologia 574: 29–46.

Sakamoto, M., 1997. Eutrophication. In. Biswas, A. K (ed.), Water resources: Environmental Planning. Management, and Development, McGraw-Hill, New York, 297–379.

Urbach, E., K. Vergin, G. Larson & S. Giovannoni, 2007. Bacterioplankton communities of Crater Lake, OR: Dynamic changes with surface food web structure and stable deep water populations. Hydrobiologia 574: 161–177.

Urbach, E., K. L. Vergin, L. Young, A. Morse, G. L. Larson & S. J. Giovannoni, 2001. Unusual bacterioplankton community structure in ultra-oligotrophic Crater Lake. Limnology and Oceanography 46: 557–572.

Hydrobiologia (2007) 574:13–27
DOI 10.1007/s10750-006-0343-5

CRATER LAKE, OREGON

Subaqueous geology and a filling model for Crater Lake, Oregon

Manuel Nathenson · Charles R. Bacon · David W. Ramsey

© Springer Science+Business Media B.V. 2007

Abstract Results of a detailed bathymetric survey of Crater Lake conducted in 2000, combined with previous results of submersible and dredge sampling, form the basis for a geologic map of the lake floor and a model for the filling of Crater Lake with water. The most prominent landforms beneath the surface of Crater Lake are andesite volcanoes that were active as the lake was filling with water, following caldera collapse during the climactic eruption of Mount Mazama ~7700 cal. yr B.P. The Wizard Island volcano is the largest and probably was active longest, ceasing eruptions when the lake was ~80 m lower than present. East of Wizard Island is the central platform volcano and related lava flow fields on the caldera floor. Merriam Cone is a symmetrical andesitic volcano that apparently was constructed subaqueously during the same period as the Wizard Island and central platform volcanoes. The youngest post-caldera volcanic feature is a small rhyodacite dome on the east flank of the Wizard Island edifice that dates from ~4800 cal. yr B.P. The bathymetry also yields information on bedrock outcrops and talus/debris slopes of the caldera walls. Gravity flows transport sediment from wall sources to the deep basins of the lake. Several debris-avalanche deposits, containing blocks up to ~280 m long, are present on the caldera floor and occur below major embayments in the caldera walls. Geothermal phenomena on the lake floor are bacterial mats, pools of solute-rich warm water, and fossil subaqueous hot spring deposits. Lake level is maintained by a balance between precipitation and inflow versus evaporation and leakage. High-resolution bathymetry reveals a series of up to nine drowned beaches in the upper ~30 m of the lake that we propose reflect stillstands subsequent to filling of Crater Lake. A prominent wave-cut platform between 4 m depth and present lake level that commonly is up to 40 m wide suggests that the surface of Crater Lake has been at this elevation for a very long time. Lake level apparently is limited by leakage through a permeable layer in the northeast caldera wall. The deepest drowned beach approximately corresponds to the base of the permeable layer. Among a group of lake filling models, our preferred one is constrained by the drowned beaches, the permeable layer in the caldera wall, and paleoclimatic data. We used a precipitation rate 70% of modern as a limiting case. Satisfactory models require leakage to be proportional to elevation and the best fit model has a linear combination of 45% leakage

Guest Editors: Gary L. Larson, Robert Collier, and Mark W. Buktenica
Long-term Limnological Research and Monitoring at Crater Lake, Oregon

M. Nathenson (✉) · C. R. Bacon · D. W. Ramsey
U.S. Geological Survey, 345 Middlefield Road, Menlo Park, CA 94025, USA
e-mail: mnathnsn@usgs.gov

proportional to elevation and 55% of leakage proportional to elevation above the base of the permeable layer. At modern precipitation rates, the lake would have taken 420 yr to fill, or a maximum of 740 yr if precipitation was 70% of the modern value. The filling model provides a chronology for prehistoric passage zones on postcaldera volcanoes that ceased erupting before the lake was filled.

Keywords Crater Lake · Geology · Filling model · Caldera

Introduction

Crater Lake occupies a caldera that is a collapse depression formed during the climactic eruption of Mount Mazama volcano ~7700 cal. yr B.P. This eruption vented ~50 km^3 (dense-rock equivalent) of magma as pumice and ash, and probably lasted no more than a few days. The geology of the lake floor is known from bathymetric surveys, sampling by dredging and coring, and observations by remotely operated vehicle (ROV) and manned submersible (Williams 1961; Nelson et al., 1988, 1994; Collier et al., 1991; Bacon et al., 2002). Dramatic improvement in understanding of lake floor morphology and geology resulted from the 2000 multibeam echo sounding survey (Gardner et al., 2001; Gardner and Dartnell 2001) in which >16 million geographically referenced (to ± 1 m) depth measurements accurate to 0.2% were made virtually to the shoreline. The maximum depth of the lake was found to be 594.0 m relative to the shoreline at 6178 feet (1883.05 m) elevation that appears on U.S. Geological Survey 1:24,000-scale topographic maps. Calculated lake area is 53.4 km^2 and volume is 18.7 km^3. Visualizations made from the 2000 data using geographic information systems allowed detailed interpretation of lake floor morphology in terms of volcanic and sedimentary processes (Bacon et al., 2002). In the present paper, we summarize findings of the interpretive study and highlight aspects we consider pertinent to other papers in this special issue.

Determining the ages of lake floor features is a key element in interpreting the geologic history of Crater Lake. Evidence from the 2000 bathymetric survey provides the basis for a quantitative model for filling the lake with water that serves as a chronometer for fossil shoreline features that track growth of volcanoes on the caldera floor. Estimating the time to fill Crater Lake after the climactic eruption presents some difficulties. The steady-state water balance is known reasonably well (Phillips 1968; Redmond 1990; Nathenson 1992); however, the depth dependence of leakage has been unknown. Phillips (1968) estimated 500–1000 years to fill the lake, assuming that much of the leakage leaves the lake 150 m or more below current lake level. Nelson et al., (1994) provided a minimum estimate of 300 years assuming 50% evaporation but ignoring leakage. Hoffman (1999) calculated a filling time of about 850 years with seepage a varying fraction of the lake volume. He also looked at the impact of reduced precipitation on the filling time from a 1000-year dry period taking place 500 years after the lake started filling and found long times (2000–3000 years) for the lake to fill. The calculation of filling time in this paper is constrained by the appearance of drowned beaches in the upper 30 m of the lake caused by reduced precipitation lasting for 10's to 100's of years. At modern precipitation rates, our calculations show that the lake would have taken 420 yr to fill, or a maximum of 740 yr if precipitation was 70% of the modern value.

Lake floor geology

Crater Lake floor morphology is easily visualized with shaded-relief perspective views. Several appear in Gardner et al., (2001), Gardner and Dartnell (2001), and Bacon et al., (2002), along with shaded relief and acoustic backscatter maps. Figure 1 is a panoramic view of the lake as photographed from the west rim of the caldera (top) with a matching perspective view (bottom) with the geology draped over the bathymetry (Ramsey et al., 2003). A geologic map with bathymetric contours on a shaded-relief base appears in Fig. 2. We describe various aspects of the geology of the lake floor with reference to Figs. 1 and 2.

Fig. 1 Panoramic photograph (by Peter Dartnell) and matching perspective view (after Ramsey et al., 2003). Geologic map has been draped over bathymetry. See Fig. 2 for key to geologic units

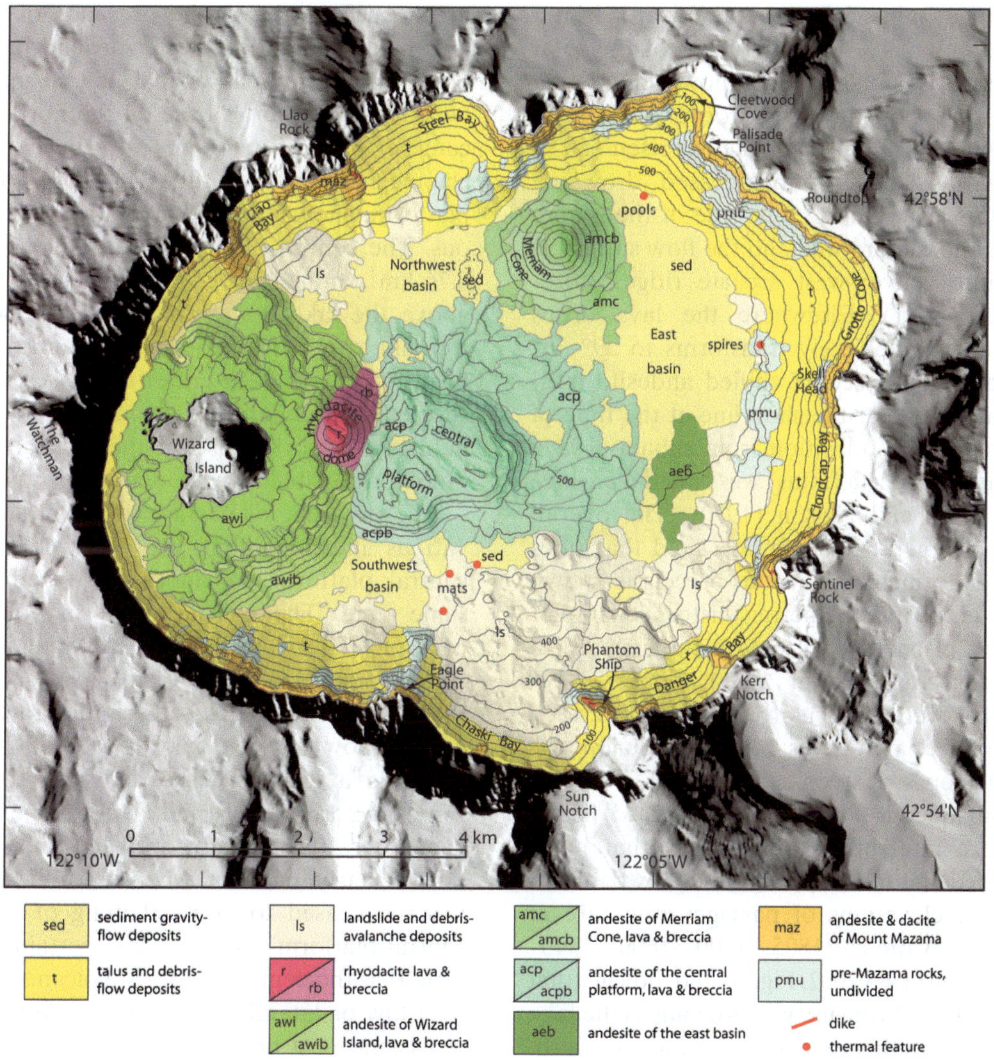

Fig. 2 Geologic map on shaded-relief bathymetric base (after Bacon et al., 2002, Fig. 7). Bathymetric contour interval 50 m relative to lake surface elevation of 6178 feet (1883.05 m)

Volcanic features

The volcanoes on the floor of Crater Lake post-date collapse of the caldera ~7700 cal. yr B.P. Virtually all postcaldera volcanic rocks are andesite of similar or related composition (Nelson et al., 1994). The sole exception is a rhyodacite dome, the youngest eruptive unit, that amounts to ~2% of the total postcaldera erupted volume of 4.1 km^3. Detailed descriptions of each of the postcaldera volcanoes are summarized here, and a table of their volumes and footprint areas is given in Bacon et al., (2002).

The Wizard Island volcano rises 750 m above the lake floor and has a volume of at least 2.6 km^3. The subaerial part of Wizard Island represents only 2.4% of the volume of the total edifice. Blocky subaerial lava flows emanating from the base of the cone form a platform extending up to 1.5 km. The lake has risen ~80 m since Wizard Island eruptions ceased, drowning much of this lava. On the west side of the island, lava abuts the caldera wall and the flow surface is buckled into convex-west arcuate ridges (foreground of Fig. 1). Elsewhere the lava drops abruptly and immediately transforms to 29°–36° sloping talus composed of chilled andesite fragments that formed at the shoreline at the time of lava effusion. The transition from subaerial lava to subaqueous breccia (talus) was called a *passage zone* by Jones and Nelson (1970). The overall structure of a gently-sloping lava field that rides over earlier-formed subaqueous breccia is known as a lava delta, after numerous historic examples and studies of ancient volcanic rocks exposed in cross section (e.g., Fuller 1931). Locally, lava streams that were particularly vigorous cascaded down the breccia slope as coherent subaqueous flows with slopes as steep as 45°.

The subaqueous flanks of the Wizard Island volcano are interrupted by benches that mark earlier lava deltas formed during the growth of this edifice. The elevations of passage zones are at consistent elevations (1805, 1700, 1600, 1540–1560 m; Bacon et al., 2002, Fig. 4) and indicate that the Wizard Island volcano was growing as the lake filled. The lake filling model presented below provides a chronology for these features. Preservation of successive passage zones as benches that would otherwise have been overridden by younger lava implies changes in source-vent location or decrease in eruption rates in comparison to lake-level rise. The breccia slopes are ready sources of sediment for supply to the deep basins by gravity flows.

The andesitic central platform volcano extends east from the Wizard Island edifice. The central platform, as its name implies, has a comparatively flat upper surface. Its flanks slope 30°–37° from the top lava surface at a ~1600-m elevation passage zone and others are present at ~1510 and ~1540–1560 m. The flanks appear to be breccia that is draped in its upper reaches by more coherent lava. The central platform differs from the Wizard Island volcano in the relief on its upper surface and in the presence of extensive lava fields in the deep water around much of its base. The sinuous flows that extend up to 2 km beyond the volcano's base on the north and east can be traced up slope into prominent lava channels or collapsed tubes on top of the central platform, showing that these flows are subaqueous. The central platform itself has a volume of 0.76 km^3, and the deep flow fields are 0.30 km^3 above the lake floor. The main east–southeast-trending channel on top of the central platform heads in an apparent crater at its west end. Two shorter but similar lava channels lead to a flow field at the north base of the central platform. Another possible lava channel leads to the south.

The presence of passage zones at similar elevations indicates that the Wizard Island and central platform volcanoes were active at the same time. The central platform volcano was constructed of lava deltas until late in its life, when formation of discrete lava channels or tubes, or perhaps increased eruption rate, allowed lava to flow into the lake without fragmenting. These lava streams descended over the breccia slopes to form fan-shaped lava flow fields on the lake floor. Its partially drowned subaerial lava flows clearly indicate when the Wizard Island volcano ceased to erupt. Timing of the end of central platform activity is less certain. It is possible that the central platform volcano was fed by overflow or lateral dike transport from the Wizard Island vent or conduit.

Merriam Cone is an andesitic volcano named by Williams (1961) after J. C. Merriam. Rising

430 m above the east basin (151 m depth below lake level), Merriam Cone has a volume of 0.34 km^3. Most of the surface of this symmetrical cone slopes 30°–32°, decreasing to ~20° near its base, and probably is composed of breccia. Radial ridges, some with lobate downhill terminations, and buttresses near the cone's base suggest lava flows. The uppermost fourth of the craterless cone appears to be lava. Surface features visible in ROV video and the character of dredge samples suggest that Merriam Cone was constructed by subaqueous eruptions, the last of which occurred in shallow water. Dredge samples are compositionally similar to the most chemically differentiated Wizard Island lava. The compositional similarity and the apparently shallow-water final eruptions suggest that Merriam Cone was active during the later stages of growth of the Wizard Island volcano.

Approximate morphologies of the central platform and Merriam Cone were known from a 1959 echo sounding survey (Byrne 1965). In the 2000 survey, a fourth probable andesitic lava field (aeb, Fig. 2) was discovered protruding into the south end of the east basin. Although there are no samples of material from this feature, its outline, slopes, and backscatter suggest sediment-covered lava flows from a vent between Kerr Notch and Sentinel Rock, now buried by debris-avalanche deposits. The unit has a minimum volume of 0.032 km^3 and may be the oldest exposed post-caldera volcanic rock.

On the east flank of Wizard Island is a rhyodacite lava dome that reaches 1854 m elevation (29 m depth) and has a volume of 0.074 km^3. Capped by relatively smooth lava, the dome has sides that slope 32°–34° and are composed of talus or breccia. Much, if not all, of the dome apparently was emplaced subaqueously. Correlative ash found in a core recovered from the surface of the central platform is bracketed by radiocarbon dates that yield a calendrical age for the dome of 4830 +460/–410 yr B.P. (Nelson et al., 1994; Bacon et al., 2002).

Caldera walls

The submerged caldera walls continue the same types of bedrock outcrops and debris slopes found above water (Fig. 1). When the ~5 × 6 km central block subsided during the climactic eruption, the unsupported walls failed by sliding, resulting in the scalloped outlines of the lake shore and caldera rim where embayments are separated by bedrock promontories. Subaerial bedrock outcrops are composed of sets of lava flows and pyroclastic deposits from vents that were active during comparatively short periods in the eruptive history of Mount Mazama. Contacts between these units commonly are expressed by benches on the caldera walls. Similar features are evident in the 2000 bathymetry, some of which have been observed and sampled by manned submersible. Many of these benches are identified as parts of the Mount Mazama edifice but the deeper ones do not have obvious counterparts and are shown as undivided pre-Mazama volcanic and intrusive rocks on the geologic map (Fig. 2). The steep triangular-faceted spur below Eagle Point (Figs. 1 and 2) may be a remnant of the footwall of a ring fault along which the central block of the caldera subsided. Observations with the submersible indicate that steep faces are bare rock, but ledges and gentle slopes are dusted with sediment.

Much of the submerged caldera wall is buried under fragmental debris aprons shown as talus on the geologic map (Fig. 2). The higher-elevation sublacustrine debris commonly is continuous with subaerial talus and has slopes of 29°–35° that systematically decrease to lower values with increasing depth in the lake. Talus cones appear to grade downslope into debris-flow or landslide material that is buried at its distal end by sediment of the deep basins. A few coarse rockfalls have been identified at the bases of steep cliffs that lack talus sources above, as off Phantom Ship.

A glaciated lava flow has been recognized in the caldera wall near Roundtop, (Bacon et al., 2002) and its extension below lake level, as interpreted from bathymetry and back-scatter imagery, is shown on the geologic map (Fig. 2) as part of the andesite and dacite of Mount Mazama continuing west towards Palisade Point. The glaciated upper surface of the lava flow appears at elevations ranging from 1838 m to 1859 m, 24–45 m below lake level. Unconsolidated, fragmental glacial deposits rest on this lava flow below the andesite of Roundtop (Bacon et al., 2002; Fig. 14), and the fragmental deposits are likely to provide a permeable pathway for leakage from Crater Lake.

Landslides

Hummocky topography and scattered large blocks on the lake floor are the expression of landslide and debris-avalanche deposits below many of the embayments in the caldera wall. The latter resemble the much larger submarine landslide deposits from volcanoes in the oceans (e.g., Moore et al., 1989). Debris-avalanche deposits are best developed below embayments in the south and southeast caldera walls in hydrothermally-altered rocks of the older part of Mount Mazama. The largest, the Chaski Bay debris-avalanche deposit, contains blocks up to ~280 m long that have traveled as much as 2–3 km from its source on the south rim to apparently run up on lava flows at the south flank of the central platform. Secondary slides have formed by mobilization of the lower reaches of the larger debris-avalanche deposits, leaving source-scar headwalls (e.g., north of Phantom Ship). The debris avalanches and secondary slides may have been triggered by earthquake shaking. Debris-avalanche deposits probably are abundant in the caldera fill below the lake floor, interlayered with pyroclastic material that accumulated during the climactic, caldera-forming eruption of Mount Mazama.

Apparent landslide deposits that lack coarse blocks on their surfaces occur west of Eagle Point and below Llao Bay, Steel Bay, and Grotto Cove. Each is essentially continuous with the talus/debris apron above it. These slide deposits appear to have formed by superposition of flow lobes and they have been mapped separately from the more uniform talus/debris surface above them. They slope from a maximum of 25° to a 3°–7° distal runout surface.

Sediment ponds and basins

The east, northwest, and southwest basins contain relatively fine-grained sediment transported from caldera wall sources by sediment gravity flows (Nelson et al., 1986). These basins, and numerous smaller sediment-filled depressions on and between lava flows and landslide deposits, have smooth, nearly flat surfaces. On the basis of seismic-reflection profiling, Nelson et al., (1986) suggested a maximum sediment thickness of 75 m in the east basin and < 50 m in the southwest and northwest basins. Sediment flows toward the basins along channels between postcaldera volcanoes and the caldera wall, ponding in local depressions along its path to the basins. Sediment ponds also occur on lava flows, on debris-avalanche deposits, below secondary slide headwall scarps, and where debris avalanches terminate against central platform lavas. The northwest basin and many of the smaller sediment ponds are dammed by low ridges (sills), and sediment apparently moves through this system to ultimate sinks in the east and southwest basins.

Geothermal phenomena

The recency of the caldera-forming eruption suggests that the lake floor should have features that reflect loss of residual heat from the magmatic and hydrothermal system beneath Mount Mazama (Bacon and Nathenson, 1996). Williams and Von Herzen (1983) found high convective heat flows in the south and northeast parts of the lake floor using oceanographic measurement techniques. Submersible and ROV investigations of these areas revealed unequivocal evidence of modern hydrothermal circulation (Dymond and Collier, 1989; Collier et al., 1991; Wheat et al., 1998). Bacterial mats associated with venting of warm water are present locally in the southern area near the northwest margin of the Chaski Bay debris-avalanche deposit (Fig. 2, "mats"). Relatively warm and solute-laden water forms pools below Cleetwood Cove (Fig. 2, "pools"). Silica spires ~10–12 m high below Skell Head are fossil subaqueous thermal-spring deposits (Fig. 2, "spires"). None of these features are large enough to have been resolved in the 2000 bathymetric survey. No doubt, there are many more thermal features undiscovered on the lake floor.

Filling of Crater Lake

The detailed bathymetry produced by the 2000 multibeam survey provides evidence of fossil shorelines that record periods of low lake level from reduced precipitation subsequent to the lake filling. The bathymetry also shows a layer of glacial deposits below Roundtop that provides a likely location for a permeable pathway for leakage from Crater Lake. These observations form anchor points in a new model for the filling of Crater Lake.

Drowned beaches

The high resolution of the multibeam bathymetry makes it possible to determine that there are drowned beaches at Crater Lake reflecting periods where the lake was significantly lower in elevation (Fig. 3). Beaches reflect stillstands from periods of reduced precipitation, and their existence constrains models for the distribution of leakage out of Crater Lake.

In order to quantify the elevations of the drowned beaches, we constructed profiles of elevation versus distance in areas that contain these features (Fig. 4). The origin of the profiles is near to but not at shoreline as the surveying boat had to stay some distance from exposed rock. We identified possible locations on the profiles where there might be a beach (vertical lines) based on

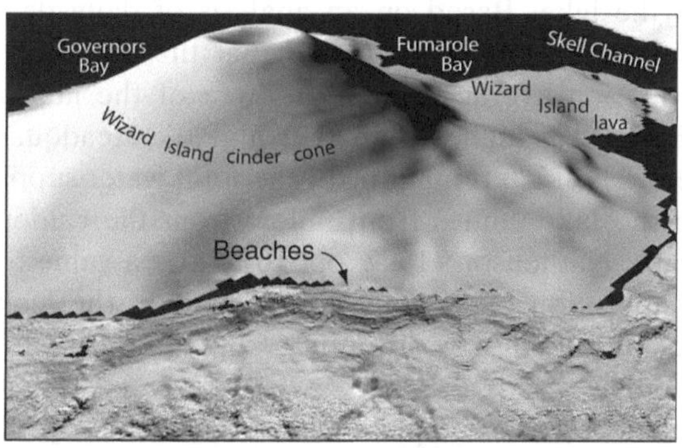

Fig. 3 Perspective view of drowned beaches on the northeast flank of Wizard Island (after Bacon et al., 2002, Fig. 12). Slumps have modified some beaches, particularly on the right half of the image. Bathymetry not shown in background (black). Subaerial terrain from USGS 10-m DEM. No vertical exaggeration

the pattern of depth versus distance and changes in the slope. Possible drowned beaches from a number of profiles were compared (Fig. 5), and those confirmed by their appearance in multiple profiles are marked with a B (Fig. 4). The Wizard Island profile (Fig. 4 top) and has a large number of possible drowned beaches. The wide, nearly level portion of the profile between 1852 and 1855 m also appears on profiles 103 and 23 of Wizard Island, but it is not found on other profiles and is probably related to local rather than lake-wide effects. Only a few of the possible beaches on the Roundtop profile (Fig. 4) are confirmed by other profiles. The contact noted on the Roundtop profile is the top of the Mount Mazama lava flow that is covered by glacial debris that we propose as an area of shallow leakage.

The summary of beaches found on multiple profiles (Fig. 5) shows nine drowned beaches ranging in elevation from 1848.5 to 1877.5 m. The relatively high elevation of the deepest beach provides a fairly strong constraint on models for where leakage occurs in the lake in response to climate change (see below). The contact found near Roundtop ranges from 1838 to 1859 m (Fig. 5) and is a likely region for some of the leakage.

In addition to the drowned beaches shown in the bathymetry, there is a wave cut platform that is a much larger feature. The photograph shown in Fig. 6 was taken between August 15 and 31, 1931 (Stephan R. Mark, NPS, written communication, 2002). In September 1931, the lake elevation was 1878.6 m (Phillips, 1968), about 4.5 m lower than the reference elevation as a result of the 1920s to 1930s drought (see water level plot in Redmond, 1990). The wave cut platform commonly is up to 40 m wide (Fig. 4, Bacon et al., 2002), much wider than any of the drowned beaches seen in the bathymetry. This difference in width implies that the lake has actually spent most of its history at an elevation of around 1879 m (1 to a few meters below the level of the last 40 years), and the recent few decades have been wetter than most of the time since Crater Lake filled. The bathymetric profile near Roundtop (Fig. 4) shows low slope in the first 7 m of the profile between elevations of 1878.8 and 1876.7 m. This is a bit lower in elevation than the

Fig. 4 Bathymetric profiles off Wizard Island and near Roundtop. Vertical lines indicate possible drowned beaches. Vertical lines marked with B's indicate features confirmed by other profiles. Slope for profiles shown along with that for angle of repose of 33°

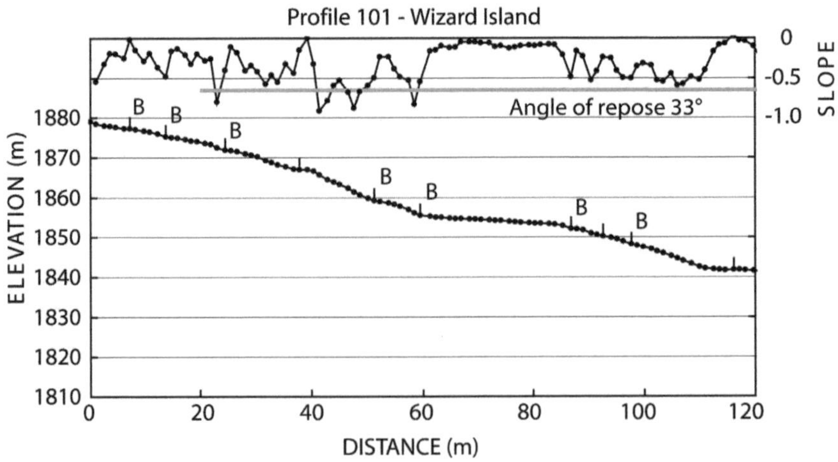

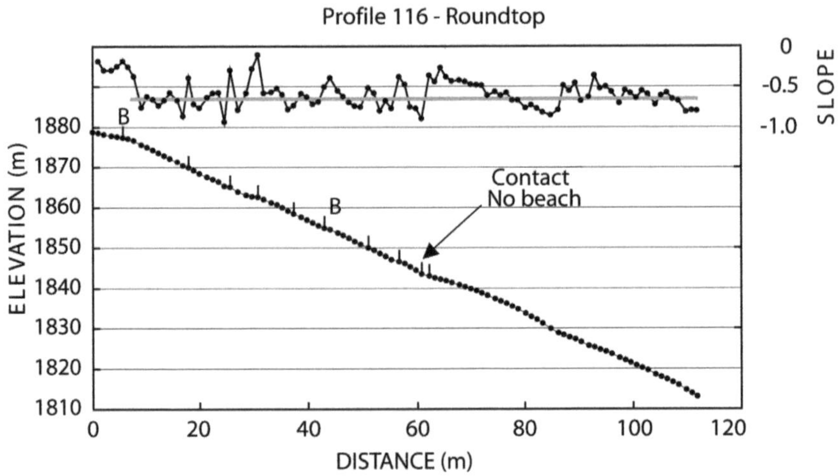

1878.6-m elevation obtained from the 1931 photo for the shallow, wave-cut platform, but there could be some uncertainty in ties to the reference elevation used to report past elevations. Elevation of the platform on other profiles generally ranges between 1876 and 1878 m.

Filling model

Crater Lake receives water from direct precipitation and inflow from the caldera walls and loses water by surface evaporation and leakage (Phillips 1968; Redmond 1990; Nathenson 1992). No streams flow out of Crater Lake. From the water balance for the years 1961–1988 (Nathenson 1992), the precipitation at the Park Headquarters rain gauge for the lake to remain at a constant level is 169.2 cm/y. The water supply to the lake from direct precipitation and inflow from the caldera walls is 224.2 cm/y over the 53.2 km^2 area of the lake. Based on an analysis of daily data, Nathenson (1992) estimated that the direct precipitation averaged over the area of the lake is about 10% higher than that at Park Headquarters. The 17% remaining of the total water supply to the lake comes from inflow from the caldera walls. Nathenson (1992, pp. 10) chose a value for evaporation of 85 cm/y and calculated the leakage of 139 cm/y by difference from the total water supply.

In order to explain the occurrence of the drowned beaches and their relatively narrow depth range, leakage through the caldera walls must vary with depth and cannot occur just at the lake bottom or at the modern lake level. A reasonable and simple model is that leakage is

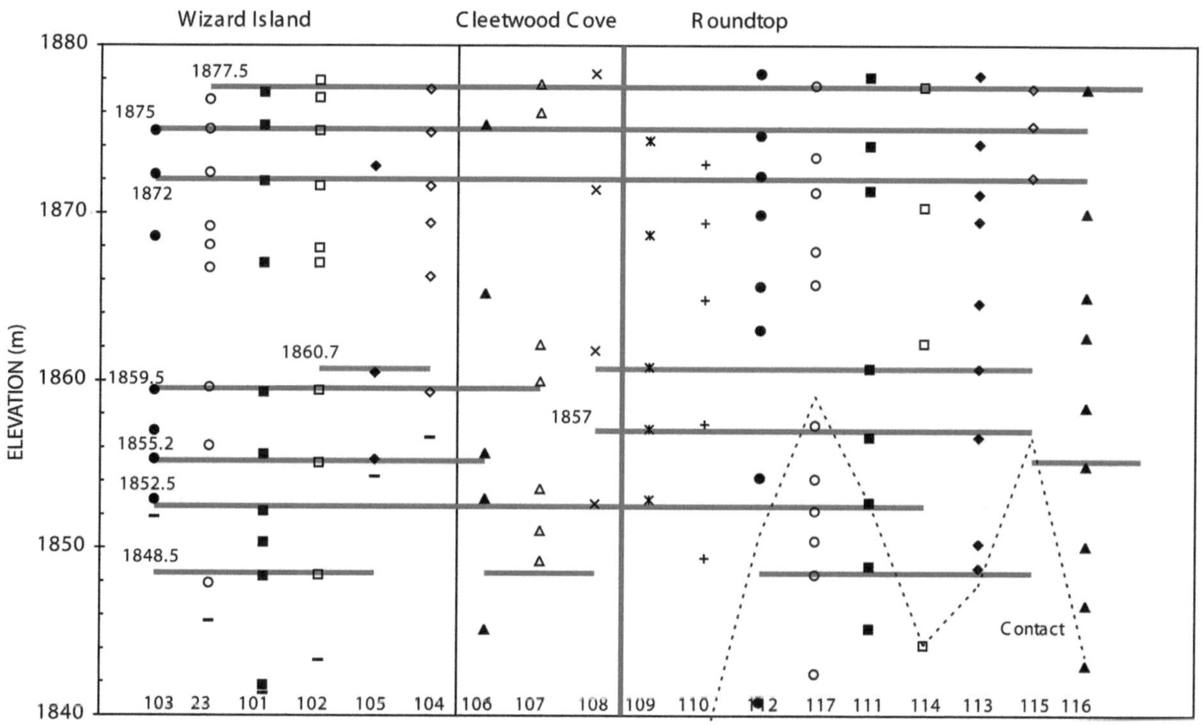

Fig. 5 Possible drowned beaches from various profiles (profile numbers at bottom of figure) on Wizard Island, Cleetwood Cove, and area near Roundtop. Chosen drowned beaches based on appearance in multiple profiles. Contact is top of lava flow covered by fragmental debris that could be an area of leakage

proportional to elevation above the bottom of the lake. Recognition that there is a thick layer of relatively permeable debris resting on glaciated lava in the northeast caldera wall above an elevation of ~1844 m suggests a variant of this model where leakage is proportional to elevation above 1844 m.

The water balance for Crater Lake for the change in lake elevation z referenced to the bottom of the lake is (Phillips 1968; Redmond 1990; Nathenson 1992):

$$Q_i + Q_p - Q_o - q_e A = A\frac{dz}{dt}$$

where $Q_i + Q_p$ is the water supply from inflow plus precipitation and is assumed constant, Q_o is the leakage from various models discussed below, q_e is the evaporation per unit area and is assumed constant, and A is the area as function of elevation $A(z)$ from the bathymetry in Fig. 2. The area of the lake is a curvilinear function of elevation that is well approximated by a series of straight lines over 50 m intervals (Fig. 7). The water balance equation is solved by arranging it as an integral over z,

$$t = \int_0^Z [A/(Q_i + Q_p - Q_o - q_e A)]dz$$

doing the integration using the trapezoidal rule, and solving for time as z increases from zero to the equilibrium level z_o. We neglect the extra volume currently taken up by lava flows and volcanoes on the floor and only fill the current volume of the lake. Because the eruptions took place during the filling, the missing volume was filled in by lava contemporaneously with filling by water.

The model with the longest filling time is where the total leakage Q_L (139 cm/y times the reference area of the lake) is out of the bottom of the lake

$$Q_o = Q_L \quad 0 \le z \le z_o,$$

Fig. 6 Photograph of Phantom Ship taken between August 15 and 31, 1931 by George Grant (Stephan R. Mark, NPS, written communication, 2002) showing wave cut platform just above shoreline. Photograph also appears in Atwood (1935, p. 147)

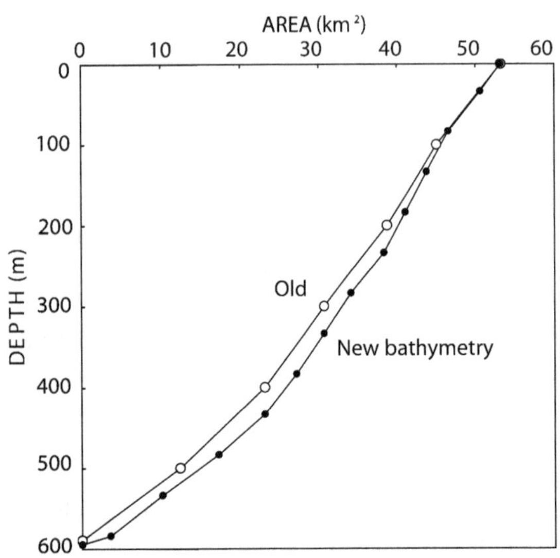

Fig. 7 Area versus depth for Crater Lake from new bathymetry from Gardner and Dartnell (2001) and old bathymetry from Byrne (1965). Note that lake area and volume given in Byrne (1965, Table 1) are not correct for his map

and the model with the shortest filling time is where there is no leakage until the lake reaches its equilibrium level z_o

$$Q_o = \begin{cases} 0 & 0 \le z < z_o \\ Q_L & z = z_o. \end{cases}$$

Given the increase in lake area with elevation from the bottom and the increasing head driving leakage as the lake fills, a reasonable model is that the leakage is proportional to elevation

$$Q_o = Q_L\, z/z_o \quad 0 \le z \le z_o.$$

For the model of leakage starting at the elevation of the glaciated lava z_s above the lake floor, the leakage function is

$$Q_o = \begin{cases} 0 & 0 \le z < z_s \\ Q_L\,(z - z_s)/(z_o - z_s) & z_s \le z \le z_o. \end{cases}$$

It turns out that in order to match the constraint from drowned beaches (see below), a combination of the last two models is needed

$$Q_o = \begin{cases} (1-\alpha)\, Q_L \dfrac{z}{z_o} & 0 \le z \le z_s \\ (1-\alpha)\, Q_L \dfrac{z}{z_o} + \alpha\, Q_L \dfrac{(z - z_s)}{z_o - z_s} & z_s \le z \le z_o \end{cases}$$

where α is the fraction leaking out in the fragmental deposits above the glaciated lava.

The time to fill the lake assuming modern values for precipitation, inflow, and leakage (Nathenson 1992) has a large range of values for the various models (Fig. 8). The calculations are only shown to within 1 m of the equilibrium level, because the function asymptotically approaches the equilibrium level requiring an infinite time to fill. However, water levels change by about half a meter each year, and the asymptote is not practically significant. In order to choose the proper model, we use the constraint provided by the drowned beaches that resulted from periods of reduced precipitation after the lake was filled. Crater Lake was filled near the end of a dry period in the Pacific northwest that started in the early Holocene (Barnosky et al., 1987). Evidence

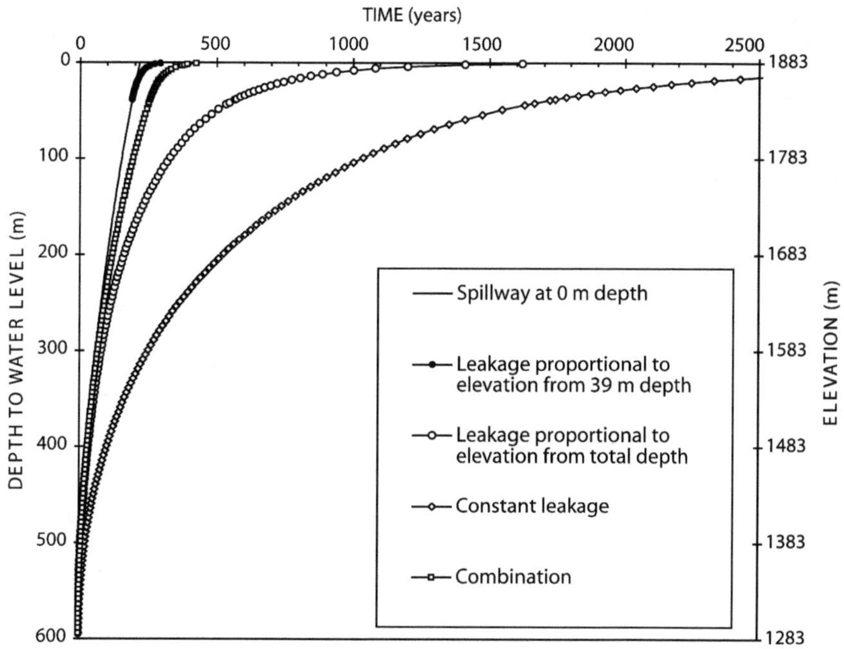

Fig. 8 Filling history for various models. Calculations stopped at 1 m below equilibrium level

for a dry period at the time of the climactic eruption is found at lower elevation in Lower Klamath Lake and Tule Lake south of Crater Lake (Nelson et al., 1994). However, studies of pollen in Crater Lake itself indicate that the period when the lake was filling was not dry at the elevation of Crater Lake, and the dry period occurred later there, between 6,000 and 5,000 C-14 years BP (Nelson et al., 1994).

In order to make quantitative estimates of the reduced precipitation to Crater Lake, we compare it to the history of Mono Lake (Stine, 1990, 1994). Droughts in California in 1928 to 1934 and 1987 to 1992 also appear in the lake-level record for Crater Lake (Redmond, 1990; Nathenson, unpublished). The behavior is not identical, but the overall character of these periods being much drier than normal is similar. Based on this comparison, we assume that the history of Mono Lake can provide data on amounts of precipitation reduction and durations that can be applied to Crater Lake. Stine (1990) reports periods of drought at Mono Lake over the last thousand years based on dated lake deposits. The drought about 1000 years BP was especially strong with inflow (precipitation) only 68% of the modern period, and one of similar magnitude occurred about 1800 years BP. Four droughts in the last thousand years had inflows ranging from 74 to 84% of the modern period. Stine (1994) estimates that the drought of 1000 years BP lasted for 220 years and the one about 600 years BP lasted for about 140 years with inflow 79% of the modern period. In order to calibrate the leakage models, we choose a minimum value for precipitation of Crater Lake as 70% of modern, and assume that this lowered precipitation produced the drowned beach at 1852.5 m (the deepest beach is somewhat suspect). This minimum value of precipitation of 70% of modern is also used to calculate one of the filling histories, assuming that it occurred over the entire time to fill the lake during an extended dry period.

The steady state values for lake level as a function of the fraction of water supply have been calculated from the water balance equation by iteration, and values are shown in Fig. 9. For the model with leakage proportional to total depth, such a large decrease in precipitation as 70–80% of modern would require drowned beaches below depths of 120 m, and we see no evidence for such deep beaches. For the model with leakage proportional to elevation above the layer of permeable debris, the deeper beaches require reductions in precipitation larger than 70%, and that seems unlikely given the climate data.

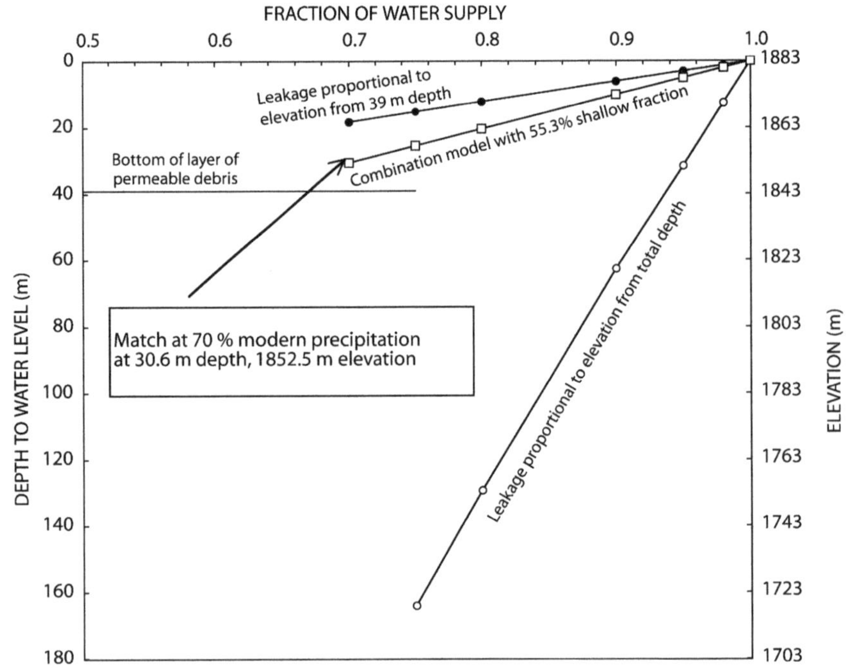

Fig. 9 Equilibrium values for depth to water level versus fraction of modern water supply for various models. Shallow fraction of 55.3% for combination model chosen to match equilibrium level of 30.6 m depth at 70% of modern precipitation

A simple linear combination of these two models with 55.3% of the leakage proportional to elevation above the layer of permeable debris and the remainder occurring proportional to total depth matches the constraint provided by the beach at 1852.5 m. Varying the minimum value for precipitation away from the 70% value would change the relative contributions of the two models, and the value for the relative percentages of leakage is reported to higher precision than justified because of the need to make calculations consistent. The particulars of the model are not unique, but the general idea that there is high-elevation leakage that depends on depth seems necessary to produce drowned beaches at various elevations as the fraction of water supply takes various values.

Phillips (1968: Fig. 5) calculated the leakage versus elevation for a number of time periods. He drew a curve through calculated points but indicated that a straight line is probably all that is justified by the calculated points of leakage versus elevation. Fitting a straight line to his points by eye, the change in seepage with lake elevation is 0.05 $m^3 s^{-1} m^{-1}$. For the combination model with 55.3% of the leakage occurring as leakage proportional to elevation above the layer of permeable debris and the remainder occurring proportional to total depth, the change in seepage with lake elevation above the layer of permeable debris is 0.035 $m^3 s^{-1} m^{-1}$. Considering the uncertainties in the calculations, this is quite good agreement.

For a reduction in precipitation of 30% for a period of 200 years such as that seen at Mono Lake, the combination model nearly reaches steady state (Fig. 10), and the water level is within 1 m of its lowest elevation for about 60 years. Although this may not have been enough time to produce a beach, if the reduction in precipitation lasted for a longer time such as during the dry period between 5,000 and 6,000 C-14 years BP, there would have been plenty of time to produce a beach. The lake level for the model with leakage proportional to elevation from 39 m depth reaches steady state faster and is within 1 m of its equilibrium level for about 115 years. The models with constant leakage and with leakage proportional to total depth do not reach steady state and cannot produce drowned beaches in the time available. For smaller reductions in precipitation, the times to reach steady state for the model with leakage proportional to elevation from 39 m depth and for the combination model are reduced, and reasonable times are available to produce beaches.

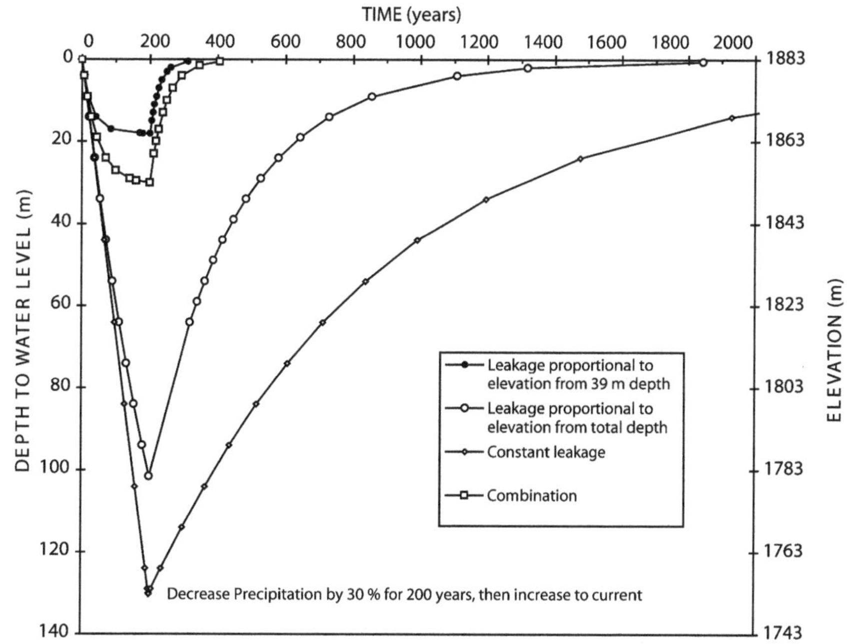

Fig. 10 Model histories for modern precipitation, then precipitation decreased by 30% for 200 years, followed by increase to modern precipitation

For the combination model, filling histories have been calculated for three values of precipitation (Fig. 11). Under the assumption that precipitation was at its modern value, the lake takes about 420 years to fill. Based on the occurrence of the wave cut platform at 1879 m, the lake may have spent most of its history at a level below the assumed equilibrium level of 1883 m, and normal precipitation would have been 95.6% of modern to yield an equilibrium level of 1879 m. Filling time only increases to about 460 years. If by chance the lake actually filled during a dry period not recognized because of poor preservation of the earliest few hundred years in the core record

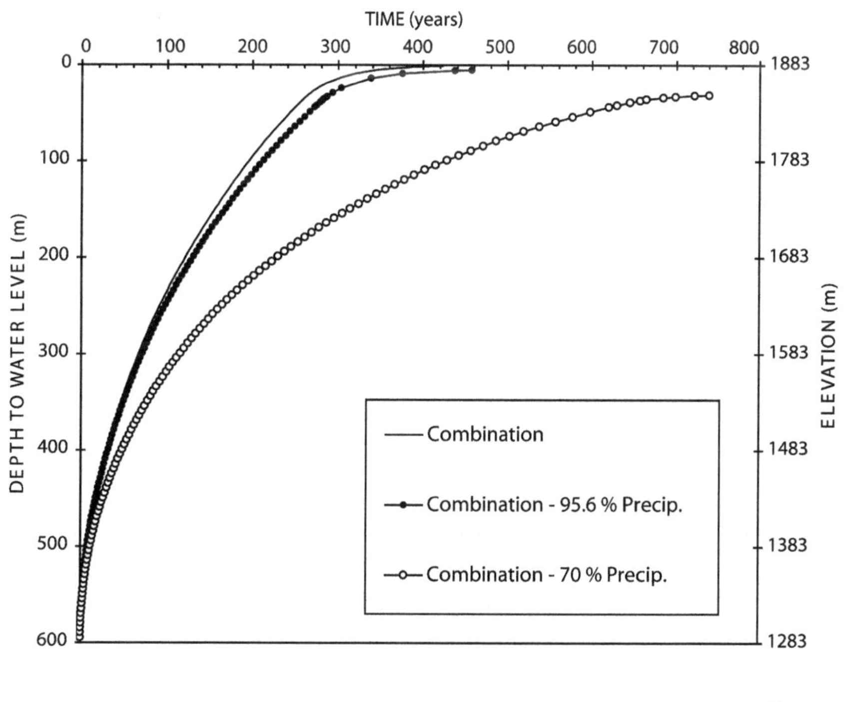

Fig. 11 Filling history for various fractions of modern precipitation for combination model with 55.3% shallow fraction. Calculations stopped at 1 m below equilibrium level

for Crater Lake (Nelson et al., 1994), a filling history with 70% of modern precipitation increases the filling time to about 740 years. The models with modern precipitation and 70% of modern have been used in Bacon et al., (2002) to bracket the times for passage zones of lava flow features that occurred while the lake was filling.

Chronology for postcaldera geologic events

The lake filling model provides a chronology for postcaldera geologic history. It is reasonable to assume that the lake began to fill virtually immediately following caldera collapse. Granted that evaporation rate probably was enhanced owing to hot caldera fill and exposure of the interior of Mount Mazama, a large amount of groundwater would have flowed into the caldera from the fractured walls. Applying the combination lake filling models to fossil shorelines (passage zones on volcanoes) indicates that andesitic volcanism ceased by 215–490 yr after caldera collapse (the time to fill to the highest elevation drowned lava delta passage zone on the Wizard Island volcano). Eruption rates calculated from postcaldera volcano volumes and filling models of $19–8.2 \times 10^6\ m^3/y$ are comparable to long-term lava effusion rates at historically active andesite-dacite volcanoes (Bacon et al., 2002, Table 2). The debris-avalanche deposits on the lake floor were emplaced sometime after much or all of the andesitic volcanism. Downslope movement of debris from the caldera walls to the deep basins is ongoing. Further details of the geologic history of Crater Lake are presented in Bacon et al., (2002).

Conclusions

The geology of the floor of Crater Lake is now reasonably well known from detailed bathymetry, sampling by submersible and dredging, and observations by submersible and ROV. The morphology of postcaldera volcanoes, including lava deltas and subaqueous breccias, reveals that all but the youngest were active during the period of lake filling. Also visible in the 2000 bathymetric survey are bedrock outcrops and talus/debris slopes of the caldera walls, debris-avalanche deposits below embayments in the walls, and sediment-filled deep basins. Thermal features, some active and others fossil, were discovered in earlier manned submersible and ROV exploration.

Models for the filling of Crater Lake with water are constrained by the presence of drowned beaches and a permeable layer in the caldera wall delineated by the 2000 bathymetric survey and suggest the lake took 420–740 years to fill. The level of Crater Lake is maintained by a balance between precipitation and inflow versus evaporation and leakage. Existence of the beaches and a broad wave-cut platform at 0–5 m depth can be explained by variations in precipitation with models for lake filling with 45% of leakage proportional to elevation plus 55% of leakage proportional to elevation above 1844 m, the elevation of the base of the permeable layer in the northeast caldera wall. The models provide a chronology for postcaldera volcanism, much of which ceased by 215–490 yr after the lake began to fill, depending of an assumed precipitation rate of 100% to 70% of the modern rate.

Acknowledgements The interpretations in this paper were possible thanks to the exceptionally detailed bathymetric survey led by James V. Gardner and to contributions to GIS analysis by Peter Dartnell, Joel E. Robinson, and James V. Gardner. Stephan R. Mark, Crater Lake National Park Historian, provided the photograph in Fig. 6. Michelle Coombs, Shaul Hurwitz, Daniel Hayba, and Kelly Redmond are thanked for helpful reviews.

References

Atwood, W. W. Jr., 1935. The glacial history of an extinct volcano, Crater Lake National Park. Journal of Geology 43: 142–168.

Bacon, C. R. & Manuel Nathenson, 1996. Geothermal resources in the Crater Lake area, Oregon. U. S. Geological Survey Open-File Report 96–663, 34 pp.

Bacon, C. R., J. V. Gardner, L. A. Mayer, M. W. Buktenica, P. Dartnell, D. W. Ramsey & J. E. Robinson, 2002. Morphology, volcanism, and mass wasting in Crater Lake, Oregon. Geological Society of America Bulletin 114: 675–692.

Barnosky, C. W., P. M. Anderson & P. J. Bartlein, 1987. The northwestern U.S. during deglaciation; Vegetational history and paleoclimatic implications. In Ruddiman, W. F. & H. E. Wright Jr. (eds), North

America and Adjacent Oceans During the Last Deglaciation. The Geology of North America, Geological Society of America K-3: 289–321.

Byrne, J. V., 1965. Morphometry of Crater Lake, Oregon. Limnology and Oceanography 10: 462–465.

Collier, R. W., J. Dymond & J. McManus, 1991. Studies of hydrothermal processes in Crater Lake, OR. Oregon State University College of Oceanography Report 90-7, 317 pp.

Dymond, J. & R. W. Collier, 1989. Bacterial mats from Crater Lake, Oregon and their relationship to possible deep-lake hydrothermal venting. Nature 342: 673–675.

Fuller, R. E., 1931. The aqueous chilling of basaltic lava on the Columbia River Plateau. American Journal of Science 5th series. 21(124): 281–300.

Gardner, J. V. & P. Dartnell, 2001. 2000 multibeam sonar survey of Crater Lake, Oregon: Data, GIS, images, and movies. U. S. Geological Survey Digital Data Series DDS-72, 1 CD-ROM.

Gardner, J. V., P. Dartnell, L. Hellequin, C. R. Bacon, L. A. Mayer, M. W. Buktenica & J. C. Stone, 2001. Bathymetry and selected perspective views of Crater Lake, OR. U. S. Geological Survey Water Resources Investigation Report 01-4046, 2 plates, scale 1:15,000, http://walrus.wr.usgs.gov/pacmaps.

Hoffman, F. O., 1999. The filling of Crater Lake. Nature Notes from Crater Lake 30: 10–13.

Jones, J. G. & P. H. H. Nelson, 1970. The flow of basalt lava from air into water — its structural expression and stratigraphic significance. Geological Magazine 107: 13–19.

Moore, J. G., D. A. Clague, R. T. Holcomb, P. W. Lipman, W. R. Normark & M. E. Torresan, 1989. Prodigious submarine landslides on the Hawaiian Ridge. Journal of Geophysical Research 94: 17,465–17,484.

Nathenson, M., 1992. Water balance for Crater Lake, Oregon. U. S. Geological Survey Open-File Report 92–505, 33 pp.

Nelson, C. H., A. W. Meyer, D. Thor & M. Larsen, 1986. Crater Lake, Oregon: A restricted basin with base-of-slope aprons of nonchannelized turbidites. Geology 14: 238–241.

Nelson, C. H., P. R. Carlson & C. R. Bacon, 1988. The Mount Mazama climactic eruption (6900 BP) and resulting convulsive sedimentation on the continent, ocean basin, and Crater Lake caldera floor. In Clifton, H. E. (ed), Sedimentologic Consequences of Convulsive Geologic Events, Geological Society of America Special Paper 229: 37–57.

Nelson, C. H., C. R. Bacon, S. W. Robinson, D. P. Adam, J. P. Bradbury, J. H. Barber Jr., D. Schwartz & G. Vagenas, 1994. The volcanic, sedimentologic, and paleolimnologic history of the Crater Lake caldera floor, Oregon: Evidence for small caldera evolution. Geological Society of America Bulletin 106: 684–704.

Phillips, K. N., 1968. Hydrology of Crater, East, and Davis Lakes, Oregon. *with a section on* Chemistry of the lakes, by A. S. Van Denburgh. U. S. Geological Survey Water-Supply Paper 1859-E, 60 pp.

Ramsey, D. W., P. Dartnell, C. R. Bacon, J. E. Robinson & J. V. Gardner, 2003. Crater Lake revealed. U. S. Geological Survey Geological Investigation Series Map I-2790, 1 plate.

Redmond, K. T., 1990. Crater Lake climate and lake level variability. In Drake E. T., G. L. Larson, J. Dymond & R. Collier (eds), Crater Lake, an ecosystem study: Pacific Div. American Association for the Advancement of Science, San Francisco, 127–141.

Stine, S., 1990. Late Holocene fluctuations of Mono Lake, eastern California. Paleogeography, Paleoclimatology, and Paleoecology 78: 333–381.

Stine, S., 1994. Extreme and persistent drought in California and Patagonia during medieval time. Nature 369: 546–549.

Wheat, C. G., J. McManus, J. Dymond, R. Collier & M. Whiticar, 1998. Hydrothermal fluid circulation through the sediment of Crater Lake, Oregon: Pore water and heat flow constraints. Journal of Geophysical Research 103: 9931–9944.

Williams, D. L. & R. P. Von Herzen, 1983. On the terrestrial heat flow and physical limnology of Crater Lake, Oregon. Journal of Geophysical Research 88: 1094–1104.

Williams, H., 1961. The floor of Crater Lake, Oregon. American Journal of Science 259: 81–83.

Hydrobiologia (2007) 574:29–46
DOI 10.1007/s10750-006-2603-9

Evaporation and the hydrologic budget of Crater Lake, Oregon

Kelly T. Redmond

© Springer Science+Business Media B.V. 2007

Abstract The hydrologic budget of Crater Lake, Oregon is investigated by taking advantage of its relatively simple geometry, climatic circumstances, and the concurrent availability of many years of traditional data. Buoy data are here utilized for the first time for this purpose. The lake gains water through precipitation and delayed runoff from the caldera sides and Wizard Island. The lake loses water through evaporation and seepage. Seepage can be estimated quite well from ice-covered precipitation-free intervals in 1985, and is 127 cm/year. Evaporation has previously been determined as a residual, but is here estimated directly from the floating buoy, with an approximate value of 76 cm/year, a downward revision from previous estimates. These losses are balanced by precipitation input, nearly all in the form of snow or snowmelt runoff. Factors contributing to the uncertainty in each of the water budget components are discussed in some detail. The buoy data corroborate previous findings based on studies of stage that evaporation is greatest on the coldest days. Seasonally, the greatest evaporation occurs in the autumn and the least in spring. Proxy records are used to extend the effective length of the buoy record. Monthly estimates of evaporation are calculated for 1950–1996 and used to deduce temporal characteristics. The standard deviation of water year precipitation is 4.6 times larger than that of evaporation. Thus the water budget is controlled more by variability of precipitation than evaporation. An additional 15 years of data since earlier studies confirm that the annual lake level variations from one September 30 to the next are highly correlated ($r = 0.96$) with Park Headquarters water year precipitation for the 42 years from 1961–2003. The lake rises 1.4 cm for every cm of measured precipitation over equilibrium value (168.6 cm) at Park Headquarters. Sources of this "magnification" are discussed.

Keywords Evaporation models · Crater lake · Climate · Water budget · High elevation

Guest Editors: Gary L. Larson, Robert Collier, and Mark W. Buktenica
Long-term Limnological Research and Monitoring at rater Lake, Oregon

K. T. Redmond (✉)
Western Regional Climate Center, Desert Research Institute, 2215 Raggio Parkway, Reno, NV 89512-1095, USA
e-mail: kelly.redmond@dri.edu

Introduction

Crater Lake, Oregon (42°56′ N, 122°6′ W) has many unique aspects that facilitate studies of its water budget. It is the second deepest lake in North America and seventh deepest in the world, and straddles the main crest of the Cascade Mountains at a reference elevation of 1882.14 m above sea level. The lake is nearly circular with an average diameter of 8.2 km and has very steep sides and only a small terrestrial collection area. The 53.2 km^2 surface of the lake itself occupies 78.5% of its own 67.8 km^2 drainage area (Phillips, 1968). Wizard Island, area 1.2 km^2, occupies just 8% of the non-lake drainage (see Larson, 2007, for map). The lake surface sits an average of about 300 m below the rim, about 600 m below the high points and 175 m below the lowest gap on the rim. Beyond the rim, essentially the drainage divide, the remainder of former Mt. Mazama falls steadily away at a typical slope of 4–8° (15° from a few high points). The lake has no surface outflow, and any surface inflow is from springs and temporary channels along the caldera walls and Wizard Island during snowmelt season. Bathymetry was remapped in 2000 (Gardner et al., 2000; Bacon et al., 2002) and the depth was revised to 593 m when the surface is at 1882 m. More detailed information on the lake can be found in Phillips (1968), Redmond (1990b), Drake et al. (1990), Larson et al. (1993), and Larson (2007).

The climate is dominated by the basic wet-winter/dry-summer pattern seen throughout the Pacific Northwest. Most precipitation-bearing systems have large dimensions and high spatial correlation of precipitation amounts on scales of tens of kilometers west of the Cascade Crest. Park Headquarters (HQ) receives 169.0 cm of precipitation annually, falling on an average of 139 days (1931–2003). Precipitation occurs frequently in winter (18 days in January bringing 16% of the annual total) and infrequently in summer (3 days in July bringing 1% of the annual total). The lake experiences very heavy snow (1931–2003 average is 1332 cm at Park HQ) and the surrounding land builds up a deep snowpack nearly every winter (average maximum depth 361 cm, extreme 640 cm in 1983 at Park HQ). Relatively little runoff occurs during the heart of the cool season (too cold) or the warm season (too dry). A strong spatial gradient in precipitation exists across the lake (about 200 cm annually on the southwest rim, about 90 cm on the northeast rim). Additional climate details can be found in Redmond (1990a, b) and web sites (e.g., Western Regional Climate Center). The lake almost never freezes completely (3 known instances since 1949), and for this work only under the most fortuitous of circumstances.

With its steep cliff-like sides and circular shape, Crater Lake acts as the rough equivalent of a giant leaky rain gage, with some incoming precipitation sticking to the sides for a few snowy months. The lake loses water downward through steady seepage and upward via evaporation. The residence time of water is about 160 years and the surface elevation has fluctuated about 5 m in

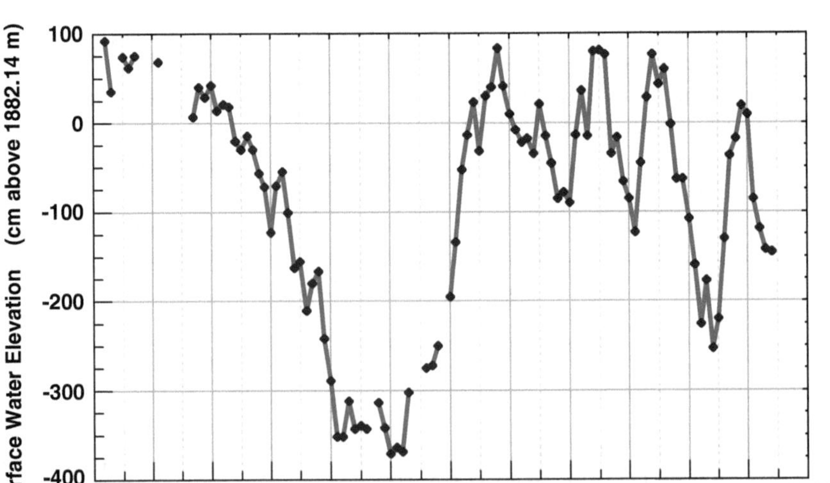

Fig. 1 Crater Lake water level, adjusted to September 30, through 2004. Referenced to nominal elevation of 1882.14 m/6175 ft

the past 110 years (Fig. 1). The lake would fill from empty in 400–500 years at modern precipitation rates (Nathenson et al., this issue). Tree ring precipitation reconstructions indicate that Crater Lake may have been higher than present in the past few centuries (Peterson et al., 1999).

The water budget of Crater Lake has been examined earlier (Phillips, 1968; Simpson, 1970; Redmond, 1990a, b; Nathenson, 1992; among others). The combination of the physical circumstances and available data give Crater Lake a unique status, allowing approaches not possible at any other lake in the world. This work extends those efforts by utilizing buoy data to provide improved annual and seasonal evaporation estimates.

Methods and data analyzed

Crater Lake Park Headquarters (HQ) has, since 1931, been a standard National Weather Service cooperative station (ID 35-1946), recording once-daily 8 a.m. measurements of precipitation, snowfall, snow depth, and max and min temperature. This site has an excellent record from a stable location, two km south-southwest of the lake, at an elevation (1974 m) about midway between the lake surface (1882 m) and the rim (2135 m). The record had significant interruptions during World War II; missing monthly values of precipitation were estimated from nearby well-correlated west-slope sites. Hourly precipitation from a recording gage (separate from the manual gage, but physically very close) is available since 1948, archived as an entirely separate data set. A precipitation storage gage was installed in 1963 on the rim above Cleetwood Cove, near the present-day parking lot. Other storage gages have operated for shorter times on Wizard Island and near the Crater Lake Lodge. Another hourly meteorology site was installed by Oregon State University (OSU) on the southwest rim, north of the visitor center. Data begin 11 December 1991, and continue (with occasional equipment malfunctions) to the present. This site measures temperature, wind speed and direction, relative humidity, solar radiation, and for part of its record, precipitation and snow depth. Since 1948, digital data are available for National Weather Service twice-daily radiosonde ascents with balloons at Medford, 88 km to the southwest. Prevailing flows frequently take the balloon toward Crater Lake. These free-air measurements consist of temperature, relative humidity, wind speed and direction, pressure, and height of the pressure level; the Crater Lake rim is about 800 mb.

Water levels have been measured by USGS at Cleetwood Cove since 14 Sept 1961, to a resolution of 0.01 foot, or 0.305 cm. Daily maximum and minimum water temperatures have also been measured here, and are now recorded at sub-daily intervals. Prior water levels were measured sporadically by visitors to the lake shore, after the first of six gages was installed in August 1896. The gage history is recounted by S. T. Harding (1953, Unpublished), and additional information is given by Phillips (1962). Most early measurements are from the warm portion of the year when the lake is accessible. Annual changes are here referenced to 30 September, near the usual low point of the half-meter annual cycle (Redmond 1990b: Fig. 12).

A buoy anchored over the deepest spot was deployed by Bob Collier of OSU on 18 Sept 1991 and has operated (with some interruptions due to equipment malfunction) through this writing. This platform, anchored to the bottom, measures hourly wind speed, and direction, air temperature, relative humidity, and water temperature near the surface and at several different depths.

Lake water budget sources and sinks

Precipitation

Precipitation ultimately provides the entire input of water to the lake. Because the lake occupies so much of its own drainage basin, most of the precipitation input occurs directly on the water, mostly as snow, with instantaneous lake level response. Winter precipitation is modest in intensity but very frequent. Summer precipitation is much less common, and showery.

The southwest rim is thought to receive 15–25% more annual precipitation (Soil Conservation Service, 1965), and the northeast rim about

30% less, than Park HQ. The storage gage on the northeast rim near Cleetwood Cove receives only about 72% as much as the Park HQ gage. This reflects the sharp west-to-east decrease in annual precipitation seen everywhere across the crest of the Cascade Range. A 40–50% decrease from the southwest to the northeast rim is quite plausible. Over the flat lake itself a precipitation minimum displaced from center toward the northeast rim would be expected. Lake Tahoe shows a similar pattern (see map in R. A. Lind & J. D. Goodridge, 1978 Unpublished). The northeast/southeast rim ratio, and thus lake/headquarters ratio, will differ from year to year. The July–June storage gage precipitation on the north rim at Cleetwood has a correlation from 1964–65 to 1987–88 with Park HQ precipitation of $r = 0.90$. The mean Cleetwood/HQ ratio is 0.715, with a standard deviation of 0.074.

Because of extreme logistical and climatic difficulties, it is not feasible to directly measure precipitation at a network of sites on or around the lake. As in all mountain regions, annual precipitation is sensitively dependent on elevation and aspect. Winds at 800 mb (rim level) over Medford (1948–1997 average) are from the south through west quadrant 61% of the time from November through March, months that account for 71% of the annual precipitation, with monthly "resultant wind" directions from 231–238 degrees of azimuth. In winter nearly every slope with a southwest aspect will locally enhance precipitation. Orographic enhancement and "lake effect" snows on the northeast caldera wall would also be expected. An unknown amount of drifted snow is transported upslope and over the rim, to be deposited on steep lee slopes or in the water. Avalanching also contributes to the winter lake budget as a form of highly efficient runoff, but this component is poorly known.

Traditional precipitation gages catch less than the true precipitation, a result of orifice aerodynamics or gage wall interception. Undercatch is common and indeed expected (Goodison, 1978; Yang et al., 1998; Doesken & Judson, 2000; Yang et al., 2001) with unshielded gages (as at Crater Lake) when precipitation is frozen, so that *measured* precipitation at Park HQ is not necessarily the full *actual* precipitation. The Park HQ gage is rather well situated with respect to nearby trees and other desired obstacles to wind, and Cascade snow is often dense; these factors reduce, but do not eliminate, undercatch. Thus, undercatch of 5–15% would not be surprising. The Park HQ gage is also not quite directly upwind with respect to the lake center.

Runoff

Very heavy snowpack develops around the narrow ring of land that drains into the lake (average 530 m horizontal distance), but little of this runs off until spring melt season. Thus a portion of basin precipitation input to the lake is delayed. By mid July, snow has melted entirely at Park HQ, but patches (depending on the year) continue to dwindle on the north slopes inside the caldera. Lake behavior implies that runoff slows very considerably by late July. The seasonal cycle in lake levels has a broad "shoulder" in spring and early summer (Redmond, 1990b: Fig. 12), after precipitation has plummeted, representing delayed lake input occurring as melt and runoff. By mid summer this has usually ended. No streams flow in or out of the basin, though 60 springs (Gregory et al., 1993) discharge into the lake and exhibit strong seasonal variation. Depending on bedding planes, some subsurface flows from inside and outside the caldera may cross the surface drainage divide. Runoff complications to assessment of instantaneous lake water budget terms should be at a minimum when everything is frozen (in winter) or when little is left to contribute (late summer, early autumn). Many of the studies reported here using Cleetwood stage data were under these circumstances.

Seepage

Seepage is thought to occur at a relatively constant rate. Redmond (1990b) used two unusual ice-covered episodes in January and December 1985 to estimate seepage at 0.347 cm/day or 127 cm/year, equivalent to an outlet stream with discharge 2.14 m^3/s. The two episodes yield values of 123 and 130 cm/year, so the uncertainty appears to be 2–3%. The young volcanic rock likely provides a large number of conduits to conduct water out of this elevated lake basin. The lake surface is far above the surrounding terrain to the east and west, and even the lake bottom is at or above the high desert to the east. Depth variations in the 20th Century (5 m) are less than 1% of the total depth. Nathenson et al. (this issue) suggest that seepage may vary by 1.5% per m when the lake is within 5–10 m of 1882.14 m because of a fractured layer near Roundtop on the northeast rim, so that seepage has remained approximately constant (within 5–8%) over the past century. The volcano is dormant, with little seismic activity in the past several decades, so it seems reasonable to assume seepage channels are stable.

Evaporation

Evaporation proceeds all year, but at a greatly variable rate. It also continues to occur while precipitation is falling. Except for a short time in summer, the lake is usually warmer than the overlying air (Redmond, 1988: Fig. 6, p. 77). Thus, free convection in the air over the lake is likely to be occurring much of the time. Relative humidity at the buoy is only about 70% over the lake in winter, and about 64% in summer. Because of its elevation and latitude, the lake feels the large scale westerly wind flow much of the year, especially in winter. Thus, forced convection is also quite likely.

In most circumstances evaporation cannot be measured directly from a large body of water. At Crater Lake, during certain times of the year, when complicating components of the water balance produce minimal contributions, one can take advantage of these simplified conditions by making use of selected stratifications of historical stage data that now extends to nearly 15,000 days. This approach has been followed in prior investigations (Redmond, 1988, 1990a, 1990b, 1993), which have shown conclusively that evaporation is not constant, either during the climatological annual cycle, or for the same part of the seasonal cycle from one year to the next.

These studies showed that for a given time of year, on colder days the lake level falls more rapidly than on warmer days (Redmond, 1993). This is true in winter and summer. Large day-to-day decreases in lake level occur often on autumn days with cool dry air overlying the thin layer of warm water left from summer. Greatest evaporation during the year quite clearly occurs on the coldest days. During the cool season, winds are stronger and enhance evaporation. Winds are likely to be stronger during stormy periods, and following winter cold frontal passages near-surface vertical stability is decreased, both of which increase evaporation. The Great Lakes literature has documented this quite well (Derecki, 1976; Derecki, 1981a, 1981b; Quinn, 1979), an effect visibly apparent in lake effect snowstorms (Eichenlaub, 1979).

A study of days with just *a little* precipitation (0.25–1.50 mm) at Park HQ was conducted. These amounts are not enough to cause runoff, should not be greatly accentuated over the rim by orographic processes, are small enough to not overwhelm the other terms, and can be approximated as small corrections to daily lake level change. They showed that the lake dropped faster on these days than on days with no precipitation, given the same temperatures. That is, the same process that brings precipitation also promotes evaporation: storms are accompanied by increased wind. During yet larger storms, precipitation effects thoroughly mask potential enhanced evaporation effects on daily stage changes.

Results

Analysis and modeling of the lake water budget

The annual lake water budget can be described in the following manner:

$$\Delta h = P_{\mathrm{L}} + R - S - E \quad (1)$$

where Δh is the lake level change, P_{L} the precipitation directly onto the lake surface, R the runoff from surrounding caldera slopes, S the seepage, E the evaporation from the lake surface.

Terms are expressed as depths, averaged over the lake area. Although (1) applies to all definitions of a year, we use the Water Year (1 October–30 September), the end of which coincides with lowest lake level, and is just prior to the start of the prolonged winter precipitation season (Redmond, 1990b: Figs. 5 and 12).

Seepage is here assumed constant. Annual evaporation can be broken into a constant part and a perturbation from the long-term mean. These perturbations are primarily functions of wind, temperature and humidity in the atmosphere, and temperature of the water, all of which vary seasonally and diurnally. The atmospheric factors affecting evaporation have a complex relationship with precipitation, especially with significant seasonal differences. We simply represent evaporation as

$$E = E_0 + e', \quad (2)$$

where E_0 is mean annual evaporation and e' is the departure from this long-term mean.

Runoff from the zone around the lake margin, extending from the water edge to the drainage divide, may be written as

$$R = dgP_{\mathrm{H}}(A_{\mathrm{R}}/A_{\mathrm{L}})(1 - L), \quad (3)$$

where P_{H} is precipitation at Park Headquarters, A_{L} is the area of the lake, A_{R} is the area of the terrestrial ring around the lake inside the drainage basin, and L is the annual fractional loss of water by evapotranspiration and sublimation in the terrestrial ring. The term d is the ratio of average precipitation in the terrestrial ring surrounding the lake to the precipitation at Park Headquarters. The term g is introduced to allow for gage undercatch; $g = 1$ implies perfect catch, $g > 1$ implies undercatch (further discussion below). The runoff amount on the right hand side of (3) is converted to the effect on water level by dividing by the area of the lake. In normalized terms, A_{L} is about 0.785 and A_{R} is about 0.215 (fraction of the total basin), so that 1 cm uniformly running off from land would raise the lake by about ($A_{\mathrm{R}}/A_{\mathrm{L}}$) cm, or 0.274 cm, and 1 cm falling over the entire basin (impervious surface, no evaporation) would raise the lake by 1.274 cm. For L the limiting cases are between 0 (no evaporation/transpiration land loss whatsoever, unlikely) and up to perhaps one-third, weighting summer and winter conditions.

The principal reason for tying lake behavior to Park HQ precipitation wherever possible is the practical need to utilize a precipitation record that is as long as possible, to perform reconstructions. This gage is the only one near that lake that spans nearly seven decades. The (unknown) average precipitation falling directly onto the lake is probably not the same as precipitation at Park HQ, but temporal variations in this quantity, especially on longer time scales, are very likely to be *proportional* to that at Park HQ.

Thus, we can write

$$P_{\mathrm{L}} = cgP_{\mathrm{H}} = cg(P_{\mathrm{H0}} + p'_{\mathrm{H}}), \quad (4)$$

where the annual time series of *measured* precipitation at Park HQ is decomposed into a long-term average, P_{H0}, and an annual departure, p'_{H}. Strictly speaking, P_{H0} is the HQ precipitation associated with long-term balance, 168.6 cm (updated through 2003 from Redmond, 1990b, and Nathenson, 1992; see below). Coefficient c is the ratio of actual basin-averaged precipitation to measured Park HQ precipitation. The value of c is probably in the range of 1.00–1.15, and d seems likely to equal or exceed c.

Combining all these terms and expanding we obtain:

$$\Delta h = dg(P_{\mathrm{H0}} + p'_{\mathrm{H}})(A_{\mathrm{R}}/A_{\mathrm{L}})(1 - L) + cg(P_{\mathrm{H0}} + p'_{\mathrm{H}}) - S - E_0 - e' \quad (5)$$

Since the terrestrial ring completely surrounds the nearly circular lake, the proportionality constants c and d are likely to be similar to each other (average precipitation in a narrow ring *around* the lake is likely to be similar to average precipitation *on* the lake). However, d could slightly exceed c, because the terrestrial ring receives more snow on the upwind arc, with surface transport from drifting, and there is likely a combined lake effect/orographic effect on the downwind side of the lake as the newly moistened air climbs out of the basin. Also, Wizard Island, whose effects are included in this ring, is very close to the snowy west shore. Rewriting (5),

$$\Delta h = 0.274(d(1-L)+c)gP_{H0} - S - E_0 \quad [A] + (0.274d(1-L)+c)gp'_H - e'[B] \qquad (6)$$

On a long-term basis term A must be equal to zero, because the long-term lake level changes average out to zero, and annual precipitation and evaporation departures in term B average out to zero, by definition. On shorter (interannual) time scales, lake level changes are then driven by term B. In (6), none of the quantities S, E_0, c, d, g, e', and L are known exactly, so there is some latitude for adjustment of the various coefficients and terms within their uncertainty. Simple order of magnitude estimates for E_0 are in the range of 70–100 cm/year. From term A, if c is assumed equal to d and S is set at 127 cm/year, long-term balance requires

$$cg(1+0.27(1-L)) = (S+E_0)/P_{H0}, \qquad (7)$$

and the range of the right hand side is from 1.17 to 1.34, representing various "magnification" effects in going from Headquarters precipitation to lake surface input.

The empirical relation between annual (water year, Oct–Sept) Park Headquarters precipitation and annual lake level change found by Redmond (1990b) has been here updated with 15 more years through 2003 (Fig. 2), with the result that

$$\Delta h = -235.82 + 1.399P_H(\text{cm}) \quad (r = 0.96), \qquad (8)$$

or,

$$\Delta h = -235.8 + 1.399P_{H0} + 1.399p'_H. \qquad (9)$$

Substitution of long term average precipitation 168.6 cm for P_{H0} and assuming constant evaporation leads to the empirical result

$$\Delta h = 1.399p'_H. \qquad (10)$$

Lake level rises 1.4 cm for every additional 1.0 cm of measured precipitation at Park HQ. Equation (9) can be now equated with term B in (6). Assuming the variance p'_H is sufficiently greater than e' (shown below), the evaporation variation can here be approximated as 0, and once

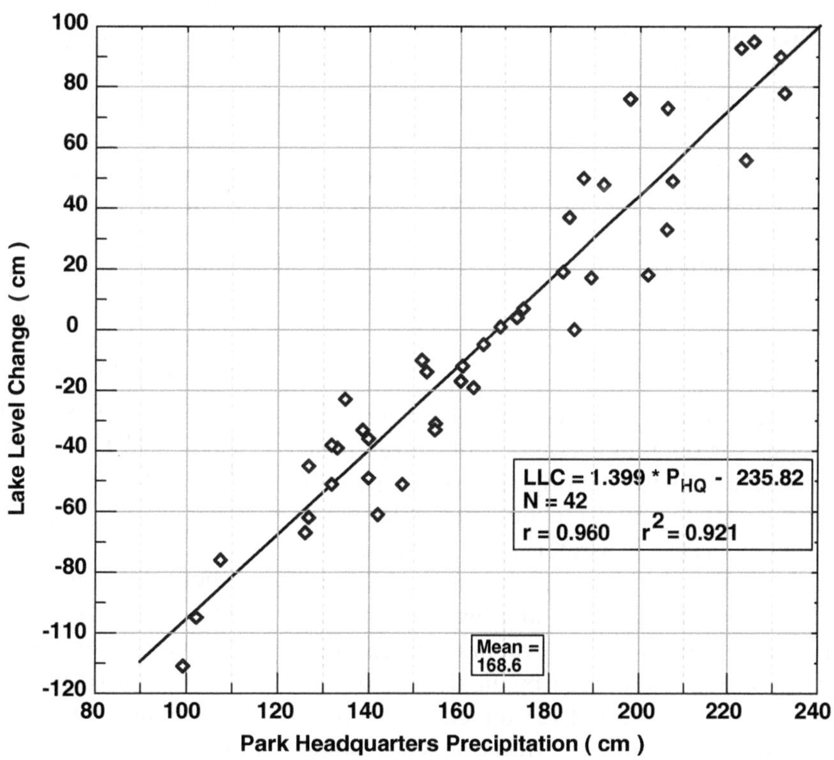

Fig. 2 Annual change in water level versus annual precipitation, from October 1 to September 30, for instrumental period of stage recorder at Cleetwood Cove, 1961–2003. Updated from Redmond (1990b)

again we approximate $c = d$. Neglecting gage undercatch ($g = 1$), then if (for example) $L = 0$, $c = 1.10$; if $L = 0.17$, $c = 1.14$; and if $L = 0.33$, $c = 1.18$. If $d = 1.1\ c$ (ring gets more than lake), c becomes 1.08, 1.12, and 1.16 respectively. These are all plausible values for c; with the steep caldera slopes it is likely that most of the interior snowpack would run off into the lake (i.e., L is likely small).

The coefficient in (10), 1.399, varies quite significantly according to how the year is defined; it is not clear why. Nathenson (1992) noted that an Oct–Sept year yields 1.45 (found by Redmond, 1990a, b, here recalculated from added data as 1.40), whereas a July–June year yields 1.26. The water year is used here because in some years the snowmelt from the previous winter is still keeping the lake high on July 1, smearing one winter's effect into two different precipitation water years if July–June is used. A mid-summer date makes it more difficult to clearly distinguish the effects of a particular winter than when a later date is chosen.

Evaporation estimates from buoy data

In 1991, an anchored meteorological buoy was deployed over the deepest part of the lake. This study has utilized hourly data from the first several years of buoy operation, 1991–1995, including about 27 full months of buoy data, with partial data for another 6 months. The buoy meteorological data was used to estimate evaporation by traditional methods.

Adams formulation

Rasmussen et al. (1995) compared several formulations for obtaining evaporation fluxes from lakes. These models were originally developed for cooling ponds for power plants, but have more general applicability. Crater Lake is somewhat similar, in that it is usually warmer than the overlying air. There are usually two components (e.g., Mahrt & Ek, 1984) to evaporative heat loss: (1) free convection, caused by air density differences in the vertical, and (2) forced convection, due largely to the mixing effects of wind. Some formulations recognize that the mixing coefficient is affected by vertical stability and use stability-dependent exchange coefficients. The formulation tried here is additive, not multiplicative, where

Heat Flux = Free Convection Flux +
Forced Convection Flux.

In the comparisons by Rasmussen et al. (1995), the criterion for "best" was based on ability to simulate the annual cycle of nine Minnesota lakes ranging in size from 0.1 to 12 km^2, as measured by regression coefficient or standard error of prediction. Based on this, a formulation by Adams et al. (1990) was selected

$$H_e = [(2.7(T_{v,w} - T_{v,a})^{1/3})^2 + (5.1A^{-0.05}W_2)^2]^{0.5}(e_s - e_a) \quad (11)$$

where H_e is the evaporative heat flux [W/m^2], $T_{v,w}$ is the virtual temperature of saturated air at water temperature, $T_{v,a}$ is the virtual temperature of the ambient air, A is the area of the lake in hectares (to account for fetch and equilibration effects), W_2 is the wind speed [m/s] at 2–m height, and e_s and e_a are the saturation vapor pressure at water temperature, and of the ambient air, respectively [mb].

Energy fluxes were converted into mass fluxes by dividing by the latent heat of vaporization and converted to equivalence in cm of water. When the air is warmer than the water, or when the free term is negatively larger than a positive forced term, the negative sign in (11) is carried through.

Hourly increments of evaporative loss are summed each day. On days with little or no runoff, a slightly different approach was used here to estimate hourly and daily lake level change as the sum of "effective precipitation" (recent plus delayed) minus evaporation minus seepage, a simplified version of (1), with ($R + P_L$) replaced by ($f\ P_H$), because R~0:

$$\Delta h_{NR} = fP_H - E - S \quad (12)$$

where Δh_{NR} is lake level change with no runoff and other terms as before. Lake-average (and for this case, basin-average) precipitation is taken to be proportional to headquarters precipitation.

The factor f accounts for the ratio of lake-basin averaged precipitation to Park Headquarters precipitation. The method used to determine *f* is to add all the budget terms for many years and compare net calculated lake level change from the first day to the last day with the observed change from lake stage differences, and then find the *f*-value that produces an exact match. Since long-term lake level change is an integral of budget sums over hundreds of thousands of hours, it is extremely sensitive to tiny changes in parameters such as this, which must be determined with more precision than we could ever obtain from first principles. Also, in reality, precipitation generation mechanisms vary from year to year, so that this ratio is not really fixed, but rather varies in some range. This approach is satisfactory as long as the resulting value is not greatly different from 1. In this case, the value is typically near 1.16.

Because runoff is not accounted for, this simple approach introduces systematic errors in estimated daily lake level changes during certain parts of the year. An attempt to account for runoff seasonality is beyond the scope of this study.

Table 1 shows the resulting values of evaporation, based on the Adams formulation, with partial months identified. With this formulation, evaporation for these particular months averages about 76 cm (37% of the total loss) and seepage averages about 127 cm (63%), with a total loss of 202.5 cm. This is not a systematic sample, and the averages are simply intended to show a sense of magnitude.

Several features are evident. The largest values are generally in late summer through late autumn, approximately August through November, when the water minus air temperature difference is greatest (Redmond, 1988: pp. 77) and also when winds are slowly increasing toward their winter maximum. Values diminish during the middle of winter, and reach their lowest values in later spring, when air is warming and lake temperatures are coolest.

Evaporation estimates when buoy data missing

With the missing days and months from the buoy it is not possible to use residuals to determine what the precipitation multiplication factor should be, because lake levels are responding to events in other months without measurements. In order to obtain systematic values for every month, and eventually to extend the record, values from long-term sites with no missing data are required. This entails finding proxy data that can be used to estimate water and air temperatures, relative humidity, and wind speed at the buoy. The proxy data must also have a high likelihood of being available every day.

Mean daily temperature [maximum (shifted back one day) plus minimum, divided by 2] at Park HQ was selected as a proxy for buoy air temperature. Cleetwood water temperature was selected as a proxy for buoy water temperature. For days without Cleetwood water temperature, a secondary proxy based on Park HQ air temperatures was developed. Relative humidity at 800 mb (approximately rim level) above Medford was selected as a proxy for buoy relative humidity. For these three, additive corrections were used. Wind speed at 800 mb above Medford was selected as a proxy for buoy wind speed, with a multiplicative correction. Climatological values were substituted for the relatively few days with missing proxy data.

Table 1 Crater Lake buoy evaporation

	Jan	Feb	Mar	Apr	May	Jun	Jul	Aug	Sep	Oct	Nov	Dec	Ann
1991									x8.13	9.94	7.80	6.73	
1992	7.15	4.87	x1.79	3.17	1.92	5.43	4.66	9.25	x10.67				
1993									8.27	7.56	10.65	7.61	
1994	5.75	6.90	4.72	3.38	x2.12			x10.12	8.33	11.41	11.57	8.59	
1995	5.59	3.67		6.16	3.81	1.93	x3.20						
Ave	6.16	5.15	4.22	3.45	1.99	4.32	4.66	9.68	8.85	9.64	10.01	7.64	75.77
Seep	10.76	9.80	10.76	10.41	10.76	10.41	10.76	10.76	10.41	10.76	10.41	10.76	126.74

x indicates partial data for that month; values have been pro-rated (based on those days with data) to a full month. Blank indicates no buoy data. Ann—sum of monthly averages. Seepage shown for comparison. Units: cm

Mean daily values were then calculated for each day in the record for the buoy with at least 20 hourly values. For all days with common data available among the paired sources, biases were determined and saved. Generally 2–3 years of simultaneous buoy and proxy data were available, but for some parts of the year only 1 year was available. For each day of the year, biases (whether additive or multiplicative) were averaged over all years (at most 4 years in this case). A moving 30-day window was then applied to smooth the annual cycle of bias. Inspection of the individual biases (on a monthly basis) from one year to the next shows relative stability in the relationships.

Figure 3 shows the long-term daily Cleetwood average water temperature for all available days between 1991–97 (generally 4–7 years of data). The buoy water temperature averages 0.4–0.8°C cooler than Cleetwood from December–April and 0.1–0.5°C warmer in June and July. Buoy water temperatures average about 3°C warmer than Park HQ mean air temperature in winter to 1°C warmer in April. Relative humidity at rim level (800 mb) from Medford radiosonde data averages just over 60% from November–April, and about 42% in August–September. The buoy relative humidity (in percent units) is about 13 units higher than the Medford relative humidity at rim level, to within 2–3 units for most of the year, with no seasonal cycle. From October–June, buoy wind speeds are 60–65% of the Medford 800 mb wind speeds, and 80–90% of those speeds from July–September.

The same calculation of evaporation as in Section 4.1 was repeated, substituting proxy data from the long-term source, adjusted to approximate buoy conditions. Since hourly data are not available from the proxy locations, estimated buoy conditions were held fixed in 24-h blocks. This overlooks possible nonlinearities associated with the daily cycle, but these are likely of no greater than second-order effect. Estimated evaporation using this technique is illustrated in Table 2.

For the 27 months in common between both tables (full monthly data), the total for the buoy-derived values is 176.82 cm, and for the proxy-derived values the total is 172.39 cm, within 3% of each other. Individual monthly differences are larger, and a few months are considerably different, but some months are very close.

For these years, seepage (127 cm) plus evaporation (76 cm) would total 203 cm, which would have to be balanced by basin average precipitation and runoff. Using the Park HQ precipitation of 169 cm, and inserting these values into (7) would imply that $c\ g\ (1 + 0.27(1\text{-}L))$ would be about 1.22. If L were about one third (loss to evapotranspiration), then the basin would receive about 103% of the headquarters value. A smaller loss ($L = 0.16$) would imply c is about

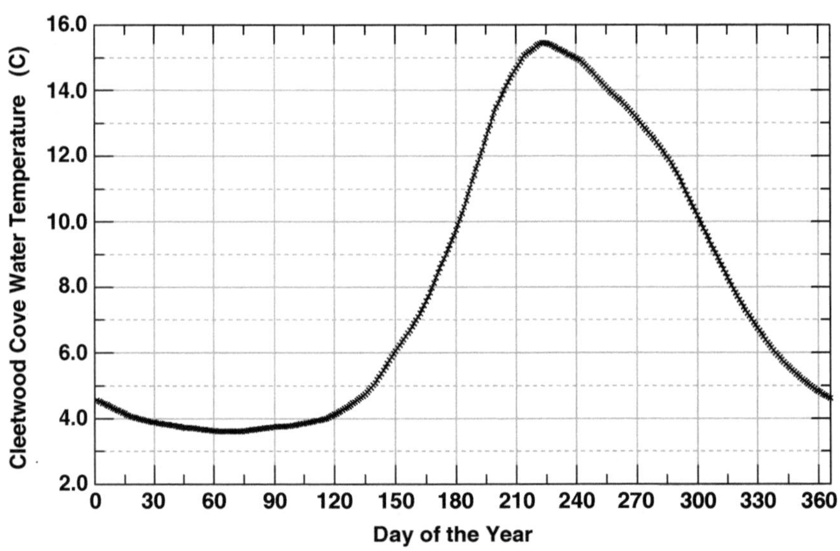

Fig. 3 Surface water temperature at Crater Lake, Cleetwood Cove USGS gage, using available data from 1991–97

Table 2 Crater Lake buoy evaporation estimated with Adams et al. (1990) formulation, using buoy data when available and simulated buoy data from proxy sources when not available

	Jan	Feb	Mar	Apr	May	Jun	Jul	Aug	Sep	Oct	Nov	Dec	Ann
1991	6.51	5.01	5.54	5.12	3.90	3.57	4.53	9.62	10.55	b11.23	b8.43	b6.80	80.81
1992	b7.52	b4.19	1.82	b1.71	b2.60	b5.74	b5.38	b9.74	11.52	7.80	9.13	8.91	76.06
1993	7.12	4.49	1.43	4.15	1.54	3.03	7.58	7.28	b7.60	b7.42	b10.90	b6.99	69.53
1994	b4.67	b5.33	b4.36	b2.39	1.03	1.47	8.96	11.66	b9.92	b9.67	b11.92	b8.08	79.46
1995	b4.38	b4.59	b5.49	b3.60	b1.74	3.62	5.03	11.22	9.94	10.27	5.99	5.84	71.71
Ave	6.04	4.72	3.73	3.39	2.16	3.49	6.30	9.90	9.91	9.28	9.27	7.32	75.51
Seep	10.76	9.80	10.76	10.41	10.76	10.41	10.76	10.76	10.41	10.76	10.41	10.76	126.74

Months with buoy values indicated by "b." Units: cm

unity, and that the park headquarters precipitation is close to the basin average precipitation. If the lake basin were to receive 110% of the headquarters value, then a loss of 60% of annual precipitation in the terrestrial ring to evapotranspiration would be implied, a rather high fraction.

Comparisons of daily variations in simulated and observed stage changes are shown in Fig. 4. Because runoff is not explicitly included for this exercise, absolute numbers are of less interest than how relative day to day changes are simulated; thus, in Fig. 4 differences are reset to zero at the start of each month. Figure 5 shows a comparison of simulated and observed levels reset only once to zero on the first day of buoy data. Intervening days have used substituted proxy data when buoy data are unavailable, in the priority order discussed earlier. Again, since no attempt is made to account for runoff, the day-by-day *changes*, rather than actual lake level itself, are the items of most interest.

Comparison of buoy evaporation with observations

Buoy estimates cannot be validated against stage data, except for days when runoff and precipitation contributions are near zero. Days were selected with simultaneous data, two-consecutive days without precipitation, low runoff, and in winter cold temperatures and deep snow on ground. During the five years, 72 winter days and 105 summer days met these criteria. A constant seepage of 0.347 cm per day was assumed and subtracted from the Cleetwood daily changes. The following comparisons result:

		Adams Buoy	Cleetwood	Buoy/ Cleetwood
Winter	72 days	11.90	10.37 cm, totaled	1.15
Summer	105 days	27.53	26.08	1.06
Total	177 days	39.43	36.46	1.08

The Adams algorithm results in somewhat higher evaporation, especially in winter. Note that the "truth" consists of a biased sample, because of the inability to utilize the same procedure on stormy days, when wind, humidity and atmospheric stability (pre- and post-frontal) might produce a different evaporation regime.

The annual evaporation of 76 cm for these 5 years is closer to previous estimates, discussed and summarized by Nathenson (1992: pp. 10 and Table 1). The annual averages given therein range from 59 cm (Phillips, 1968), to 69 cm (Simpson, 1970), to 83 cm (Meyers, 1962), to 97 cm (Nathenson, 1992), to an earlier value by Redmond of 120 cm (1990b). The latter estimate stands out and from this analysis now appears to be excessively high. All of these estimates are based on various assumptions about the remaining water budget terms. Nathenson proposed a consensus value of 85 cm, ±10%, not greatly different from the estimate of 76 cm given here, the only one based on hourly meteorological measurements directly on the lake surface.

Proxy-based extension of the evaporation record

The comparison between proxy-assisted and buoy-only values gives enough confidence to further extend the record with this technique.

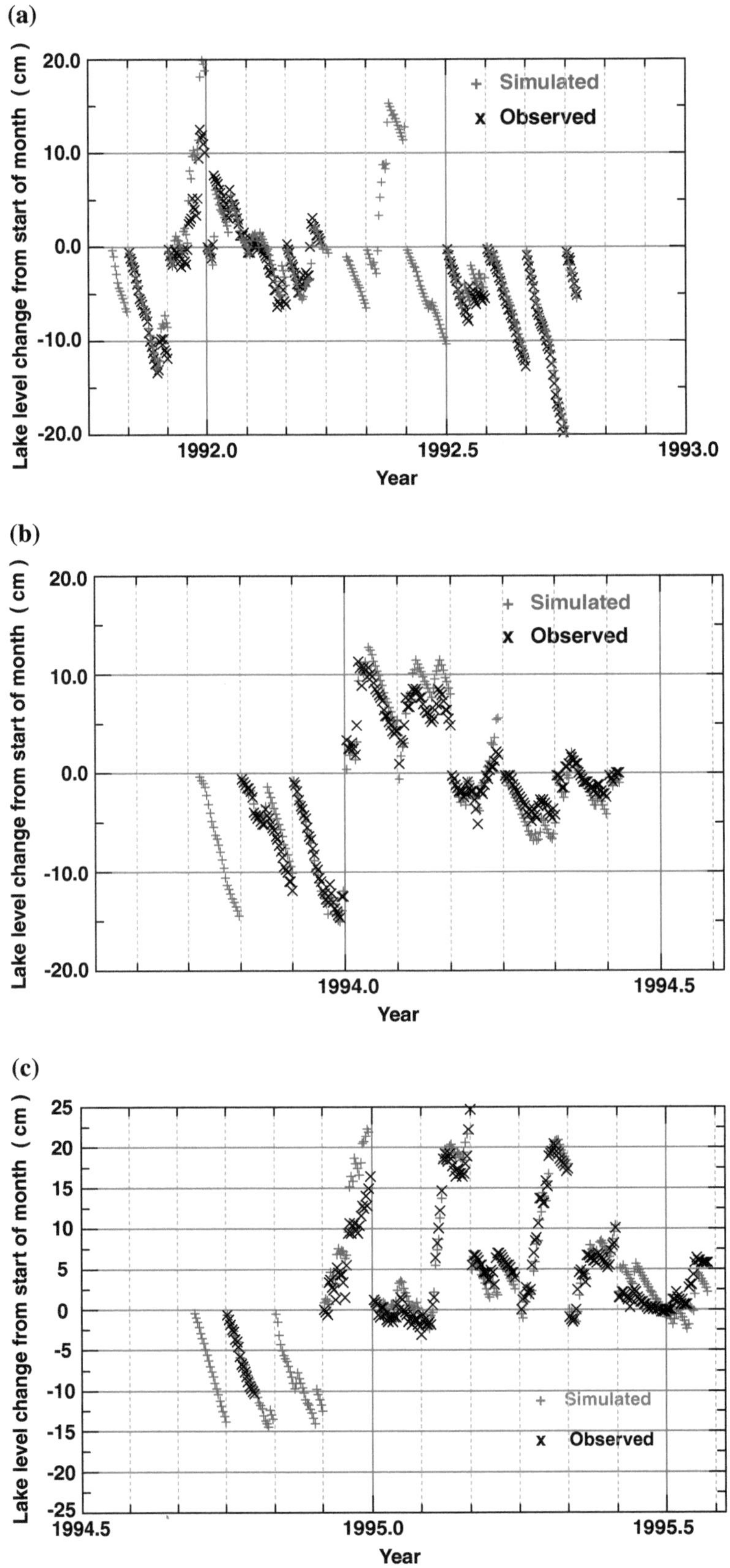
(a)
Lake level change from start of month (cm)
20.0
10.0
0.0
-10.0
-20.0
1992.0
1992.5
1993.0
Year
Simulated
Observed
(b)
Lake level change from start of month (cm)
20.0
10.0
0.0
-10.0
-20.0
1994.0
1994.5
Year
Simulated
Observed
(c)
Lake level change from start of month (cm)
25
20
15
10
5
0
-5
-10
-15
-20
-25
1994.5
1995.0
1995.5
Year
Simulated
Observed

◀ **Fig. 4** Simulated versus observed daily lake level changes, using f = 1.166, reset to zero the first day of each month and accumulated through the month. Only relative changes are of interest. Winters are (**a**) 1991–1992 and 1992–1993, (**b**) 1993–1994, and (**c**) 1994–1995

Accordingly, the same approach was subsequently applied for the period from 1950 through 1996, taking advantage of the long upper air record from Medford. Prior to 1961 there are no lake temperatures, so a climatological estimate for each day was used. Summary statistics for calendar years (to show one more year) are shown in Table 3 for 1950–1996. The overall mean is 72.4 cm. For Water Years 1950–51 through 1995–96, mean evaporation is 72.5 cm, standard deviation 7.7, for a coefficient of variation of 10.6%. For the same period at Park HQ, mean precipitation is 171.2 cm, standard deviation 32.2, and coefficient of variation of 20.9%. Figure 6 shows the relative temporal variations of water year observed precipitation and calculated evaporation during this 47 year period, and confirms the importance of water supply variations compared with water loss variations in affecting the lake level change.

Statistical summaries of evaporation for each month in the period 1950–1996 are shown in Fig. 7. The time series of calculated individual monthly evaporation is shown in Fig. 8, with a running 12-month mean.

It is very difficult to find long-term time series of evaporation or evapotranspiration, especially for similar circumstances, to compare these temporal statistics. From 1950 through 1996, the calculated evaporation values ranged from 54.7 cm in the very wet (224 cm) El Nino water year of 1983 up to 95.6 cm in the dry water year (136 cm) of 1967, a range of 76–132%. (The wet/dry association of the two extreme years may just be chance—see below.) Hostetler & Bartlein (1990) estimate annual values of 90, 88, 92, 97, and 103 cm from 1981–1985 for Malheur–Harney

Fig. 5 Tracking of water elevation changes, comparison between estimate and Cleetwood-determined observations, reset to zero only in October 1992

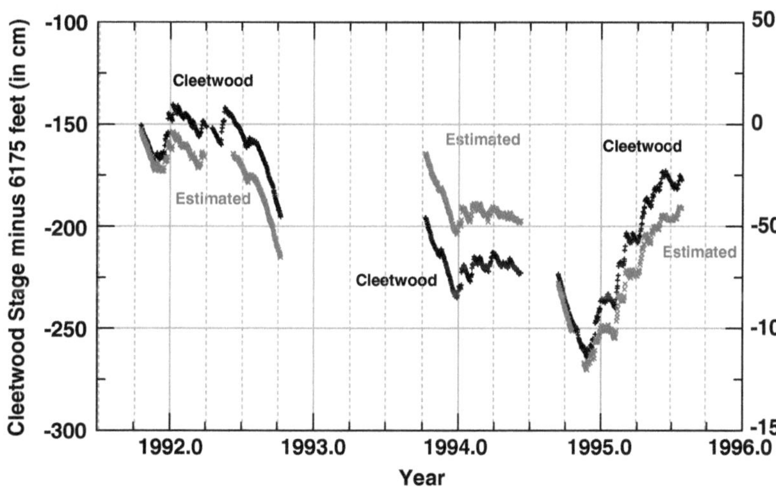

Table 3 Statistics of estimated proxy-based evaporation for calendar years. 1950–1996

	Jan	Feb	Mar	Apr	May	Jun	Jul	Aug	Sep	Oct	Nov	Dec	Ann
Ave	6.47	5.40	5.00	3.64	2.53	2.68	4.92	8.38	9.29	8.71	8.48	6.93	72.42
SD	2.60	2.41	2.35	2.00	1.68	1.74	2.33	2.97	3.11	3.01	2.96	2.69	8.56
C. V.	0.40	0.45	0.47	0.55	0.66	0.65	0.47	0.35	0.34	0.35	0.35	0.39	0.12
Max	9.33	8.46	9.70	5.94	4.31	5.73	8.96	13.53	14.32	12.11	11.92	9.91	94.30
Min	3.78	2.62	1.43	1.71	1.02	1.09	1.98	5.52	4.89	4.43	5.73	3.75	49.82
Years	47	47	47	47	47	47	47	47	47	47	47	47	47

Ave is average, SD is standard deviation, C.V. is coefficient of variation, Max is the highest extreme, Min is the lowest extreme, Years is the number of years used. Units: cm

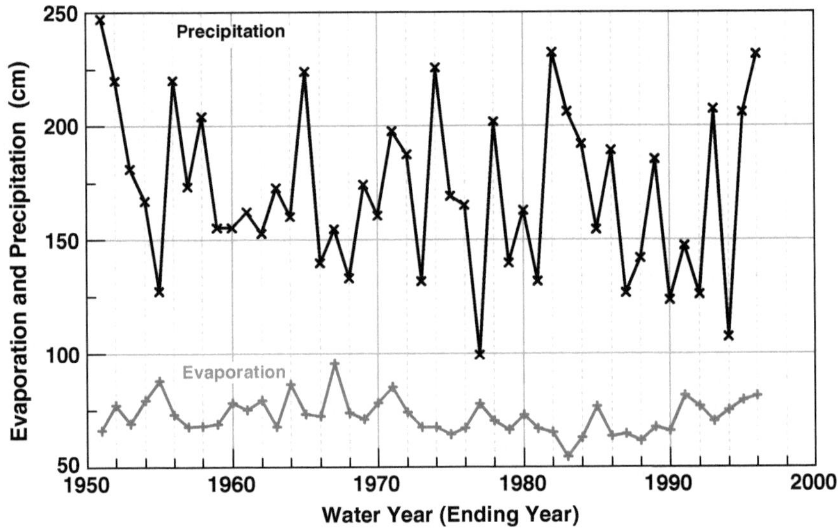

Fig. 6 Relative variability and annual means of annual (calendar year) Park Headquarters precipitation, and of calculated evaporation, 1950–1996 (Adams formulation). Precipitation mean 170.9 cm, standard deviation 36.22 cm; evaporation mean 72.44 cm, standard deviation 7.73 cm

Lake in eastern Oregon's high desert 200 km east, a range of nearly 10% around the mean of an admittedly short sample. Their shallow lake system has a summer maximum in evaporation, is at a lower altitude (1249 m) and at the same altitude as surrounding land, and thus may not be a good analog for Crater Lake. Notably, the summer climate of 1983 was highly unusual in terms of frequent precipitation, high humidity, and failure of then-recently sharply increased levels in many playa lakes in the northern Great Basin to fall as much as usual during July, and Crater Lake shows its lowest annual evaporation in the entire record that year.

Perhaps a better comparison would be with the Great Lakes. Although much lower in elevation, like Crater Lake they have an autumn–early winter evaporation maximum. Derecki (1976) reports an average of 91 cm evaporation for Lake Erie from 1937–1968, standard deviation 12.2 cm, coefficient of variation 0.13, and a range of 68–111 cm, or 75–122% in these 32 years. Derecki (1981a, b) gives an average of 48.3 cm for Lake Superior from 1942–1975, standard deviation 5.6 cm, coefficient of variation 0.12, and a range of 40.4–62.6 cm, or 83–130% in these 34 years. This lake does freeze.

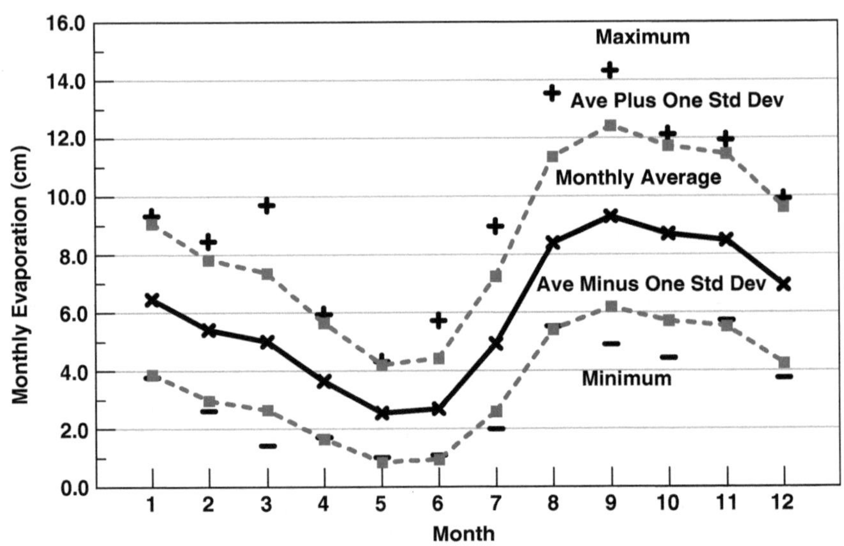

Fig. 7 Statistical summary of monthly estimated evaporation, 1960–1996

Fig. 8 Calculated evaporation, 1950-96, for individual months and for 12-month running mean, based on proxy data from Medford radiosonde and Park Headquarters climate station

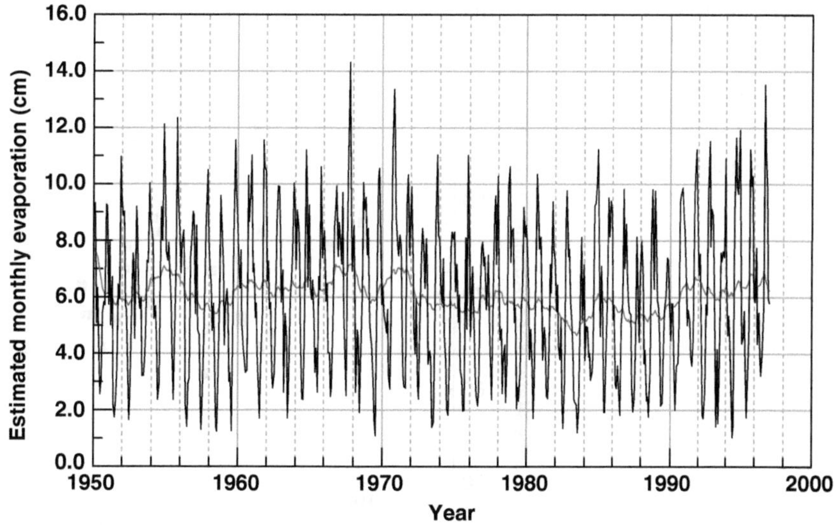

Pan evaporation is often used to estimate evaporation from both lakes and ponds and from terrestrial vegetative cover. R. A. Lind & J. D. Goodridge (1978, Unpublished) used warm season pan evaporation at Tahoe City for 1958–1977, and estimated (constant) values for winter (November–April, when the pan is frozen), and applied a pan coefficient of 0.70 to arrive at a value of 68 cm for Lake Tahoe. They found a standard deviation over these 20 years of 3.6 cm, a coefficient of variation of 0.052, and a range from 61.9–76.1 cm, or 91–112%. They assigned 24.6% of the annual mean evaporation to November–April, whereas the above estimates indicate this should be closer to 50% and has significant variability, so the Tahoe variations are not large enough. Lake Tahoe (1898.6 m when full) almost exactly the same elevation as Crater Lake (1882.14 m).

For contrast, in a non-lake environment, a 40-year set of calculated annual potential evapotranspiration values for central Illinois shows about an 8% coefficient of variation for annual values (data supplied by Ken Kunkel, Midwest Regional Climate Center).

Using a value for f of 1.166, constant seepage, and the calculated evaporation, lake level estimates were determined for every day from 1950 through 1996. Because runoff is not accounted for, as mentioned earlier it is best to restrict comparisons to mid-winter and late summer months. An examination of many individual months shows a large number of close estimates during these 564 months. Using only annual changes from one September 30 to the next, a period when runoff issues should not confound the comparison, and since 1961 when the Cleetwood gage was installed, the correlation between estimated and observed changes is $r = 0.85$ for the 34 years. This value, based on summing 365 days, is not as good as the correlation of $r = 0.96$ between annual precipitation at Park HQ and annual lake level change in Fig. 2, but it is not clear which relation one should expect to be better. Three of the years have lake level change differences (estimated minus observed) of more than 30 cm, accounting for most of the poorer fit, and none of the 34 years have an incorrect sign.

The relationships between precipitation and evaporation are likely to vary by time scale, from daily to seasonally to annually, so that all of the simplifying assumptions made in prior studies are open to question. Any relation of annual precipitation to annual evaporation, if there is one, is likely to be complicated. The annual October–September proxy-based evaporation values are plotted in Fig. 9 against Park HQ precipitation for the same period. There is essentially no correlation ($r = -0.16$) on annual time scales between precipitation and estimated evaporation for these 46 years, and it is not intuitive that there should be any such relationship.

Fig. 9 Relation between water year precipitation at Park Headquarters and calculated water year evaporation from lake surface, 1950–1951 through 1995–1996

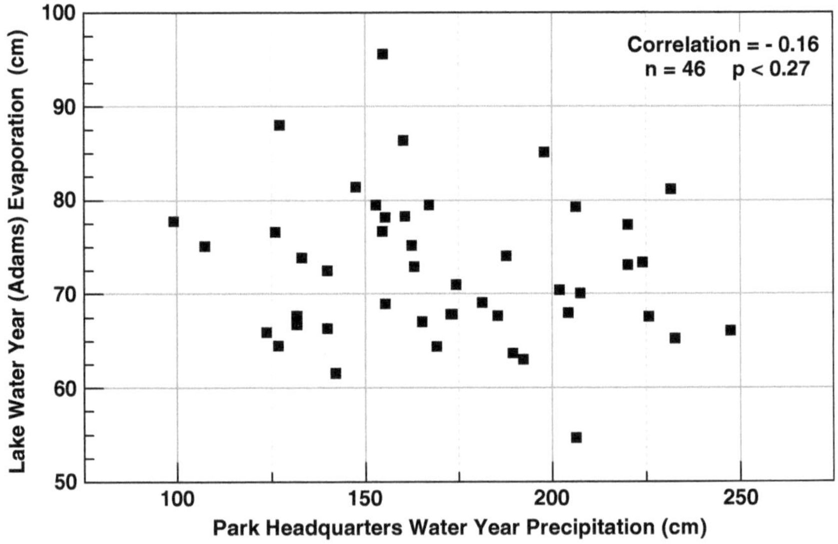

Discussion

Each additional cm of annual precipitation measured at Park HQ corresponds to an eventual lake rise of 1.4 cm. What is the source of this magnification? There appear to be at least four multiplicative factors: (a) The terrestrial ring contributes runoff, (b) the terrestrial ring could receive more precipitation (including drift) than the lake average ($d > 1$), c) the lake could receive more precipitation than the actual precipitation at Park HQ ($c > 1$), and 4) actual precipitation at the gage could be more than what the gage catches ($g > 1$). If two-thirds of water in the terrestrial ring runs off, this factor would be 1.18 (likely range 1.14–1.23). The terrestrial ring could easily receive 1.05 times the lake average (likely range 1.0–1.1). The lake could easily receive 1.05 times the actual precipitation at Park HQ (likely range 0.9–1.1). The actual precipitation could easily exceed the measured precipitation by 1.07 (likely range 1.05–1.15). The product of these four factors can readily account for the 1.4 magnification.

The sum of S and E (about 203 cm/year) differs from the constant term, 235.8, in (8). This constant does not necessarily represent the water supply to the lake, if there is gage undercatch at Park HQ. If the rain contribution to annual Park HQ precipitation was measured within 1–2% of the true value, but snow was undermeasured by 10%, a simple exercise shows that this nonlinearity in bias could account for the discrepancy. An examination of records from co-located shielded and unshielded gages at Park HQ shows that the elevated shielded gage reports about 1.18–1.22 times more precipitation over the past five years than the unshielded gage at the ground/snow surface.

Within the uncertainties in the various ratios and parameters, the buoy data have been extremely useful in providing better estimates of lake evaporation. Additional years of data could be further exploited. It seems unlikely that the other components of the lake budget will be independently estimated any time soon to the accuracy needed to confidently close the lake water budget, especially the lake averaged precipitation. Systematic snow surveys around the rim of the lake near April 1 during an average winter to assess snow water content would be very helpful. The relatively large annual variability in water supply compared with loss variability, the lack of correlation between these quantities, and the ability of a small gage with a surface area of only 324.3 cm^2 to serve as a surrogate for a basin that is 2.1 billion times larger, largely explain why annual lake level change is so closely correlated with annual Park HQ precipitation.

The Crater Lake buoy is located over the deepest portion of the lake, well away from the nearest shoreline. There are likely systematic variations associated with changes in atmospheric profiles of wind, temperature and humidity along the overwater trajectory, and with growth of water waves as the air column moves from the shoreline to the lake center, as well as effects associated with airflow descending and ascending the steep caldera walls. In other large lakes with high autumn evaporation, the upwind side experiences more evaporation since the air is not yet moistened or warmed, although in this case the speed may be temporarily reduced in the lee of the steep sides. The buoy conditions thus do not necessarily represent exactly the atmospheric or water conditions everywhere on the entire lake; however, it appears that its present position (Station 13, Larson, 1990) is probably close to optimal if a single buoy is used. A three-dimensional atmospheric flow model would be needed to investigate these effects. Sufficiently fine resolution is now available with current atmospheric models that this approach would be preferred in future improvements.

Acknowledgements Much of this work was originally supported by the Crater Lake Limnological Studies Program at Oregon State University. I am very thankful for much help, advice, and data received over the years from Gary Larson, Bob Collier, Greg Crawford, and Mark Buktenica, Crater Lake park staff, and from many others at Oregon State University and elsewhere. Comments by Manuel Nathenson and Steve Hostetler and one anonymous reviewer were very helpful and greatly appreciated.

References

Adams, D. E., D. J. Cosler & K. R. Helfrich, 1990. Evaporation from heated water bodies: predicting combined forced plus free convection. Water Resources Research 26: 425–435.

Bacon, C. R., J. V. Gardner, L. A. Mayer, M. W. Buktenica, P. Dartnell, D. W. Ramsey & J. E. Robinson, 2002. Morphology, volcanism and mass wasting in Crater Lake, Oregon. GSA Bulletin 114: 675–692.

Derecki, J. A., 1976. Multiple estimates of Lake Erie evaporation. Journal of Great Lakes Research 2: 124–149.

Derecki, J. A., 1981a. Operational estimates of Lake Superior evaporation based on IFYGL findings. Water Resources Research 17: 1453–1462.

Derecki, J. A., 1981b. Stability effects on Great Lakes evaporation. Journal of Great Lakes Research 7: 357–362.

Doesken, N. J. & A. Judson, 2000. The Snow Booklet: a Guide to the Science, Climatology, and Measurement of Snow in the United States. Colorado State University, Department of Atmospheric Sciences.

Drake, E. T., G. L. Larson, J. Dymond & R. Collier, (eds), 1990. Crater Lake: an Ecosystem Study. American Association for the Advancement of Science, Pacific Division, Allen Press, Lawrence, KS.

Eichenlaub V., 1979. Weather and Climate of the Great Lakes Region. University of Notre Dame Press, Notre Dame, IN.

Gardner, J. V., L. A. Mayer & M. Buktenica, 2000. Cruise Report, R/V *Surf Surveyor* Cruise S1–00-CL, Mapping the Bathymetry of Crater Lake, Oregon. USGS Open-File Report 00–405.

Gregory, S. V., G. L. Larson, C. D. McIntire & M. Buktenica, 1993. Water chemistry of caldera springs. In Larson G. L., C. D. McIntire & R. W. Jacobs (eds), Crater Lake Limnological Studies Final Report. Technical Report NRTR–93/03. Cooperative Park Studies Unit, College of Forestry, Oregon State University, Corvallis OR, 131–174.

Goodison, B. E., 1978. Accuracy of Canadian snow gauge measurements. Journal of Applied Meteorology 27: 1542–1548.

Hostetler, S. W. & P. J. Bartlein, 1990. Simulation of lake evaporation with application to modeling lake level variations of harney-malheur Lake, Oregon. Water Resources Research 26: 2603–2612.

Larson, G. L. 1990. Status of the ten-year limnological study of Crater Lake, Crater Lake National Park. In Drake E. T., G. L. Larson, J. Dymond & R. Collier (eds), Crater Lake: An Ecosystem Study. Pacific Division, American Association for the Advancement of Science, San Francisco, CA, 7–18.

Larson, G. L., C. D. McIntire & R. W. Jacobs, 1993. Crater Lake Limnological Studies Final Report. Technical Report NRTR-93/03. Cooperative Park Studies Unit, College of Forestry, Oregon State University, Corvallis OR.

Larson, G. L., (2007). Preface: History of Crater Lake Program.

Mahrt, L., & M. Ek, 1984. The influence of atmospheric stability on potential evaporation. Journal of Climate and Applied Meteorology 23: 222–234.

Meyers, J. S., 1962. Evaporation from the 17 Western States. Includes Section on Evaporation Rates (J.T. Nordenson). USGS Professional paper 272-D.

Nathenson, M. 1992. Water balance for Crater Lake, Oregon. USGS Open-file report 92–505. USGS, Menlo Park, CA.

Nathenson, M., C. R. Bacon & D. W. Ramsey, (this issue). Subaqueous Geology and a Filling Model for Crater Lake, Oregon.

Peterson, D. L., D. G. Silsbee & K. T. Redmond, 1999. Detecting long-term hydrological patterns at Crater Lake, Oregon. Northwest Science 73: 121–130.

Phillips, K. N., 1962. Station Analysis for All Records of Stage, 1878–1962, Crater Lake near Crater Lake, Oregon. USGS manuscript furnished to Crater Lake National Park, March 30, 1987.

Phillips, K. N., 1968. Hydrology of Crater, East, and Davis Lakes, Oregon. USGS Water Supply Paper 1859-E.

Quinn, F. H., 1979. An improved aerodynamic evaporation technique for large lakes with application to the International Field Year fot the Great Lakes. Water Resources Research 15: 935–940.

Rasmussen, A. H., M. Hondzo & H. G. Stefan, 1995. A test of several evaporation equations for water temperature simulations in lakes. Water Resources Bulletin 31: 1023–1028.

Redmond, K. T., 1988. Climate, climate variability, and Crater Lake. In Larson G. L. (ed.), Crater Lake Limnological Studies, 1987. Cooperative Park Studies Unit, Oregon State University, Corvallis, OR, or available from author, 69–79.

Redmond K. T., 1990a. Crater Lake climate studies. In Larson G. L. (ed.), Crater Lake Limnological Studies, Annual Report, 1989. Cooperative Park Studies Unit, Oregon State University, Corvallis, OR, 49–56.

Redmond, K. T. 1990b. Crater Lake climate and lake level variability. In Drake E. T., G. L. Larson, J. Dymond & R. Collier (eds), Crater Lake: An Ecosystem Study. Pacific Division, American Association for the Advancement of Science, San Francisco, CA: 127–144.

Redmond, K. T. 1993. Climate variability at Crater Lake National Park and its effect on water level. In Larson G. L., C. D. McIntire & R. W. Jacobs (eds), Crater Lake Limnological Studies Final Report. Technical Report NRTR-93/03. Cooperative Park Studies Unit, College of Forestry, Oregon State University, Corvallis OR, 39–61.

Simpson, H. J., 1970. Tritium in Crater Lake, Oregon. Journal of Geophysical Research 75: 5195–5207.

Soil Conservation Service, 1965. Mean Annual Precipitation, State of Oregon, 1930–1957. Single-page map.

Western Regional Climate Center, NOAA/NCDC. <www.wrcc.dri.edu> (historical data, Oregon).

Yang D., B. E. Goodison, J. R. Metcalfe, V. S. Golubev, R. Bates, T. Pangburn & C. Hanson, 1998. Accuracy of NWS 8" Standard nonrecording precipitation gauge: results and application of WMO intercomparison. Journal of Atmospheric and Oceanic Technology 15: 54–68.

Yang, D., B. Goodison, J. Metcalfe, P. Louie, E. Elomaa, C. Hanson, V. Golubev, T. Gunther, J. Milkovic & M. Lapin, 2001. Compatibility evaluation of national precipitation gage measurements. Journal of Geophysical Research 106: 1481–1491.

Hydrobiologia (2007) 574:47–68
DOI 10.1007/s10750-006-0345-3

CRATER LAKE, OREGON

Long-term observations of deepwater renewal in Crater Lake, Oregon

G. B. Crawford · R. W. Collier

© Springer Science+Business Media B.V. 2007

Abstract We examine observations of key limnological properties (primarily temperature, salinity, and dissolved oxygen), measured over a 14-year period in Crater Lake, Oregon, and discuss variability in the hypolimnion on time scales of days to a decade. During some years (e.g., 1994–1995), higher-than-average wintertime deep convection and ventilation led to the removal of significant amounts of heat and salt from the hypolimnion, while dissolved oxygen concentrations increase. In other years, such as the winter of 1996–1997, heat and salt concentrations increase throughout the year and dissolved oxygen levels drop, indicating conditions were dominated by the background geothermal inputs and dissolved oxygen consumption by bacteria (i.e., minimal deep convection). Over the entire 14 year period, no statistically significant trend was observed in the annual hypolimnetic heat and salt content. Measurements from several thermistors moored in the hypolimnion provide new insight into the time and space scales of the deep convection events. For some events, cool water intrusions are observed sequentially, from shallower depths to deeper depths, suggesting vertical mixing or advection from above. For other events, the cooling is observed first at the deepest sensors, suggesting a thin, cold water pulse that flows along the bottom and mixes more slowly upwards into the basin. In both cases, the source waters must originate from the epilimnion. Conditions during a strong ventilation year (1994–1995) and a weak ventilation year (1996–1997) were compared. The results suggest the major difference between these 2 years was the evolution of the stratification in the epilimnion during the first few weeks of reverse stratification such that thermobaric instabilities were easier to form during 1995 than 1997. Thus, the details of surface cooling and wind-driven mixing during the early stages of reverse stratification may determine the net amount of ventilation possible during a particular year.

Keywords Hypolimnion · Vertical mixing · Nutrient upwelling · Ventilation · Cold-water intrusions · Interannual variability · Long-term monitoring · Climate change

Guest Editors: Gary L. Larson, Robert Collier, and Mark W. Buktenica
Long-term Limnological Research and Monitoring at Crater Lake, Oregon

G. B. Crawford
Oceanography Department, Humboldt State University, 1 Harpst St., Arcata, CA 95521-8299, USA
e-mail: gbc3@humboldt.edu

R. W. Collier
College of Oceanic and Atmospheric Sciences, Oregon State University, 104 Ocean Administration Building, Corvallis, OR 97331-5503, USA

Introduction

Crater Lake is a caldera lake situated at an elevation of 1882 m above sea level. The lake sits atop Mount Mazama, a dormant volcano in the Cascade Mountains of Oregon, U.S.A. Originally formed by a climactic eruption roughly 7700 years ago (Bacon & Lanphere, 1990; Bacon et al., 2002), Crater Lake is the deepest lake in the United States (594 m) and the seventh deepest in the world. As part of ongoing limnological studies at the lake (see Larson, 1996), we have collected a series of physical, chemical, and biological data with the long-term goal of understanding the relationships between physical forcing and biogeochemical cycles in the lake. In this paper, we examine relationships between physical and chemical data in an extended time series, spanning as much as 14 years (1987–2001). The observations demonstrate episodic deep ventilation events and we discuss how these might relate to long-term nutrient budgets in the lake.

The primary data we discuss below are annual vertical profiles of temperature, salinity, and dissolved oxygen, along with detailed thermistor recordings at 18 depths along a mooring located in the deepest part of the North Basin of the lake (Fig. 1); some brief discussion of nitrate measurements at the same location are also included. We will focus our attention on the deep waters that do not directly (or fully) participate in the upper lake's annual thermal stratification cycle. These waters are, however, the major reservoir of fixed nitrogen stored in the system, which is derived from the steady microbial oxidation of particulate organic matter raining down from the euphotic zone. The rate of vertical exchange or ventilation of the deep water eventually controls the availability of new nitrogen in the euphotic zone. Nitrogen appears to be the limiting nutrient

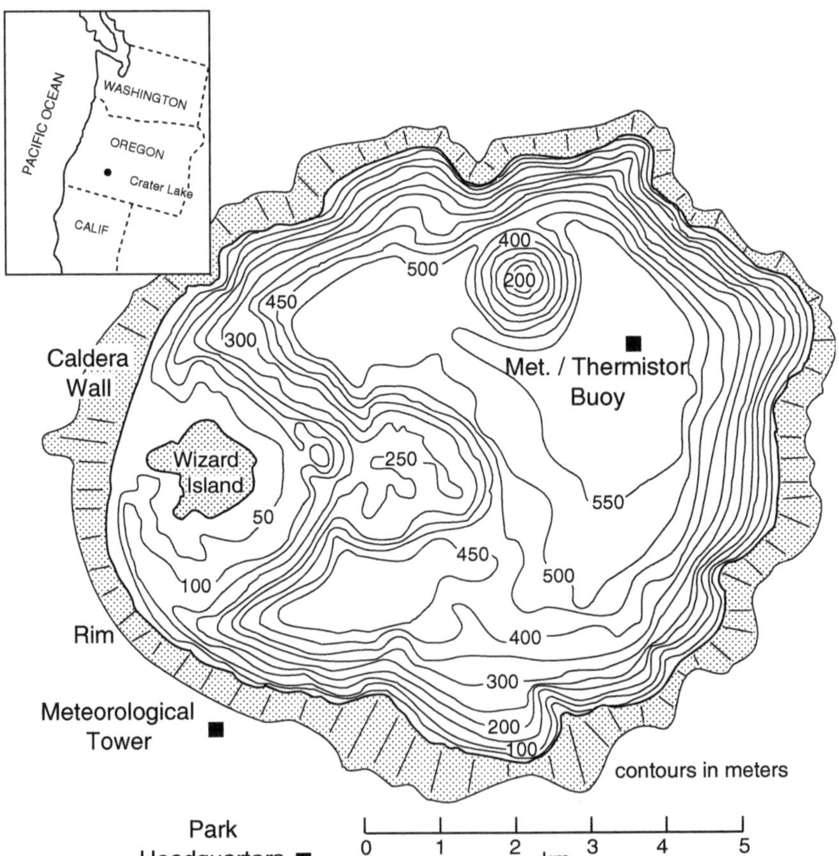

Fig. 1 The location (inset) and bathymetry of Crater Lake, showing approximate location of the observation sites. The CTD casts, thermistor mooring, and primary meteorological station are all located near the deepest part of the North Basin; a second meteorological station is located on the southwest rim of the caldera

in this ultraoligotrophic system (Larson et al., 1996; Groeger, 2007).

The present study provides a number of new key results, including: an assessment of the importance of rotation in the lake; observations of the greatest annual hypolimnetic ventilation thus far recorded (1995); observations corresponding to no (or minimal) annual hypolimnetic ventilation (e.g., 1997); long-term data sets that allow assessment of decadal variations; and high temporal resolution measurements of thermal changes in the hypolimnion that provide new insight to the mixing and exchange properties associated with deep convection.

Background

The lake has a surface area of approximately 53.2 km^2 and a volume of 17.3 km^3 (Phillips, 1968). Surrounded by steep caldera walls, the lake occupies 79% of its drainage basin. Annually-averaged lake level has been nearly constant over at least the past century, as water enters the lake primarily by direct deposition of snow and leaves by evaporation and seepage (Redmond, 1990; Nathenson, 1992). The morphology of the lake (Fig. 1) includes two sub-basins, separated by a sill of about 425–450 m depth; the larger, deeper, North Basin has a maximum depth of 594 m, while the smaller and more shallow South Basin reaches a maximum depth of about 495 m. With extremely low nutrient inputs and organic production (e.g., McIntire et al., 1996), the lake is one of the clearest in the world.

Mixing processes in the lake have been discussed in some detail by McManus et al. (1993). Near-surface temperatures typically range from about 2 to 3°C in late winter or early spring up to as much as 18°C in late summer. In classical limnological terms, then, the lake is dimictic, in that there are two periods when the upper lake (~200 m) becomes well-mixed near 4°C. The hypolimnion, however, receives small, but not insignificant, fluxes of heat and solutes (1 Watt m^{-2} and 5 μg m^{-2} s^{-1}, respectively). Upward diffusion of these fluxes gives rise to a deep-lake salinity gradient of 10^{-2} mg l^{-1} m^{-1} and a hyperadiabatic temperature gradient of 10^{-4}°C m^{-1} (Neal et al., 1971; Williams & Von Herzen, 1983; McManus et al., 1993). The resulting deep water column is slightly stable, with N^2 typically ~10^{-7} s^{-2} (where N is the buoyancy frequency; Crawford & Collier, 1997). Geothermal sources in the South Basin account for most of the benthic heat and salt inputs. Previous observations over an annual cycle (McManus et al., 1993) have shown a steady accumulation of heat and salt in the deep part of the North and South Basins through much of the year due to the hydrothermal inputs, but partial ventilation of the deep waters during winter (typically late December through early March) tends to remove much of this excess heat and salt and 'reset' the deep water temperature and salinity.

The process details associated with the annual partial mixing and ventilation of the hypolimnion with surface waters still remain elusive, as in many other deep lakes. In Lake Baikal, for example, a variety of deep-water ventilation processes have been identified, including thermobaric instabilities (Carmack & Weiss, 1991), thermal bars (Shimaraev et al., 1993; Holland et al., 2001), hydrothermal springs (Kipfer et al., 1996), and bottom boundary layer mixing (Ravens et al., 2000). Recently, cold water intrusions in Lake Baikal have been described (Wuest et al., 2005), but the mechanism for their generation is uncertain. In Lake Issyk-kul, deep water renewal is believed to be caused by cold water density plumes traveling along sunken river channels (Peeters et al., 2003). In the case of Crater Lake, Crawford & Collier (1997) have discussed several different potential processes and suggest that either thermobaric instabilities generated by internal waves or seiches (Walker & Watts, 1995) or by breaking internal waves near a sloping bottom (e.g., Ericksen, 1985; Wuest et al., 1995; Ledwell & Hickey, 1995). Other secondary processes will not be considered in our analysis but may also play some role in the mixing at Crater Lake. For instance, the high cliff walls (wall height/lake width ~0.05) can modify the wind forcing and insolation of the lake in both space and time; slumping of sediment and snow from the cliff walls may also influence mixing (e.g., Hamblin et al., 1999).

Instrumentation and data

Temperature and conductivity profiles have been measured in Crater Lake using a Seabird SEACAT model SBE19 profiler, modified for low conductivity environments, from 1987 to 1998 and with SEACAT model SBE19plus profiler from 1998 through 2001. The sensors are calibrated regularly. The profiles presented here were all taken from the North Basin, in the deepest part of the lake, at a site commonly referred to as "Station 13" (which is based on an early sampling grid developed for Crater Lake). Measurements have also been undertaken occasionally at other locations in the lake, but the sampling history at Station 13 remains the longest and most continuous. Lake access, and hence in situ profiling, is generally limited to summer months (July–September) due to large snowfall accumulations on the caldera walls and rim, although a very few profiles are available from surveys during other seasons.

Temperature resolution for the CTD measurements is +/– 0.0001°C and the absolute accuracy of the temperature sensor is estimated to be about 0.005°C. Salinity is determined from conductivity using a relationship developed for Crater Lake, based on detailed analysis of the chemical composition of the lake water (McManus et al., 1992). Typically, profiles are measured at the center of the North Basin once a month or more during the summer. The salinity data show some high frequency noise (rms ~0.1 mg l^{-1}), which is believed to be primarily instrument noise. After standard processing of the CTD data to 1 m bin averages, the data are then smoothed using a triangular weighting function of width 10 m, which removes the bulk of these high frequency fluctuations. In order to be consistent, the temperature profiles are also smoothed in the same way, although temperature variations are far less noisy than the salinity data (compared to the main, larger-scale profile features).

We also report measurements made from internally-recording thermistors (Alpha-Omega, MDR 9102) moored in the North Basin since 1992, also located at Station 13. The thermistors are mounted at 18 depths on this mooring, with vertical spacings of typically 10–25 m in the upper lake and 40–50 m in the deep lake. The two deepest thermistors were positioned at approximately 545 and 585 m, with depth variability on the order of +/– 15 m or so from deployment to deployment. The thermistors have been calibrated every 1–2 years, with uncertainties estimated to be typically +/– 0.002°C. The data sampling period is 10 min; the instruments are typically recovered and re-deployed twice a year, in early July and late September.

Water samples for dissolved oxygen water were collected in vanDorn bottles at intervals between 10 and 50 m. Oxygen concentrations were determined using a whole-flask Winkler titration method (Carpenter, 1965a, b; Grasshoff et al., 1983). The titration is controlled by a Brinkman Dosimat titrator interfaced with a microcomputer and monitored amperometrically using a Pt electrode (Knapp et al., 1990). The overall precision of the oxygen measurement based on duplicate sample collection and analysis and is better than 0.5%. Solubility for dissolved oxygen was based on the work of Benson & Krause (1984); mean atmospheric pressure at the lake surface was calculated from the altitude-dependent equation given by Mortimer (1981).

Two meteorological stations were established in late 1991, one on a tower on the southwest caldera rim and another on a buoy at Station 13, near the thermistor mooring (Fig. 1). Both of these stations measure wind speed and direction (RM Young 05103), air temperature, and relative humidity (Vaisala HMP35). The buoy includes a thermistor for surface water temperature and the rim station includes a barometer (Vaisala PTA427) and pyranometer (LiCor LI-200SZ). All meteorological instruments were controlled, sampled and logged by Campbell Scientific data loggers (CR10X) and data were retrieved by through a modem to an RF network.

Scaling

Assessing the importance of rotation

For small lakes with sufficient stratification, first-mode basin-scale internal waves appear as internal seiches. If the lake is very large or the stratification is sufficiently weak, the Coriolis effect becomes a dominant factor and seiches are not a free-wave response. The relative importance of the Coriolis effect in a lake can be estimated in terms of the Burger number, S_1. For a flat-bottomed, two-layer system, S_1, is given by:

$$S_1 = \frac{c_1}{Lf}$$

where L is the characteristic horizontal dimension of the lake, f is the Coriolis frequency (also known as the inertial frequency or planetary vorticity),

$$f = 2\Omega \sin \lambda$$

Ω is the angular rotation rate of the earth (2π/86400 s), λ is the latitude (42.93°N), c_1 is the first-mode internal wave phase speed,

$$c_1 = \sqrt{\frac{g' h_1 h_2}{(h_1 + h_2)}} = \sqrt{\frac{g(\rho_2 - \rho_1) h_1 h_2}{\rho_0 (h_1 + h_2)}}$$

h_1 is the thickness of the top (epilimnion) layer and h_2 is the thickness of the bottom layer, ρ_1 and ρ_2 are representative densities for those two layers, respectively, g is gravitational acceleration (taken as 9.81 m/s^2), ρ_0 is an average water density (taken as 1000 kg m^{-3}), and g' is the reduced gravity. When the $S_1 >> 1$, the Coriolis force is unimportant and internal wave setup generated by the wind may lead to a first-mode internal seiche; when S_1 approaches 1 or less, the Coriolis force is important in the lake and the internal wave response tends to be in the form of Kelvin and Poincare waves. Note that S_1 is approximately 1.4 c_1, when the phase speed is expressed in m s^{-1}.

In order to determine the Burger number, we need to calculate density, which requires salinity profiles. Using CTD data from 1989 through 1997, a monthly climatology was developed for the salinity profile (Fig. 2). The number of available salinity profiles for the monthly averages was greatest (7–8) for June through September, since the CTD sampling program is most intensive (and lake access is most reliable) during this period. Profiles from the rest of the year have been made infrequently (April 1989: three profiles; May 1989: three profiles; January 1990: two profiles; January 1997: two profiles). For months where no data are available, we linearly interpolate in time based on the closest months where data are available. As can be seen in Fig. 2, the salinity profile changes very little through the year. The greatest variations occur in the upper lake: in winter, the epilimnion deepens and freshens to about 103.8 mg l^{-1}; in summer, heating and evaporation can lead to a thin, stable, warm and slightly salty (up to 104.8 mg l^{-1}) surface layer.

For most freshwater lakes of shallow to moderate depths, the water column stability can be characterized in terms of the vertical gradient of the in situ density, which in turn depends primarily on temperature and salinity. The issue of stability in deep, freshwater lakes is a little more complex because of the influence of pressure on density and the temperature of maximum density (T_{md}). In situations where the water column is weakly stratified, the in situ density will increase with depth primarily because of the pressure effect. In such a case, vertical perturbations will change the pressure of water parcels, such that their densities are comparable to the background density. Thus, a stability estimate based on the in situ density gradient alone will tend to be an overestimate. Potential density referenced to the surface of the lake, $\rho_\theta(S,\theta,p = 0)$, where θ is potential temperature, is sometimes used, but this approach is also unreliable as a predictor of stability at depth when temperatures are between 0 and 4°C (a consequence of the temperature and pressure effects on density, which can make the water column metastable). We choose instead to use quasi-density, ρ_N, defined as (Peeters et al., 1996)

$$\rho_N(z) = \rho(T, S, 0) \left[1 - \frac{1}{g} \int_z^0 N^2 \mathrm{d}z \right]$$

where

$$N^2 = -\frac{g}{\rho} \frac{\partial \rho}{\partial z} - \frac{g^2}{c^2}$$

is the square of the buoyancy frequency, ρ is the in situ density, c is the speed of sound in water,

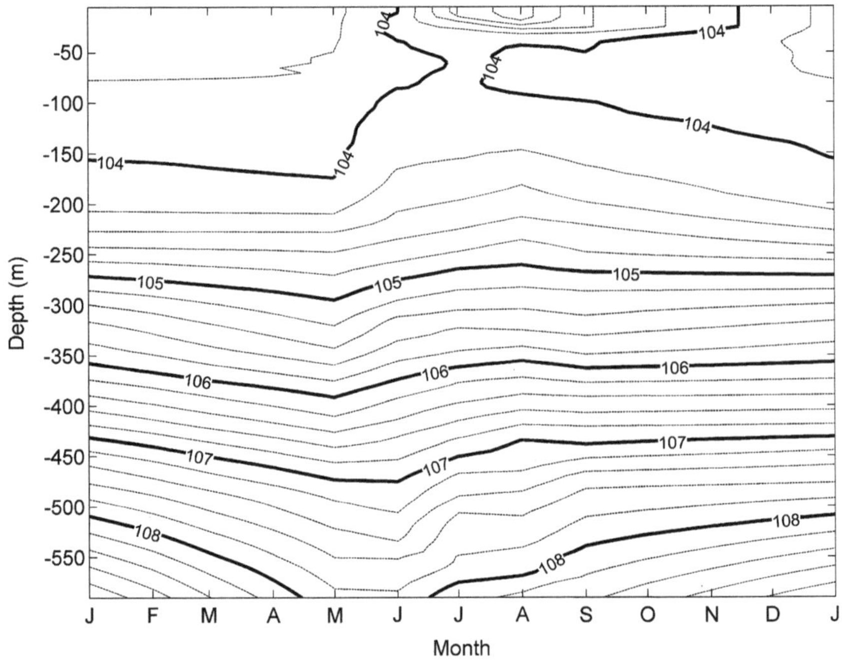

Fig. 2 Salinity climatology (mg l^{-1}) derived from available CTD data (1989–1997). Evaporation effects lead to slightly salty surface waters from June through October, but coincident warming keeps associated density profile stable during this time. Partial ventilation in winter reduces hypolimnetic salinities in mid-winter through early spring

and z is 0 at the surface and negative below the surface. We calculate N^2 using the equation of state for fresh water in Chen & Millero (1986). The quantity ρ_N should be more reliable for evaluating stability of the water column under weak stratification for deep freshwater lakes and temperatures around 0–4°C. We note that calculations for N^2, and associated "bulk" calculations of dynamical quantities c_1 and S_1, should be meaningful when vertical perturbations are not too large. As one anonymous reviewer of this manuscript pointed out, if there is enough downwelling of the epilimnion to generate thermobaric instabilities, there is no restoring force in that region of the water column and the dynamics become much more complicated.

We next consider how to calculate the Burger number. For a nearly-circular lake of uniform depth, L is given by the lake radius (Antenucci & Imberger, 2001). The surface area for Crater Lake, which is roughly circular, is taken as 53.2 km^2, so we take L to be the radius, approximately 4.11 km. The characterization of the layer depths within the lake, however, requires some additional consideration. Frequently h_1 is defined as the shallowest depth where the temperature is cooler than the surface temperature by some fixed amount (e.g., 0.1°C) or the vertical temperature gradient increases above a critical value. However, these approaches are not applicable to this lake, which has subtle temperature and density structures associated with surface water temperatures below 4°C. Here we choose to define h_1 using a more robust, integral form:

$$h_1 = \frac{\int_0^{590\text{m}} z \frac{\partial \rho_N}{\partial z} A(z) \mathrm{d}z}{\int_0^{590\text{m}} \frac{\partial \rho_N}{\partial z} A(z) \mathrm{d}z}$$

where we integrate from the bottom of the lake up to the surface (because of the steep sides to the lake, we also scale the calculations using the hypsometric curve for the lake basin, $A(z)$, in m^2). For a characteristic total depth for the lake, $h_1 + h_2$, we use 590 m. The reduced gravity, g', is calculated based on average values of ρ_N in the upper (h_1) and lower (h_2) layers; we take an average density ρ_0 of 1000 kg m^{-3}.

Wind forcing

As a measure of the strength of the wind forcing, we consider the friction velocity, $\vec{u}_*$, which is estimated from the "law of the wall" similarity theory:

$$|\vec{u}_*|\vec{u}_* = C_z \frac{\rho_a}{\rho_0} \left|\vec{U}\right| \vec{U}$$

$$C_z = \frac{C_{10}}{\left(1 - \frac{C_{10}^{1/2} \log(10/z_u)}{\kappa}\right)^2}$$

where $\vec{U}$ is the wind velocity measured at a height z_u (m) above the lake surface, ρ_a is the air density (taken as 1.225 kg m^{-3}), C_{10} is the drag coefficient at 10 m height (taken here to be 0.0013), and κ is von Karman's constant (taken as 0.4).

Characterizing interannual variability

To compare changes in the deep lake heat and salt budget, we assume horizontal homogeneity and integrate using a hypsographic curve for Crater Lake to examine the relative heat content (relative to 0°C) and the salt content of the hypolimnion, calculated as

$$HC = \int_{-z_1}^{-z_2} \rho c_p T(z) A(z) \mathrm{d}z$$

$$SC = \int_{-z_1}^{-z_2} S(z) A(z) \mathrm{d}z$$

where HC is the heat content (*J*), SC is the salt content (*g*), $T(z)$ is the temperature profile (°C), $S(z)$ is the salinity profile (mg l^{-1}), $A(z)$ is the lake area as a function of depth (m^2), ρ is the density (taken as a constant 1000 kg m^{-3}), c_p is the specific heat capacity (taken as a constant 4.184 J kg^{-1}°C^{-1}), and z_1 and z_2 are the depths that span the hypolimnion. In this case, we take the hypolimnion to be at depths greater than 350 m, as suggested by McManus et al. (1993); we take 540 m as the lower limit of depth calculations because some CTD casts only extended to this depth. This region appears to be relatively well insulated from surface fluxes, except occasionally during mixing in the winter.

It is tempting to estimate net fluxes to the deep lake for other quantities, such as dissolved oxygen and nitrate concentrations, but their budgets are more complex and less well constrained than the budgets for heat and salt. This is especially true for nitrate, since POM and DOM represent major reservoirs and transport vectors for nitrogen and carbon, which are not represented by the simple mixing model that can represent temperature and salinity. The deepwater respiration rate is expected to vary as a function of the flux of organic material from the surface waters and the rate of nitrification should also relate to this POC flux (Dymond et al., 1996; McManus et al., 1996). Although some of the data suggest that the nitrate and oxygen respond to extreme ventilation events, a reasonable budget for these chemical tracers must include the biogeochemical processes that dominate their cycles (Dymond et al., 1996; Fennel et al., 2007) and is beyond the scope of this discussion.

In order to quantify the relative amount of ventilation from year to year, we define a ventilation anomaly, VA, given as:

$$VA = -\left[(HF / 1\ \mathrm{W\ m^{-2}}) + (SF / 5\ \mu\mathrm{g m^{-2} s^{-1}})\right]/2$$

where HF and SF are the annually averaged heat and salt fluxes to the deep lake from 350 m to 540 m. These fluxes are each normalized by the net geothermal heat and salt fluxes, respectively; these normalized fluxes are equally weighted in calculating VA. For VA ≥ 0.5, we characterize the degree of ventilation as 'strong'; for VA ≤ –0.5, we characterize the ventilation as 'weak' (i.e., deep lake heat and salt changes are dominated by geothermal inputs); for intermediate values of VA (–0.5 to 0.5), we define the ventilation as 'average'.

Results

The importance of rotation

Two examples of winter time series of h_1, the epilimnion temperature, T_1 (averaged from the surface to h_1), and S_1, for 1994/95 and 1996/97 are displayed in Fig. 3. As late summer proceeds to winter, the epilimnion depth increases from 20–30 m in mid-September to 275–300 m in early January (Fig. 3a). From January through March,

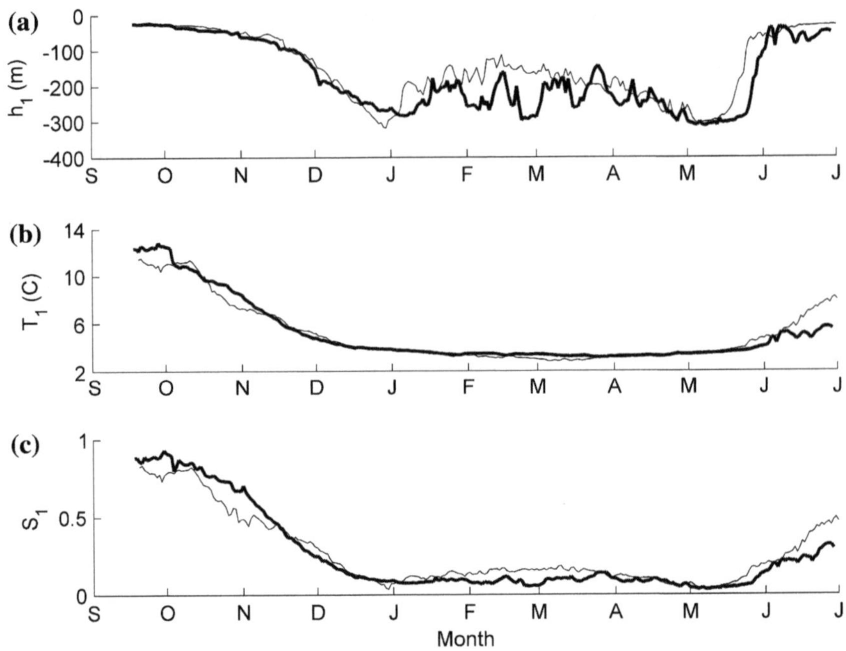

Fig. 3 (**a**) Depth of epilimnion, h_1, in m; (**b**) epilimnion temperature, T_1; (**c**) Burger number, S_1. Thick lines correspond to winter of 1994/95; thin lines correspond to winter of 1996/97

there are a number of times when h_1 decreases again as reverse stratification sets in, although the values in 1997 are more shallow (typically $h_1 \sim$ 125–200 m) and less variable during this period than in 1995 (when $h_1 \sim$ 150–290m). This difference reflects the more extensive reverse stratification in 1997, as we discuss later. From April until mid-May, as the water column begins to warm up, convection and wind-driven mixing keep the stratification low in the upper water column and h_1 remains fairly deep (~200–300 m). By mid- to late May, the surface temperatures warm to greater than 4°C (Fig. 3b), stratification again begins to build, and h_1 returns to the shallow seasonal thermocline.

The Burger number always remains less than 1 in the lake, varying from about 0.9 in late summer to 0.03–0.24 in mid-winter (Fig. 3c). Thus, the Coriolis force is always important and internal seiches are not a natural mode of oscillation in the lake. The Burger number also mimics the epilimnion temperature (Fig. 3b) over much of the year, except when the temperature gets below 4°C and colder surface temperatures generally indicate greater stratification and therefore higher Burger numbers (The Burger number is effectively a function of both h_1 and T_1).

Interannual variability

Figure 4 presents a sequence of summer temperature profiles (CTD) for the deep portion of the lake (below 200 m) over the period 1987–2001. We have chosen to focus on data from September (late summer) of each year because internal wave activity in the hypolimnion is much less significant than during other periods. The CTD casts are measured routinely each year at this time and almost always extend to at least 540 m. There were, however, 2 years (1995 and 1997) where the September profiles did not extend as deeply, so for these cases we have used profiles from the same location in mid-August.

To facilitate comparison of year-to-year changes, profiles for two consecutive years are plotted in each panel. The first year is plotted as a thick line; the subsequent year is plotted as a thin line. Overall, the profiles from year to year are similar. Temperatures below 300 m increase with depth, with a hyperadiabatic temperature gradient of roughly 2–4 × 10^{-4} °C m^{-1}, although sections of some profiles show regions of zero or negative temperature gradients (e.g., 1990, 1991); in addition, the profile for 1988 shows a more complex thermal structure between 400 and 500 m

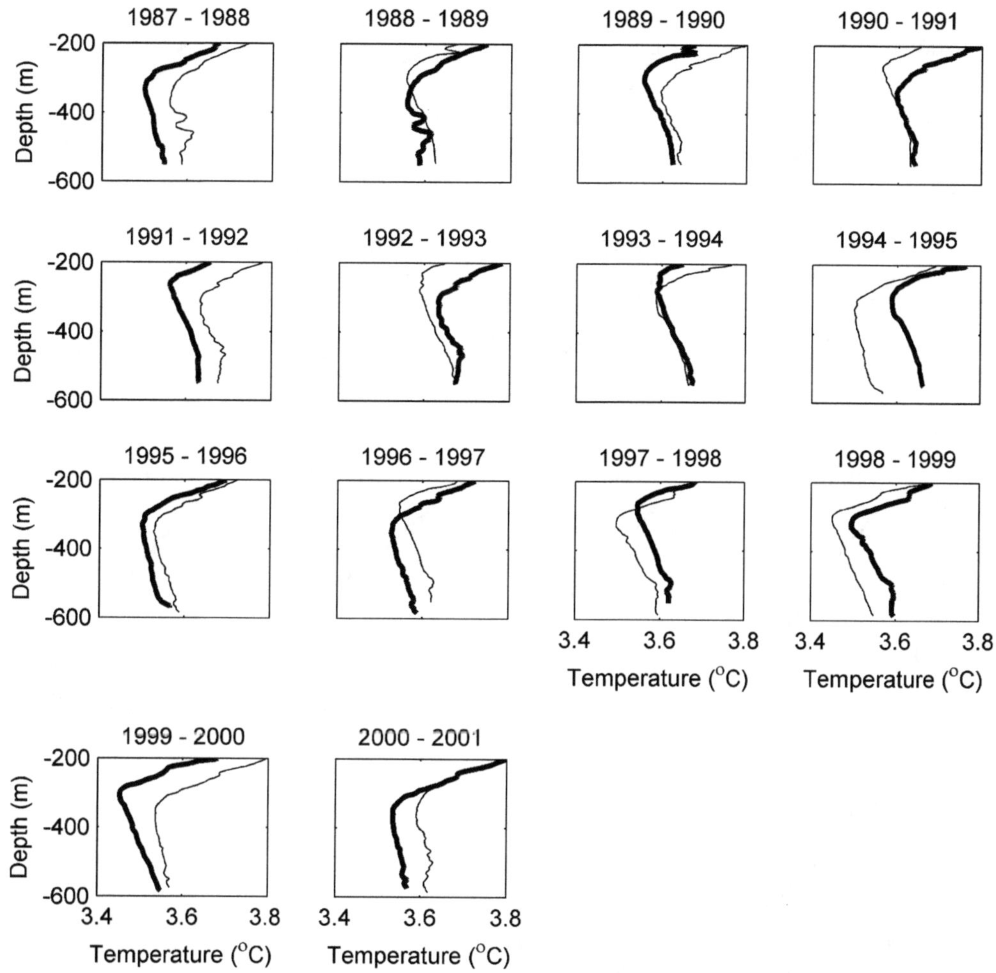

Fig. 4 Interannual variability in deep lake thermal structure at Station 13, North Basin. Temperature profiles are based on mid-September CTD casts (except 1995 and 1997, which are based on mid-August profiles). Each panel displays the temperature profile for a given year and the subsequent year (thick line and thin line, respectively). Each profile has been smoothed in the vertical using a centered, 10 m-wide triangular weighting function. The maximum depth attained during casts varied from year to year. Uncertainties are roughly +/– 0.001°C

compared to other profiles. This structure is likely a consequence of horizontal transport of geothermal heat from the South Basin that has not yet mixed vertically. These features, which are not uncommon in September, are usually smoothed out within days to a few weeks (McManus et al., 1993).

There are some significant differences from year to year. In 1988, the deep lake temperatures are roughly 0.05°C higher than in 1987. Similar net heating is seen from 1991 to 1992, 1995 to 1996, 1996 to 1997, 1999 to 2000, and 2000 to 2001. On the other hand, a substantial decrease of –0.10 to –0.15°C occurred from 1994 to 1995, and of about –0.05 to –0.07°C from 1997 to 1998 and again from 1998 to 1999. Generally, these increases or decreases in temperature from year to year are fairly uniform with depth below about 350 m, although the thermal gradient does change markedly from about 4×10^{-4} °C m^{-1} in 1999 to about 2×10^{-4} °C m^{-1} in 2000; this reduced thermal gradient is still present in the later summer of 2001.

Figure 5 presents the sequence of summer salinity profiles for the same periods as in Figure 4. Profiles were smoothed using the same method as for temperature. All of the profiles show the same general increase in salinity with depth, with an

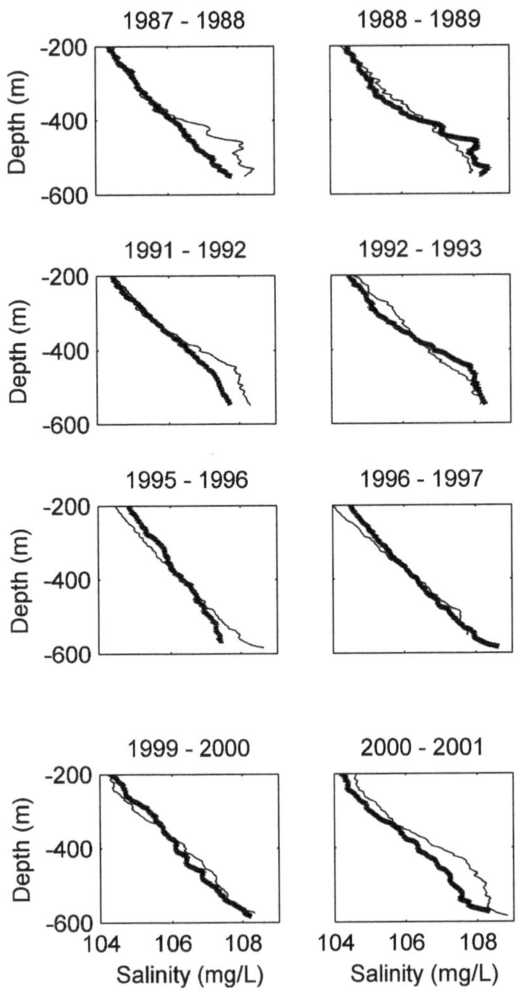

Fig. 5 Interannual variability in deep lake salinity structure at Station 13, North Basin. Salinity profiles are based on mid-September CTD casts (except 1995 and 1997, which are based on mid-August profiles). Each panel displays the salinity profile for a given year and the subsequent year (thick line and thin line, respectively). Uncertainties are about +/– 0.02 mg l^{-1} (McManus et al., 1992)

overall gradient of typically 8–10 × 10^{-3} mg l^{-1} m^{-1} (again, the 1988 profile shows more structure than other years in the deep lake; 2001 also shows a significant bulge of more saline water between 400 and 590 m). Salinities clearly increase in the deep lake from 1987 to 1988 (by typically 0.7 mg l^{-1}) and from 1991 to 1992 (by typically 0.5 mg l^{-1}) and decrease between 1994 and 1995 (–0.7 to –1.0 mg l^{-1}) and again between 1997 and 1998 (–0.3 to –0.5 mg l^{-1}). Relatively little change is observed from 1995 to 1996, and from 1996 to 1997, although there are slight differences in the salinity gradients.

Time series calculations of heat and salt content are shown in Fig. 6. Over the 14 year span, the heat and salt content of the hypolimnion increased by 1 × 10^{15} J and 3 × 10^{6} kg, respectively. However, no significant trend (at the 95% confidence level) is observed in either the hypolimnetic heat or salt content. We therefore conclude that there is no evidence for global warming effects in the hypolimnion over this 14-year period. There are, of course, significant interannual variations, including: an increase in heat content (8 × 10^{14} J) and salt content (20 × 10^{5} kg) from 1987 to 1988; a relatively large decrease in heat and salt content (–15 × 10^{14} J and –24 × 10^{5} kg, respectively) from 1994 to 1995; a large decrease in heat content (–13 × 10^{15} J) from 1997 to 1999,

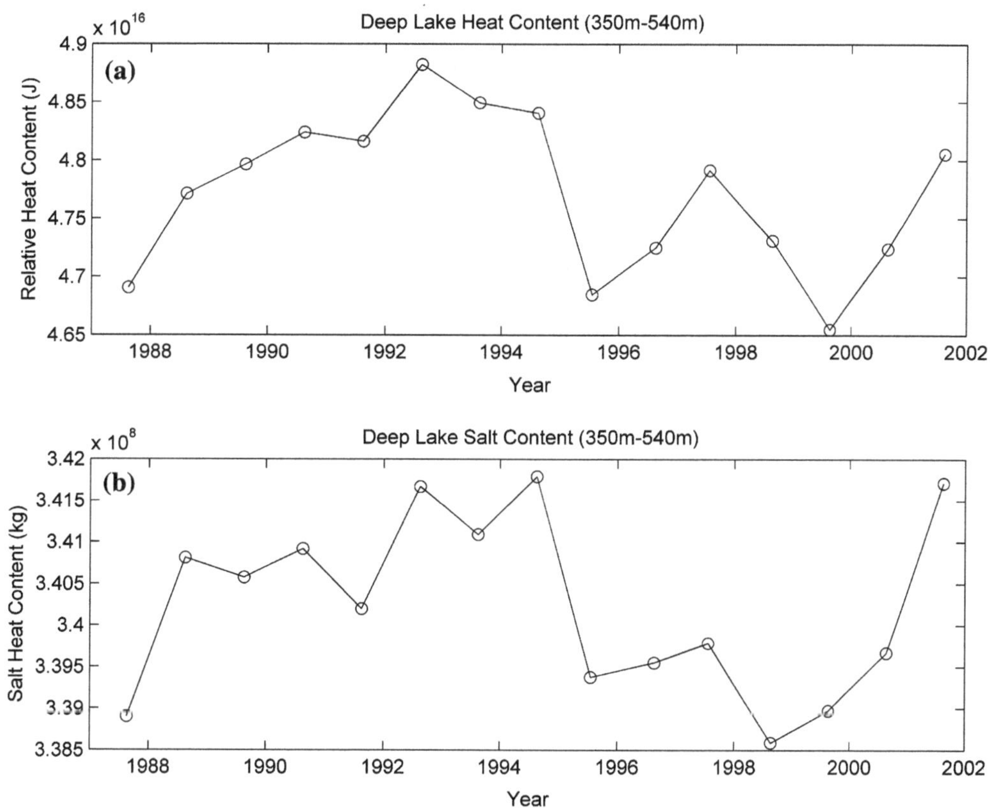

Fig. 6 Relative heat content (upper panel) and salt content (lower panel) of the hypolimnion, based on summertime CTD data from Station 13, North Basin. Calculations are based on the depth range 350–540 m, assuming horizontal homogeneity

with a more limited decrease in salt (-6×10^5 kg); roughly equivalent re-heating between 1999 and 2001 (14×10^{15} J), with a coincident large increase in salt content (28×10^5 kg).

We use the differences in these annual budgets between years to estimate the net annual heat and salt flux into the hypolimnion, as shown in Fig. 7a, b. A net value of zero heat flux over a year corresponds to an annual balance between geothermal heat input, heat loss to surface waters (and ultimately the atmosphere) through vertical mixing, and any heat loss through seepage. Similarly, a zero net salt flux through a year corresponds to an overall balance between chemical inputs with geothermal fluids, vertical transport of salt to the shallower low salinity epilimnion, and loss through seepage. Figure 7c shows a time series of the ventilation anomaly, VA, as defined previously. Based on this methodology, 1995, 1998 and 1999 were seen as strong ventilation years, while 1988, 1992, 1997, 2000 and 2001 were weak ventilation years.

The year-to-year comparisons of annual dissolved oxygen profiles are shown in Fig. 8. Generally the profiles decrease with depth, primarily as a consequence of bacterial oxidation of organic matter. McManus et al. (1996) also identify a minor contribution due to oxidation of reduced inorganic species introduced to the lake via subsurface hydrothermal springs. Overall the profiles are relatively consistent from year to year in the face of ongoing POC fluxes and oxidation—some ventilation of these deep waters must occur nearly every year in order to maintain the oxygen levels at these relatively steady and elevated levels. There is some interannual variability: measurable increases in dissolved oxygen (strong ventilation) occurred between 1994 and 1995 (as much as 5–8 μM l^{-1}), and again between 1998 and 1999. The strongest decrease occurred between

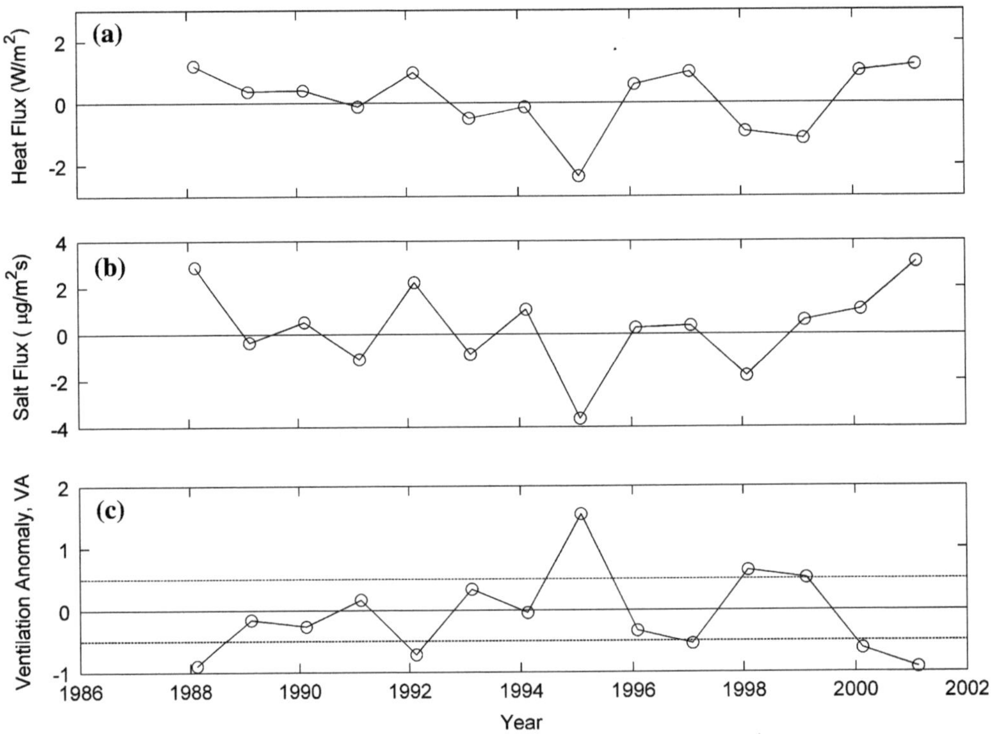

Fig. 7 Net annual heat flux (**a**) and salt flux (**b**) to the hypolimnion, based on summertime CTD data from Station 13, North Basin; (**c**) deep lake ventilation anomaly, VA, as described in text. Values of 0.5 or greater are considered to indicate strong ventilation; values of –0.5 or less indicate weak ventilation

1996 and 1997 (as much as 5 μM l^{-1}), while smaller decreases (2–3 μM l^{-1}) occurred in 1992, 2000 and 2001. These results are consistent with our ventilation anomaly results, reflecting increases in deep lake dissolved oxygen concentrations in strong ventilation years and decreases in weak ventilation years.

Dissolved nitrate is also removed from the deep lake by the vertical exchange processes that introduce new oxygen (McManus et al., 1996; Larson et al., 2007). Nitrate is usually undetectable in the upper lake during the summer and increases consistently below 200 m to a maximum concentration of nitrate (~0.015 mg-N l^{-1}) at around 500 m. There is considerable noise in the data from profile to profile (data not presented here), since the concentrations and resolution are close to the analytical detection limit. The lowest nitrate concentrations at 500 m, 0.012 mg-N l^{-1}, were observed in 1995 after the strongest ventilation year, suggesting that significant deepwater nitrate had been exchanged with depleted upper lake water. Apart from these low extremes, there is no systematic relationship between nitrate and the ventilation anomaly.

Detailed temperature observations in the deep lake

Figure 9 presents a 10-year temperature time series from the two deepest lake thermistors on the mooring in the North Basin. In order to account for variability in thermistor depths from deployment to deployment, we have interpolated the temperature data to specific depths depths: 545 and 585 m (when the deepest thermistor was shallower than 585 m, we still identify it as 585 m). We refer to the associated temperature time series as T_{545} and T_{585}, respectively. Figure 9a shows measurements from the deeper thermistor (T_{585}), while Fig. 9b shows the difference between the deeper and shallower thermistor. Overall, the warmest recorded deep lake temperatures occurred in December, 1992 (3.73°C). From the mid-1990s until 2000, these

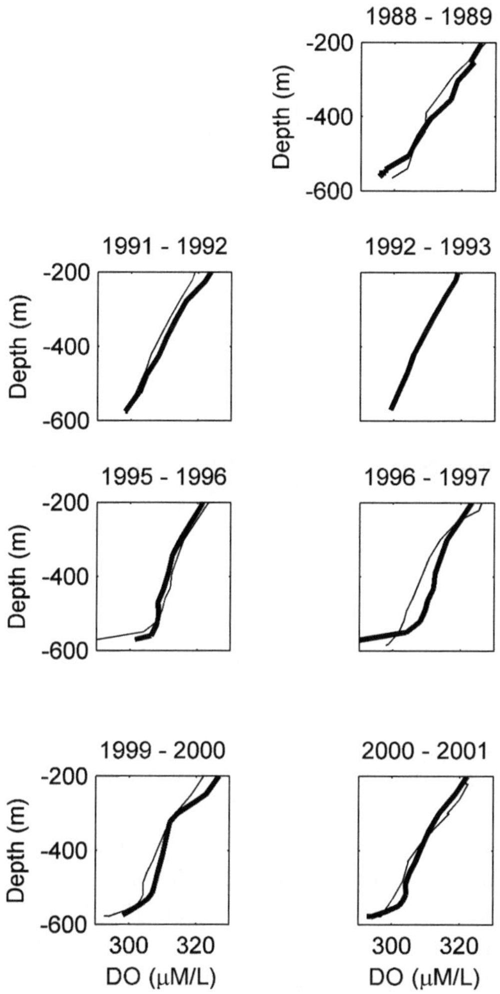

Fig. 8 Interannual variability in deep lake dissolved oxygen concentrations at Station 13, North Basin. Profiles are based on mid-September bottle casts. Depth spacing varied between 10 and 50 m. Each panel displays the dissolved oxygen profile for a given year and the subsequent year (thick line and thin line, respectively). No oxygen data are available from 1987 or 1993. Relative uncertainties are estimated to be typically about 0.5% (McManus et al., 1996)

temperatures were cooler, averaging around 3.5–3.6°C, with short-lived (hours to days) "bursts" as low as 3.18°C (early 1999). For the next 2 years, the deep lake warmed overall, approaching its 1992 maximum at the end of 2001.

The annual cycle of temperature in the deep lake is also clearly reflected by the deep thermistor. From spring through fall there is a net warming in the deep lake (Fig. 9a) and this is associated with an increase in salinity due to hydrothermal inputs, primarily in the South Basin (McManus et al., 1993). In general, the increase in salinity more than compensates for the heat input, maintaining a slightly stable hypolimnion through much of the year ($N^2 \sim 10^{-7}\ s^{-2}$). A slight cooling trend appears in the early winter, presumably due to vertical mixing with slightly cooler waters from above, and generated by internal wave action or boundary mixing as the lake stratification decreases and the effects of wind forcing can propagate more deeply. In late winter and early spring, when the surface temperature of the upper half of the lake is below 4°C, episodic events bring colder, fresher water from the overlying surface lake, effectively 'resetting' the deep lake conditions.

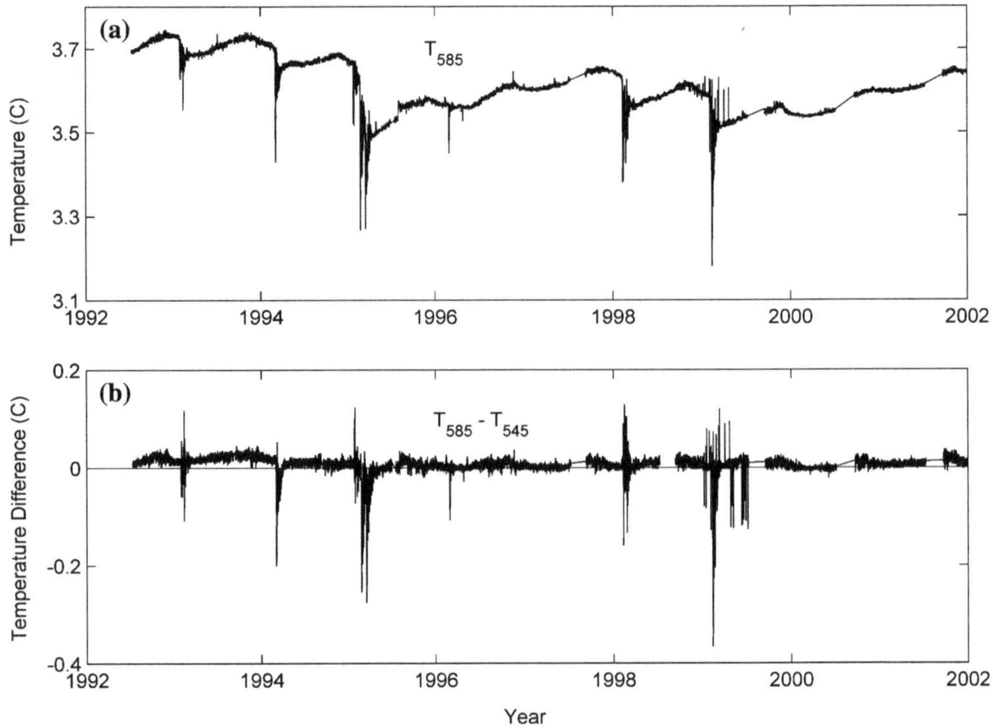

Fig. 9 (**a**) Decadal time series of deep lake thermistor, interpolated to 585 m depth. (**b**) Time series of temperature difference for data interpolated to 585 m and to 545 m (i.e., T_{585}–T_{545})

Generally, the geothermal flux maintains the deep temperature (585 m) about 0.01°C greater than that 40 m shallower (slightly positive values in Fig. 9b). However, particularly during the episodic cooling events in winter, the deeper lake has been observed to be substantially warmer (up to 0.13°C) or cooler (as much as –0.37°C) for periods ranging from hours to several weeks.

We interpret the winter cooling and ventilation of the deep lake, in qualitative terms, as a result of both turbulent diffusion and, episodically, direct advection of surface waters downward. Carmack & Weiss (1991) suggested this may occur in Lake Baikal; an analogous process happens in the ocean as well (see, for example, Marshall & Schott, 1999; Jiang & Garwood, 1995). Admittedly, a clear delineation between these two processes is not always possible from the available data on a single vertical mooring. Here we consider "local" mixing to be mixing of adjacent water masses, which evolves on a relatively long time scale (days to weeks) and probably occurs vertically and to some extent horizontally (enhanced at boundaries). We consider "nonlocal" mixing refer to mixing due to waters from the upper lake that result from thermobaric instabilities (Carmack & Weiss, 1991; Crawford & Collier, 1997), convect vertically on relatively rapid time scales (hours to days) and penetrate into the deep lake largely as distinct "packets" or "pulses" of cool water. Such pulses of water may be observed descending at the mooring, as shown in the data presented in Crawford & Collier (1997), or they may merely show up near the bottom of the thermistor chain in this deepest part of the lake.

The details of deep lake temperature variability in winter can provide some indication of the nature of these episodic cooling events. Figure 10 displays temperature time series from individual thermistors (without vertical interpolation or extrapolation), at depths ranging from 390 to 590 m.

We choose to contrast observations from a strong ventilation year (1995; Fig. 10b) and a weak ventilation year (1997; Fig. 10c). We also show some additional observations from 1993, an average ventilation year, which provide a broader context for the single episodic event previously

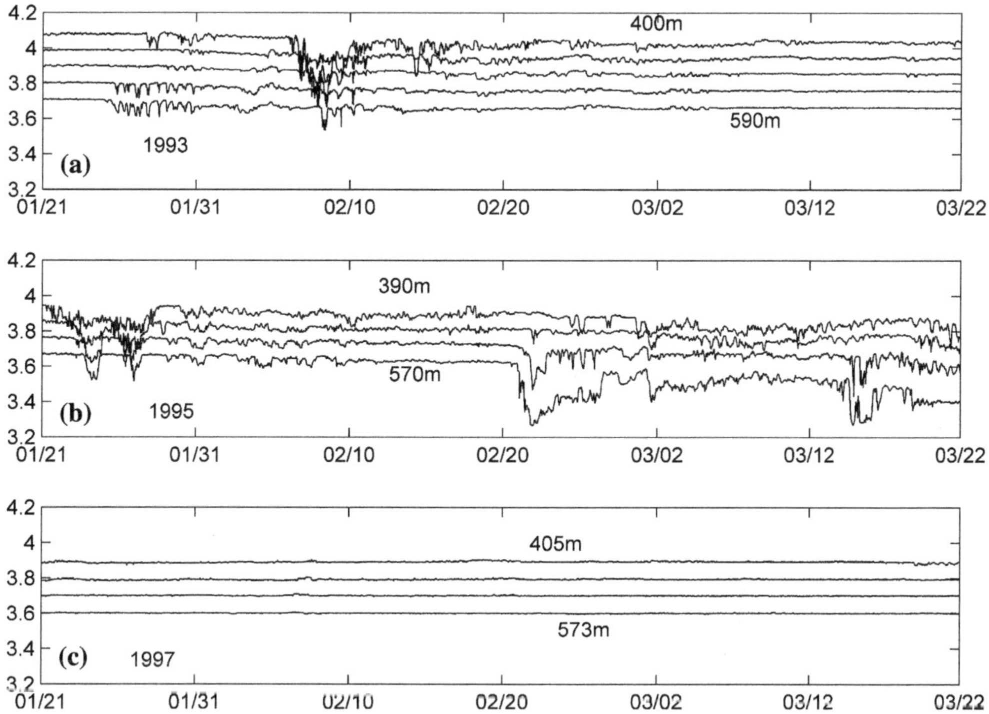

Fig. 10 Deep lake temperature time series from 21 January to 22 March (60 d) for three winter seasons: (**a**) 1993 [590, 550, 500, 450, and 400 m]; (**b**) 1995 [570, 520, 455, and 390 m]; (**c**) 1997 [573, 539, 472, and 405 m]. For each year, data from the deepest thermistor are plotted in the usual manner; an additional 0.1°C offset is added to each subsequently more shallow thermistor data set, to facilitate comparisons (Thus, the deepest thermistor in each case corresponds to the coldest temperatures depicted)

discussed in detail by Crawford & Collier (1997; Fig. 10a).

In these three data sets, we observe two relatively distinct types of episodic cooling events: those that appear to descend vertically at the mooring, and those that arrive in the deep lake without passing the mooring thermistors at intermediate depths. One example of the first type occurred between 7 and 11 February 1993 (Fig. 10a) and was described by Crawford & Collier (1995). The data suggest the vertical descent of cool water at the mooring, as cool temperatures (~3.6°C) are observed sequentially, starting at about 200 m on 5 February and reaching 590 m on 8 February (Fig. 3; Crawford & Collier, 1997). Over the next several days, the hypolimnion temperatures readjust to close to, but slightly cooler than, the values present before this February event. The net cooling between 400 and 550 m due to this event was approximately 0.03°C (although there was no net cooling at 590 m). Other examples of this type of vertical cooling event occur on 23 January and 26 January, 1995 (Fig. 10b).

Three examples of cold water intrusions arriving horizontally are also displayed. The first occurs between 25 and 31 January 1993 (Fig. 10a). Cold water fluctuations appear at 590 and 550 m (on the order of 3.63°C, or about 0.1°C colder than ambient conditions at those depths), but not between 400 and 500 m, where the temperature holds steady at about 3.68°C. Before this period, temperatures colder than 3.63°C were limited to depths shallower than 360 m. Thus, we interpret that the cold water seen at depth must have originated at a depth shallower than 360 m and moved to the deep waters of the North Basin horizontally. (The deepwater temperature fluctuations oscillate between about 3.63 and 3.71°C at a relatively regular period of about 11.5 h. A superinertial period like this usually indicates the presence of a Poincare wave, however the data in

Fig. 10a suggest the temperature oscillations are due primarily to advection of horizontal inhomogeneities in the temperature field.) Temperatures then remain relatively stable at about 3.68°C between 31 January and 2 February, suggesting that the net cooling in the deep lake due to the episodic cooling event(s) between 26 and 31 January was about 0.03 and 0.05°C at 545 and 585 m, respectively.

The coldest intrusion occurred around 21 February, 1995 (Fig. 10b). The ambient temperature at the bottom of the lake was about 3.63°C. Temperatures begin to decrease dramatically at 570 m early on 21 February, reaching 3.26°C by the end of that day. At 520 m, the temperature cools less dramatically from 3.62 to 3.56°C through the first half of 21 February and then takes a dramatic drop to a minimum of about 3.36°C by the end of the day, which is still 0.10°C warmer than the coolest temperature at 570 m. Unlike the 1993 case, however, this intrusion arrives first at the deepest thermistor (570 m), then about 12 h later at 520 m; little or no response was seen at 455 or 390 m. Again, this cold water intrusion waters was likely transported to the mooring horizontally, as a cold tongue of water along the bottom. The coldest waters indicated here must have originated within the upper 50 m or so of the lake. The most likely explanation is that cool waters from the upper half of the lake descended along one side of the lake wall, were downwelled sufficiently to generate a thermobaric instability, and eventually arrived as a cold "pulse" of water at the bottom of the lake. The coldest waters appear at the very bottom of the lake, associated with the core of this hypothetical water pulse.

The third dramatic cooling burst begins on March 14, similar to the event three weeks earlier. Cool water first appears at 573 m, reaching 3.27°C early on 15 March; temperatures at 539 m start to cool about 6 h after they do at 573 m, and only reach about 3.32°C (0.05°C warmer than the coldest temperature at 573 m). Temperatures at 405 and 472 m hover around 3.45–3.58°C during this whole time (actually warming slightly, then cooling after the deeper cooling subsides).

Deep lake temperatures (405–573 m depth) during the late winter/early spring period in 1997 show no significant variability (Fig. 10c). Temperatures remain quite steady throughout this period at about 3.60°C. There is no evidence of episodic cooling this year that penetrates to the deepest part of the lake.

The bottom-trapped cold water intrusions observed in 1993 and 1995 are reminiscent of the cold water intrusions observed in Lake Baikal recently reported by Wuest et al. (2005). In their study, they used thermistor chains attached to the lake bottom, with 10 m vertical resolution, while our resolution at depth is about 50 m. Wuest et al. (2005) estimate a plume height of 100 m. In Crater Lake, these plumes are of comparable height ranges, extending to 50–150 m above the bottom. The temperature anomalies in Lake Baikal are on the order of –0.1°C; in Crater Lake, we have seen anomalies as great as –0.37°C. As in our case, Wuest et al. (2005) do not have enough information to verify the dynamical processes, but they also suggest thermobaric instabilities formed at the side boundaries of the lake are ultimately responsible for these cold water instrusions.

Dynamical considerations: stratification and wind forcing

Why does the degree of deep convection and ventilation vary from year to year? Some insight may be gained by comparing the evolution of temperature profiles from years with distinctly different amounts of ventilation. In Fig. 11, we present monthly temperature profiles for two different winter deployment periods representing strong (1994/95; VA = 1.6) and weak (1996/97; VA = –0.5) ventilation years.

In early October (Fig. 11a) for both 1994 and 1996, the temperature profiles were very similar, showing substantial thermal stratification in the upper 100 m, with an epilimnion depth, h_1, of 10–15 m. The epilimnion temperature, T_1, at this time was 13.3°C for 1994 and 11.3°C for 1996; in the deep half of the lake, the temperatures were about 0.1°C warmer in 1994 than in 1996. As surface cooling progresses to November (Fig. 11b), T_1 in 1994 and 1996 was about 8.8°C (MLD = 30 m) and 7.7°C (h_1 = 15 m), respectively. By December, however, T_1 had cooled to 5.0°C in 1994 (h_1 = 90 m), while in 1996 T_1 was

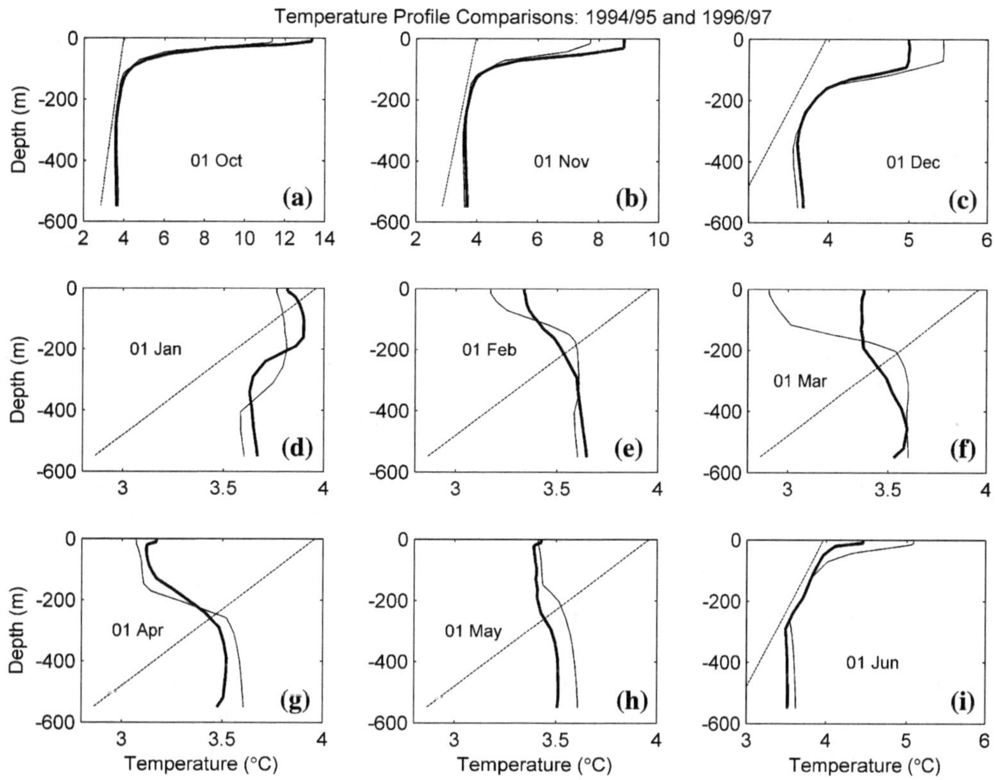

Fig. 11 Monthly temperature profiles spanning the fall-through-spring period of 1994/95 (thick line) and 1996/97 (thin line). Profiles are one month apart, ranging from 1 October (**a**) to 1 June (**i**), as indicated. Each profile is averaged over 4 days. The 1994/95 data represent a year with significant deep convection and ventilation (VA = 1.6; see Fig. 5c); the 1996/97 data represent a year when there was weak ventilation (VA = –0.5). The dashed line in each profile represents the temperature of maximum density line

5.4°C (h_1 = 70 m), suggesting slightly more cooling during November 1994 than November 1996 (Fig. 11c).

By January of both years, the surface temperatures are below 4°C and reverse stratification had begun to set in (Fig. 11d). There are, however, a few subtle differences between years. During the transition from 1 December 1994 to 1 January 1995, the waters are relatively isothermal to about 200 m depth, with a thermocline between about 200 and 250 m. From 1 December 1996 to 1 January 1997, there appears to be enhanced mixing of epilimnion and metalimnion waters, such that the epilimnion extended as deep as 225 m and the metalimnion to 400 m.

By 1 February, the structure of the two profiles is quite different in the upper 300 m (Fig. 11e): in 1995, the surface temperature is approximately 3.3°C and the thermal gradient in the metalimnion is relatively weak (1×10^{-3} °C m^{-1}) down to about 300 m, implying relatively weak stability; in 1997, the surface waters are measurably cooler, and the metalimnion thermal gradient is stronger (4×10^{-3} °C m^{-1}) and only extends to about 175 m. Overall, however, the heat contents reflected by the two profiles are relatively similar. In addition, the first deep convection event(s) from 1995 identified Fig. 10b has taken place by this time, leading to cooling in the hypolimnion. No such cooling is seen in the 1997 profile. By March 1 (Fig. 11f), the 1997 profile shows colder epilimnion temperatures (as low as 2.9°C) and even greater reverse stratification than February (up to 7×10^{-3} °C m^{-1} in the metalimnion). At depths greater than about 200 m, there is little change compared to 1 February. In contrast, the 1 March 1995 profile is more isothermal in the upper 200 m (~3.4°C), with a broad, weak (~1×10^{-3} °C m^{-1}) thermocline extending to 400 m; at almost all depths, the temperatures are measurably cooler compared to February. In the lower half of the lake, most of this cooling is likely a response to the deep convection event(s) through February, as seen in Fig. 10b.

The temperature profiles for April 1995 and 1997 (Fig. 11g) are qualitatively similar, with a thick (~125–150 m), cold epilimnion (~3.2°C). The metalimnion is still slightly more broad and the hypolimnion is even colder in 1995 (~3.5°C) than in 1997 (~3.6°C), as the deep convective events of March 1995 (Fig. 10b) evolve and mix. By May (Fig. 11h), the epilimnion waters are warming and convecting (T_1 is still less than 3.5°C), maintaining a thick epilimnion (h_1 ~ 225 m for 1995; 175 m for 1997); by June (Fig. 11i), surface temperatures are greater than 4°C and surface stratification has begun to build again.

For episodic cooling to occur in the hypolimnion in winter, the upper lake waters need to be colder than the deep waters, which are typically around 3.6–3.7°C. Overall, the near-surface temperature was cold for similar periods of time in both years: less than 3.6°C for 121 days in 1995 and 125 days during 1997. The most significant difference in thermal structure between the 2 years is that the February and March 1997 data show a cooler, shallower epilimnion and therefore stronger reverse stratification, than during the same time in 1995. This period generally appears to be the critical period when deep convection and ventilation is observed.

What drives the hypolimnetic mixing and episodic cooling? Our primary hypothesis is that wind-driven forcing during periods of low stratification and relatively cool (but not too cool) surface temperatures are responsible. One possible mechanism is wind-driven transport of the epilimnion to one side of the lake (downwind if $S_1 >> 1$, 90° to the right of the wind in the Northern Hemisphere if $S_1 < < 1$). If the wind-driven transport results in enough downwelling of cold epilimnion waters such that conditions become hydrostatically unstable (i.e., a thermobaric instability develops), then cool water convection may occur.

While we do yet not have an adequate three-dimensional, nonhydrostatic dynamical model in place to investigate this hypothesis, we can at least examine how deeply the epilimnion needs to be depressed to generate a thermobaric instability. For now, we will ignore the salinity effect in the lake and consider only temperature. Furthermore, in our discussion we shall ignore any potential mixing between the epilimnion waters and those below.

Figure 12 shows a time series from 1995 of mean daily u_*^2 (vector-averaged), h_1, and the depth, h_{tb}, to which the epilimnion needs to be displaced to generate a thermobaric instability (in other words, the depth at which $T_1 = T_{md}$, the temperature of maximum density). The period of interest here is the same as in Fig. 10b (21 January–22 March, 1995). The epilimnion temperature hovers around 3.4°C from mid-January to the end of February, then decreases to about 3.2°C by about March 12 (Fig. 3c). The value for h_1 averages around 200 m, varying by about +/– 30 m, for the entire period (Fig. 12c), and is frequently close to the depth necessary to generate thermobaric instabilities (e.g. between 1 and 10 February; between 22 February and 2 March).

Estimates of the friction velocity provide a measure of wind stress (Fig. 12a). We choose, somewhat arbitrarily, to consider storm periods where the peak value of the square of the friction velocity is 1.5×10^{-4} m^2 s^{-2}. From January through the end of March, 1995, there were four such strong storm periods: 6–16 January (storm I [not shown]), 30 January to 2 February (Storm II), 12–19 February (Storm III, comprising two distinct, but closely spaced bursts), and 4–14 March (Storm IV, comprising one short burst of winds, followed by five days of strong winds). Table 1 indicates estimates of the amount of downwelling required to generate thermobaric instabilities for each storm period. After three of the four storms (I, III and IV), there is clearly a cold water intrusion showing up in the deep lake, as described earlier; there is also thermal variability in the deep lake after Storm II, but it is unclear if this is in response to this storm or a remnant from Storm I. For Storm I, the necessary downwelling distance to generate an instability was relatively small (0–50 m) and the response at the mooring appeared to be a downward propagating cold-water intrusion. For Storm III and IV, the required downwelling

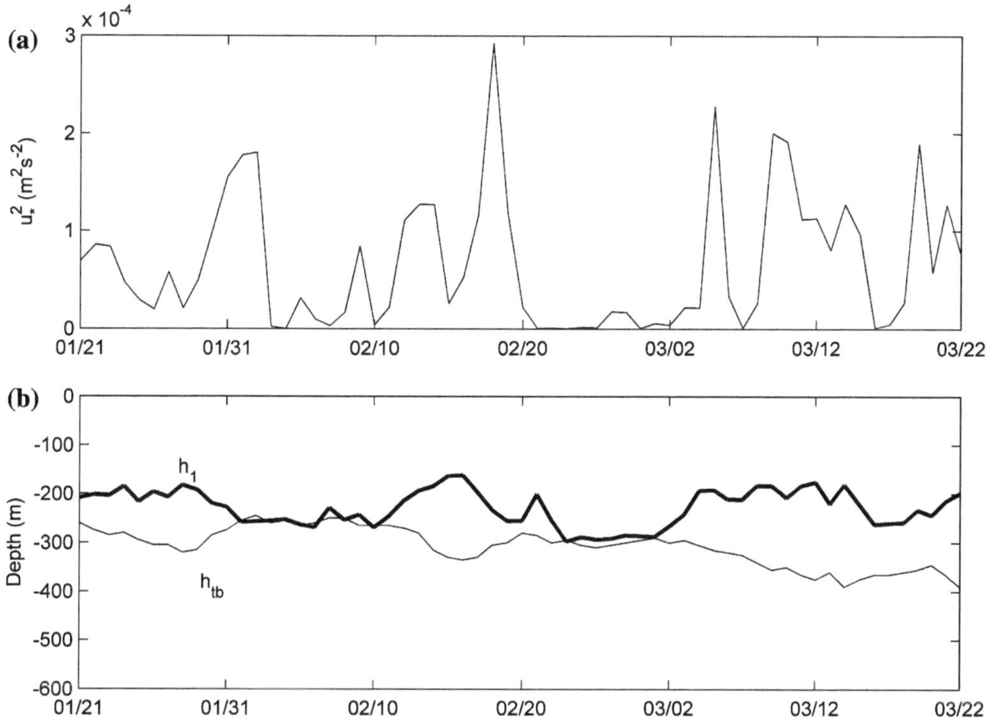

Fig. 12 Time series for 21 January to 22 March 1995: (**a**) mean-squared (vector averaged) friction velocity; (**b**) depth of epilimnion, h_1 (thick line) and depth h_{tb} (thin line) to which epilimnion would need to be depressed to create a thermobaric instability

distance was larger (60–200 m) and the cold water appeared as a horizontal intrusion at the deep lake mooring.

Figure 13 shows mean daily u_*^2 (vector-averaged), h_1, and h_{tb} for 1997, corresponding to the period in Fig. 10c. While h_1 and h_{tb} are very similar (~200 m) in mid-January, the subsequent cooling and increased stratification both shallows the epilimnion depth and deepens h_{tb} substantially. By early March, h_{tb} is actually deeper than the bottom of the lake (indicated by the gaps in the associated line in Fig. 13b).

There are three strong storm periods during this winter. A brief storm occurred in late January (31 January through 1 February; Storm V), followed by roughly 35 days of low-to-moderate winds and cooling conditions (leading to the relatively cold, more shallow epilimnion and a consequent increase in reverse stratification. Two large storms then followed (9–13 March [Storm

Table 1 Storms, required downwelling distance of epilimnion to create thermobaric instabilities, and apparent deep lake response for January to March, 1995 and 1997

Year	Storm identifier	Period	h_{tb}–h_1 (m)	Apparent deep lake response at mooring
1995	Storm I	6 Jan–16 Jan	~0	Vertical intrusion
	Storm II	30 Jan–2 Feb	0–50	Indeterminate
	Storm III	12 Feb–19 Feb	60–170	Horizontal intrusion
	Storm IV	4 Mar–14 Mar	100–200	Horizontal intrusion
1997	Storm V	31 Jan–2 Feb	100–150	Negligible
	Storm VI	9 Mar–13 Mar	*	Negligible
	Storm VII	16 Mar–20 Mar	*	Negligible

* indicate periods when the required downwelling was greater than the lake depth (i.e., thermobaric instabilities could not be formed)

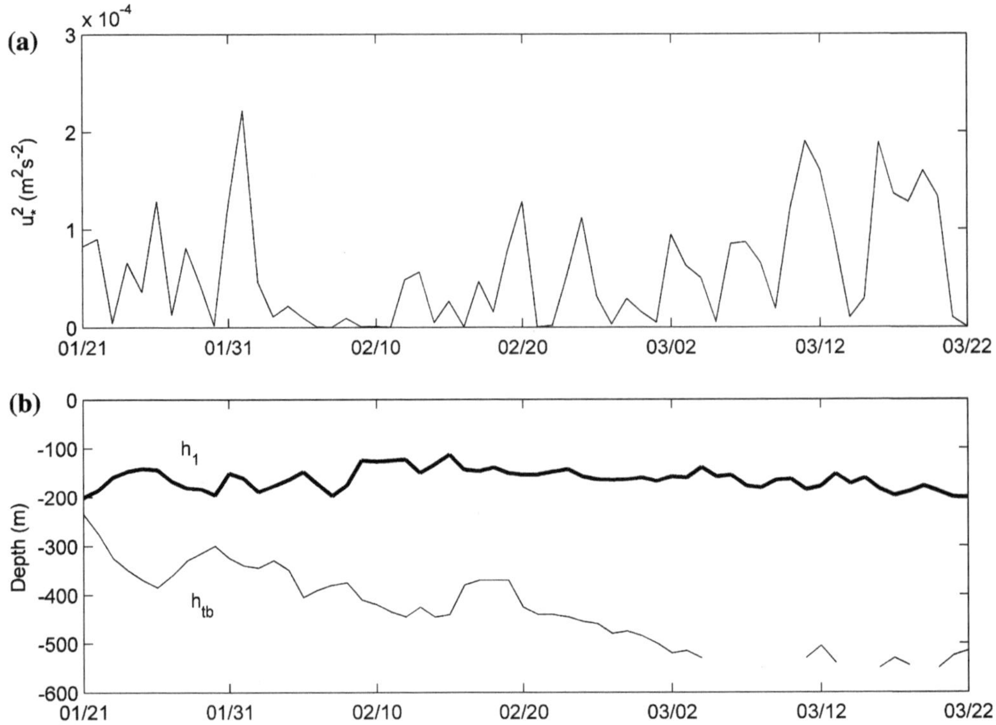

Fig. 13 Time series for 21 January to 22 March 1997: (**a**) mean-squared (vector averaged) friction velocity; (**b**) depth of epilimnion, h_1 (thick line) and depth h_{tb} (thin line) to which epilimnion would need to be depressed to create a thermobaric instability. Periods when h_{tb} is not plotted in the time series reflect periods when this calculated instability depression depth is greater than the lake depth

VI] and 16–20 March [Storm VII]). There is no indication of cold water intrusions in response to any of these storms (Fig. 10c). For Storm VI and VII, thermobaric instabilities were unlikely to form, because the epilimnion would need to be downwelled all the way to the bottom (or deeper). For Storm V, the downwelling distance is 100–200 m, comparable to Storm IV in 1995, but there was no response apparent in the deep lake. We hypothesize that the strong winds did not blow long enough to generate enough downwelling in this case.

Discussion

Variations in the net annual exchange of hypolimnetic waters with the overlying waters are manifested in annual variations of heat, salt, dissolved oxygen, and (occasionally) nitrate. Year-to-year changes in the temperature and salinity profiles are well-correlated and provide a sense of the extent of deep ventilation. Variations in oxygen profiles also reflect variations in net annual ventilation, although they are also likely subject to variations in particulate organic carbon oxidation rates. Nitrate profiles remain relatively steady from summer to summer, with most of the variability lying within measurement error, although during winters with much ventilation the hypolimnetic waters may show measurable depletion as late as the following summer. From these data, we have developed a ventilation anomaly using a combination of salinity and temperature changes from year to year. This anomaly provides a quantitative measure of the degree of ventilation of the hypolimnetic waters from year to year.

The moored thermistor data provide additional insight into the process of this mixing and ventilation and the variations from year to year. Geothermal heating is observed in the deep lake through summer and fall; slight cooling of deep waters generally occurs in early winter, most likely due to wind-driven turbulent diffusion of colder waters from above. In many, though not all, years, significant cooling events are observed between January and March. Many deep mixing events appear to propagate vertically to the deep lake, although the most dramatic cooling events observed reach the deepest part of the lake from the sides, suggesting formation of these downwelling waters closer to the lake edges.

Based on the examination of observations from two very different deep mixing years (1994–1995 and 1996–1997), thermobaric instabilities remain the most likely process for the generation of the observed deep mixing events. We believe the most critical issues for determining the extent of annual deep ventilation are the initial formation and evolution of the cold, buoyant surface layer during the reverse stratification period. In particular, we suggest that:

1. If the surface layer remains relatively thick and only slightly colder than hypolimetic waters (~3.6–3.7°C), then thermobaric instabilities may develop that appear at the mooring as vertically propagating cold water intrusions;
2. If the epilimnion is somewhat colder (say, 3.2–3.4°C) and winds are sufficiently strong and long enough in duration (perhaps a few days), downwelling and instabilities may occur at one side of the lake and cold water intrusions may arrive horizontally in the deep at the mooring;
3. If the epilimnion is very cold (< 3.0°C), then the required downwelling distance is too great for instabilities to occur and no cold water intrusions will be observed.

Acknowledgements We wish to thank Craig Stevens, Martin Schmid, and an anonymous reviewer, all of whom made several suggestions which improved this paper.

References

Antenucci, J. P. & J. Imberger, 2001. Energetics of long internal gravity waves in large lakes. Limnology and Oceanography 46: 1760–1773.

Bacon, C. R. & M. A. Lanphere, 1990. The geological setting of Crater Lake, Oregon. In Drake E. T. et al. (eds), Crater Lake: An Ecosystem Study. American Association for the Advancement of Science: 19–27.

Bacon, C. R., J. V. Gardner, L. A. Mayer, M. W. Buktenica, P. Dartnell, D. W. Ramsey & J. E. Robinson, 2002. Morphology, volcanism, and mass wasting in Crater Lake, Oregon. Geological Society of America Bulletin 114: 675–692.

Benson B. B. & D. Krause Jr., 1984. The concentration and isotopic fractionation of oxygen in freshwater and seawater in equilibrium with the atmosphere. Limnology and Oceanography 29: 620–632.

Carmack, E. C. & R. F. Weiss, 1991. Convection in Lake Baikal: an example of thermobaric instability. In Chu, P. C. & J. C. Gascard (eds), Deep convection and deep water formation in the oceans. Elsevier: 215–228.

Carpenter, J. H., 1965a. The Chesapeake Bay Institute technique for the Winkler dissolved oxygen method. Limnology and Oceanography 10: 141–143.

Carpenter, J. H., 1965b. The accuracy of the Winkler method for dissolved oxygen analysis. Limnology and Oceanography 10: 135–140.

Chen, C. T. & F. J. Millero, 1986. Precise thermodynamic properties for natural waters covering only the limnological range. Limnology and Oceanography 31: 657–662.

Crawford, G. B. & R. W. Collier, 1997. Observations of deep mixing in Crater Lake, Oregon. Limnology and Oceanography 42: 299–306.

Dymond, J., R. Collier, J. McManus & G. Larson, 1996. Unbalanced particle flux budgets in Crater Lake, Oregon: Implications for edge effects and sediment focusing in lakes. Limnology and Oceanography 41: 732–743.

Ericksen, C. C., 1985. Implications of ocean bottom reflection for internal wave spectra and mixing. Journal of Physical Oceanography 15: 1145–1156.

Fennel, K., R. W. Collier, G. L. Larson, G. B. Crawford & E. Boss, 2007. Seasonal nutrient and plankton dynamics in a physical-biological model of Crater Lake. Hydrobiologia 574: 265–280.

Grasshoff, K., M. Ehrhardt & K. Kremling, 1983. Methods of Seawater Analysis (2nd ref. ed.). Verlag Chemie GmbH, Weinheim.

Groeger, A. W., 2007. Nutrient limitation in Crater Lake, Oregon. Hydrobiologia 574: 205–216.

Hamblin, P. F., C. L. Stevens & G. A. Lawrence, 1999. Simulation of vertical transport in mining pit lake. Journal of Hydraulic Engineering 125: 1029–1038.

Holland , P. R., A. Kay & V. Botte, 2001. A numerical study of the dynamics of the riverine thermal bar in a deep lake. Environmental Fluid Mechanics 1: 311–332.

Jiang, L. & R. W. Garwood Jr., 1995. A numerical study of three-dimensional dense bottom plumes on a southern ocean continental slope. Journal of Geophysical Research 100: 18471–18488.

Kipfer, R., W. Aeschbach-Hertig, M. Hofer, R. Hohmann, D. M. Imboden, H. Baur, V. Gobulev & J. Klerkx, 1996. Bottom-water formation due to hydrothermal activity in Frolikha Bay, Lake Baikal, eastern Siberia. Geochemica et Cosmochimica Acta 60: 961–971.

Knapp, G. P., M. Stalcup & R. J. Stanley, 1990. Automated oxygen titration and salinity determination. WHOI

Technical Report, WHOI-90-35, Woods Hole Oceanographic Institution.
Larson, G. L., 1996. Development of a 10-year limnological study of Crater Lake, Crater Lake National Park, Oregon, USA. Journal of Lake and Reservoir Management 12: 221–229.
Larson, G. L., C. D. McIntire, M. Hurley & M. W. Buktenica, 1996. Temperature, water chemistry, and optical properties of Crater Lake. Journal of Lake and Reservoir Management 12: 230–247.
Larson, G. L., R. L. Hoffman, C. D. McIntire, M. W. Buktenica & S. F. Girdner, 2007. Water quality and optical properties of Crater Lake, Oregon. Hydrobiologia 574: 69–84.
Ledwell, J. R. & B. M. Hickey, 1995. Evidence for enhanced boundary mixing in the Santa Monica basin. Journal of Geophysical Research 100: 665–679.
Marshall, J. & F. Schott, 1999. Open-ocean convection: observations, theory and models. Reviews of Geophysics 37: 1–64.
McIntire, C. D., G. L. Larson, R. E. Truitt & M. K. Debacon, 1996. Taxonomic structure and productivity of phytoplankton assemblages in Crater Lake, Oregon, Crater Lake National Park, Oregon, USA. Journal of Lake and Reservoir Management 12: 259–280.
McManus, J., R. W. Collier, C. -T. Chen & J. Dymond, 1992. On the physical properties of Crater Lake, OR: determination of a conductivity and temperature dependent expression for salinity. Limnology and Oceanography 37: 41–53.
McManus, J., R. W. Collier & J. Dymond, 1993. Mixing processes in Crater Lake, Oregon. Journal of Geophysical Research 98: 18295–18307.
McManus, J., R. Collier, J. Dymond, C. G. Wheat, G. Larson, 1996. Spatial and temporal distribution of dissolved oxygen in Crater Lake, Oregon. Limnology and Oceanography 41: 722–731.
Mortimer, C. H., 1981. The oxygen content of air-saturated fresh waters over ranges of temperature and atmospheric pressure of limnological interest. Mitteilungen-Internationale Vereingung fuer Theoretische und Angewandte Limnologie 22: 23.
Nathenson, M., 1992. Water balance for Crater Lake, Oregon. USGS Open File Rept. 920595.
Neal , V. T., S. J. Neshyba & W. W. Denner, 1971. Temperature microstructure in Crater Lake, Oregon. Limnology and Oceanography 16: 695–700.
Peeters, F., G. Piepke, R. Kipfer, R. Hohmann & D. M. Imboden, 1996. Description of stability and neutrally buoyant transport in freshwater lakes. Limnology and Oceanography 41: 1711–1724.
Peeters, F., D. Finger, M. Hofer, M. Brennwald, D. M. Livingstone & R. Kipfer, 2003. Deep-water renewal in Lake Issyk-Kul driven by differential cooling. Limnology and Oceanography 48: 1419–1431.
Phillips, K. N., 1968. Hydrology of Crater Lake, East Lake, and Davis Lakes, Oregon. U.S. Geol. Surv. Water-Supply Pap. 1859-E.
Ravens, T. M., O. Koscis, A. Wuest & N. Granin, 2000. Small-scale turbulence and vertical mixing in Lake Baikal. Limnology and Oceanography 45: 159–173.
Redmond, K. T., 1990. Crater Lake climate and lake level variability. In Drake E. T. et al. (eds), Crater Lake: An Ecosystem Study. AAAS: 127–142.
Shimaraev, M. N., N. G. Granin & A. A. Zhdanov, 1993. Deep ventilation of Lake Baikal due to spring thermal bars. Limnology and Oceanography 38: 1068–1072.
Walker, S. J. & R. G. Watts, 1995. A three-dimensional numerical model of deep ventilation in temperate lakes. Journal of Geophysical Research 100: 22,711–22,731.
Williams, D. L. & R. P. Von Herzen, 1983. On the terrestrial heat flow and physical limnology of Crater Lake, Oregon. Journal of Geophysical Research 88: 1094–1104.
Wuest , A., D. C. Van Senden, J. Imberger, G. Piepke & M. Gloor, 1995. Comparison of diapycnal diffusivity measured by tracer and microstructure techniques. Dynamics of Atmospheres and Oceans 24: 27–39.
Wuest, A., T. M. Ravens, N. G. Granin, O. Kocsis, M. Schurter & M. Sturm, 2005. Cold intrusions in Lake Baikal: direct observational evidence for deep-water renewal. Limnology and Oceanography 50: 184–196.

Hydrobiologia (2007) 574:69–84
DOI 10.1007/s10750-006-0346-2

Thermal, chemical, and optical properties of Crater Lake, Oregon

Gary L. Larson · Robert L. Hoffman · David C. McIntire · Mark W. Buktenica · Scott F. Girdner

© Springer Science+Business Media B.V. 2007

Abstract Crater Lake covers the floor of the Mount Mazama caldera that formed 7700 years ago. The lake has a surface area of 53 km^2 and a maximum depth of 594 m. There is no outlet stream and surface inflow is limited to small streams and springs. Owing to its great volume and heat, the lake is not covered by snow and ice in winter unlike other lakes in the Cascade Range.The lake is isothermal in winter except for a slight increase in temperature in the deep lake from hyperadiabatic processes and inflow of hydrothermal fluids. During winter and spring the water column mixes to a depth of about 200–250 m from wind energy and convection. Circulation of the deep lake occurs periodically in winter and spring when cold, near-surface waters sink to the lake bottom; a process that results in the upwelling of nutrients, especially nitrate-N, into the upper strata of the lake. Thermal stratification occurs in late summer and fall. The maximum thickness of the epilimnion is about 20 m and the metalimnion extends to a depth of about 100 m. Thus, most of the lake volume is a cold hypolimnion. The year-round near-bottom temperature is about 3.5°C. Overall, hydrothermal fluids define and temporally maintain the basic water quality characteristics of the lake (e.g., pH, alkalinity and conductivity). Total phosphorus and orthophosphate-P concentrations are fairly uniform throughout the water column, where as total Kjeldahl-N and ammonia-N are highest in concentration in the upper lake. Concentrations of nitrate-N increase with depth below 200 m. No long-term changes in water quality have been detected. Secchi disk (20-cm) clarity varied seasonally and annually, but was typically highest in June and lowest in August. During the current study, August Secchi disk clarity readings averaged about 30 m. The maximum individual clarity reading was 41.5 m in June 1997. The lowest reading was 18.1 m in July 1995. From 1896 (white-dinner plate) to 2003, the

Guest Editors: Gary L. Larson, Robert Collier, and Mark W. Buktenica
Long-term Limnological Research and Monitoring at Crater Lake, Oregon

G. L. Larson (✉)
USGS Forest and Rangeland Ecosystem Science Center, 3200 SW Jefferson Way, Corvallis, OR 97331, USA
e-mail: gary_l._larson@usgs.gov

R. L. Hoffman
USGS Forest and Rangeland Ecosystem Science Center, 777 NW 9th Street, Suite 400, Corvallis, OR 97330, USA

D. C. McIntire
Department of Botany and Plant Pathology, Oregon State University, Corvallis, OR 97331, USA

M. W. Buktenica · S. F. Girdner
Crater Lake National Park, Crater Lake, OR 97604, USA

average August Secchi disk reading was about 30 m. No long-term changes in the Secchi disk clarity were observed. Average turbidity of the water column (2–550 m) between June and September from 1991 to 2000 as measured by a transmissometer ranged between 88.8% and 90.7%. The depth of 1% of the incident solar radiation during thermal stratification varied annually between 80 m and 100 m. Both of these measurements provided additional evidence about the exceptional clarity of Crater Lake.

Keywords Crater Lake · Water quality · Secchi disk · Optical properties

Introduction

Larson et al. (1996) evaluated the water quality and optical properties of Crater Lake, a deep ultraoligotrophic caldera (594 m) lake in the southern Cascade Mountain Range of Oregon State, for the period between 1983 and 1991. The lake was thermally stratified in late summer and early fall. The maximum thickness of the epilimnion was about 20 m and the metalimnion extended to a depth of about 100 m. Thus, most of the lake volume was a cold hypolimnion. Secchi disk clarity in summer and early fall typically varied between the high 20-m and low 30-m range. The lake was slightly basic, but alkalinity and conductivity were slightly elevated near the lake bottom (Dymond & Collier, 1990). The entire water column was well oxygenated. Nitrogen and phosphorus were low in concentration. These results were consistent with observations made at infrequent intervals between 1896 and 1982 (Larson et al., 1996).

The objectives of the present paper are to assess lake characteristics for long-term changes and to extend the period of observation to 2000, or in the cases of some variables, to 2003. An additional objective is to evaluate the nitrate-N concentration in an inlet spring (Spring 42) suspected of being contaminated by a sewage system on the caldera rim and impacting the productivity of the lake (Larson, 1984).

Methods

Study site

The lake is located in Crater Lake National Park in south-central Oregon at a latitude of 42°56′ N and longitude of 122°06′ W. The lake covers the floor of the caldera formed after the eruption and subsequent collapse of Mount Mazama, approximately 7700 years ago (Fig. 1; Bacon et al., 2002). The lake has a maximum depth of 594 m, a mean depth of 325 m, and a surface area of 53.2 km^2. The shoreline is 31 km in length and the diameter varies between 8 km and 10 km. There are no surface outlet streams, and thus, water is lost from the system by seepage and evaporation. The caldera functions like a large rain gauge (catchment area of about 15 km^2). The surface elevation of the lake varies about 0.5 m annually, but exhibits long-term changes in elevation (Redmond, 1990). The lake is rarely capped by snow and ice in winter as are other Cascade Range lakes.

The surface elevation of the lake dropped about 4 m between 1910 and 1931 and then returned to a reference elevation of 1,882 m by the early 1950s. The lake exhibited level fluctuations of about +/–1 m from the reference level through the mid-1980s and then dropped about 3 m by

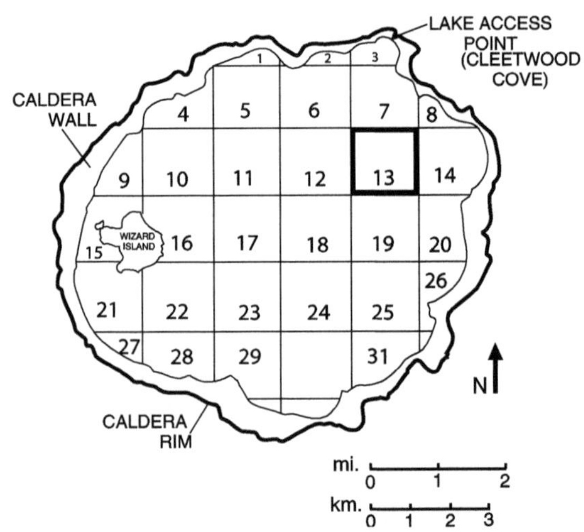

Fig. 1 Grid system of Crater Lake stations established by Hoffman (1969). The main monitoring station (13) is 594 m deep

1994. The lake level returned to near the reference level in 2000.

Nutrient inputs into the lake are derived from runoff from the caldera wall, precipitation, and seepage from underwater hydrothermal springs (Collier et al., 1993; Nelson et al., 1996). Crater Lake is an ultraoligotrophic lake with low nutrient concentrations and a high Secchi disk transparency, usually between 28 m and 33 m from June to September (Larson & Buktenica, 1998). Additional information about the geology of the caldera, and chemical and physical properties of Crater Lake is available in publications by Byrne (1965), Phillips (1968), Bacon & Lanphere (1990), Barber & Nelson (1990), Nathenson & Thompson (1990), McManus et al. (1993), Nelson et al. (1994), Larson et al. (1996), Bacon et al. (2002), and Nathenson et al. (2007).

Sampling and laboratory methods

All water column temperature, chemical, and optical data were obtained in the deepest basin in the lake at Station 13, a location approximately 3 km south of Cleetwood Cove (see Larson et al., 1996; Fig. 1). Details of sampling methods and laboratory procedures, including those for sampling and processing nitrate-N samples from Spring 42, were described by Larson et al. (1996). Physical and chemical data were derived from samples obtained between 1983 and 2003, with the exceptions of transmissometer (1991–2000) and spectraradiometer (1996–2003) data. No photometer readings were recorded from 1992 to 1995. The Licor spectraradiometer data were compared with the older photometer data by reprocessing using the light response curves for the Kahl photometer filters. Data for the red filter (Kahl) and red band (Licor) were not used because of unresolved issues related to light contamination (Hargreaves et al., 2007).

Statistical methods

For depth profiles the database was queried to obtain mean values and associated standard errors for 10 chemical variables at the water surface (0–2 m depth), and at depths of 5, 10, 20, 60, 100, 200, 300, 400, 500, and 550 m below the surface. The analysis of long-term change was confined to the 13 sets of samples obtained in August (1988–2000) when the lake exhibited thermal stratification.

Relationships between sample year and the August values for each of 11 water quality variables were analyzed using simple linear regression (NCSS 2000) (Hintze, 1998). Level of significance was $P \leq 0.05$. This statistical method was used because the 11 variables exhibited very little interannual variability. Ten variables (i.e., temperature, pH, alkalinity, conductivity, total phosphorus, orthophosphate-P, nitrate-N, total Kjeldahl-N, ammonia-N, and light transmission) were analyzed statistically at the surface, 100 and 550 m. Secchi disk clarity was analyzed using the mean on each sampling date (usually three observations by two observers). Four variables at one or more depths (i.e., pH, 100 m; nitrate-N, 0 and 100 m; ammonia-N, 100 and 550 m; percent transmission, 2 and 550 m), were not normally distributed and were natural-log transformed prior to analysis. Regression analysis (Hintze, 1998) was used to evaluate the relationship between snow depth in May on the caldera rim and the concentration of nitrate-N in Spring 42 in August from 1983 through 2003 ($P \leq 0.05$).

An additional statistical test for analysis of the water column chemical data was used to examine spatial and temporal changes in water masses. In this case, water masses were defined by a cluster analysis using the program CLUSB4. The CLUSB4 algorithm, a non-hierarchical divisive procedure, provided a minimum variance partition of the data (McIntire, 1973). Before partitioning the data into discrete clusters, the program standardized each of the 10 chemical variables by subtracting the mean from each observation and dividing the difference by the standard deviation. Therefore, this transformation corrected for the unit differences among the variables and gave each variable equal weight. After the transformations, the data were partitioned into a four-cluster structure. Results of this analysis were reported in the form of a cluster diagram that illustrated the temporal and spatial patterns and discontinuities in the chemical composition of the Crater Lake water column.

Results

Water quality

Depth profiles

Temperature profiles for the Crater Lake water column were based on measurements obtained by a conductivity–temperature–depth probe between June 1988 and September 2000. The standard error of the mean for all data obtained during the measurement period indicated that there was very little annual variation in the profiles for the months illustrated in Fig. 2. In winter and spring, temperature was relatively uniform throughout the water column. Values varied between 2.35°C and 4.11°C in January, and between 2.55°C and 3.62°C in April and May over the 13-year sampling period. With the onset of thermal stratification in June, the mean water temperature at the lake surface was 8.87°C. As the development of an epilimnion continued throughout the summer, the mean temperature at the lake surface gradually reached a maximum value of 15.48°C in August. Mean water temperatures at the lake surface in September and October were 14.38 and 11.21°C, respectively.

The maximum depth (about 20 m) of the epilimnion occurred in August. Limited data obtained in 1988, 1989, and 1994 and continuous moored thermister data since 1992 (Crawford & Collier, 2007) indicated that thermal stratification continued through the month of October and then the mixed layer deepened and cooled through the rest of the year.

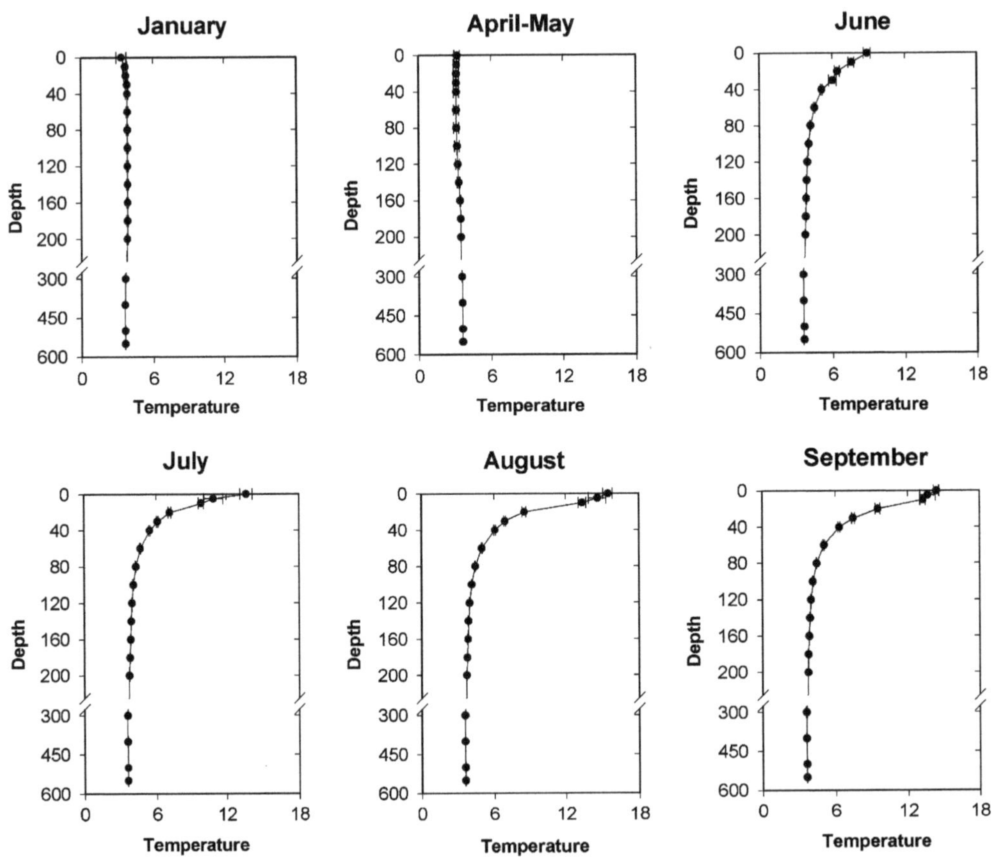

Fig. 2 Temperature depth profiles for selected months from the water surface to a depth of 550 m at Station 13. The profiles were based on all temperature measurements at each of 11 depths in each month between June 1988 and September 2000. Each point plotted represents the mean value and associated standard error for each depth

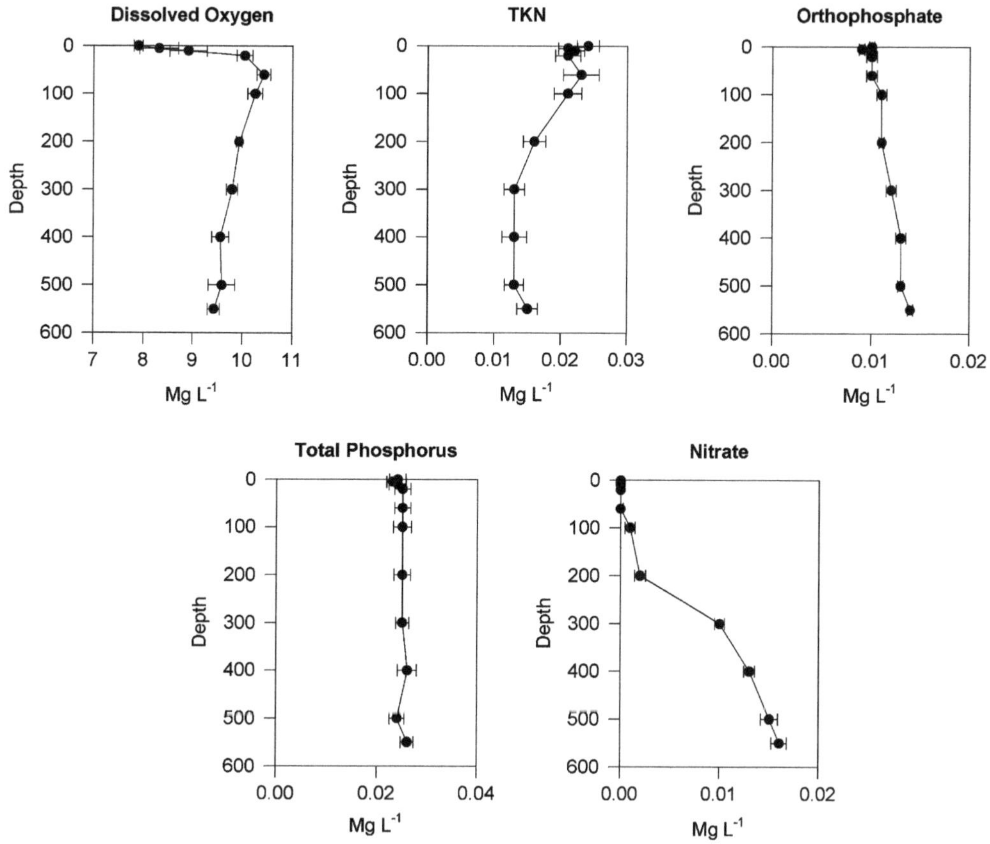

Fig. 3 Depth profiles for dissolved oxygen, total Kjeldahl-N, orthophosphate-P, total phosphorus, and nitrate-N from the lake surface to a depth of 550 m at Station 13. The analysis was confined to the 13 sets of samples obtained in August (1988–2000) when the lake exhibited thermal stratification. Each point plotted represents the mean value and associated standard error for each depth

Dissolved oxygen concentration exhibited an orthograde curve, which is typical of deep oligotrophic lakes with relatively warm surface temperatures and low productivity during thermal stratification (Fig. 3). Total Kjeldahl-N was highest in the euphotic zone (surface to a depth of about 100 m) and decreased with increasing depth between 100 m and 550 m below the surface. Concentrations of orthophosphate-P and total phosphorus were fairly uniformly distributed throughout the water column, whereas the nitrate concentration was near the limit of detection (0.001 mg l^{-1}) in the upper 200 m of the lake and gradually increased with depth to a maximum mean of 0.016 mg l^{-1} at 550 m. Silicon, alkalinity, and conductivity increased slightly from the surface waters to the bottom of the lake, whereas pH exhibited a corresponding decrease throughout the water column (Fig. 4). Standard error values associated with the mean values of each of the chemical variables, with the exception of ammonia-N, indicated that there was relatively little annual variation in the chemical properties of the water column during the period of thermal stratification.

Based on regression analyses of temporal variations of 11 water quality variables measured at the surface, 100 and 550 m, and Secchi disk clarity readings, only 5 of 31 analyses were statistically significant (Table 1). These results suggested that there were no long-term changes in the water quality of the lake. The significant samples (temperature at 550 m; pH at 0 m; conductivity at 100 m; nitrate-N at 550 m; and ammonia-N at 100 m) all had negative regression

Table 1 Regression results for water quality variables measured during the month of August 1988–2000 ($N = 13$): except TP at 0 m measured 1988–1991, 1994–2000 ($N = 11$); temperature and % transmission at 2 and 100 m measured 1988, 1990–1991, 1993–2000 ($N = 11$); and temperature and % transmission at 550 m measured 1990–1991, 1993–2000 ($N = 10$)

Variable	Depth	Mean	SD	Range	R2	Slope	*P*-value
Temperature (°C)	2	15.56	1.5788	12.51–17.74	0.0249	–0.0646	0.643
	100	4.10	0.1265	3.94–4.34	0.0251	–0.0052	0.642
	550	3.60	0.0476	3.53–3.66	0.5174	–0.0103	0.019
pH (standard units)	0	7.73	0.0751	7.6–7.8	0.3929	–0.0121	0.022
	100	7.74	0.0506	7.7–7.8	0.0286	–0.0003	0.581
	550	7.60	0.1080	7.5–7.8	0.0192	0.0038	0.651
Alkalinity (mg l^{-1})	0	7.13	0.0902	6.92–7.26	0.0031	0.0013	0.856
	100	7.10	0.0966	6.87–7.23	0.0019	0.0011	0.887
	550	7.30	0.0900	7.09–7.41	0.0055	–0.0017	0.810
Conductivity (μS cm^{-1})	0	114.62	1.8187	110.9–116.9	0.1702	–0.1926	0.161
	100	114.91	1.5240	112.0–117.1	0.5576	–0.2922	0.003
	550	118.78	2.1271	115.5–121.5	0.2014	–0.2451	0.124
Total phosphorus (mg l^{-1})	0	0.0225	0.0044	0.016–0.030	0.1922	–0.0005	0.177
	100	0.0231	0.0054	0.015–0.036	0.2500	–0.0007	0.082
	550	0.0249	0.0045	0.019–0.035	0.2611	–0.0006	0.074
Orthophosphate-P (mg l^{-1})	0	0.0098	0.0010	0.008–0.011	0.0286	0.0000	0.581
	100	0.0102	0.0010	0.009–0.012	0.2517	–0.0001	0.077
	550	0.0135	0.0011	0.012–0.016	0.0231	0.0000	0.620
Nitrate-N (mg l^{-1})	0	0.0002	0.0004	0.000–0.001	0.2630	–0.1186	0.073
	100	0.0006	0.0080	0.000–0.008	0.0238	–0.0483	0.615
	550	0.0158	0.0030	0.011–0.023	0.4496	–0.0005	0.012
Total Kjeldahl-N (mg l^{-1})	0	0.0222	0.0094	0.000–0.030	0.2674	0.0012	0.070
	100	0.0199	0.0108	0.000–0.030	0.0357	0.0005	0.537
	550	0.0139	0.0084	0.000–0.020	0.0823	0.0006	0.342
Ammonia-N (mg l^{-1})	0	0.0015	0.0023	0.000–0.005	0.0460	–0.0001	0.482
	100	0.0013	0.0023	0.000–0.007	0.3381	–0.2662	0.037
	550	0.0013	0.0023	0.000–0.007	0.1327	–0.1668	0.221
Transmission (percent)	2	87.25	2.974	80.68–90.92	0.0001	–0.0001	0.976
	100	88.48	0.4316	88.02–89.46	0.0005	0.0025	0.947
	550	89.15	1.6833	85.17–90.78	0.0674	0.0015	0.469
Secchi disk depth (m)	observed	28.0	4.156	22.3–36.4	0.0004	0.0214	0.948

slopes that were very nearly zero. No long-term changes were detected for Secchi disk clarity (Table 1).

Identification of water masses

Water masses in Crater Lake were defined by a cluster analysis of water samples obtained between May 1986 and August 2000. This analytical approach examined the patterns of a group of chemical variables collectively, and identified spatial and temporal patterns of homogeneity and discontinuity in the water column of the lake. The analysis included all samples that had a complete set of data for the 10 chemical variables plotted in Figs. 3 and 4. In this case there were 364 samples representing 15 years (1986–2000) and 8 months of the year. However, sampling was unbalanced, as August was the only month represented in each of the 15 years. Sampling at different months of the year occurred from 1986 to 1992; subsequently samples were obtained only during the month of August.

Results of the cluster analysis, as presented in a diagram depicting a 4-cluster structure (Fig. 5), illustrate the vertical distribution and temporal persistence of the concentrations of water quality variables in the lake. The darkened areas in the diagram represent samples that were excluded from the analysis because of missing data. The 4-cluster structure identified a group of samples (Cluster 3) that were collected mostly in the spring and fall of 1989 and the summer of 1990. This cluster had a slightly higher mean concentration of

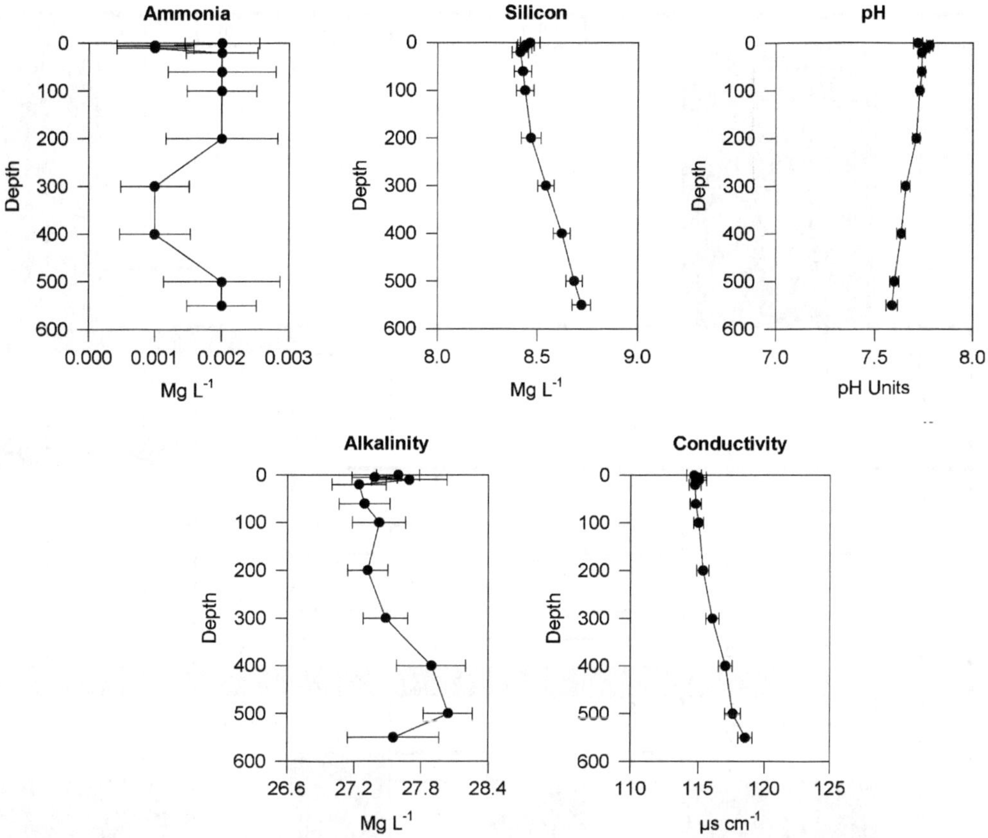

Fig. 4 Depth profiles for ammonia-N, silicon, pH, alkalinity, and conductivity from the lake surface to a depth of 550 m at Station 13. The analysis was confined to the 13 sets of samples obtained in August (1988–2000) when the lake exhibited thermal stratification. Each point plotted represents the mean value and associated standard error for each depth

dissolved oxygen and a four-times higher mean concentration of ammonia (Table 2) than the more prominent group of samples in the upper layer of the water column (Cluster 2). Also, there were differences in mean concentrations of total Kjeldahl-N and total phosphorus between samples in

Table 2 Mean values of chemical variables used in a cluster analysis of samples taken from the water column of Crater Lake from May 1986 to August 2000

Variable	Cluster 1	Cluster 2	Cluster 3	Cluster 4
Dissolved oxygen	9.654	9.404	10.870	9.970
Total Kjeldahl-N	0.011	0.021	0.017	0.017
Orthophosphate-P	0.014	0.010	0.011	0.014
Total phosphorus	0.028	0.023	0.032	0.030
Nitrate-N	0.014	0.001	0.001	0.003
Ammonia-N	0.002	0.002	0.009	0.002
Silicon	8.644	8.401	8.358	8.386
pH	7.579	7.751	7.662	7.630
Alkalinity	7.228	7.069	6.967	7.048
Conductivity	117.558	114.961	114.676	114.028

The values correspond to the 4-cluster structure of the data

Depth	0 m	5 m	10 m	20 m	60 m	100 m	200 m	300 m	400 m	500 m	550 m
May-86	4			4	4	4	4	1	1		1
Jun-86	4			4	4	4	4	1	1		1
Jul-86	4			4	4	4	4	1	1		1
Aug-86	2			4	4	4	4	1	1		1
Jan-87	4			4	4	4	4	4	1		1
Apr-87	4			4	4	4	4	4	4		1
Jun-87	4			4	4	4	4	4	4		1
Jul-87	4			4	4	4	4	4	4		4
Aug-87	2			4	4	4	4	4	1		1
Sep-87	2			4	4	4	4	4	4		1
Jun-88	4	2	2	2	2	2	3				
Jul-88	2	2	2	2	2	2	4	4	4	1	
Aug-88	2	2	2	2	2	2	4	4	1	1	1
Sep-88	2	2	2	2	2	2	3	4	1	1	1
Oct-88	2	2	2	2	2	2	2	4			
Apr-89	4	3	3	3	3	3	3	4	1	1	1
May-89	3	3	3	3	3	3	3	4	1	1	1
Jun-89	4	4	2	4	2	2	4	4	4	1	1
Jul-89	2	2	2	2	2	2	2	1	1	1	1
Aug-89	2	2	2	2	2	3	2	1	1	1	1
Sep-89	2	2	2	3	3	3	3	1	1	1	1
Jan-90	2	2	2	2	2	2	2	1	1	1	1
Jun-90	2	2									
Jul-90	2	3	3	2	2	2	3	3	1	1	1
Aug-90	2	2	2	2	2	2	2	1	1	1	1
Sep-90	2	2	2	3	2	3		1	1	1	1
Jun-91	2			2		2		1			1
Jul-91	2			2		2		1			1
Aug-91	2			2		2		4			1
Sep-91	2			2		2		1			1
Aug-92		2	2	4	4	4		1	1	4	1
Sep-92	2	2	2	2	2	2	4	1	1	4	2
Aug-93		2	2	2	2	2	4	1	1	1	1
Aug-94	2	2	2	2	2	2	2	1	1	1	1
Aug-95	2	2	2	2	2	2	2	1	1	1	1
Aug-96	2	2	2	2	2	2	2	1	1	1	1
Aug-97	2	2	2	2	2	2	2	2	2	1	1
Aug-98	2	2	2	2	2	2	2	2	2	1	1
Aug-99	2	2	2	2	2	2	2	4	1	1	1
Aug-00	2	2	2	2	2	2	2	4	1	1	1
Depth	0 m	5 m	10 m	20 m	60 m	100 m	200 m	300 m	400 m	500 m	550 m

Fig. 5 Results of a 4-cluster analysis of selected water quality variables (see Table 2) from the lake surface to a depth of 550 m from May 1986 to August 2000. All samples were collected at Station 13

Clusters 2 and 3. Cluster 4 identified a group of samples that were collected in the upper layer of the water column, primarily in 1986 and 1987. Samples represented by Cluster 4 were unrelated to season, as the group included samples collected in January, April–May, June–August, and September–October. Cluster 1 samples were primarily collected from 300 m to 550 m. These samples had the highest means for conductivity, alkalinity, and nitrate-N.

Spring 42 nitrate-N

The nitrate-N concentration in Spring 42 in August was selected to illustrate temporal variation and possible long-term trends (Fig. 6).

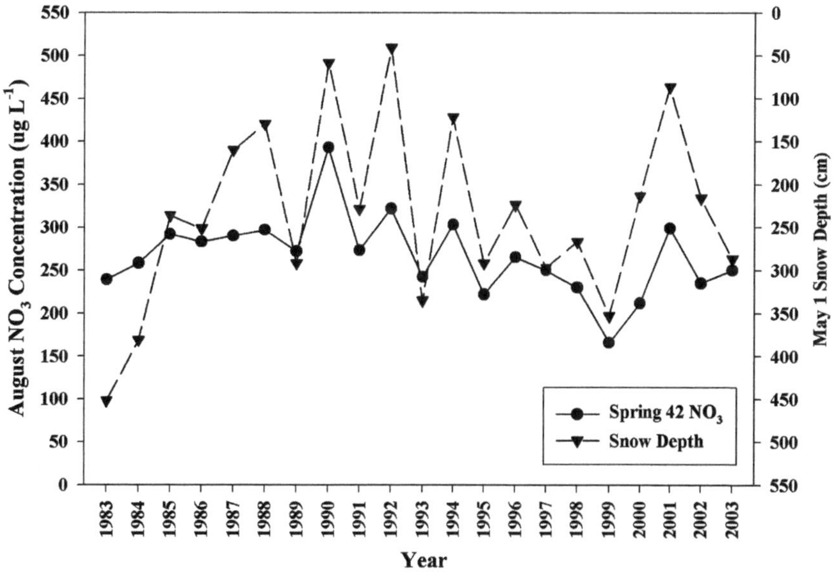

Fig. 6 Temporal changes in August concentration of nitrate-N in Spring 42 and the snow depth on the caldera rim on May 1 from 1983 to 2003

Nitrate-N concentrations increased from 1983 to 1990, decreased to 1999, and then increased to 2003. The mean concentration for the period between 1983 and 2003 was 266.3 μg l^{-1} (SD = 46.5). Temporal variation of August nitrate-N concentrations, however, corresponded negatively to the depth of the snow pack on the caldera rim in May (Fig. 7).

Optical properties

Secchi disk clarity

Clarity measurements were primarily recorded from June to September between 1978 and 2003

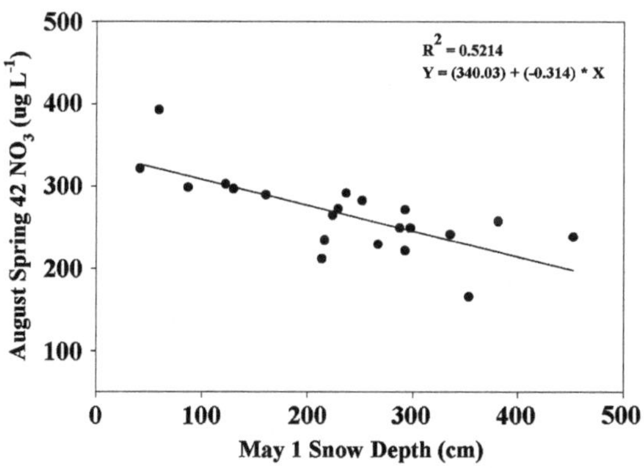

Fig. 7 Relationship between snow depth on the caldera rim on May 1 and the August concentration of nitrate-N in Spring 42 (1983–2003)

(Fig. 8). Average clarity measurements exhibited seasonal variability. Clarity measurements for months with 10 or more observations were highest in June and then decreased in August and September when the lake was thermally stratified.

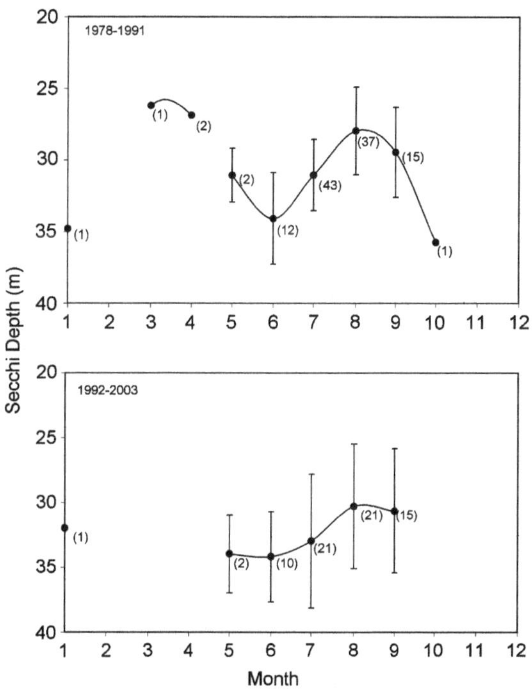

Fig. 8 Temporal changes of average Secchi disk clarity readings at Station 13 for 1978–1991 (Larson et al., 1996) and 1992–2003. Numbers in parentheses are sample sizes and error bars represent ±1 SD

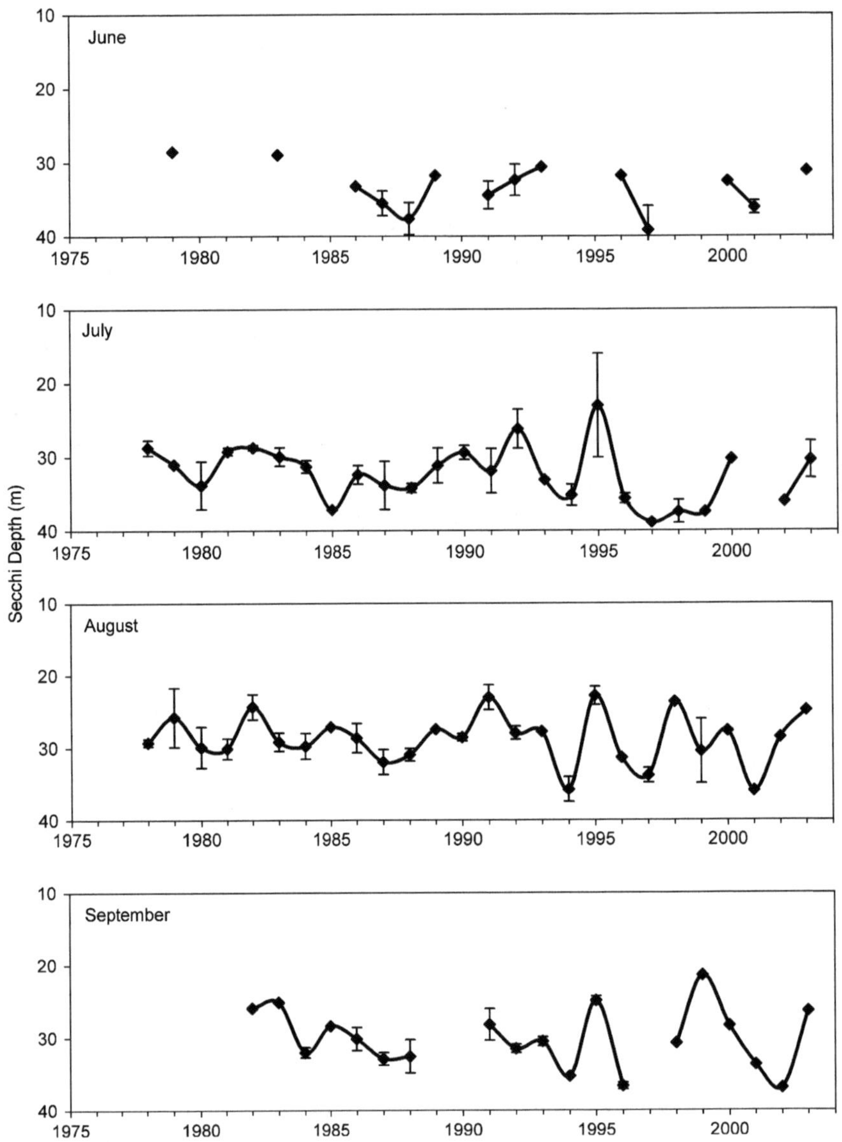

Fig. 9 Temporal patterns of Secchi disk clarity readings for June, July, August, and September at Station 13. Error bars represent ±1 SD

Secchi disk clarity readings exhibited considerable interannual and seasonal variability between 1992 and 2003 (Fig. 9). June clarity readings typically were >30 m and were near 40 m in 1997. Secchi readings averaged 33.0 m in July, 30.3 m in August, and 30.6 m in September. Secchi disk readings recorded between 1978 and 1991 (Larson et al., 1996) generally were consistent with the 1992–2003 recordings, but there was less interannual variation. The deepest single reading was 41.5 m on 25 June 1997, whereas the shallowest reading was 18.1 m on 31 July 1995 after a torrential summer storm (rain and hail) that caused extensive mud flows into the lake. Collectively, the average of Secchi disk clarity readings recorded in August from 1978 to 2003 (28.8 m) was consistent with observations made prior to 1978 (Fig. 10). The average Secchi disk clarity for August measurements from 1896 to 2003 was 29.5 m (Fig. 10). Statistically, there were no indications of increasing or decreasing long-term trends in the Secchi disk clarity readings (Table 1).

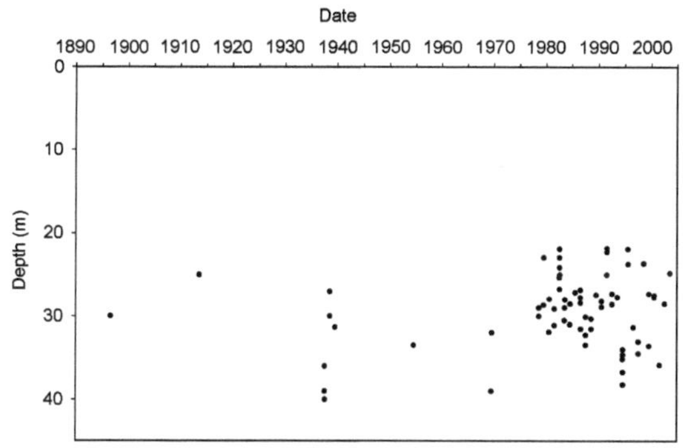

Fig. 10 Mean daily Crater Lake Secchi disk clarity readings in August from 1896 to 2003

Light penetration

Optical properties of the water column to a depth of 160–170 m were recorded with a Licor spectraradiometer beginning in 1996. In June, July, August, and September, the depth of 1% incident solar radiation for the green and clear bands was between 80 m and 100 m, and the blue band was between 100 m and 125 m (Fig. 11). Photometer data collected between 1985 and early 1992 yielded 1% light penetration depths for the green filter (80.9 m, $N = 38$); the blue filter (98.0 m, $N = 32$) and the clear filter (79.0 m, $N = 39$).

Transmissometer

Average beam percent transmissions from a depth of 2–200 m for June–September for the period 1991–2000 ranged from 87.9% to 90.3% (Fig. 12). The variability and lower clarity driven by biological and lithogenous particles. The deeper lake experiences less variability in particle concentrations and light transmission. The average transmissions between the depths of 202 m and 550 m ranged between 89.3% and 90.8% (Fig. 12). The lowest percent transmissions occurred in the upper 100 m of the water column and near the lake bottom. Although seasonal and interannual variations of beam transmissions were observed, there were no long-term trends of declining or increasing transmissions (Table 1).

Discussion

Water quality characteristics of Crater Lake from 1992 to 2000 were consistent with those reported for 1983–1991 (Larson et al., 1996). In winter the lake was essentially isothermal, with the exceptions of reverse stratification near the lake surface and a slight increase in temperature in the deep lake (~0.07°C) due to geothermal warming (Collier et al., 1993). The initial phase of thermal stratification began in June and the upper lake was fully stratified in August and September. The epilimnion was shallow (about 20 m thick) and the metalimnion extended to a depth of about 100 m. Most of the lake volume, therefore, was a cold hypolimnion (mean = 3.64°C ± 0.0006 SE, 102–550 m).

McManus et al. (1993) demonstrated that the lake mixes in winter and spring to 200–250 m by wind energy and convection. They also showed that the concentration of dissolved oxygen in the deep lake decreased during summer and fall, but that the concentration increased by winter and spring. The replenishment of dissolved oxygen in the deep lake during winter and spring was caused by infrequent down-welling of cold water that originated from the lake surface (Crawford & Collier, 1997). The down-welling waters also caused deep lake water with relatively high concentrations of nitrate-N to up-well into the upper portion of the lake. Although the frequency and impact of upwelling events are not well understood (Crawford & Collier, 2007), there is some evidence that a major upwelling event(s) occurred in 1986. This period of upwelling may have been responsible for the slightly elevated concentrations of nutrients in Cluster 4 throughout most of the water column in 1986 and 1987 (Fig. 5). Moreover, the samples in Cluster 4 from the upper part of the lake roughly corresponded to a period (1986–1989) of maximum chlorophyll concentration in the lake (McIntire et al., 2007). Therefore, we suggest that the slightly higher concentrations of plant nutrients detected in the samples of Cluster 4 may have been associated with a period of relatively high autotrophic production in the upper 200 m of the lake.

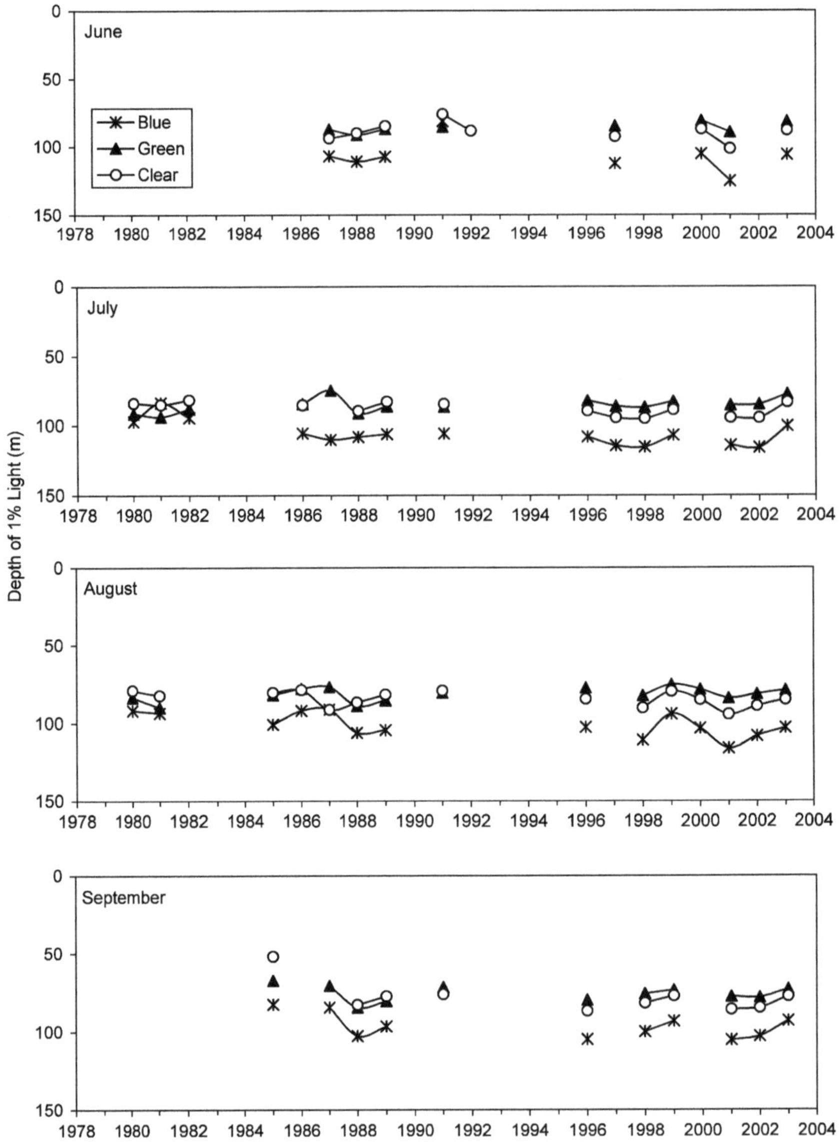

Fig. 11 Temporal changes in the depth of 1% incident solar radiation for blue, green, and clear light in June, July, August, and September at Station 13. A Kahl model 268WA350 was used in the early 1980s, followed by a Kahl model 68WB365TD in the late 1980s and early 1990s. A Licor spectraradiometer was used from the mid–1990s to 2003

Unlike the cluster analysis that grouped related samples based on a set of water quality variables, regression analysis was used to evaluate each variable for temporal change. The regression analyses suggested that there were no long-term changes in water quality variables, except for a few instances where slight variations resulted in negative trends that were statistically significant (Table 1). These significant temporal decreases were small and were not biologically meaningful. It should be noted that a significant *P*-value can be obtained even when the difference between the estimated and null hypothesis parameters is small and the parameters are, for practical purposes, equivalent (Berger & Sellke, 1987). In this context the *P*-value can be misleading and/or uninformative (see Anderson et al., 2000). The cluster analysis showed the distribution of Cluster

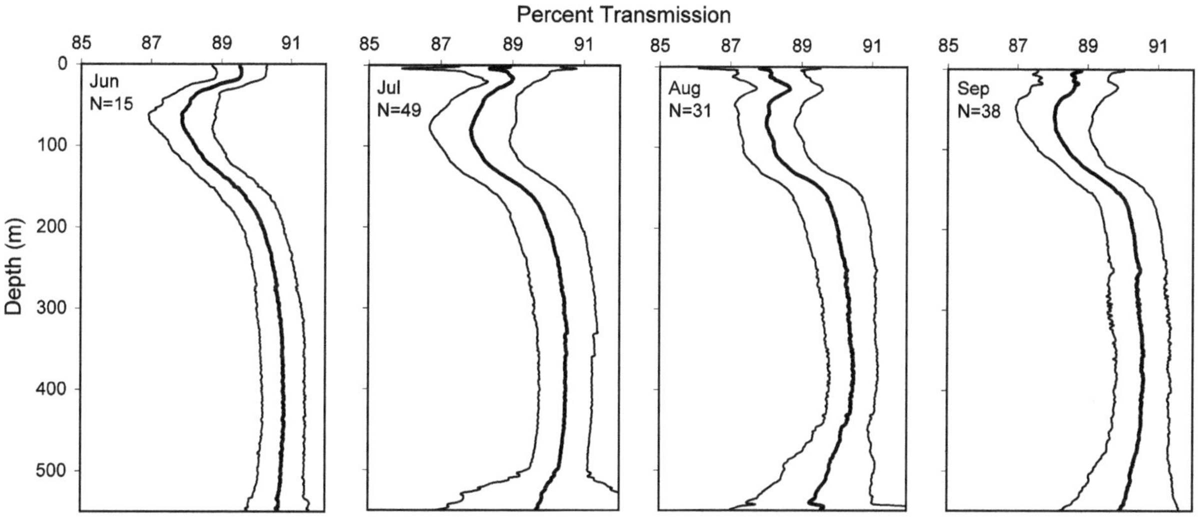

Fig. 12 Average percent transmission and standard deviations by month at Station 13 from 2 m to 550 m

4 throughout most of the water column during 1986 and 1987 was also based on very small differences among the four clusters (Table 2; Fig. 5). Thus, the two methods of analyses provided insight into the influences that very small changes in water quality might have had on the productivity of the lake during the mid to late 1980s.

Nelson et al. (1996) developed a chemical solute mass balance model for Crater Lake that showed that the major chemical inputs derived from the caldera springs and from atmospheric deposition did not equal the mass output in seepage from the lake. In an earlier study, Collier et al. (1993) documented the amount of solutes from hydrothermal fluids entering the system at the lake bottom. Based on this information, Nelson et al. (1996) was able to balance the chemical solute model for Crater Lake, which provided a theoretical framework for our observed long-term stability of the chemical solute mass balance of the lake (Phillips, 1968).

Wetzel (1983) and Horne & Goldman (1994) suggest that certain physical and chemical conditions typify oligotrophic lakes. In Crater Lake, these included high transparency, orthograde oxygen and nitrate-N depth profiles, and low nitrate-N concentrations in the epilimnion. The relatively high concentrations of total phosphorus in Crater Lake are in the range usually associated with mesotrophic lakes (Wetzel, 1983), but since the lake is strongly nitrogen limited (Groeger,

Table 3 Concentrations of selected chemical variables in near surface water samples of non-caldera eutrophic (E), mesotrophic (M), and oligotrophic (O) lakes in the Cascade Range of Oregon (Johnson et al. (1985)

Trophic status	N	Alkalinity (mg l^{-1})	Conductivity (μS cm^{-1})	pH (su)	Total phosphorus (μg l^{-1})
E	2	15	29–50	8.4–9.5	24–61
M	3	13–18	32–37	7.3–9.3	28–47
O	13	<1–16	3–50	6.3–7.6	2–72
East	1	105	310	7.9	16
Paulina	1	340	560	8.3	45
Crater[1]	1	26–30	80–121	7.1–7.9	13–43

[1] Present study

East Lake (M), Paulina Lake (M), and Crater Lake (O) are caldera lakes receiving hydrothermal fluids. N refers to number of lakes and su = standard units

2007), the phosphorus concentrations do not reflect the trophic status of the lake. Furthermore, the alkalinity and conductivity of Crater Lake are much higher than expected for other lakes in the Cascade Range of Oregon (Table 3). The relatively high alkalinity and conductivity of Crater Lake and two other near-by caldera lakes (East Lake and Paulina Lake) are associated with inputs of hydrothermal fluids (Table 3; Collier et al, 1993). Therefore, some of the physical and chemical properties of Crater Lake are unique and do not conform to the range of criteria typically associated with oligotrophic lakes.

The Crater Lake program began in 1983 in response to Secchi disk clarity readings recorded between 1978 and 1982 that were about 25% shallower than readings in 1937 and 1969. There was considerable concern that these shallow readings were caused by deteriorating lake water quality (Larson, 1984). A primary suspect for the apparent change in the water quality characteristics of Crater Lake was sewage contamination (nitrate-N) of an intracaldera spring (Number 42) located below a sewage system on the caldera rim. The sewage system was disconnected in 1991 when sewage from visitor facilities on the caldera rim was drained to a treatment plant about 2 km from park headquarters on the lower flanks of Mount Mazama. Furthermore, the old septic gallery system was removed from the caldera rim in 1992. Annual water samples collected from Spring 42 during the period from 1983 to 2003 indicated that the nitrate-N concentration did not change after the sewage system was removed (Fig. 6). There was, however, considerable interannual variation in the nitrate-N concentration, but this appeared to be primarily a function of spring water discharge as indicated by the correlation between the depth of the snow pack on May 1 and the August nitrate-N concentrations in Spring 42 (Fig. 7). Thus, the available information suggests that sewage contamination was not a reason for the elevated nitrate-N in the spring. The current hypothesis is that the elevated nitrate-N concentration in the spring is a natural phenomenon (Stan Gregory, personal communication).

Secchi disk clarity of Crater Lake varies seasonally and annually in response to a variety of biological and geological sources of particulate matter in the upper portion of the water column. Although the shallow readings recorded between 1978 and 1982 were of concern based on deeper readings in earlier years (Larson, 1984), the readings recorded between 1983 and 2003 substantiated that variability was "normal". There were no long-term variations in the readings to suggest that the clarity of the lake was changing. Larson & Buktenica (1998) evaluated the Secchi disk clarity database and demonstrated that variations in depth measurements are very sensitive to variations in particle densities and compositions, differences in observers, and the surface conditions of the lake (e.g., smooth, boat waves, or rippled by wind energy). Large variations in Secchi disk clarity measurements in such a clear lake are the result of the hyperbolic relationships between particle densities and disk clarity that inhibit the ability of observers to see the 20 cm disk at a range of >30 m (Larson & Buktenica, 1998). It is of interest to note that a 28 m reading on 19 July 1995 was followed by a reading of 18.1 m on 31 July 1995 after a torrential summer storm (rain and hail) caused extensive mud flows into the lake. These results were similar to observations after extensive mud slides in 1938 (i.e., Secchi disk depths of 27–30 m, as compared to 36–40 m readings in 1937). Based on available information, Crater Lake remains one of the clearest lakes in the world (Hargreaves, 2003). Nonetheless, correct interpretations of the dynamics of Secchi disk clarity readings will remain an important goal of the long-term limnological program.

Comparing and interpreting long-term changes in light penetration into lakes when different instruments were used is problematic. Although Goldman (1988) addressed this problem for the long-term limnological investigations of Lake Tahoe, we were unable to quantitatively compare and interpret the data collected from the two Kahl instruments and the Licor spectraradiometer relative to any long-term changes in light penetration. Incident solar radiation (e.g., clear and blue light) penetrated to exceptionally great depths in Crater Lake and showed no obvious systematic long-term changes. Hargreaves et al. (2007) use observed relationships over 13 years between blue-filtered photometer data and Secchi disk depth, to predict longer-term changes in UV absorption of the water column.

The exceptionally high clarity of the lake was also substantiated by the transmissometer data. The relatively high turbidity in the epilimnion may have been associated with high densities of phytoplankton (McIntire et al., 2007) and undetermined abiotic particles originating from the atmosphere, intracaldera springs and streamlets, and nearshore erosion of lake sediments. Beneath the epilimnion the lake decreased in turbidity and then increased near the lake bottom from the deposition of particles from the water column and the flux of particles from the edges of the lake that move down along the steep slopes into the deep lake (Dymond et al., 1996).

In summary, Crater Lake is an ultraoligotrophic system with exceptional optical properties that had shown relatively stable water quality properties over the long period of observation. Although there were concerns during the early 1980s that the optical properties and water quality of the lake had declined in comparison to sparse historical information, the larger data set discussed here suggests that natural variations within any particular year mask our abilities to detect any minor long-term changes that might be occurring.

Acknowledgements We thank James Larson, Shirley Clark, Ray Herrmann, Robert Benton, David Morris, Al Hendricks, Charles Lundy, Jon Jarvis, James Milestone, Jerry McCrea, Mac Brock, John Salinas, Ed Starkey, and a host of permanent and seasonal park employees for their support and contributions to the long-term project. We also thank Robert Collier, Jack Dymond, Kelly Redmond, Peter Nelson, Manuel Nathensen, Charles Bacon, Hans Nelson, and members of the peer review panels, especially panel chair Stanford Loeb. Douglas Larson provided Secchi disk clarity data from 1978 to 1983.

References

Anderson, D. R., K. P. Burnham & W. L. Thompson, 2000. Null hypothesis testing: problems, prevalence, and an alternative. Journal of Wildlife Management 64: 912–923.

Bacon, C. R. & M. A. Lamphere, 1990. The geological setting of Crater Lake, Oregon. In Drake, E. T., G. L. Larson, J. Dymond & R. Collier (eds), Crater Lake: An Ecosystem Study. Pacific Division, American Association for the Advancement of Science, San Francisco, CA, 19–27.

Bacon, C. R., J. V. Gardner, L. A. Mayer, M. W. Buktenica, P. Dartnell, D. W. Ramsey & J. E. Robinson, 2002. Morphology, volcanism, and mass wasting in Crater Lake, Oregon. Geological Society of America Bulletin 114: 675–692.

Barber Jr., J. H. & C. H. Nelson, 1990. Sedimentary history of Crater Lake caldera, Oregon. 1990. In Drake E. T., G. L. Larson, J. Dymond & R. Collier (eds), Crater Lake: An Ecosystem Study. Pacific Division, American Association for the Advancement of Science, San Francisco, CA, 29–39.

Berger, J. O. & T. Sellke,1987. Testing a point null hypothesis: the irreconcilability of *p* values and evidence. Journal of the American Statistical Association 82: 112–122.

Byrne, J. V., 1965. Morphology of Crater Lake, Oregon. Limnology and Oceanography 10: 462–465.

Collier, R., J. Dymond & J. McManus, 1993. Studies of hydrothermal processes. In Larson, G., C. D. McIntire & R. W. Jacobs (eds), Crater Lake Limnological Studies, Final Report, Technical Report NPS/PNROSU/NRTR – 93/03: 205–213.

Crawford, G. B. & R. W. Collier, 1997. Observations of a deep mixing event in Crater Lake, Oregon. Limnology and Oceanography 42: 299–306.

Crawford, G. B. & R. W. Collier, 2007. Long-term observations of deepwater renewal in Crater Lake, Oregon. Hydrobiologia 574: 47–68.

Dymond, J. & R. Collier, 1990. The chemistry of Crater Lake sediments: definition of sources and implications for hydrothermal activity. In Drake, E. T., G. L. Larson, J. Dymond & R. Collier (eds), Crater Lake: An Ecosystem Study. Pacific Division, American Association for the Advancement of Science, San Francisco, CA, 41–60.

Dymond, J., R. Collier & J. McManus, 1996. Unbalanced particle flux budgets in Crater Lake, Oregon: implications for edge effects and sediment focusing in lakes. Limnology and Oceanography 41: 732–743.

Goldman, C. R., 1988. Primary productivity, nutrients, and transparency during the early onset of eutrophication in ultra-oligotrophic Lake Tahoe, California-Nevada. Limnology and Oceanography 33: 1321–1333.

Groeger, A., 2007. Nutrient Limitation in Crater Lake, Oregon. Hydrobiologia 574: 205–216.

Hargreaves, B. R., 2003. Water column optics and penetration of UVR. In. Helbling E. W & H. E. Zagarese (eds), UV Effects in Aquatic Organisms and Ecosystems, Comprehensive Series in Photochemical and Photobiological Sciences. Royal Society of Chemistry, Cambridge, UK, 59–105.

Hargreaves, B. R., S. F. Girdner, M. W. Buktenica, R. W. Collier, E. Urback & G. L. Larson, 2007. Ultraviolet Radiation and Bio-optics in Crater Lake, Oregon. Hydrobiologia 574: 107–140.

Hintze, J. L., 1998. Number Cruncher Statistical System (NCSS 2000). Kaysville, UT.

Hoffman, F. O., 1969. The horizontal distribution and vertical migration of the limnetic zooplankton in Crater Lake, Oregon. M.S. Thesis, Oregon State University, Corvallis. 60 pp.
Horne, A. J. & C. R. Goldman, 1994. Limnology.2nd edn. McGraw-Hill, Inc. NY, 592 pp.
Johnson, D. M., R. R. Petersen, D. R. Lycan, J. W. Sweet, M. E. Neuhaus & A. L. Schaedel, 1985. Atlas of Oregon Lakes. Oregon State University Press, Corvallis, Oregon, 317 pp.
Larson, D. W, 1984. The Crater Lake study: detection of possible optical deterioration of a rare, unusually deep caldera lake in Oregon, USA. Verhandlungen. Internationale Vereinigung fur theoretische und angewandte Limnologie 22: 513–517.
Larson, G. L. & M. W. Buktenica, 1998. Variability of Secchi disk readings in an exceptionally clear and deep caldera lake. Archiv fur Hydrobiologie 141: 377–388.
Larson, G. L., C. D. McIntire, M. Hurley & M. W. Buktenica, 1996. Temperature, water chemistry, and optical properties of Crater Lake. Lake and Reservoir Management 12: 230–247.
McIntire, C. D., 1973. Diatom associations in Yaquina Estuary, Oregon: a multivariate analysis. American Naturalist 129: 97–121.
McIntire, C. D., G. L. Larson & R. E. Truitt, 2007. Taxonomic Composition and Production Dynamics of Phytoplankton Assemblages in Crater Lake, Oregon. Hydrobiologia 574: 179–204.
McManus, J., R. W. Collier & J. Dymond, 1993. Mixing processes in Crater Lake, Oregon. Journal of Geophysical Research 98: 18295–18307.
Nathenson, M. & J. M. Thompson, 1990. Chemistry of Crater Lake, Oregon, and nearby springs in relation to weathering. In Drake, E. T., G. L. Larson, J. Dymond & R. Collier (eds), Crater Lake: An Ecosystem Study. Pacific Division, American Association for the Advancement of Science, San Francisco, CA, 115–126.
Nathenson, M., C. R. Bacon & D. W. Ramsey, 2007. Subaqueous Geology and a Filling Model for Crater Lake, Oregon. Hydrobiologia 574: 13–27.
Nelson, C. H., C. R. Bacon, S. W. Robinson, D. P. Adam, J. P. Bradbury, J. H. Barber Jr., D. Schwartz & G. Vagenas, 1994. The volcanic, sedimentologic, and paleolimnologic history of the Crater Lake caldera floor, Oregon: evidence for small caldera evolution. Geological Society of America Bulletin 106: 684–704.
Nelson, P. O., J. F. Reilly & G. L. Larson, 1996. Chemical solute mass balance of Crater Lake, Oregon. Lake and Reservoir Management 12: 248–258.
Phillips, K. N., 1968. Hydrology of Crater Lake, East Lake, and Davis Lake, Oregon. US Geological Survey Water Supply Paper 1859-E, 60 pp.
Redmond, K. T., 1990. Crater Lake climate and lake level variability. In Drake, E. T., G. L. Larson, J. Dymond & R. Collier (eds), Crater Lake: An Ecosystem Study. Pacific Division, American Association for the Advancement of Science, San Francisco, CA, 127–141.
Wetzel, R. G., 1983. Limnology. 2nd edn. Saunders College Publishing, PA, 767 pp.

Hydrobiologia (2007) 574:85–105
DOI 10.1007/s10750-006-0347-1

CRATER LAKE, OREGON

The extent and significance of petroleum hydrocarbon contamination in Crater Lake, Oregon

Daniel R. Oros · Robert W. Collier · Bernd R. T. Simoneit

© Springer Science+Business Media B.V. 2007

Abstract In order to evaluate hydrocarbon inputs to Crater Lake from anthropogenic and natural sources, samples of water, aerosol, surface slick and sediment were collected and analyzed by gas chromatography-mass spectrometry (GC-MS) for determination of their aliphatic and aromatic hydrocarbon concentrations and compositions. Results show that hydrocarbons originate from both natural (terrestrial plant waxes and algae) and anthropogenic (petroleum use) sources and are entering the lake through direct input and atmospheric transport. The concentrations of petroleum hydrocarbons range from low to undetectable. The distributions and abundances of *n*-alkanes, polycyclic aromatic hydrocarbons (PAH) and unresolved complex mixture (UCM) from petroleum are similar for all surface slick sampling sites. The estimated levels of PAH in surface slicks range from 7–9 ng/m^2 which are low. Transport of petroleum-derived hydrocarbons from the lake surface has resulted in their presence in some sediments, particularly near the boat operations mooring (total petroleum HC = 1440 μg/kg, dry wt. compared to naturally derived *n*-alkanes, 240 μg/kg, dry wt.). The presence of biomarkers such as the tricyclic terpanes, hopanes and steranes in shallow sediments further confirms petroleum input from boat traffic. In the deep lake sediments, petroleum hydrocarbon concentrations were very low (16 μg/kg, dry wt.). Very low concentrations of PAH were detected in shallow sediments (17–40 μg/kg at 5 m depth near the boat operations) and deep sediments (3–15 μg/kg at 580 m depth). The individual PAH concentrations in sediments (μg/kg or ppb range) are at least three orders of magnitude less than reported threshold effects levels (mg/kg or ppm range, test amphipod *Hyalella azteca*). Therefore, no adverse effects are expected to occur in benthic biota exposed to these sediments. Boating activities *are* leaving a detectable level of petroleum in surface waters and lake sediments but these concentrations are *very low*.

Keywords Boating · Hydrocarbon contamination · PAH · Petroleum · Surface slicks · Sediments

Guest Editors: Gary L. Larson, Robert Collier, and Mark W. Buktenica
Long-term Limnological Research and Monitoring at Crater Lake, Oregon

D. R. Oros (✉)
San Francisco Estuary Institute, 7770 Pardee Lane, 2nd Floor, Oakland, CA 94621, USA
e-mail: daniel@sfei.org

R. W. Collier · B. R. T. Simoneit
College of Oceanic and Atmospheric Sciences, Oregon State University, Corvallis, OR 97331, USA

Introduction

Crater Lake is the primary resource feature of Crater Lake National Park. The lake fills the caldera of Mt. Mazama, and has a total surface area of 53.2 km^2 and a volume of 17.3 km^3 (Phillips, 1968). It is situated at an elevation of 1882 m above sea level and is the deepest lake in the U.S. with two semi-enclosed basins, one in the north-eastern portion of the lake (North Basin, 580 m) and the second in the south-western section of the lake (South Basin, 485 m). It is one of the clearest bodies of water in the world and attracts tourists and research scientists worldwide. National concern about the clarity of Crater Lake was demonstrated in September 1982 when Congress approved Public Law 97–250 authorizing and directing the Secretary of the Interior to conduct a 10-year limnological study of Crater Lake and to immediately implement such actions as may be necessary to retain the lake's natural pristine water quality.

The objective of the present investigation is to conduct the first comprehensive assessment of the levels and distribution of petroleum hydrocarbons in Crater Lake water and sediments. Hydrocarbons in the lake can originate from a variety of natural (terrestrial plant waxes, algal productivity, etc.) and anthropogenic (petroleum combustion, biomass burning, etc.) sources and can enter the lake through a variety of pathways (direct input, runoff, long range atmospheric transport, etc.). Because Crater Lake has no surface outlet, it may be particularly susceptible to anthropogenic pollution inputs. The Park Service and its concessionaire currently operate four tour boats, two research boats and three skiffs on Crater Lake. These boats, along with other anthropogenic as well as natural sources, introduce unknown quantities of hydrocarbons into the water. Outboard engines release their oil-enriched exhaust at and beneath the water surface. Particulate matter and volatile combustion products from inboard engine exhaust enter the water directly. Although careful measures are taken to deter all petroleum contamination, small amounts of uncombusted lubricating oil and gasoline are unavoidably introduced into the lake during repairs, fueling and pumping of bilge from engine compartments.

The environmental effect of using marine engines for visitor tours and park operations is an ongoing concern that has not been fully evaluated. Qualitatively, hydrocarbon contamination is not apparent in the lake; limited visible fuel slicks are generally localized around operating boat exhausts. However, research on the levels of hydrocarbons is needed by the Park Service in order to make informed decisions on current and future boat use on Crater Lake. In this case, the Park Service's long-range plan considers the option of not having boats in Crater Lake. Whatever the outcome, the Park Service will continue to do everything possible to meet the goals of the park mandate to keep the system unaltered.

Petroleum hydrocarbon pollution and contamination of the environment, especially of water bodies such as estuaries and lakes, are of major regulatory concern. Petroleum pollution can be an obvious phenomenon (e.g., crude oil spill), whereas in low-level chronic cases it is not as clearly obvious. For this study, pollution is defined as a consequence of the anthropogenic introduction of substances in excess of their natural concentrations that results in a detrimental and detectable impact on the environmental system. Contamination is defined as merely an excess of a substance above its natural concentration without a detrimental effect. This distinction is generally evaluated on a case-by-case basis. We propose that the present status of Crater Lake is best characterized in terms of minor contamination from petroleum hydrocarbons.

Nevertheless, the concern with petroleum hydrocarbon contamination of water bodies is important. Petroleum is a complex mixture of tens of thousands or more compounds, including low concentrations of the polycyclic aromatic hydrocarbons (PAH) which impart the carcinogenic and mutagenic properties to the total mixture (Farrington & Meyers, 1975; Farrington, 1980). It should also be emphasized that the PAH, which are the primary health concern, can be derived from other thermal combustive processes as major products, besides being present at trace levels in petroleum. In addition, the volatile and more water-soluble petroleum components cause detrimental effects on fish reproduction and behavior, and on water quality (e.g., Cranwell, 1975).

Below, we present background information summarizing features of the lake that are particularly significant to the hydrocarbon study.

Natural hydrocarbon sources

Terrestrial

Crater Lake is an ultra-oligotrophic lake, thus natural hydrocarbon contributions to the water column and surface are mainly from atmospheric input of detritus from higher plants. The primary hydrocarbons contributed by terrestrial vascular plants include high molecular weight epicuticular waxes (i.e., lipids, on leaves and needles), terpenes (bark and tree resins) and other particles containing lipids (spores, pollen, etc.). Lipids derived from higher plants may be characterized by a number of features (Hatcher et al., 1982; Simoneit, 1978; Simoneit et al., 1980). The *n*-alkanes in the range from n-C_{20} to n-C_{40} show a strong predominance of odd-carbon-numbered over even-numbered homologous compounds. This predominance is especially apparent from n-C_{25} to n-C_{35}, with a strong preference of the n-C_{27}, n-C_{29}, and n-C_{31} alkanes. Even carbon numbered aliphatic alcohols with 24–36 carbon atoms are also relatively common, especially in plant waxes. The most prominent fatty acids generally are palmitic acid (C_{16}), stearic acid (C_{18}), and C_{18} monounsaturated acids.

Aquatic

Natural hydrocarbons in the lake may originate from aquatic sources such as phytoplankton, zooplankton, bacteria, macrophytes, zoobenthos and fish. Phytoplankton are the most important producers of organic matter in the aquatic environment. Within the water column algae (e.g., diatoms and dinoflagellates) may contribute saturated and unsaturated hydrocarbons having both straight and branched chains. Algae synthesize *n*-alkanes in the range from n-C_{14} to n-C_{32}, where often n-C_{15} or n-C_{17}, or both are the predominating alkanes (Simoneit et al., 1980; Tissot & Welte, 1984). Zooplankton and other micronekton also contribute hydrocarbons to the water column, primarily by reproduction, excretion, feeding activities and by the decomposition of detrital organic matter. Lipids from zooplankton include wax esters (consisting of long-chain alcohol, C_{12}–C_{18}, and fatty acid, C_{12}–C_{18}, constituents) and pristane (2,6,10,14-tetramethyl-penta-decane). Pristane, but not phytane, is a major component found in marine zooplankton body fat and may be used for maintaining buoyancy in the water column (Blumer et al., 1963). Bacteria in the water column and in the sediments can contribute functionalized hopanoid biomarkers (e.g., hopanols, hopenes) which are pentacyclic terpenoids derived primarily from their membranes (Simoneit, 1978; Peters & Moldowan 1993).

Recent research utilizing manned submersibles showed that there are inputs of hydrothermal fluids into the bottom of Crater Lake (Collier et al., 1991). Sublacustrine hydrothermal springs and concomitant organic matter alteration by a magmatic intrusion-heating source can contribute hydrothermally derived hydrocarbons to the water column (Tiercelin et al., 1993). However, measurements by Collier et al. (1991) showed no evidence of hydrothermally generated hydrocarbons seeping into the lake.

Atmospheric

Crater Lake is subject to aerosol fallout from pollen, natural particles and charcoal. Natural aerosols are normally composed of particles with adsorbed organic compounds from vegetation sources such as high molecular weight epicuticular plant waxes, fatty acids (C_{12}–C_{30}, higher plants) and biomarkers such as terpenes (conifer resins) (Simoneit, 1989). The pollen fallout during early summer represents a significant organic matter input to the lake surface. Most of this organic matter is degraded and only traces are preserved in the lake sediments.

Anthropogenic hydrocarbon sources

Environmental effects

The major environmental concerns with regards to petroleum hydrocarbon contamination of water bodies such as lakes and estuaries are the detrimental effects on aquatic biota and water quality (Cranwell, 1975; Rowland et al., 2001). Biodegradation of lower chronic levels of petroleum hydrocarbons on the water surface and in the water column results in increased turbidity from particulate matter, with a concomitant increase in micronekton productivity (e.g., Bidleman et al., 1990; Edgerton et al., 1987; Jackivicz & Kuzminski, 1973; Marcus et al., 1988). Chronic petroleum contamination persists in the environment due to incomplete degradation, bioconcentration and bioaccumulation. The cosolubility of petroleum hydrocarbons in natural lipids (fats) aids this preservation. Descending biowaste products such as fecal pellets and other particulate matter (e.g., pollen) in the water column ultimately result in an overall incremental build-up of petroleum hydrocarbons in the sedimentary sinks. Once buried in sediments the petroleum hydrocarbons, including the PAH, are preserved (Cerniglia & Heitkamp, 1989; Wilcock et al., 1996).

The interaction of hydrocarbon fuel dispersion from marine engine use with the aquatic environment has been amply reviewed (e.g., Edgerton et al., 1987; English et al., 1963a, b; Jackivicz & Kuzminski, 1973). A more recent study has confirmed the presence of gasoline compounds in Lake Tahoe (Boughton & Lico, 1998). Mainly the gasoline anti-knock additive methyl *tert*-butyl ether (MTBE) was found as deep as 30 m below the lake surface during the summer boating season.

Marine engine use (tours and research)

The Park Service and its concessionaire currently operate four tour boats, two research boats and three skiffs on Crater Lake. Boat operation facilities are located at Cleetwood Cove and on Wizard Island (Fig. 1). The lake shore terminal at Cleetwood Cove contains floating docks, a small ticket sales counter, a manually operated gasoline storage tank and a restroom facility. During the summer, approximately 700 people a day hike down the 1.1 mile long Cleetwood Trail to view the lake and/or take a guided boat tour around the lake. The estimated number of boat tours for the 1995 summer season was 656 trips.

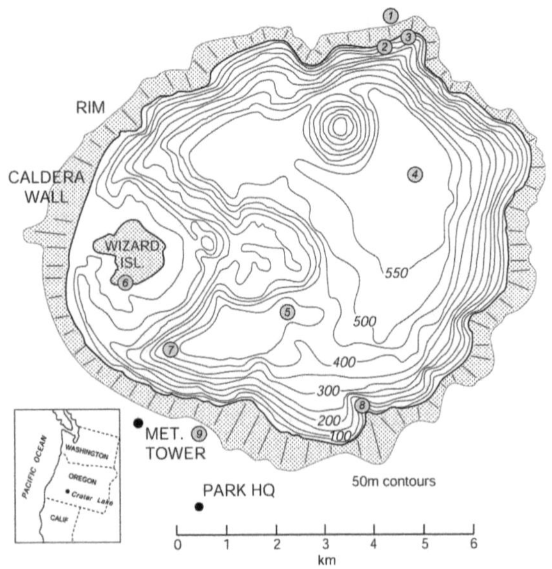

Fig. 1 Map showing locations of hydrocarbon sampling sites at Crater Lake. Sites indicated include: (1) lake access point/parking lot; (2) Cleetwood Cove boat dock; (3) Cleetwood Cover boat mooring; (4) North Basin (east of mid-lake mooring); (5) SE deep basin sediments; (6) Wizard Island boat dock; (7) helicopter crash site; (8) Phantom Ship; (9) air samples collected on the caldera rim near the Lodge

The concessionaire owns and operates the four 60-passenger boats, which provide 2-hour tours around the lake, and a skiff with a small outboard motor. The tour boats are powered by unleaded gasoline (Texaco: Rogue Valley Fuels, Klamath Falls, OR) using inboard engines (2 boats: Ford Redline 460 inch^3 = 7.5 l; 1 boat: Ford Redline 351 inch^3 = 5.7 l; 1 boat: Chevrolet Crusader 350 inch^3 = 5.7 l). Each tour boat contains two gasoline tanks with a maximum capacity of 65 gallons (246 l) per tank. Engine exhaust for these boats exits below the water line. At the end of every tour operation day, excess water (possibly containing lubricating oil) from the boat engine compartment is released into the lake through bilge pumping.

The Park Service's primary research vessel was replaced in 1994. The new vessel is equipped with inboard engines rather than outboard engines to minimize release of unburned fuel and lubricating oils into the lake. The research vessel, "Neuston" is equipped with two inboard engines (2 × 5.0 l V8 Volvo). Two smaller boats equipped with outboard engines are also used: the "Whaler"

(70 HP Johnson outboard); and the "Livingston" (9.9 HP Johnson outboard). The two-cycle outboard engines use a premixed 16:1 unleaded gasoline to lubricating oil mixture for operation.

A 2000 gallon (7570 l) tank storing the gasoline to fuel all boats operating on Crater Lake is located about 0.25 miles west of the Cleetwood Cove parking area, adjacent to Rim Drive. The gasoline is gravity fed to a 300 gallon (1136 l) tank located close to the Cleetwood Cove boat dock. The gasoline from the 300 gallon tank is transferred using a hand operated pump to a fuel dispenser located at the boat dock. The total amount of gasoline delivered to Cleetwood Cove for boat use in 1995 was 7629 gallons (28,876 l); 93% of the total was used for tour boat operations. This fuel delivery system was redesigned and upgraded in 1999 to reduce the risk of fuel spills.

Wizard Island has two boat landings, two concession-owned boat houses and one Park Service boat house. The boat houses, which are on the south side of the island, are used primarily for storage of boats during the winter season, for maintenance, and for scientific research purposes during summer lake operations (July through early September). Hazardous chemical storage lockers containing various petroleum products (e.g., lube oil, gasoline, grease) are located in both boat houses. Approximately 50 gallons (190 l) of gasoline are stored at the Park Service facility to support research operations from the island. Other materials (rags, paper towels, etc.) that may have come into contact with petroleum products are stored in aluminum containers with lids.

Atmospheric input

Crater Lake is subject to aerosol fallout from urban particles and charcoal besides the natural particles and pollen. Local vehicular traffic (mostly visitor automobiles during summer and Park Service snow plows during winter) on the caldera rim is also a source of atmospheric input. According to the Park Service, the number of tourist visits, hence automobile traffic, to Crater Lake National Park has increased gradually over the last two decades (1986 at 428,000; 1995 at 543,000).

Natural aerosols are normally composed of particles with adsorbed organic compounds from vegetation sources (Simoneit, 1989). Organic aerosols derived from anthropogenic sources in urban areas by combustion processes and vehicular emissions (petroleum, heating and cooking oils, etc.) and charcoal also from biomass burning (e.g., wild fires), are composed mainly of petroleum hydrocarbons and minor amounts of PAH (Simoneit, 1984). Some PAH are of environmental concern because they are genotoxic and carcinogenic to many organisms (Cerniglia, 1984; Heitcamp & Cerniglia, 1987).

The results of monitoring aerosols at Crater Lake during the period of March 1993 through February 1994, showed that the atmospheric particle composition is fairly evenly split between sulfates (urban output, coal/oil fired power plants, refining and smelting activities), organics (biogenic natural emissions, smoke, industrial and urban emissions) and soot (vehicular exhaust, smoke) (National Visibility Monitoring Program, NVMP, 1995). Dirty days (dirtiest 20% of total observations) in spring, autumn and winter showed increased nitrate concentrations (automobiles, any combustion source), which were virtually absent from median (median 20%) and clean (cleanest 20%) days. Visibility conditions at Crater Lake are relatively uniform throughout the year with a visibility decrease during the autumn season often due to forest fires. According to the Inter-Agency Monitoring of Protected Visual Environments (IMPROVE) results for site visibility, Crater Lake ranks as the 6th cleanest site out of 42 total sites nationally (NVMP, 1995).

The park service is responsible for maintaining road accessibility within the National Park. During the winter season, the northern park entrance remains closed to all automobile traffic. Vehicle access to the park during this period is through the south entrance along Oregon State Highway 62. During periods of heavy snowfall, two diesel powered push plows and one diesel operated rotary snowplow are usually operating on a daily basis to keep park roads open. These vehicles can contribute particulate matter from vehicular exhaust into the atmosphere and on the caldera rim. The use and maintenance of roadways in the Park will always contribute some amount of hydrocarbons to the local particulate matter in the atmosphere and caldera rim runoff. The design and operation of these facilities needs to include an element considering these hydrocarbons.

Atmospheric transport and deposition of metals, such as iron associated with fine dust particles, has been suggested by Collier et al. (1990). The vertical profiles for iron, manganese and lead collected during the summer of 1984 showed the existence of a surface maximum, trapped in the seasonal thermocline, which decreases rapidly below 75 m. The atmospheric transport and deposition of metals associated with particulate matter to the lake surface water may enhance primary production if the deposited metals are essential for phytoplankton growth and metabolism (e.g., Fe or Mn for photosynthesis).

The atmospheric transport and deposition of pollen into Crater Lake occurs annually during the months of June and July. Pollen particles tend to float and accumulate along the shoreline at Cleetwood Cove and other protected embayments forming "yellow pudding" masses alongside and underneath rocks. Pollen contributes natural hydrocarbons (waxes) as it degrades in the environment.

Acute hydrocarbon inputs

On September 23, 1995 a single-engine helicopter (Aerospatiale AS-350 B-1 Astar) crashed into Crater Lake. Park service officials estimated that the helicopter was carrying approximately 70–90 gallons (265–340 l) of fuel, 2–3 gallons (~10 l) of lubricating oil, and 2 quarts (~2 l) of hydraulic fluid at the time of impact. Rescue and surface water clean-up operations were begun within hours after the crash. A boom was deployed to skim off petroleum hydrocarbons on the lake surface. Surface water samples and miscellaneous debris were also collected at the crash site. The vehicle was not recovered.

This case represents a "unique" circumstance that characterizes a class of uncontrolled/accidental input that can occur in association with human uses of the Park. Potential vehicle accidents (including boats on the lake, automobiles, aircraft) are rare due to appropriate regulation of their use. The helicopter crash is probably representative of the pollutant impacts that might be expected from other worse-case accidents (such as grounding/break-up of a tour boat).

Materials and methods

Sample collection and treatment

A variety of analytical techniques and interpretive tools were applied to examine the character of fossil fuel hydrocarbons as well as natural sources of hydrocarbons in surface water and sediments in terms of their structural and compositional makeup (Bidleman et al., 1990; Dimock et al., 1980; Eglinton et al., 1975; Marcus et al., 1988; Simoneit & Aboul-Kassim, 1994; Voudrais & Smith, 1986). Samples representative of the different inputs were taken to assess hydrocarbon concentrations and compositions. In order to evaluate hydrocarbon inputs from boating (anthropogenic) and natural (biogenic) sources to Crater Lake, a preliminary baseline sampling survey and a final analytical sampling survey of various lake environments were conducted. Surface slick (surface microlayer), water column and sediment samples were collected for the determination of their aliphatic and aromatic hydrocarbon concentrations and compositions. Air samples were also collected on the caldera rim to determine the characteristics of atmospheric hydrocarbon inputs to the lake. A map showing the hydrocarbon sampling sites is given in Fig. 1.

Chemical analyses of samples for hydrocarbons were conducted after suitable extraction and fractionation by using gas chromatography (GC) and gas chromatography-mass spectrometry (GC-MS) instrumentation at Oregon State University (OSU) and at the Geochemical and Environmental Research Group (GERG) laboratory at Texas A & M University. This study discusses the characterization of petroleum hydrocarbons present in selected samples from Crater Lake. Most of the total extracts consist primarily of natural lipid components with traces of petroleum hydrocarbons.

Gasoline and lubricating oil (Napa SAE 30) used primarily for tour boats were obtained from the concessionaire at the Cleetwood Cove boating facility. The premixed fuel for the two-cycle motor boats was also sampled. The hydrocarbon fluid samples were collected in preheated (< 350°C) 10.0 ml vials with teflon lined caps. Immediately following collection, the samples were stored at 4°C and then transported to OSU and GERG for the determination of source hydrocarbons by GC and GC-MS analyses.

At GERG, source fluid samples for hydrocarbon analysis were first subsampled by diluting an aliquot with a known amount of methanol. An aliquot of the diluted sample (< 5.0 ml) was then taken and spiked with internal standard in 5.0 ml of water. The spiked sample was then analyzed by GC and GC-MS for BTEX (benzene, toluene, ethylbenzene and xylenes) volatiles which were determined by purge and trap procedures based on EPA Method 8020 (EPA, 1986). PAH were extracted using the procedure outlined in GERG Standard Operating Procedure (SOP) 8902 Rev 4 and analyzed according to protocol in GERG SOP 9406 Rev 1. The following parent and alkylated PAH compounds were analyzed: C_0, C_1, C_2, C_3, C_4-naphthalenes; biphenyl; acenaphthylene; acenaphthene; C_1, C_2, C_3-fluorenes; C_0, C_1, C_2, C_3, C_4-phenanthrenes; C_0, C_1, C_2, C_3, C_4-anthracenes; C_0, C_1, C_2, C_3-dibenzothiophenes; C_1-fluoranthenes; C_0, C_1-pyrenes; benz[*a*]anthracene; C_0, C_1, C_2, C_3, C_4-chrysenes; benzofluoranthenes; benzopyrenes; perylene; indeno[1,2,3-*cd*]pyrene; dibenz[*a,h*]anthracene; and benzo[*g,h,i*]perylene. Data was also reviewed for accuracy and met the quality assurance (QA) criteria as specified in the SOP for BTEX and PAH products based on the methods above.

Water column samples were collected at approximately 1.0 m below the surface using preheated (< 350°C) 500 ml narrow mouth glass bottles (amber with teflon lined lids). Immediately following collection, the unfiltered water samples were stored at 4°C and then transported to GERG for chemical analysis of BTEX compounds by EPA Method 8020.

Samples of water surfaces impacted by fuel residues and with natural slicks were acquired using 20 × 26 cm pre-cleaned quartz fiber filter sheets (annealed at 350°C for a minimum of 4 h) (Simoneit & Aboul-Kassim, 1994). Each filter was used to blot the surface film six times by alternating sides during collection. This assumes that the area blotted (0.31 m^2/filter) adsorbed the surface slick or surface microlayer, typically 100 μm thickness, uniformly from that area. Sample filters were then placed in pre-heated (< 350°C) "Qorpak" wide-mouth jars (with teflon lined caps), spiked with approximately 15 ml of a chloroform/methanol (2:1) solvent mixture to stop microbial alteration and degradation, then stored at 4°C for transport to OSU and GERG. At OSU, the sample filters were solvent extracted four times each with 50.0 ml aliquots of methylene chloride (CH_2Cl_2). The CH_2Cl_2 fraction containing the hydrocarbons was then separated from water in a separatory funnel, filtered and evaporated under aspirator vacuum to approximately 1.0 ml. The crude extracts were fractionated by liquid chromatography into aliphatic and aromatic hydrocarbons using a column of silica gel and alumina with gradient solvents as eluent. Total extract and sample fractions (positives and blanks) were concentrated using a rotary evaporator and then under a stream of filtered N_2 gas to a final volume (< 100 μl) necessary for hydrocarbon detection by GC. Surface slick samples were also submitted to GERG for chemical analysis using the GERG SOP for chemical extraction and determination of PAH.

Lake sediments were collected using a Soutar box corer and by SCUBA diver. Samples were taken from mid-lake at approximately 580 m depth and at various locations and depths in Cleetwood Cove. For hydrocarbon analysis only the top 2.0 cm of undisturbed surface sediment was sampled. All sediment samples were placed in Kapak bags, sealed and then stored on dry ice. At the OSU laboratory, the wet sediments were extracted using ultrasonic agitation for three periods of 15 min each in 200 ml of 3:1 CH_2Cl_2/methanol solvent mixture. The extractions were carried out in organically clean 500 ml pyrex beakers. The solvent extract was filtered using a Gelman Swinney filtration unit containing an annealed glass fiber filter for the removal of insoluble particles then followed by isolation of the CH_2Cl_2 soluble organic fraction from water using a separatory funnel. The filtrate was first concentrated on a rotary evaporator and then using a stream of filtered N_2 gas to an extract volume of approximately 4.0 ml. The volume was then adjusted to 4.0 ml exactly by addition of CH_2Cl_2 then analyzed for hydrocarbons by GC and GC-MS. Sediment samples were also submitted to GERG for chemical analysis using the GERG SOP for chemical extraction and determination of PAH.

Air sampling at Crater Lake rim during the winter (May) and summer seasons (July) was done over 48 h periods to collect ambient aerosol particulate matter. A standard high volume air sampler (GCA/Precision Scientific) with a flow rate of 40 ft^3/min (1.13 m^3/min) was positioned in an open meadow (snow field in winter) approximately 300 m south of Crater Lake lodge. After sample collection, a small portion of the sampling filter (2.5 × 3.5 cm) was removed for total carbon analysis (as volatilizable and black soot carbon) and the remainder placed in organically clean Qorpak wide mouth jars (with teflon lined caps), preserved with 5.0 ml of chloroform and then stored at 4°C until chemical analysis. In the laboratory, the air sample filters were extracted three times using ultrasonic agitation for 15 min each in 200 ml of CH_2Cl_2. The extractions were carried out within the filter storage jars. The solvent extract was filtered using a Gelman Swinney filtration unit containing an annealed glass fiber filter for the removal of insoluble particles. The filtrate was first concentrated on a rotary evaporator and then using a stream of filtered nitrogen gas to an extract volume of approximately 4.0 ml. The volume was then adjusted to 4.0 ml exactly by addition of CH_2Cl_2. An aliquot of the total extract was then subjected to GC and GC-MS analyses.

The organic carbon analysis of the particulate matter on the aerosol filters consists of a two-step laser combustion method (Johnson et al., 1981; Birch & Cary, 1996). The CO_2 generated first from the volatilizable organic matter is quantified and then that from the black carbon (soot).

GC and GC-MS analyses

Samples were fingerprinted by maintaining the same conditions with high resolution GC on a Hewlett-Packard (HP) Model 5890A GC, equipped with a split/splitless injector and a flame ionization detector (FID). The samples were analyzed in the split or splitless modes using a fused capillary column (30 m × 0.25 mm i.d., 0.25 μm film thickness, DB-5, J & W Scientific) with helium as carrier gas and operating conditions as follows: FID 300°C, injector 300°C, the oven temperature was programed from 45 to 300°C at 4°C/min after 15 min and held isothermal at 300°C for 20 min. The analog signal was monitored with an HP 3393A integrator or with an HP Chemstation Software program. The data is presented as plots of relative response versus time. Identification was based on comparison of the retention times with standard reference compounds. The following standard mixtures were injected on GC: (1) Wisconsin diesel range hydrocarbons (AccuStandard Inc.); (2) a series of *n*-alkanes ranging from C_{10} to C_{36}; and (3) pristane and phytane.

The GC-MS analyses were conducted on an HP 5973 MSD mass spectrometer interfaced directly with a HP Model 6890 GC and equipped with a 30 m × 0.25 mm i.d. fused silica capillary column. The operating conditions were as the same as given above. The GC-MS data were acquired and processed with HP Chemstation Software equipped with a GC-MS Data Library. The data is presented as plots of relative intensity of the total ion counts or individual fragment ions (called mass fragmentograms) versus time. Compound assignments were made from individual mass spectra and GC retention times and with comparison to authentic standards where possible. Laboratory blanks and positive samples were analyzed as controls for this method.

Hydrocarbon parameters

The following organic geochemical parameters are used for interpreting the data:

Makeup of natural and anthropogenic organic components (homologous alkane series, Simoneit, 1978).

Carbon preference index (CPI): The ratio of odd-carbon-numbered to even-carbon-numbered *n*-alkane peaks in a given sample (Simoneit, 1978). The CPI helps in differentiating biogenic from petrogenic *n*-alkanes in organic environmental samples. In particular, it is useful for making estimates of terrestrial plant wax contribution versus fossil fuel contamination. Vascular plants synthesize epicuticular waxes as odd number *n*-alkane hydrocarbons usually in the C_{25}–C_{33} range (CPI > 1.0). In crude oils, the high molecular weight *n*-alkanes inherited from terrestrial plants are normally diluted by hydrocarbons from kerogen degradation which increases the even number *n*-alkane concentrations, resulting in a CPI of around 1.0. An even-over-odd *n*-alkane predominance, although much less common, is associated with organisms such as bacteria and diatoms.

The *n*-alkanes were detected in the total extract or separated hydrocarbon fractions by the GC retention index or in GC-MS data by the mass fragmentogram plot of the *m/z* 85 key fragment ion.

Isoprenoid hydrocarbons: The isoprenoid hydrocarbons pristane (Pr) and phytane (Ph) can be used together as specific indicators for the presence of petroleum residues (Peters & Moldowan, 1993). They are mature biomarkers generally found in all crude oils and are stable in the environment. Pr and Ph have specific chemical structures which are unique to their source and together they are not synthesized by contemporary biota. However, a high concentration of pristane alone (>*n*-heptadecane) can be derived from zooplankton.

Unresolved complex mixture (UCM): The broad "hump" as observed in chromatograms is associated with the heavier compounds in petroleum and lubricating oils. It results from an unresolved "complex" mixture of branched and cyclic hydrocarbons, which generally indicates a petrogenic hydrocarbon input from heavier hydrocarbon fractions of petroleum. The UCM is always present in unburned petroleum emissions, however, its chemical components cannot be fully determined. Its major input vector into environmental systems is from engine lubricating oils.

Petroleum biomarkers: Biomarkers or molecular markers are indicator compounds that can be used for defining the sources of organic matter in the environment (e.g., Simoneit, 1978, 1986). As applied here, biomarkers characteristic of petroleum products are characterized to confirm such an origin for extractable organic matter. The petroleum biomarkers are triterpenoid hydrocarbons (17α(H)-hopanes) and steroid hydrocarbons (steranes and diasteranes), which are minor but unique components in petroleum products such as lubricating oils (Bieger et al., 1996; Peters & Moldowan, 1993; Rogge et al., 1993a). The hopanes were detected in the GC-MS data by the mass fragmentogram plot of the *m/z* 191 key fragment ion and similarly the steranes and diasteranes are found in the key fragment ion plots of *m/z* 217 and 218.

Polycyclic aromatic hydrocarbons: PAH can be derived from three sources in an environment such as Crater Lake. First, combustion emissions from vehicles using petroleum-derived fuels and lubricants contain PAH with a relatively high amount of alkyl substituents (e.g., Marcus et al., 1988). The typical indicators used are the phenanthrene/anthracene (P/A) series. This signature is distinguishable from the PAH emitted by the second source, biomass burning, where the P/A series would show an enriched content of C_2 and C_4 homologs (i.e., pimanthrene and retene, alteration products from conifer resin compounds) (Ramdahl, 1983; Simoneit, 1998). The third source is high temperature combustion which emits the higher molecular weight PAH as described for many urban areas (Neff, 1979). These PAH (e.g., pyrene, chrysene, etc.) were detectable in some of the samples analyzed from Crater Lake. However, the source emission for this third category overlaps with the first.

BTEX: These are the volatile aromatic ($< C_{12}$) petroleum products which include benzene, toluene, ethylbenzene and xylenes and other derivatives. BTEX products are generally found in gasoline as additive hydrocarbons to improve fuel efficiency and are extremely volatile once released in the environment. The BTEX compounds have some limited solubility in water.

Results

Petroleum hydrocarbon use

The primary petroleum products used in the park include gasoline, diesel, heating and lubricating oils. Diesel and heating oil are the main petroleum products used in Munson Valley, especially during the winter season. On the lake, gasoline is used to fuel watercraft (tour and research boats, skiffs) and lubricating oil is used for boat engines and for making up the fuel-oil mixture (50:1–100:1) necessary to operate outboard motors. Boat fueling and tour and research operations unavoidably introduce petroleum hydrocarbons and their combustion residues to the lake.

Analyses conducted

Sampling sites were chosen to best represent the distributions and concentrations of hydrocarbons, BTEX and PAH products in Crater Lake (Fig. 1). All samples were subjected to hydrocarbon and PAH analyses while water column samples and petroleum fluids were further tested for BTEX products. The concentrations and distributions of petroleum hydrocarbons present in Crater Lake from engine emissions and other combustion processes (PAH) are reported in Table 1. Analytical data for the *n*-alkane, UCM and isoprenoid hydrocarbon constituents for all the samples collected are reported in Table 2.

Table 1 Total PAH compounds in Crater Lake samples

	Total PAH[a]
Surface slicks	(ng/m^2)
Cleetwood Cove-boat mooring	1.7–8.5
East of mid-lake mooring	0.7–7.6
Phantom ship-north side	0–7.7
Wizard island-south bay	0.5–7.2
Sediments	(μg/kg)
Cleetwood Cove (5 m water depth)	17–40
North basin (580 m water depth)	3–15
Petroleum fluids	(mg/kg)
Lubricating oil	159
Gasoline	13500

[a] PAH analysis consists of the following compounds: C_0, C_1, C_2, C_3, C_4-naphthalenes; biphenyl, acenaphthylene; acenaphthene; C_0, C_1, C_2, C_3-fluorenes; C_0, C_1, C_2, C_3, C_4-phenanthrenes; C_0, C_1, C_2, C_3, C_4-anthracenes; C_0, C_1, C_2, C_3-dibenzothiophenes; C_0, C_1-fluoranthenes; C_0, C_1-pyrenes; benz[*a*]anthracene; C_0, C_1, C_2, C_3, C_4-chrysenes; benzofluoranthenes; benzopyrenes; perylene; indeno[1,2,3-*cd*]pyrene; dibenz[*a,h*]anthracene; and benzo[*g,h,i*]-perylene (data from GERG, Texas A & M University). See Table 3 for individual compound analyses

Water column hydrocarbons

Crater Lake water column samples from the Cleetwood Cove boat mooring, the mid-lake science mooring, and near Wizard Island were subjected to BTEX analysis for determination of petroleum input from gasoline spillage and boat exhaust. The BTEX products were not found at concentrations above the procedural blanks used for this analysis (< 5.0 μg/l). Thus, the BTEX compound concentrations in the water column are not a significant component or contributor of hydrocarbons. BTEX products, once applied to a water surface from a gasoline spill or from boat exhaust, are very volatile and have limited solubility in cold water. Therefore, they are not likely to concentrate significantly in the water column. Analyses for heavier petroleum hydrocarbons dissolved in the water column, such as PAH, were not conducted because their solubilites are low and they generally have high octanol-water partition coefficients (Callahan et al., 1979), hence a strong affinity for adsorbing to water column particulate organic matter. Their concentrations in the surface slicks were also found to be low suggesting that water column concentrations would be below the method detection limits. Petroleum hydrocarbons in the water column are adsorbed to particulate material, which ultimately sinks to the lake sediments.

Surface slick hydrocarbons

Aliphatic and higher molecular weight hydrocarbons (C_{10}–C_{30}) are hydrophobic and therefore concentrate at the water–air interface forming a slick (film) with the natural lipids, which also accumulate there. These surface films (also termed surface microlayer, upper 100 μm) are important for concentrating lipophilic higher molecular weight compounds such as petroleum hydrocarbons and natural lipids (Morris & Culkin, 1975). The larger-chain surface-active molecules tend to concentrate at the water–air interface and mixing of the water surface by wind and currents displaces the shorter-chain, more hydrophilic compounds downward. Thus, the more water soluble and volatile petroleum components (e.g., BTEX) partition into the water

Table 2 Analytical data of the *n*-alkane, UCM and isoprenoid hydrocarbon constituents in Crater Lake samples

Sample			C_{range}	C_{max}	CPI[a]	Pr/Ph	Plant wax[b]	Petroleum	Total	UCM	Total	Total
Date-ID	Sampling site	Description					*n*-alkanes	*n*-alkanes	*n*-alkanes		PAH	HC[c]
Air							(ng/m^3)	(ng/m^3)	(ng/m^3)	(ng/m^3)	(ng/m^3)	(ng/m^3)
5/06/96–1	Crater Lake rim	48 h/1 filter	12–29	25	3.3	ND	94	139	233	ND	ND	ND
7/09/96–1	Crater Lake rim	48 h/2 filters	15–29	23	1.8	ND	118	155	274	ND	ND	ND
7/11/96–1	Crater Lake rim	48 h/2 filters	14–31	16	1.4	1.7	430	227	657	ND	ND	ND
Surface slicks							($\mu g/m^2$)	($\mu g/m^2$)	($\mu g/m^2$)	($\mu g/m^2$)	($\mu g/m^2$)	($\mu g/m^2$)
7/27/95–1	Cleetwood Cove	Interior of boat dock	16–34	22	1.2	1.4	50	630	680	3440	ND	4070
7/27/95–3	Cleetwood Cove	Outboard engine exhaust	14–35	22	1.8	2	400	790	1190	2190	ND	2980
7/27/95–5	Cleetwood Cove	Slick with foam	14–36	28	0.5	1.2	280	220	500	1680	ND	1900
7/27/95–13	Wizard Island[d]	Tour boat wake at dock	16–26	18	1	1.9	300	1610	1910	7140	ND	8750
7/27/95–16	Spring 42	2 filters	16–25	22	1.4	2.8	340	1060	1410	ND	ND	ND
8/11/96–1	Wizard Island	4 filters/GERG	NA	NA	NA	NA	NA	NA	NA	NA	7.18	ND
8/11/96–3	Phanthom ship	4 filters/GERG	NA	NA	NA	NA	NA	NA	NA	NA	7.65	ND
8/11/96–5	Mid-lake mooring	4 filters/GERG	NA	NA	NA	NA	NA	NA	NA	NA	7.56	ND
8/11/96–7	Cleetwood mooring	4 filters/GERG	NA	NA	NA	NA	NA	NA	NA	NA	8.52	ND
9/26/95–1	Helicopter crash site	2 filters	15–28	28	0.3	0.7	1200	840	2030	3920	ND	4760
9/26/95–2	Phanthom ship	2 filters	14–29	28	0.1	1.8	5050	710	5760	4330	ND	5040
9/26/95–5	Mid-lake mooring	2 filters	14–30	28	0.1	2	5120	830	5950	4870	ND	5700
9/26/95–6	The Palisade	2 filters/3 m to shore	16–28	28	0.1	1.1	5860	370	6230	ND	ND	ND
Sediments							($\mu g/kg$)	($\mu g/kg$)	($\mu g/kg$)	($\mu g/kg$)	($\mu g/kg$)	($\mu g/kg$)
9/08/95–27	Cleetwood Cove	5 m depth/diver collected	15–33	24	0.7	2.2	243	505	748	ND	39.6	ND
9/08/95–28	Cleetwood Cove	10 m depth/diver collected	15–33	29	2.1	1.4	210	290	500	ND	ND	ND
9/08/95–29	Cleetwood Mooring	5 m depth/diver collected	15–33	29	1.1	1.2	236	1439	1675	ND	ND	ND
9/08/95–30	Cleetwood Mooring	10 m depth/diver collected	15–33	29	1	1.5	115	503	618	ND	ND	ND
9/08/95–31	North Basin	580 m depth/box core	16–33	21	2.6	ND	13	16	29	ND	15.5	ND
Soils							($\mu g/kg$)	($\mu g/kg$)	($\mu g/kg$)	($\mu g/kg$)	($\mu g/kg$)	($\mu g/kg$)
7/27/95–6	Cleetwood Cove	Next to gas pump	12–26	13	1	2.8	6E+06	4E+07	5E+07	ND	ND	ND
7/27/95–15	Spring 42	Mud composite	18–30	25	4.7	ND	1E+05	5E+04	1E+05	ND	ND	ND
7/27/95–17	Sinot memorial	Mud composite	17–36	27	2.4	ND	2E+04	2E+04	4E+04	4E+05	ND	ND
7/27/95–25	Watchman overlook	Lake side of road	17–36	29	3.6	ND	3E+04	2E+04	5E+04	ND	ND	ND
Petroleum products							(mg/l)	(mg/l)	(mg/l)	(mg/l)	(mg/l)	(mg/l)
7/27/95–2	Cleetwood Cove	Gasoline	to C_{10}	ND	ND	ND	ND	ND	ND	ND	ND	ND
7/27/95–4	Cleetwood Cove	Gasoline/lube oil mixture	to C_{10}	ND	ND	ND	ND	ND	ND	ND	ND	ND
7/27/95–22	Cleetwood Cove	Purple lube oil/whaler	12–22	12	0.5	1.1	2E+06	5E+09	5E+06	1E+07	ND	ND
8/11/96–11	Cleetwood trail	Diesel	10–25	16	1	ND	ND	ND	ND	ND	ND	ND
Surface debris							($\mu g/m^2$)	($\mu g/m^2$)	($\mu g/m^2$)	($\mu g/m^2$)	($\mu g/m^2$)	($\mu g/m^2$)
9/26/95–3	Phanthom ship	White foam	14–36	18	1.2	1	90	480	570	1870	ND	ND
9/26/95–4	Phanthom ship	White foam	17–28	21	1.6	ND	2940	2700	5640	ND	ND	ND
9/26/95–7	Helicopter crash site	Pilot book/page with fuel	12–19	13	0.4	3.6	1E+08	2E+09	2E+09	2E+09	ND	ND
9/26/95–8	Helicopter crash site	Pilot book/cover with fuel	12–18	13	0.2	1.2	1E+11	8E+10	2E+11	2E+11	ND	ND

Table 2 continued

Sample			C_{range}	C_{max}	CPI[a]	Pr/Ph	Plant wax[b]	Petroleum	Total	UCM	Total	Total
Date-ID	Sampling site	Description					*n*-alkanes	*n*-alkanes	*n*-alkanes		PAH	HC[c]
Snow												
7/27/95–18	Sinot memorial	Surface scraping	15–30	20	0.5	3	ND	ND	ND	ND	ND	ND
7/27/95–20	Southeast Shore	Surface scraping	17–35	20	0.9	ND	ND	ND	ND	ND	ND	ND
7/27/95–26	Watchman overlook	Surface scraping	16–31	20	0.7	2.4	ND	ND	ND	ND	ND	ND
Biology							(µg/kg)	(µg/kg)	(µg/kg)	(µg/kg)	(µg/kg)	(µg/kg)
7/27/95–8	Cleetwood Cove	Aquatic moss beneath dock	17–33	17	2.1	ND	1E+07	8E+06	2E+07	ND	ND	ND
7/27/95–11	Cleetwood beach	Pollen pudding at water line	15–29	21	4.2	0.4	1E+05	8E+04	2E+05	ND	ND	ND
7/27/95–24	Traffic Stop/rim village	Mountain hemlock (waxes)	24–36	34	0.9	ND	4E+07	7E+07	1E+08	ND	ND	ND

[a] CPI determined from C16 through C33 *n*-alkanes

[b] Remaining concentration after subtraction of the petroleum *n*-alkane concentration (calculation cited from Simoneit et al., 1991)

[c] Total HC: Total hydrocarbon concentration as sum of PAH, *n*-alkanes, and UCM. PAH determined by GC-MS, *n*-alkanes and UCM determined by GC

[d] Sample collected in wake of tour boat

Abbreviations: NA, not analyzed; ND, not detected (thus not determined)

column or evaporate and are therefore depleted in the water-air interface slick. Therefore, sampling of surface slicks is a way to analyze an enriched *upper limit* in concentration of hydrophobic petroleum hydrocarbons superimposed on the natural background lipids.

The primary petroleum hydrocarbons found in Crater Lake surface slicks have been identified as exhaust products and motor lubricating oil residues from internal combustion engines. The results of the chemical analyses show that normal and isoprenoid alkanes and an envelope (UCM, hump) of unresolved branched and cyclic hydrocarbons, typical of petroleum products, were detected in surface slicks at different sites of Crater Lake. The GC fingerprint for gasoline hydrocarbons, which has a characteristic peak pattern for naphthenic compounds in the low molecular weight range ($< C_{12}$), and for lubricating oil, with a pronounced UCM, are used as indicators for direct input (unburned) of petroleum products. The presence of *n*-alkanes from C_{17} to C_{26} (with CPI = 1) is also indicative of petroleum-derived hydrocarbons from this source.

Slick samples were collected at various sites to best represent the overall spatial distribution of hydrocarbons on the lake surface. Analytical data of the hydrocarbons collected from sampling sites with highest traffic from boating activities (Cleetwood Cove, Wizard Island), sites along or near the boat touring route (The Palisades, Phanthom Ship, Spring 42) and a background site (East of Mid-lake Mooring) are shown in Table 2 (also see Fig. 1). For comparison purposes, the hydrocarbon data from surface slicks collected within days after the helicopter crash are used to determine the impact of the crash on surface hydrocarbon levels.

Typical examples of GC-MS traces of surface slicks containing outboard engine exhaust products and hydrocarbons from natural sources are shown in Figs. 2 and 3. A complete summary of the hydrocarbon concentrations and distributions are shown in Table 2. Surface slicks generally show *n*-alkanes present as a homologous series ranging in chain length from *n*-C_{14}–*n*-C_{36} with C_{max} at C_{18}, C_{21}, C_{22} and C_{28}. The presence of odd-carbon number hydrocarbons with chain length $>C_{25}$ indicates a minor contribution from

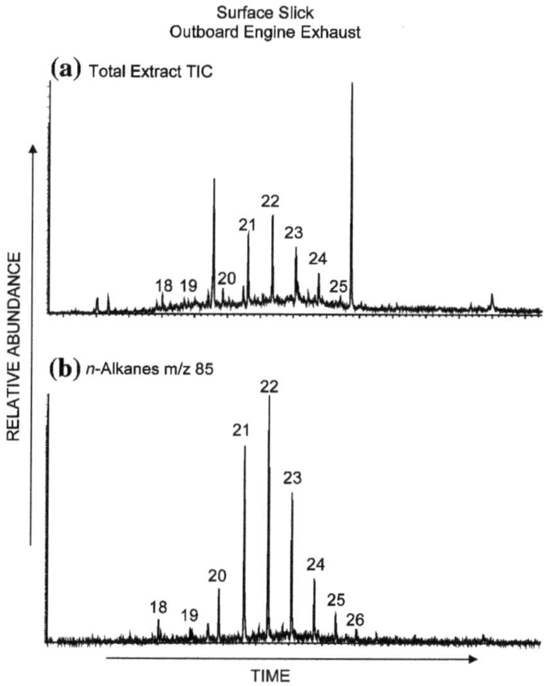

Fig. 2 GC-MS data of a surface slick sampled immediately behind an outboard engine exhaust: (**a**) total ion current trace of the hydrocarbon fraction showing resolved peaks of *n*-alkanes and other lipids; (**b**) *n*-alkane hydrocarbons (detected in data of A by the key fragment ion *m/z* 85) with C_{max} at C_{22}. Numbers refer to carbon chain length of components

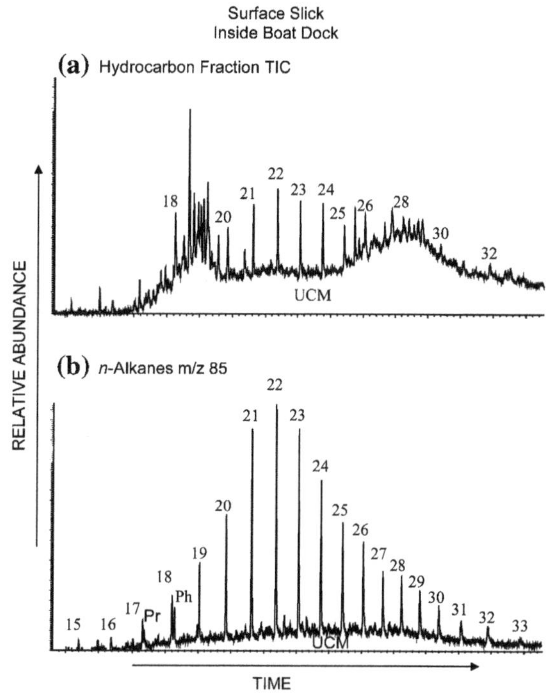

Fig. 3 GC-MS data of fractionated surface slick extract collected at the interior section of the Cleetwood Cove boat dock: (**a**) total ion current trace of the total hydrocarbon fraction showing resolved peaks of *n*-alkanes and other lipids; and (**b**) *m/z* 85 key fragment ion for *n*-alkane hydrocarbons with C_{max} at C_{22}. The *n*-alkane distribution is similar to that found in outboard engine exhaust. Numbers refer to carbon chain length of components, Pr = pristane, Ph = phytane, UCM = unresolved complex mixture

natural sources (higher plant wax). The CPIs range from 0.1 to1.8 and reflect the contribution from both natural and anthropogenic hydrocarbon sources. The presence of an UCM and *n*-alkanes with a CPI = 1.0 (from C_{14} to C_{24}) confirms that these samples also contain lube oil and engine exhaust products. The occurrence of both pristane and phytane support the petroleum related origin of most of the *n*-alkanes and the UCM in the surface slick samples.

Within several days after the helicopter crash in Crater Lake, surface slick samples and crash debris were collected for analysis. The GC trace of the hydrocarbons extracted from the recovered log book page, which had a characteristic fuel odor, showed a fingerprint for petroleum components typical of kerosene or Jet A fuel (Mayfield & Henley, 1991). Hydrocarbons were also present in natural foam samples collected near Phantom Ship after the crash, however, they are not necessarily derived from the aircraft fuels (as opposed to the sampling activity). The presence of the *n*-alkanes, C_{14} and C_{15}, indicated that a trace of unburned helicopter jet fuel was present in this surface slick. The absence of alkanes $< C_{14}$ may have been due to evaporative losses that begin immediately after hydrocarbon exposure at the air-water interface. The helicopter crash introduced petroleum hydrocarbons (*n*-alkanes and UCM) to the lake surface and water column from fuel, lubricating oil and hydraulic fluid leakage. However, the measured level of total surface slick petroleum hydrocarbons as *n*-alkanes (840 $\mu g/m^2$) does not differ significantly from a site sampled during the same period as a background reference (Mid-lake near mooring: 830 $\mu g/m^2$) or from a surface slick collected 2 months earlier at a nearby site (Spring 42: 1060 $\mu g/m^2$). Thus, the hydrocarbons released during the helicopter crash were difficult to detect in surface slicks within several days after the crash.

The plant wax concentrations measured as *n*-alkanes in surface slicks increased greatly during the summer months of July, 1995 (average at 270 μg/m^2, n = 5) through September (average at 4310 μg/m^2, n = 4). This increase reflects the seasonal release of pollen from the surrounding forests and new plant growth, with deposition of detritus to the lake surface by atmospheric transport. The accumulation of pollen is very noticeable along the shoreline of Cleetwood Cove and Wizard Island during the months of July through September. A visual observation of the shorelines and surface waters conducted in October showed a decrease in the abundance of pollen at these sites.

The concentrations and distributions of total PAH products on the lake surface as a thin film are essentially constant (between 0.5 and 7.2 ng/m^2 measured at Wizard Island and 1.7–8.5 ng/m^2 at Cleetwood Cove, Table 1). The slightly higher PAH concentration evident in the surface slick from Cleetwood Cove may be due to increased tour boat activities at this location. The absence of the C_2 and C_4 alkylated phenanthrenes, corresponding to pimanthrene and retene, respectively, indicates that these PAH are derived from engine exhaust and not from wood smoke in aerosol particle fallout.

A comparison of the concentrations and distributions of the total petroleum hydrocarbons (petroleum *n*-alkanes, UCM, and PAH) for the surface slick sampling sites is given in Table 2. The petroleum *n*-alkane concentrations range from 630 to 1610 μg/m^2 and have the highest concentration at Wizard Island. The Wizard Island sample was collected in the wake of a tour boat, hence this sample represents an upper limit concentration for engine exhaust products (petroleum *n*-alkanes, UCM, and PAH) in a surface slick. At the other sampling sites the petroleum *n*-alkane concentrations have lower values. The UCM concentration at Wizard Island (7140 μg/m^2) is almost twice the levels found at other surface slick sampling sites during the same period (range 1680–3440 μg/m^2) and following the helicopter crash (late September, range 3920–4870 μg/m^2). The Wizard Island sample reflects the tour boat activity of docking, idling and launching which contributes more engine exhaust products to the water surface. An additional contributing factor to the higher UCM and petroleum *n*-alkane concentrations may be the physical setting of Wizard Island's natural harbor, which can accumulate hydrocarbons by the different wind conditions resulting in decreased surface slick dispersal away from the area.

The total PAH concentrations were estimated by summing the individual PAH concentrations reported by GERG (Table 3). The PAH concentrations and distributions for all four surface slicks sampled show that naphthalene concentrations exceed the higher molecular weight pyrolytic PAH concentrations. Naphthalene is a major component of gasoline and thus may represent unburned fuel.

Sediment hydrocarbons

A mass fragmentogram showing the abundances and distributions of petroleum hydrocarbons in sediments collected by diver from the tour boat mooring area in Cleetwood Cove is shown in Fig. 4. The major peaks in the total ion current trace are natural lipid compounds. The relative concentrations of the *n*-alkanes are resolved better by the *m/z* 85 fragmentogram and the specific key fragment ions for the petroleum biomarkers, tricyclic terpanes and hopanes, are plotted over the more limited elution range shown (Fig. 4b, c) for detection.

In Table 2, the *n*-alkanes for all sediment samples range from C_{15}–C_{33}, with C_{max} at C_{21}, C_{24}, and C_{29}, which indicates the presence of natural (C_{21} from algae, C_{29} from plant wax) and petroleum (C_{24}) derived hydrocarbons. The relatively high content of even carbon numbered *n*-alkanes versus the odd, i.e., CPI near 1.0, further indicates the presence of petroleum residues from combustion engine exhaust. This *n*-alkane contribution from petroleum was calculated as previously described (Simoneit et al., 1991). Biomarker signatures (i.e., minor polycyclic hydrocarbons with specific structures in petroleum) are used to confirm the petroleum source. The 17α(H)-hopanes and the extended tricyclic

Table 3 PAH compounds analyzed and their concentrations

PAH[a]	Surface slicks				Sediments[b]	
	Phantom ship (ng/m^2)	Wizard Island (tourboat wake) (ng/m^2)	East of mid-lake (ng/m^2)	Cleetwood Cove (ng/m^2)	Deep Lake (0–2 cm) (μg/kg dry wt)	Cleetwood Cove (μg/kg dry wt)
Naphthalene	2.29 e	2.76 e	2.20 e	2.21 e	2.8 e	6.7
C_1-Naphthalenes	2.04 e	1.19 e	1.19 e	1.59 e	1.4 e	2.6 e
Biphenyl	0.18 e	0.56	0.71	1.03	0.3 e	1
Acenaphthylene	0.30 e	0.35 e	0.21 e	0.33 e	0.3 e	0.05 e
Acenaphthene	0.31 e	0.38 e	0.50 e	0.71	0.7 e	0.5 e
Fluorene	0.41 e	0.31 e	0.58 e	0.26 e	0.2 e	0.5 e
Phenanthrene	0.34 e	0.42 e	0.45 e	0.61 e	0.4 e	2.3 e
Anthracene	0.38 e	0.16 e	0.34 e	0.15 e	0.4 e	0.4 e
Dibenzothiophene	0.34 e	0.12 e	0.18 e	0.26 e	0.3 e	0.9
Fluoranthene	0.27 e	0.28 e	0.45 e	0.34 e	0.3 e	3.2
Pyrene	0.30 e	0.20 e	0.36 e	0.34 e	0.5 e	3.1 e
Benzo[a]anthracene	0.07 e	0.06 e	0.06 e	0.07 e	0.2 e	0.8
Chrysene	0.08 e	0.09 e	0.02 e	0.12 e	0.2 e	1.8 e
C_1-chrysene	ND	ND	ND	ND	0.5 e	ND
C_2-chrysenes	ND	ND	ND	ND	2.9	ND
C_3-chrysenes	ND	ND	ND	ND	1.0 e	ND
Benzo[b]fluoranthene	0.02 e	0.06 e	0.06 e	0.10 e	0.2 e	3 e
Benzo[k]fluoranthene	0.06 e	0.02 e	0.02 e	0.03 e	0.1 e	0.9 e
Benzo[a]pyrene	0.06 e	0.03 e	0.06 e	0.07 e	0.5 e	1.5 e
Benzo[e]pyrene	0.06 e	0.04 e	0.06 e	0.05 e	0.6 e	1.7 e
Perylene	0.04 e	0.05 e	0.05 e	0.12 e	0.3 e	0.6 e
Indeno[1,2,3-c,d]pyrene	0.03 e	0.02 e	0.03 e	0.06 e	0.3 e	4.2
Dibenzo[a,h]anthracene	0.04 e	0.06 e	0.02 e	0.03 e	0.1 e	0.2 e
Benzo[g,h,i]perylene	0.02 e	0.02 e	0.02 e	0.05 e	1 e	3.6 e
Total PAH (estimated)[c]	≤8	≤7	≤8	≤9	3–15	17–40

Data qualifiers: e = qualitative estimate, value below statistical detection limit (2s of blank). ND = not detected. Other PAH analyzed but not detected in any samples include: $C_{2,3,4}$-Naphthalenes; $C_{1,2,3}$-Fluorenes; $C_{1,2,3,4}$-Phenanthrenes/Anthracenes; $C_{1,2,3}$-Dibenzothiophenes; C_4-Chrysenes

[a] Data from GERG, Texas A & M University

[b] A total of 5 other sediment samples were analyzed for Total PAH—none were detected (ND, see Table 2)

[c] Total PAH is estimated as the sum of the individual PAH. The minimum of the range tabulated is the sum of only the PAH detected above the detection limit, whereas the high estimate includes all quantitative and qualitative values

terpanes (*m/z* 191 mass fragmentogram) are present at low levels decreasing in concentration to the deeper sediment samples. The biomarker distribution matches with that reported for sediment samples from other geographic areas (e.g., Simoneit & Kaplan, 1980) confirming the petroleum product source. The steranes and diasteranes, another petroleum biomarker group, of the same sediment samples provide secondary confirmation of a petroleum source. These compounds occurred at trace levels and were only detectable in the shallow sediments.

A sediment sample from a deep part of the lake (North Basin 580 m depth) was extracted and analyzed for hydrocarbon content. The total extract was analyzed by GC-MS (Fig. 5). The major peaks in the total ion current trace are natural fatty acids (shown as methyl ester derivatives and confirmed by the *m/z* 74 key ion plot, Fig. 5b), docosanol, and elemental sulfur (S_8). Hydrocarbons are minor constituents (Fig. 5c) and are of a natural origin from plant wax (>C_{25}) and degraded algal (plankton) lipids (< C_{25}). There is no UCM present and the hopane and sterane biomarkers from petroleum are not detectable. The total hydrocarbons attributable to petroleum in this deep sediment sample amount to < 20 μg/kg. Significant petroleum residues from engine exhaust are detectable in the sediments of Cleetwood Cove, but the values are low

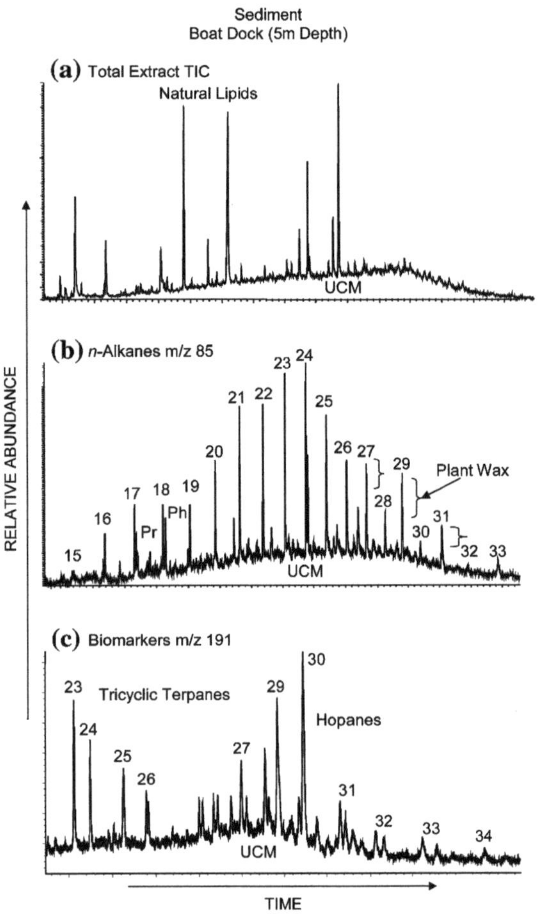

Fig. 4 GC-MS data of a total hydrocarbon fraction from the extract of sediment collected at the Cleetwood Cove boat mooring in 5 m water depth: (**a**) Total ion current trace showing natural lipids; (**b**) *m/z* 85 key fragment ion for *n*-alkanes, Pr = pristane, Ph = phytane; (**c**) *m/z* 191 key fragment ion for tricyclic terpane and hopane biomarkers from petroleum. The biomarkers confirm the presence of petroleum in this sediment sample. Carbon numbers are indicated, UCM = unresolved complex mixture

compared to other areas (Table 4). It should be pointed out that the guideline cutoff for non-polluted (by oil and grease, assumed equivalent to petroleum residues) harbor sediment is <1000 mg/kg (<10^6 µg/kg) and moderate pollution is from 1000 to 2000 mg/kg (EPA, 1977).

The UCM was minor or not found at levels above the instrument detection limits in the sediments. Both plant wax and petroleum *n*-alkane concentrations decrease with depth at both Cleetwood Cove sampling sites possibly due to the variability in the sediment particle grain sizes and their organic carbon content. For example, in the water column, hydrophobic compounds such as *n*-alkanes will tend to bind (partition) with suspended particulate organic materials (e.g., sediments, phytoplankton, etc.), while in sediments, hydrophobic compounds preferentially adsorb to the fine grained (<0.63 µm, silt and clay) fractions of sediments and not to the larger grain sized fractions (>0.63 µm, sand to gravel) (Readman et al., 1984). Because the petroleum *n*-alkanes are found at greater concentrations than plant wax *n*-alkanes at both Cleetwood Cove sites and at each sampling depth, the organic

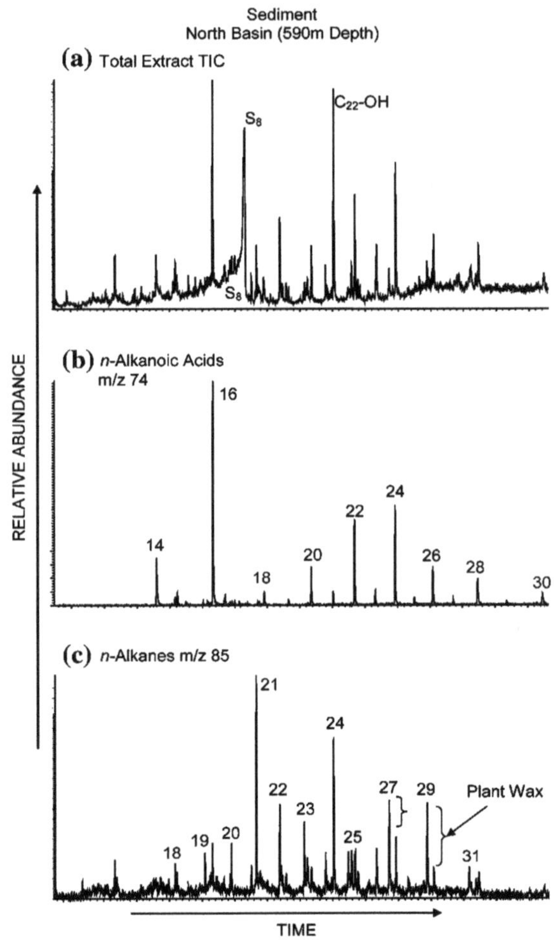

Fig. 5 GC-MS data of a total extract of sediment from North Basin in Crater Lake (590 m water depth): (**a**) Total ion current trace showing resolved peaks of natural components; (**b**) *m/z* 74 key fragment ion for fatty acids (as methyl esters); and (**c**) *m/z* 85 key fragment ion for *n*-alkanes. Numbers refer to carbon chain length of components, S_8 = elemental sulfur, OH = *n*-alkanol

carbon content of these sediments is not considered the determining factor for petroleum contaminant distribution. This was especially obvious in Cleetwood Cove boat mooring sediments (5 m depth) where the petroleum *n*-alkanes are six times more concentrated than plant wax *n*-alkanes.

The results of PAH analyses for surface sediments collected with a box corer nearshore at Cleetwood Cove (5 m depth) and at North Basin (580 m depth) show PAH concentration levels between 17 and 40 μg/kg and 3–15 μg/kg, respectively. Total PAH concentrations in the shallow water sediments are higher than in the deep lake sediments (Table 4). These compounds tend to accumulate once deposited in sediments due to slower degradation when compared with the water column or lake surface.

Atmospheric hydrocarbons

Because the hydrocarbon contributions to the lake surface may originate from a variety of natural and anthropogenic sources, air samples were collected to determine the levels of hydrocarbons in atmospheric particulate matter. The *n*-alkanes ranged from C_{12} to C_{31} with a C_{max} at C_{16}, C_{23} or C_{25} and significant odd-carbon number predominance. Pristane and phytane were not detectable and the UCM was minor. This indicates a dominant signature from natural sources versus urban aerosol (Simoneit, 1984, 1989; Simoneit & Mazurek, 1982). The data show that petroleum and plant wax *n*-alkanes were relatively low (0.2 μg/m^3) with a higher concentration (0.7 μg/m^3) in July. During the third air-sampling period the concentrations of plant wax *n*-alkanes increase as much as four times compared to the other samples. This increase reflects air sampling variability that occurs at this location caused by changes in wind stress and direction. The general weather conditions were recorded as follows: May 5–7, low -5°C, high 6°C, SW wind at 3 m/s; July 7–9, low 14°C, high 24°C, SW wind at 2 m/s; and July 9–11, low 12°C, high 20°C, East wind at 2 m/s. The different wind direction from the east on July 9–11, may contribute to the increased hydrocarbon levels observed during that air sampling period.

Table 4 The concentrations of total petroleum hydrocarbons and PAH in recent sediments of Crater Lake compared with other locations

Sampling site	Petroleum HC (mg/kg dry wt of sediment)	PAH (mg/kg dry wt) of sediment)
Crater Lake, Cleetwood Cove, 5 m water depth	0.3–1.4	< 0.04
Crater Lake 580 m water depth (North Basin)	0.016	< 0.015
Rhone River Estuary, Mediterranean Sea[a]	167	3
Mediterranean Sea 100 m water depth[a]	21	3
Coburn Mountain Pond, ME[b]	20	
South Orkney Island, Antarctica[c]	0.4	0.04

[a] Bouloubassi and Saliot (1993), moderately polluted

[b] Ho et al. (1991), contaminated from regional atmospheric deposition

[c] Cripps (1994), low impacted region, typical background

During the July 9–11 period, when the total *n*-alkanes neared 0.7 μg/m^3, the petroleum *n*-alkane component also increased, confirming that petroleum derived hydrocarbons were adsorbed and transported with atmospheric particulate matter collected during air sampling events. The samples reflect the high levels of atmospheric particulate matter from both natural background and vehicular emissions. The high concentration of natural background may be the result of stronger local updraft winds, which strip plant wax from vegetation and also resuspend soil detritus, and the components of vehicular emissions may reflect local traffic near the sampling site or longer distance transport of urban aerosol.

The concentrations of extractable (volatilizable carbon) and elemental carbon (black soot for example from diesel engine exhaust) in Crater Lake air particulate matter are given in Table 5. It indicates that the extractable/volatilizable carbon concentration is high for the second July sample probably due to the different wind direction during the sampling period. The elemental carbon concentrations remained steady. Since petroleum *n*-alkane concentrations increased and elemental carbon concentrations remained steady through the sampling period, these results show

Table 5 The concentrations of extractable/volatilizable and elemental carbon in aerosols at Crater Lake

Sample	Date	Collection time (h)		Extractable/Volatilizable[a] Carbon (μg/m^3)	Elemental Carbon[b] (μg/m^3)
1	5/08/96	48	1 filter	2.2	0.31
2	7/09/96	48	1st filter	2.5	0.22
3	7/11/96	48	1st filter	4.7	0.22
Average				3.1	0.25

[a] Carbon analyzed as volatilizable carbon (Birch and Cary, 1996; Johnson et al., 1981)

[b] Carbon remaining analyzed as black soot (elemental C) (Birch and Cary, 1996; Johnson et al., 1981)

that petroleum hydrocarbon components may be transported concurrently with particulate matter from higher plant sources, which is consistent with the molecular data discussed above. Furthermore, since petroleum hydrocarbon components present in air particles may derive from both gasoline and diesel engine exhaust, the petroleum hydrocarbon and elemental carbon concentrations and distributions may also reflect the input of petroleum components to air from increased vehicle traffic (tourism) during the time of sampling (May and July).

Discussion

The objective of the present investigation was to conduct the first comprehensive assessment of the levels and distributions of petroleum hydrocarbons in Crater Lake water and sediments. The results of the hydrocarbon analyses show the presence of both anthropogenic and natural hydrocarbons in Crater Lake. Petroleum hydrocarbon abundances greater than background levels have been found in some environmental samples taken but concentrations are generally very low or below our current detection limits. Removal processes such as evaporation or the formation of aerosols cause gasoline (aromatic and naphthenic compounds and *n*-alkanes $<C_{12}$) to disappear quickly from the surface. The presence of higher molecular weight PAH in surface slicks (100 μm film thickness) is essentially uniform in the areas sampled (estimated range 0.5–8.5 ng/m^2), independent of boat traffic. The relative distributions and abundances of *n*-alkanes, PAH and UCM from petroleum are similar for all surface slick sampling sites, however, the slick sample recovered at Wizard Island, has the highest petroleum contribution because it was collected in the wake of a tour boat immediately after boat mooring, idling and launching activities in that more enclosed area.

Boating activities introduce petroleum hydrocarbons and their combustion residues to the lake as evident from chemical analysis of environmental samples. It is well documented that petroleum products and their combustion residues (e.g., PAH) are persistent in the environmental compartments of soil, water and sediment (Bieger et al., 1996; Howard et al., 1991). A comparison of the estimated PAH levels present in two of the sediment samples shows that PAH concentrations are lower at depth in Crater Lake (Table 3). The wide range of uncertainty in each analysis reflects that almost all concentrations of PAH are at or near the detection limits. The individual PAH concentrations in sediments (μg/kg or ppb range) are ~3 orders of magnitude less than reported threshold effects levels (mg/kg or ppm range, test amphipod *Hyalella azteca*) (Buchman, 1999). Therefore, no adverse effects are expected to occur in benthic biota exposed to these PAH contaminated sediments.

In nearshore sediments collected at Cleetwood Cove and at the Cleetwood Cove boat mooring area, the impact of the increased boating activity is easier to detect. Concentrations of petroleum *n*-alkanes in sediments collected at 5 m depth at the Cleetwood Cove boat mooring area (1.44 mg/kg) exceed the petroleum and plant *n*-alkane concentrations at the other sites. Another important observation is that petroleum *n*-alkane concentrations are greater than plant *n*-alkane concentrations in all near shore sediments that were sampled. Aliphatic petroleum hydrocarbons are at low levels in the sediments of North Basin (580 m water depth, 0.016 mg/kg). The UCM normally associated with major petroleum inputs was not significant in these sediments. However, the presence of biomarkers such as the tricyclic terpanes, hopanes and steranes further confirms petroleum product input to these sediments.

Aerosol fallout is identified as a potentially minor contributor of petroleum hydrocarbons to surface water via atmospheric deposition because Crater Lake is located away from major urban areas. Extractable and elemental carbon analyses on aerosols do show that hydrocarbons associated with particles are present. There is a seasonal input of hydrocarbons from higher plants during spring through summer. Increased tourism to Crater Lake National Park may also contribute to petroleum *n*-alkane concentrations during summer from motor vehicle exhaust sources.

In order to quantitatively estimate the amounts of petroleum hydrocarbons released by combustion emissions from boating it is necessary to understand gasoline engine emission characteristics which include fuel consumption, fine particulate emission rates and bulk content of organic and elemental carbon of the particles. In Crater Lake, all boats are equipped with gasoline engines. Since the emission characteristics of the boats were not determined in this study, it is necessary to refer to external studies which model petroleum hydrocarbon emissions from similar internal combustion engines. Hildemann et al. (1991) determined the average mass emission rates for a variety of catalyst and non-catalyst equipped gasoline engines of automobiles similar to those of the boats at Crater Lake. Rogge et al. (1993b) further applied the same mass emission rates to model the levels of petroleum-derived combustion aerosols released to the atmosphere. By assuming that all tour boat combustion emissions are entering the water surface and are homogeneously distributed, it is possible to determine a mass balance for elutable organic matter in boat exhaust emissions.

Application of a mass balance calculation showed that approximately 0.50 μg/m^2 of the total hydrocarbons found on Crater Lake surface water is attributable to petroleum from daily boating activities. It should be noted that the natural hydrocarbons from plant waxes ranged from 50 to 5900 μg/m^2 in the surface slicks (Table 2). This estimate of petroleum concentration is based on surface slick PAH concentrations. Since a variety of organic compounds are produced by gasoline engine combustion, conservative marker compounds such as PAH may be used as tracers to determine the levels of petroleum combustion emissions from boat exhaust. The actual exhaust emission contribution to surface slick is unknown due to primary physical processes (dispersion, dissolution, evaporation) and removal processes (adsorption, biochemical degradation, solubility). Slick formation is also coupled with variations in wind direction and speed which determine the concentrations and distributions of petroleum and natural hydrocarbons on the lake surface, and ultimately in the water column and sediments. Those areas of Crater Lake where boat traffic is highly concentrated (traffic lanes, tourist loading/unloading docks, fueling areas, mooring areas where cold engine start up occurs) are mostly likely to show water and sediment petroleum hydrocarbon contamination. Also, the increased number of tourist visits to Crater Lake National Park over the last two decades presents an additional major source of petroleum hydrocarbons to the lake surface by atmospheric transport and deposition of automobile engine exhaust particles.

Acknowledgements We would like to thank Chris Moser and Dale Hubbard (COAS, Oregon State University), and Mark Buktenica, Scott Girdner, Ashly Gibson and Brad Hecht (National Park Service) for their assistance in the field. Comments and suggestions by Mark Buktenica, Roy Irwin and Mac Brock to clarify and improve this report are greatly appreciated. This project was supported by an USDI-NPS grant (CA-9000-8-0006) to Robert W. Collier and Bernd R.T. Simoneit.

References

Bidleman, T. F., A. A. Castleberry, W. T. Foreman, M. T. Zaranski & D. W. Wall, 1990. Petroleum hydrocarbons in the surface water of two estuaries in the southeastern United States. Estuarine, Coastal and Shelf Science 30: 91–109.

Bieger, T., J. Hellou & T. A. Abrajano, 1996. Petroleum biomarkers as tracers of lubricating oil contamination. Marine Pollution Bulletin 32: 270–274.

Birch, M. E. & R. A. Cary, 1996. Elemental carbon-based method for monitoring occupational exposures to particulate diesel exhaust. Aerosol Science and Technology 25: 221–241.

Blumer, M., M. M. Mullin & D. W. Thomas, 1963. Pristane in zooplankton. Science 140: 974.

Boughton, C. J. & M. S. Lico, 1998. Volatile organic compounds in Lake Tahoe, Nevada and California, July-September 1997. U.S. Geological Survey Fact Sheet FS-055-98.

Bouloubassi, I. & A. Saliot, 1993. Investigation of anthropogenic and natural organic inputs in estuarine sediments using hydrocarbon markers (NAH, LAB, PAH). Oceanologica Acta 16: 145–161.

Buchman, M. F., 1999. NOAA Screening Quick Reference Tables, NOAA HAZMAT Report 99-1, Seattle, WA, Coastal Protection and Restoration Division, National Oceanic and Atmospheric Administration.

Callahan, M. A., M. W. Slimak, N. W. Gabel, I. P. May, C. F. Fowler, J. R. Freed, P. Jennings, R. L. Durfee, F. C. Whitmore, B. Maestri, W. R. Maber, B. R. Holt & C. Gould, 1979. Environmental fate of 129 priority pollutants, Vol. 2. EPA-440/4-79-029b: 95–98.

Cerniglia, C. E., 1984. Microbial metabolism of polycyclic aromatic hydrocarbons. Advances in Applied Microbiology 30: 31–71.

Cerniglia, C. E. & M. A. Heitkamp, 1989. Microbial degradation of polycyclic aromatic hydrocarbons (PAH) in the aquatic environment. In Varanassi U. (ed.), Metabolism of Polycyclic Aromatic Hydrocarbons in the Aquatic Environment. CRC, Boca Raton, FL, USA, 41–68.

Collier, R., J. Dymond, J. McManus & J. Lupton, 1990. Chemical and physical properties of the water column at Crater Lake, Oregon. In Drake E.T., G. L. Larson, J. Dymond & R. W. Collier (eds), Crater Lake an Ecosystem Study. Allen Press, Lawrence, KS, 69–79.

Collier, R., J. Dymond & J. McManus, 1991. Studies of hydrothermal processes in Crater Lake, OR. Oregon State University, College of Oceanography Report #90-7. Submitted to the National Park Service, PNW Region, Seattle, WA.

Cranwell, P. A., 1975. Environmental organic chemistry of rivers and lakes, both water and sediment. In Eglinton, G. (ed.), Environmental Chemistry, Vol. 1. The Chemical Society, London, 22–54.

Cripps, G. C., 1994. Hydrocarbons in the Antarctic marine environment: Monitoring and background. International Journal of Environmental Analytical Chemistry 55: 3–13.

Dimock, C. W., J. L. Lake, C. B. Norwood, R. D. Bowen, E. J. Hoffman, B. Kyle & J. G. Quinn, 1980. Field and laboratory methods for investigating a marine gasoline spill. Environmental Science and Technology 14: 1472–1475.

Edgerton, S. A., R. W. Coutant & M. V. Henley, 1987. Hydrocarbon fuel spill dispersion on water: A literature review. Chemosphere 16: 1475–1487.

Eglinton, G., B. R. T. Simoneit & J. A. Zoro, 1975. The recognition of organic pollutants in aquatic sediments. Proceedings of the Royal Society, London, B189, 415–442.

English, J. N., G. N. McDermott & C. Henderson, 1963a. Pollutional effects of outboard motor exhaust—laboratory studies. Water Pollution Control Federation Journal 35: 923–931.

English, J. N., E. W. Surber & G. N. McDermott, 1963b. Pollutional effects of outboard motor exhaust—field studies. Water Pollution Control Federation Journal 35: 1121–1132.

Environmental Protection Agency (EPA), 1977. Guidelines for the pollutional classification of Great Lakes harbor sediments. U.S. EPA, Region V, April, 1977.

Environmental Protection Agency (EPA), 1986. Office of Solid Waste and Emergency Response. Test Methods for Evaluating Solid Waste, Vol. 1B, Laboratory Manual, Physical/Chemical Methods, Nov., 1986. Washington, DC.

Farrington, J. W., 1980. An overview of the biogeochemistry of fossil fuel hydrocarbons in the marine environment. In Petrakis, L. F. & T. Weiss (eds), Petroleum in the Marine Environment, Advances in Chemistry Series 185. ACS, Washington, DC: 1–22.

Farrington, J. W. & P. A. Meyers, 1975. Hydrocarbons in the marine environment. In Eglinton, G. (ed.), Environmental Chemistry, Vol. 1. The Chemical Society, London, 109–136.

Hatcher, P. G., B. R. T. Simoneit, F. T. MacKenzie, A. C. Neumann, D. C. Thorstenson & S. M. Gerchakov, 1982. Organic geochemistry and pore water chemistry of sediments from Mangrove Lake, Bermuda. Organic Geochemistry 4: 93–112.

Heitcamp, M. A. & C. E. Cerniglia, 1987. Effects of chemical structure and exposure on the microbial degradation of polycyclic aromatic hydrocarbons in freshwater and estuarine ecosystems. Environmental Toxicology and Chemistry 6: 535–546.

Hildemann, L. M., G. R. Markowski & G. R. Cass, 1991. Chemical composition of emissions from urban sources of fine organic aerosol. Environmental Science and Technology 25: 744–759.

Ho, E. S., P. A. Meyers & S. Pettingill, 1991. Geolipid content of sediments from an isolated lake: evidence for diagenetic alteration of source indicators. In Berthelin J. (ed.), Diversity of Environmental Biogeochemistry. Elsevier, Amsterdam: 67–74.

Howard, P. H., R. S. Boethling, W. F. Jarvis, W. M. Meylan & E. M. Michalenko, 1991. Handbook of Environmental Degradation Rates. CRC Press, Boca Raton, FL, USA.

Jackivicz, T. P., Jr. & L. N. Kuzminski, 1973. The effects of the interaction of outboard motors with the aquatic environment—a review. Environmental Research 6: 436–454.

Johnson, R. I., J. J. Shaw, R. A. Cary & J. J. Huntzicker, 1981. An automated thermal-optical method for the analysis of carbonaceous aerosol. In Macias, E. S. & P. H. Hopke (eds), Atmospheric Aerosol: Source/Air Quality Relationships. Amer. Chem. Soc., Symp. Ser. 167, Washington, DC, 223–233.

Marcus, J. M., G. R. Swearingen, A. D. Williams & D. D. Heizer, 1988. Polynuclear aromatic hydrocarbon and heavy metal concentrations in sediments of coastal South Carolina marinas. Archives of Environmental Contamination and Toxicology 17: 103–113.

Mayfield, H. T. & M. V. Henley, 1991. Classification of jet fuels using high resolution gas chromatography and pattern recognition. In Monitoring Water in the 90's: Meeting New Challenges, ASTM Special Technical Publication #1102, Philadelphia, PA, 579–597.

Morris, R. J. & F. Culkin, 1975. Environmental organic chemistry of oceans, fjords and anoxic basins. In Eglinton, G. (ed.), Environmental Chemistry, Vol. 1. The Chemical Society, London, 81–108.

National Visibility Monitoring Program (NVMP) 1995. Integrated report of optical, aerosol and scene monitoring data. Crater Lake National Park, March, 1993 through February, 1994, Interagency Monitoring of Protected Visual Environments (IMPROVE), 33 pp.

Neff, J. M., 1979. Polycyclic Aromatic Hydrocarbons: Evaluations of Sources and Effects, National Academy Press, Washington, DC.

Peters, K. E. & J. M. Moldowan, 1993. The Biomarker Guide: Interpreting Molecular Fossils in Petroleum and Ancient Sediments, Prentice-Hall Inc., Englewood-Cliffs, NJ.

Phillips, K. N., 1968. Hydrology of Crater, East, and Davis Lakes, Oregon. U.S. Geological Survey, Water-Supply Paper 1859-E.

Ramdahl, T., 1983. Retene—a molecular marker of wood combustion in ambient air. Nature 306: 580–582.

Readman, J. W., R. F. C. Mantoura & M. M. Rhead, 1984. The physico-chemical speciation of polycyclic aromatic hydrocarbons (PAH) in aquatic systems. Fresenius Z Analytical Chemistry 319: 126–131.

Rogge, W. F., L. M. Hildemann, M. A. Mazurek, G. R. Cass & B. R. T. Simoneit, 1993a. Sources of fine organic aerosol. 2. Noncatalyst and catalyst-equipped automobiles and heavy-duty diesel trucks. Environmental Science and Technology 27: 636–651.

Rogge, W. F., M. A. Mazurek, L. M. Hildemann, G .R. Cass & B. R. T. Simoneit, 1993b. Quantitation of urban organic aerosols on a molecular level: Identification, abundance and seasonal variation. Proceedings of the Fourth International Conference on Carbonaceous Particles in the Atmosphere, Atmospheric Environment 27A: 1309–1330.

Rowland, S., P. Donkin, E. Smith & E. Wraige, 2001. Aromatic hydrocarbon "Humps" in the marine environment: Unrecognized toxins? Environmental Science and Technology 35: 2640–2644.

Simoneit, B. R. T., 1978. The organic geochemistry of marine sediments. In Riley, J. P. & R. Chester (eds), Chemical Oceanography, Vol. 7, 2nd edn. Academic Press, New York, 233–311.

Simoneit, B. R. T., 1984. Organic matter of the troposphere-III: Characterization and sources of petroleum and pyrogenic residues in aerosols over the western United States. Atmospheric Environment 18: 51–67.

Simoneit, B. R. T., 1986. Cyclic terpenoids of the geosphere. In Johns, R. B. (ed.), Biological Markers in the Sedimentary Record. Elsevier Science Publishers, Amsterdam, 43–99.

Simoneit, B. R. T., 1989. Organic matter of the troposphere-V: Application of molecular marker analysis to biogenic emissions into the troposphere for source reconciliations. Journal of Atmospheric Chemistry 8: 251–275.

Simoneit, B. R. T., 1998. Biomarker PAHs in the Environment. In Neilson, A. H. (ed.), The Handbook of Environmental Chemistry, Vol. 3, Part 1, PAH and Related Compounds. Springer-Verlag, Berlin, Heidelberg, 176–221.

Simoneit, B. R. T. & T. A. Aboul-Kassim, 1994. Detection of fuels on water in port of an estuary: Final report. Pacific States Marine Fisheries Commission.

Simoneit, B. R. T. & I. R. Kaplan, 1980. Triterpenoids as molecular indicators of paleoseepage in Recent sediments of the Southern California Bight. Marine Environmental Research 3: 113–128.

Simoneit, B. R. T. & M. A. Mazurek, 1982. Organic matter of the troposphere-II. Natural background of biogenic lipid matter in aerosols over the rural western United States. Atmospheric Environment 16: 2139–2159.

Simoneit, B. R. T., H. I. Halpern & B. M. Didyk, 1980. Lipid productivity of a high Andean lake. In Trudinger P. A., M. R. Walter & B. J. Ralph (eds), Biogeochemistry of Ancient and Modern Environments. Australian Academy of Science, Canberra, 201–210.

Simoneit, B. R. T., J. N. Cardoso & N. Robinson, 1991. An assessment of terrestrial higher molecular weight lipid compounds in aerosol particulate matter over the south Atlantic from about 30–70°S. Chemosphere 23: 447–465.

Tiercelin, J. J., J. Boulègue & B. R. T. Simoneit, 1993. Hydrocarbons, sulphides, and carbonate deposits related to sublacustrine hydrothermal seeps in the North Tanganyika Trough, East African Rift. In Parnell, J., H. Kucha & P. Landais (eds), Bitumens in Ore Deposits. Springer-Verlag, Berlin, 96–113.

Tissot, B. P. & D. H. Welte, 1984. Petroleum Formation and Occurrence: A New Approach to Oil and Gas Exploration. Springer-Verlag, New York.

Voudrias, E. A. & C. L. Smith, 1986. Hydrocarbon pollution from marinas in estuarine sediments. Estuarine, Coastal and Shelf Science 22: 271–284.

Wilcock, R. J., G. A. Corban, G. L. Northcott, A. L. Wilkins & A. G. Langdon, 1996. Persistence of polycyclic aromatic compounds of different molecular size and water solubility in surficial sediment of an intertidal sandflat. Environmental Toxicology and Chemistry 15: 670–676.

Hydrobiologia (2007) 574:107–140
DOI 10.1007/s10750-006-0348-0

CRATER LAKE, OREGON

Ultraviolet radiation and bio-optics in Crater Lake, Oregon

B. R. Hargreaves · S. F. Girdner · M. W. Buktenica · R. W. Collier · E. Urbach · G. L. Larson

© Springer Science+Business Media B.V. 2007

Abstract Crater Lake, Oregon, is a mid-latitude caldera lake famous for its depth (594 m) and blue color. Recent underwater spectral measurements of solar radiation (300–800 nm) support earlier observations of unusual transparency and extend these to UV-B wavelengths. New data suggest that penetration of solar UVR into Crater Lake has a significant ecological impact. Evidence includes a correlation between water column chlorophyll-*a* and stratospheric ozone since 1984, the scarcity of organisms in the upper water column, and apparent UV screening pigments in phytoplankton that vary with depth. The lowest UV-B diffuse attenuation coefficients ($K_{d,320}$) were similar to those reported for the clearest natural waters elsewhere, and were lower than estimates for pure water published in 1981. Optical proxies for UVR attenuation were correlated with chlorophyll-a concentration (0–30 m) during typical dry summer months from 1984 to 2002. Using all proxies and measurements of UV transparency, decadal and longer cycles were apparent but no long-term trend since the first optical measurement in 1896.

Keywords Ultraviolet radiation · Plankton · Optics · UV-B · Stratospheric ozone

Guest Editors: Gary L. Larson, Robert Collier, and Mark W. Buktenica
Long-term Limonological Research and Monitoring at Crater Lake, Oregon

B. R. Hargreaves (✉)
Department of Earth & Environmental Sciences, Lehigh University, 31 Williams Drive, Bethlehem, PA 18015, USA
e-mail: brh0@lehigh.edu

S. F. Girdner · M. W. Buktenica
Crater Lake National Park, Crater Lake, OR 97604, USA

R. W. Collier
COAS, Oregon State University, 104 Ocean. Admin. Bldg, Corvallis, OR 97331, USA

E. Urbach
Department of Microbiology, Oregon State University, 220 Nash Hall, Corvallis, OR 97331, USA

Present Address:
E. Urbach
eMetagen, L.L.C., 3591 Anderson St., Suite 207, Madison, WI 53704, USA

G. L. Larson
USGS Forest and Rangeland Ecosystem Science Center, 3200 Jefferson Way, Corvallis, OR 97331, USA

Introduction

Current interest in ultraviolet radiation (UVR) in Crater Lake, Oregon, follows naturally from its well-known transparency (Larson, 2002) and the greater incident UVR and lake transparency found at high elevations (Laurion et al., 2000; Sommaruga, 2001). Crater Lake is famous for its depth (594 m, Bacon et al., 2002), visual clarity, and deep blue color. It is sub-alpine in elevation at 1800 m yet it rarely freezes during winter; it has an average radius of 4.1 km and is enclosed by a volcano's caldera, whose steep walls shelter it from strong winds and limit hydraulic exchange largely to snowmelt, direct precipitation, evaporation, and outseepage (Redmond, 1999). Optical studies of water transparency in Crater Lake began in 1896 when a white dinner plate was lowered into the water until it disappeared at 30 m (Diller, 1897, cited in Larson et al., 1996a). Black and white Secchi disk measurements were made sporadically since then until the late 1970s, when regular measurements began. An underwater photometer equipped with colored filters and matching deck cell was used to characterize water transparency in 1940 (Utterback et al., 1942) and again in the late 1960s (Larson et al., 1996a; Larson, 2002). Concerns about pollution and degraded water quality in the 1970s led to an improved sewage disposal at the tourist facilities in 1975 and the complete removal of sewage in 1991 (Larson, 2002), and to a federally funded monitoring program beginning in 1983 (Larson et al., 1996a; Larson, 2002). Routine optical measurements with modern instruments began in 1987 with a beam transmissometer (25 cm, 660 nm beam) attached to an automated CTD profiler (conductivity, temperature, depth); a UV scanning radiometer was added in 1996, and a chlorophyll fluorescence sensor was added in 1999.

Crater Lake is not only visually clear; it is also remarkably transparent to solar UVR (280–400 nm). In the 1930s Crater Lake water was examined to determine scattering of UVR in a laboratory comparison with purified water and deep ocean water (Pettit, 1936), only a decade after the first measurements of solar ultraviolet radiation (UVR) in the atmosphere using a photoelectric sensor (Coblentz & Stair, 1936), and more than a decade before the penetration of solar UV-B radiation (280–320 nm) into natural waters was reported using a photoelectric sensor (Johnson [=Jerlov], 1946; Jerlov, 1950; Højerslev, 1994). Crater Lake underwater UVR was first investigated in the 1960s. Tests of newly designed scanning underwater radiometers were the basis for a series of summer measurements in Crater Lake during 1964–69 (Tyler, 1965; Smith & Tyler, 1967; Tyler & Smith, 1970; Smith et al., 1973). Spectral measurements from 1969 in Crater Lake were also compared with similar measurements in Lake Tahoe, another clear deep lake (Smith et al., 1973). These authors and Pettit (1936) observed similarities between the upper 20 m of Crater Lake water and highly purified water and noted the importance of scattering of short wavelengths to the color of Crater Lake. In an effort to characterize spectral absorption and attenuation by pure water from data published to date, Smith and colleagues used the 1967 Crater Lake optical measurements (360–700 nm) as the basis for equating its transparency to the clearest waters of the Sargasso Sea (Smith & Tyler, 1976) and established absorption and attenuation spectra (200–800 nm) for pure freshwater and seawater that would be used for many years (Smith & Baker, 1981). Morel & Prieur (1977) used Crater Lake reflectance data in an assessment of scattering and absorption in both pure and natural waters to infer the content of ocean water by remote sensing.

The high elevation (1800 m) and frequently clear summer skies combine with the UV transparent water and shallow upper mixed layer (usually < 10 m deep and persistent from summer through early fall, Larson et al., 1996a) to create an unusually broad and stable depth gradient of UVR exposure for aquatic organisms. Research on underwater UVR and its potential for biological impact has increased in response to concerns about depletion of stratospheric ozone and climate change (Schindler & Curtis, 1997; Pienitz & Vincent, 2000). Several reports on the distribution of organisms in Crater Lake suggest a possible inhibitory role for UVR (McIntire et al., 1994; McIntire et al., 1996; Larson et al., 1996b), especially above 40 m. UVR measurements in Crater Lake resumed in 1996 when a submersible wavelength-scanning radiometer (300–800 nm) was added to the monitoring program and used to acquire incident and underwater solar spectra at a range of depths. Visiting scientists made additional optical measurements in UV and visible wavelengths starting in 1999. A recent report confirmed the unusual UV-B transparency of Crater Lake and derived estimates of UV-B attenuation by pure water and phytoplankton from measurements in its surface waters (Hargreaves, 2003).

Our approach here was first to characterize the spectral properties of near-surface and deeper water in Crater Lake, emphasizing UVR wavelengths, and to compare these with other clear lakes and ocean waters. We then used bio-optical signals and daily stratospheric ozone levels to examine factors controlling UVR attenuation with depth and over time, to develop proxies for estimating UVR attenuation, and to evaluate the impact of UVR on the Crater Lake ecosystem. Finally, we used the complete record of direct measurements and proxy estimates of UVR attenuation to look at long-term trends, and to speculate about the impact of future climate change.

Optical background

In studies of underwater light in aquatic ecosystems it is customary to characterize the transparency or attenuation of natural waters instead of underwater irradiance because transparency and attenuation are persistent properties from which underwater UVR irradiance can be calculated for any given time and depth. A useful measure of UVR transparency in natural waters is $K_{d,\lambda}$, the spectral diffuse attenuation coefficient for downwelling irradiance $E_{d,\lambda}$ (Baker & Smith, 1979). K_d can be used to characterize water transparency and factors controlling transparency or to reconstruct underwater solar spectra as a function of depth. K_d is calculated either from discrete measurements of E_d made at several depths or from a set of E_d values recorded over a continuous range of depths. An average K_d is often calculated for a range of depths considered to be optically mixed (e.g. the upper mixed layer or epilimnion) but depth-specific K_d values can also be derived from underwater measurements to reveal how attenuation varies with depth.

Valid spectral applications of K_d values are difficult to obtain unless the wavebands are narrow enough to be "spectrally neutral" (where $K_{d,\lambda}$ varies little across the waveband). An example of a broad yet spectrally neutral waveband in clear water is 400–500 nm. In contrast, when underwater K_d is calculated in clear water for the PAR (400–700 nm, PAR = photosynthetically active radiation) waveband, attenuation of solar radiation varies strongly with depth even though the water is uniformly mixed (Kirk, 1994b). This is because spectral variation in $K_{d,\lambda}$ leads to shifts with depth in relative proportions of different wavelengths within the waveband. $K_{d,PAR}$ in a uniformly mixed body of clear water is much greater at the surface than it is deeper because at the surface the strongly attenuated red part of the solar spectrum contributes to the average K_d; at deeper depths only the weakly attenuated blue and violet wavelengths are still present and only these contribute to average K_d. Another example of a problematic broadband application is when the entire UV-B range of wavelengths is used to calculate a single $K_{d,UVB}$, yielding values that are difficult to compare or interpret (Hargreaves, 2003). Underwater spectral radiometers useful for K_d determinations have moderate bandwidths in the range of 10 nm or less (Kirk et al., 1994), although significant spectral shifts can occur with moderate bandwidth sensors at the shorter UV-B wavelengths (Patterson et al., 1997).

Other sensor properties can influence spectral K_d measurements. For downwelling irradiance (E_d) a sensor with an accurate cosine response to the angle of incident photons is needed. Accurate determinations of $K_{d,\lambda}$ are possible when a sensor is not accurately calibrated as long as the same sensor is used in all measurements, travels in a vertical plane during displacement over precisely determined depths, and maintains stability of its wavelength sensitivity, calibration, and cosine response to the angular distribution of light. In practice a correction for a dark offset signal and response to changing temperature may need to be incorporated into the measurement protocol (Kirk et al., 1994). Internal radiation sources (fluorescence and Raman scattering, also called "inelastic scattering") can interfere with attempts to relate $K_{d,\lambda}$ to other optical properties when these contribute a significant fraction of the detected irradiance (Haltrin et al., 1997; Gordon, 1999). Measurement of spectral reflectance ratios (either irradiance reflectance, E_u/E_d, or radiance reflectance, L_u/E_d) can suggest the wavelengths and depths where such interference is occurring (Haltrin et al., 1997) but must account for self-shading of the upwelling signal when a large instrument package is deployed (Dierssen & Smith, 2000).

For depth-specific measurements the equation is

$$K_d = \log_e(E_{d,Z_1}/E_{d,Z_2})/(Z_2 - Z_1) \quad (1)$$

where E_d is downwelling cosine irradiance measured at two depths, Z_1 and Z_2 (Kirk, 1994a, b). When E_d is measured continuously with depth by a UV profiling instrument the equation becomes

$$E_{d,Z_2} = E_{d,Z_1} e^{-K_d(Z_2 - Z_1)} \quad (2)$$

The value of K_d in equation (2) is typically estimated for a specific wavelength (λ) by regression analysis, solving for the slope of the straight line formed by plotting $\log_e(E_{d,Z})$ versus depth, Z, after correcting $E_{d,Z}$ for dark signal & other noise (Hargreaves, 2003). With either method an average value for $K_{d,\lambda}$ is attributed to a specific depth range. The value of $K_{d,\lambda}$ will be approximately constant throughout depths that are uniformly mixed but can increase or decrease somewhat with depth until an equilibrium extent of diffuseness develops (Gordon, 1989). Variations in K_d near the surface of well-mixed water are related to surface waves (Zaneveld et al., 2001) and to changes in diffuseness determined by sky conditions and sun angle (up to 20–25% for UV wavelengths, Hargreaves, 2003). It is because of the response of K_d to diffuseness and light angle that K_d has been called an apparent optical property (AOP) of the water body, in contrast to inherent optical properties discussed below.

K_d and other optical measurements respond to the concentration of particulate and dissolved matter and can be used to investigate factors controlling transparency of natural waters. In addition to K_d, (an AOP) these include inherent optical properties of the water, or IOPs (Tyler & Presiendorfer, 1962), properties controlled by the composition of the water and not influenced by the light field. IOPs include the beam absorption coefficient, a, the beam scattering coefficient, b, and their sum, the beam attenuation coefficient, c. Direct measurement of c in the water column has been common for years using the beam transmissometer, typically with a red (e.g. 660 nm) light source. When measured at a long wavelength where CDOM absorption is negligible, variations in c are correlated with the concentration of particles because of their impact on scattering. Particulate organic carbon (POC), microbial biomass, or phytoplankton cells are the dominant particles in many aquatic systems (Boss et al., 2007). At the typical wavelength of 660 nm used in transmissometers to measure beam c, the relatively constant absorption of water (c_{w660} = 0.411 m^{-1}, varying slightly with temperature, Morel, 1974; Pegau et al., 1997; Pope & Fry, 1997) can be subtracted to yield the particulate beam attenuation coefficient, c_{p660}.

Field measurements of a and b are relatively rare in visible wavelengths and extremely scarce in UV wavelengths (but see Boss et al., 2007). The value of a is the optical sum of absorption by dissolved and particulate constituents of natural waters in combination with a_w, absorption by H_2O. The primary contributor to absorption by dissolved constituents in natural waters is colored dissolved organic matter (CDOM); a_{cdom} is typically measured using a laboratory spectrophotometer after particles are removed from the water sample by filtration. Values for a_{CDOM} can also be estimated from measurements of DOC concentration if DOC-specific absorption can be estimated as well. Although suspended mineral particles can sometimes make a large contribution to attenuation, especially in shallow water or near inflow from glaciers or rivers, the optically important particles in lakes are typically phytoplankton. Spectral absorption by phytoplankton can be measured in a spectrophotometer by concentrating a water sample onto a glass fiber filter. While primarily used for visible wavelengths (Yentsch & Phinney, 1989; Mitchell, 1990; Lohrenz, 2000), the technique has also been used for UV wavelengths (Ayoub et al., 1996; Sosik, 1999; Helbling, et al., 1994; Belzile et al., 2002; Hargreaves, 2003; Laurion et al., 2003).

Indirect measures of optical properties can be predictive of phytoplankton abundance and $K_{d,UV}$ in low-CDOM systems. The concentration of the primary photosynthetic pigment in phytoplankton (chlorophyll *a*) can be detected in vivo by its red absorption peak (676 nm) or by fluorescence measurements (F_{chl} emission peak at 683 nm) in the water column. Solar-stimulated fluorescence from phytoplankton pigments can also be detected by spectral reflectance meters after correction for Raman scattering. In natural waters the c_{p660} signal described above primarily responds to particle concentration because of scattering at 660 nm but when the particles are predominantly biotic, c_{p660} is expected to covary also with absorption and also attenuation at other wavelengths. None of these indirect measures is likely to be useful alone in predicting UV attenuation over a range of depths because of photoacclimation: the deeper phytoplankton adjust to dim light by increasing the efficiency of light utilization, the concentration of chlorophyll per cell, and the absorption per unit of chlorophyll, and decreasing the proportion of UV-screening pigments (MacIntyre et al., 2002). Summing F_{chl} and c_{p660}, with proper adjustment of their relative contribution, might provide a useful index of changing UV attenuation with depth when direct measures of UV attenuation are unavailable.

Another optical measurement that should be related to $K_{d,UV}$ in UV-transparent systems is Secchi depth (Z_{SD}). Measurement of Z_{SD} has been used for many years as a simple transparency index of water quality (Larson et al., 1996a). The depth at which a 20 cm white disk is barely visible under ideal conditions (flat surface, no reflections from the surface, and adequate solar radiation) depends on a combination of scattering that obscures the image of the underwater disc and absorption that diminishes the light reaching the disk from the surface. The inverse of Secchi depth ($1/Z_{SD}$, unit m^{-1}) has been shown to correlate with $[K_d + c]$ where K_d and c are measured for the appropriate range of wavelengths dependent on the combination of human vision and peak transmission wavelengths (Tyler, 1968; Preisendorfer, 1986) and depth-averaged from the surface to Z_{SD}. Human visibility of black objects underwater has been shown to vary inversely with beam attenuation in green wavebands (530 nm, Davies-Colley, 1988; Zaneveld & Pegau, 2003) but the blue waveband is likely to be more important in the case of very clear water such as Crater Lake. Because phytoplankton contribute to both scattering and absorption in blue wavelengths and typically have UV-absorbing protective pigments in a high UVR environment, blue attenuation and $1/Z_{SD}$ should be correlated with UV attenuation when the latter is affected by phytoplankton. In other studies where phytoplankton and suspended mineral sediments control transparency, Secchi depth has been correlated with K_d measurements for the PAR waveband and with the concentration of suspended sediments (e.g. Jassby et al., 1999). In systems where the relative contributions to optical attenuation by phytoplankton and suspended mineral particles are variable, the relationships among c_{p660}, $1/Z_{SD}$, K_d, and phytoplankton concentration would be expected to vary somewhat, with K_d less responsive to increases in scattering than the other two measurements.

Methods

Site characterization

Crater Lake is located in Crater Lake National Park in southwest Oregon, USA, at an elevation (lake surface) of 1883 m. Most of the measurements reported here were made at Station 13 (42.95° N, 122.08° W) near the deepest part of the lake (589 m, Larson et al., 1996a, 594 m, Bacon et al., 2002). Because of the steep caldera walls rising to an average elevation of 2100 m, the lake watershed area projected to a flat horizontal surface is relatively small, only 14.7 km^2 or 28% of the 53.2 km^2 surface area of the lake (Larson et al., 1996a).

Crater Lake data archives

The Long Term Limnological Monitoring Program database maintained by the Crater Lake National Park staff was the source for Secchi depth, photometer profiles, weather data, LI-COR radiometer profiles, chlorophyll-a data, and CTD profiles with beam transmissometer and fluorometer data (Larson et al., 1996a). The longest record is for Secchi depths; these data were screened to meet specific criteria, and with the exception of the first record (the white dinner plate lowered by Diller in 1896, Diller, 1897), all Z_{SD} data represent 20 cm disk data recorded during non-stormy weather with good visibility through the water surface.

Measuring and modeling solar radiation and spectral diffuse attenuation

Published measurements of Crater Lake $K_{d,380}$ from the 1960s were obtained from several sources (Smith and Tyler, 1967; Tyler & Smith, 1970; Smith et al., 1973). A LI-COR scanning radiometer (model LI-1800 uw) was the primary instrument for the new measurements of UVR irradiance and attenuation reported here (performance reviewed in Kirk et al., 1994). The self-contained programmable scanning radiometer records downwelling cosine irradiance at 2 nm intervals from 300 nm to 800 nm with a bandwidth of 8 nm. LI-COR post-collection software provided immersion corrections to maintain accuracy both above and below the water surface. To create a depth-series of spectra the instrument was programmed to scan every two minutes and was then lowered on the sunny side of a small vessel near mid-day to specific depths and held for timed intervals. On many occasions an incident PAR signal was recorded on deck during the underwater scans in order to detect changing sky conditions. For each depth-series of spectra the data were evaluated at specific wavelengths to compute spectral K_d for each pair of depths (typically at 5 or 10 m vertical spacing). The data were carefully examined for anomalies (identified as outliers in both spectral and depth plots of $K_{d,\lambda}$) caused by surface waves or clouds; the occasional anomalies were eliminated either by interpolation between adjacent wavelengths or depths. The depth assigned to each K_d was the average for the upper and lower pair of E_d measurements. The LI-1800 uw was factory calibrated on 26 May 1995, 18 July 2000, and 23 January 2002.

On 20 August 2001 two other UV radiometers (from Biospherical Instruments) were also used to record water depth and temperature and up to 20 fixed wavelengths of downwelling irradiance (E_d) and upwelling radiance (L_u) using filter-based diode sensors at a rate of 5 spectra per second. A PUV-2500 profiling UV radiometer recorded seven downwelling channels (305, 313, 320, 340, 380, 395 nm with nominal 8 nm bandwidth and PAR, 400–700 nm) plus upwelling radiance in the chlorophyll-*a* natural fluorescence waveband (center = 683 nm). A PRR-800 profiling reflectance radiometer recorded 19 channel pairs of upwelling radiance and downwelling irradiance (340–710 nm) and PAR. The PRR-800 was lowered in its normal orientation to measure E_d and L_u and also lowered inverted to record upwelling irradiance E_u. The data from the Biospherical instruments were binned at 2 m depth intervals using $\log_e$ averages. To detect internal radiation sources that could interfere with interpretation of diffuse attenuation measurements we measured spectral reflectance ratios for a range of depths using the PRR-800 radiometer. To reduce near-surface noise cause by waves and ripples we used both running averages of K_d (combining several adjacent depths) and polynomial regression of Ln(Ed) versus depth from which K_d was then calculated.

On this date we also computed incident irradiance using a radiative transfer model (RTBASIC, Biospherical Instruments, Inc.; see Madronich, 1993 and Biospherical Instruments, 1998, for more details) for comparison with the LI-1800 uw incident spectra. The parameters for the 8-stream disort model were set to account for noon PDT conditions near the time that optical profiles were collected: current solar zenith angle (34.4°), barometric pressure (774 mbar based on elevation above sea level), nominal albedo (5%) and current column ozone (313 DU, URL: http://toms.gsfc.nasa.gov/). To compute a spectrum with a bandwidth comparable to the LI-1800 uw (8 nm) and comparable reporting interval (2 nm), model values with 1 nm bandwidth were first generated at 2 nm intervals from 296 to 804 nm. These initial model values were then converted to represent an 8 nm bandwidth by calculating geometric means spanning 8 nm around each 2 nm interval.

To calculate PAR (400–700 nm) from LI-1800 uw data, values reported as W nm^{-1} m^{-2} were summed and then multiplied by 2 nm per record to get the energy in the band, W m^{-2}. Quantum PAR irradiance (μMol m^{-2} s^{-1}) at the surface was calculated from W m^{-2} by multiplying by 4.60 (Kirk, 1994b). To interpolate to an intermediate wavelength (e.g. 305 nm from 304 nm and 306 nm measurements) we used the geometric mean of measured irradiance. To reduce surface noise caused by waves and ripples the scans near the surface were repeated and averaged, or computed using the incident scan as the upper value after reducing it by 5% for nominal reflectance.

Other optical and bio-optical measurements

Bio-optical signals (most of which were described in Larson et al., 1996a) were recorded by a Sea-Tech transmissometer (25 cm 660 nm) and Wetlabs, Inc. WetStar chlorophyll-*a* fluorometer integrated with a SeaBird CTD profiler. Beam attenuation c_{660} was calculated from 2 m binned transmittance data ($c_{660} = -\ln(T)*100/25$), then converted to c_{p660} by subtracting c_{w660}, the beam attenuation coefficient for pure water (0.411 from $c_w = a_w + b_w$; a_w from Pope & Fry, 1997; b_w from Morel, 1974). Previous estimates of c_{w660} (Smith & Baker, 1981; Zaneveld & Bartz, 1984; Bishop, 1986) used to calculate c_{p660} have been superseded by much-improved measurements of a_w (Pope & Fry, 1997).

Particles and whole water were analyzed on several occasions for organic carbon content. On two dates in 1999 (corresponding to c_{p660} profiles) samples were collected from three depths in 30 l. Niskin samplers (acid-washed), transferred to 20 l acid-washed carboys and 4–8 l were filtered through pre-combusted GF/F filters. Filters were analyzed on a Carlo Erba NA1500 Carbon/Nitrogen/Sulfur Analyzer using GF/F filter blanks and cystine standards. Whole water samples were also collected similarly (Urbach et al., 2001) on three dates from 12 depths and analyzed for total organic carbon (high temperature combustion using Shimadzu TOC-5000). On two dates the measurements of particulate organic carbon (POC g C/m^3) at three depths were correlated with c_{p660} (binned at 2 m intervals) yielding an average relationship where $c_{p660} = 0.019*$[POC μM C] ($r^2 = 0.53$). Because of the small number of samples and low r^2 value we used the relationship reported with greater precision by Boss et al. (2007) where $c_{p660} = 0.032*$[POC μM C]–0.024 ($r^2 = 0.996$) to calculate POC for all depths from c_{p660}. We adjusted the offset term each month for slight variations in transmissometer baseline in order to match the deep particulate signals to 0.38 μM C for depths 300–500 m in all summer months.

On 20 August 2001 spectral absorption of particulate samples was analyzed. Water (500–1000 ml) from five depths was filtered on GF/F filters (Whatman, with nominal retention of diameters greater than about 0.7 μm). Optical density was measured with the filter attached to a quartz disk in a Shimadzu 1601 UV spectrophotometer using a modified Quantitative Filter Technique (QFT, Yentch & Phinney, 1989; Mitchell, 1990) adapted for UV wavelengths (Helbling et al., 1994; Ayoub et al., 1996; Sosik, 1999) and corrected for path-length amplification using the method of Lohrenz (2000).

Filter photometer depth profiles began with Utterback et al. (1942), using a custom-built underwater photometer in July 1940 (they employed Schott BG12 filters, URL http://www.us.schott.com/, and a Weston cell, described by Barnard, 1938; the combined response curve has a peak at 450 nm with half-maximum responses at 390 nm and 490 nm). During 1968–1991 several Kahl filter photometers with similar properties to the instrument described above (Kahlisco, Inc. underwater and deck sensors equipped with clear, red, green, and blue filters over a photodetector cell, see Larson, 1972 for added details) were used to collect light profiles to a typical depth of 145 m. The blue filter data were converted to $K_{d,blue}$ by calculating transmittance ($E_{d,z}/E_{d,o}$) from raw deck (E_{do}) and underwater ($E_{d,Z}$) data and then converting this using $K_{d,blue} = \ln(E_{d,o}/E_{d,z})/(Z_2–Z_1)$. $K_{d,blue}$ was then averaged from 10 m to the Secchi depth (40 m if no Secchi depth). Irregular photometer $K_{d,blue}$ values near the surface were excluded as needed.

Proxies for $K_{d,UV}$

Least squares linear regressions were used to develop proxy $K_{d,UV}$ values from measured Z_{SD} (using $1/Z_{SD}$), $K_{d,blue}$, and $K_{d,380}$ averaged over the 0–40 m or the Secchi depth. To develop an empirical model for $K_{d,380}$ based on chlorophyl-*a* fluorescence and c_{p660} we used trial and error to adjust two parameters to minimize residuals in comparison with $K_{d,380}$ determined from LI-1800 uw spectral scans. To compute $K_{d,320}$ and $K_{d,380}$ from $1/Z_{SD}$, $K_{d,blue}$ was computed as an intermediate step. When historical data were measured over a different range of depths than 0–40 m we used recent patterns of K_d versus depth to calculate adjusted values.

Results

Spectral measurements of incident and underwater solar radiation

In the extremely clear water of Crater Lake the wavelengths penetrating most deeply were in the blue waveband (400–500 nm). Incident and underwater spectra recorded on 20 August 2001 (a date when the surface waters were unusually transparent) are shown on a logarithmic irradiance scale for a range of depths in Fig. 1a. Incident irradiance measurements made with the LI-COR LI-1800uw scanning radiometer were validated by comparing with model data for incident irradiance under ambient conditions (also plotted); in most of the UV range (310–400 nm) the ratio of Measured:Modeled ranged from 87% to 108% and averaged 96%. Underwater the solar radiation diminished with increasing depth; both short and long wavelengths were attenuated more rapidly than those in the blue range (400–500 nm), resulting in a broad peak from 400 nm to 500 nm at 80 m and a narrower peak from 460 nm to 480 nm at 160 m. The limit of detection of the LI-COR LI-1800-uw scanning radiometer (Kirk et al., 1994) is reached between 10^{-4} Watts m^{-2} nm^{-1} and 10^{-5} Watts m^{-2} nm^{-1} depending on the wavelength. For comparison with relative attenuation depths reported for other natural waters Fig. 1b shows curves for both 10% and 1% attenuation depths across the UV and part of the visible spectrum derived from the data plotted in Fig. 1a. Figure 1a shows that underwater irradiance is attenuated fairly evenly by wavelengths from 390 nm to 490 nm (also evident in the low slope of Fig. 1b for wavelengths > 390 nm). The most rapid attenuation with depth was at wavelengths longer than 600 nm.

Spectral upwelling irradiance, E_u (Fig. 2a, measured on the same date as Fig. 1 data with a PRR-800 radiometer equipped with 18 filter wavebands (lowered in inverted orientation so that cosine irradiance sensors faced downward) suggests the quality of light that would be visible from above the surface. The peak waveband is 412 nm (violet), with a nearly constant intensity from 380 nm to 490 nm (UV-A through blue). Another peak at 683 nm (the red peak emission wavelength for chlorophyll-*a* fluorescence) is evident at all depths but more distinct below 30 m. Figure 2b shows the reflectance ratio of upwelling radiance (L_u) to downwelling irradiance (E_d) at different depths (PRR-800 was used in its normal orientation), with a regular spectral pattern of declining L_u:E_d at increasing wavelengths observed for shallow depths, but with increasing depth there is a dramatic rise in reflectance for longer wavelengths (>590 nm at 15 m depth, >565 nm at 25 m, and >490 nm at 120 m).

Spectral diffuse attenuation, validation, and variations over space and time

Spectral diffuse attenuation ($K_{d,\lambda}$) is shown in Fig. 3 calculated from the spectral irradiance data shown in Fig. 1 (K_d was averaged from multiple scans for specific depth ranges and the standard errors of the mean for these are indicated by error bars). One additional depth range (10–18 m) not appearing in Fig. 1a was calculated from Biospherical Instruments, Inc. (BSI) PRR-800 and PUV-2500 profiling radiometers. In the UV and blue wavebands K_d spectra were lowest near the surface and rose with increasing depth down to the deep chlorophyll maximum (DCM = 130 m). The K_d pattern at wavelengths longer than 550 nm varied little with depth when measured

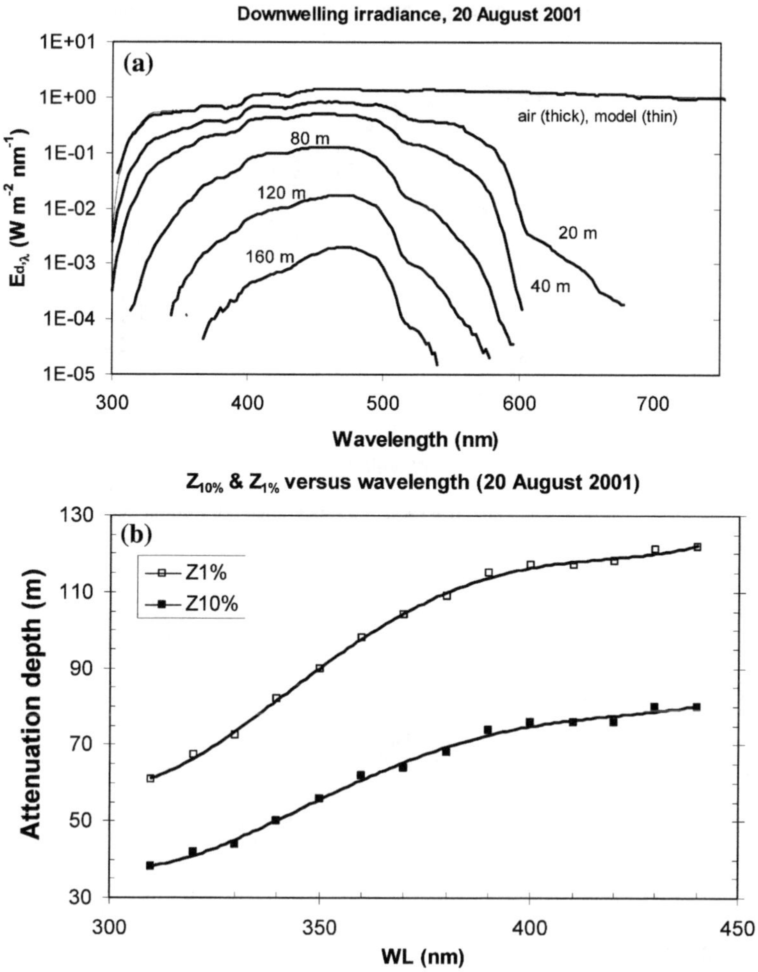

Fig. 1 (**a**) Crater Lake solar irradiance spectra (304–750 nm in air; 300–750 nm at 20–160 m), 20 August 2001 (air scan 12:04 PDT). LI-COR LI-1800 uw spectroradiometer (8 nm bandwidth; cosine response to downwelling irradiance). Incident E_d irradiance (thick line) plotted along with model E_o irradiance (thin line, distinct from the measurements only at the lowest and highest wavelengths) from RTBasic (Biospherical Instruments; 8-stream option, 2 nm bandwidth; model output adjusted using 8 nm running averages of model log (E_d) to account for the 8 nm bandwidth of the LI-1800uw data; ozone = 313 DU from satellite; P_{atm} = 774 torr from elevation). Ratios of measured:modeled data averaged 98% for 308–800 nm, and 96% for 308–400 nm (range 87–109%). (**b**) Attenuation depths (10% and 1%) from spectral irradiance in cFigure 1a. For PAR, $Z_{10\%}$ = 50 m, $Z_{1\%}$ = 100 m, $Z_{0.1\%}$ = 140 m. For UV-B (320 nm), $Z_{10\%}$ = 37 m, $Z_{1\%}$ = 62 m. Polynomial regressions fitted to the points: $Z_{10\%} = 4.9534E\text{–}07 \cdot WL^4 – 7.6076E\text{–}04 \cdot WL^3 + 4.3388E\text{–}01 \cdot WL^2 – 1.0855E{+}02 \cdot WL^1 + 1.0082E{+}04$; $Z_{1\%} = 8.2806E\text{–}07 \cdot WL^4 – 1.2583E\text{–}03 \cdot WL^3 + 7.0974E\text{–}01 \cdot WL^2 – 1.7559E{+}02 \cdot WL^1 + 1.6128E{+}04$

near the surface but at depths deeper than 20–30 m and shallower at the longest visible wavelengths the K_d values declined with increasing depth (not plotted). LI-COR K_d spectra for shallow depths (not plotted) were similar to the plotted K_d spectrum from the BSI radiometers but were noisier because fewer measurements were used in the calculations.

Variations of K_d with depth for specific wavelengths on 20 August 2001 appear in Fig. 4. The LI-COR UV wavelengths were selected to match the filter wavebands (320, 340, 380 nm and PAR) of two BSI radiometers (all three were deployed within about 1 h of solar noon). The curves show a region of uniform and low $K_{d,UV}$ from 0 to 10 m (the thermally mixed layer

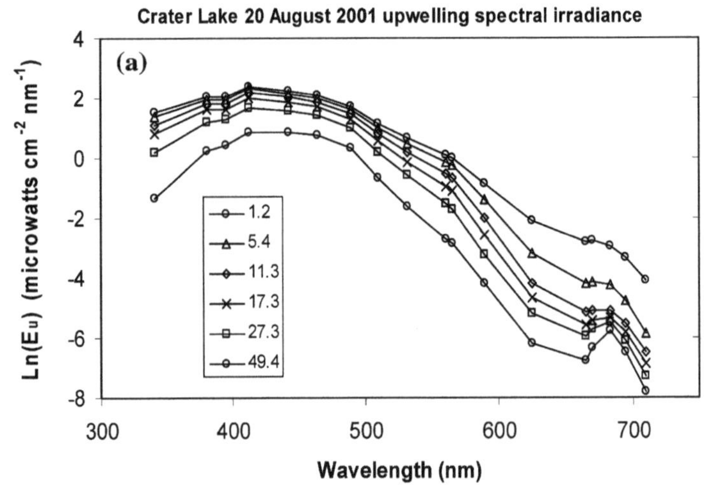

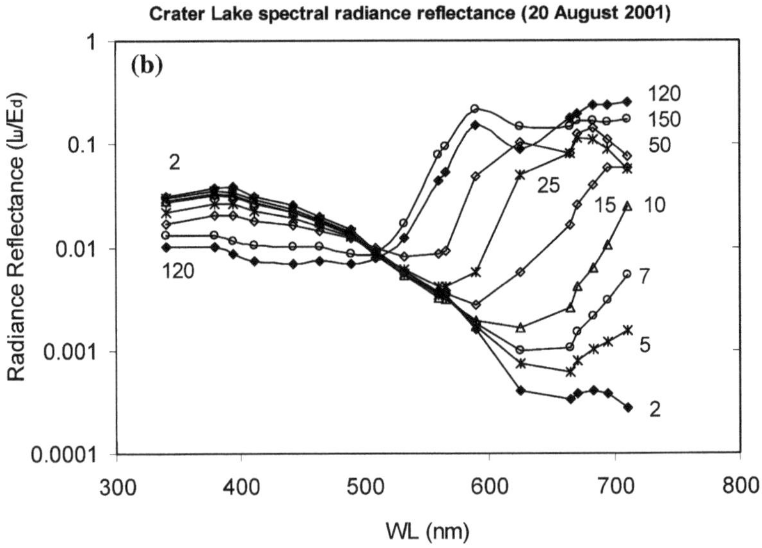

Fig. 2 (**a**) Crater Lake upwelling irradiance, 20 August 2001 (Biospherical Instruments PRR-800 reflectance profiler lowered inverted, units μW cm^{-2} nm^{-1}). Depth (m) is indicated in the legend. Peak near 400 nm is source of deep blue color when viewed from above. Red peak is fluorescence from phytoplankton (683 nm). (**b**) Crater Lake radiance reflectance, 20 August 2001 (Biospherical Instruments PRR_800, data binned at 1 m intervals). Near the surface the spectrum is similar to upwelling irradiance. At greater depths the reflectance in UV and blue wavelengths declines because absorbance coefficients increase with depth faster than backscatter coefficients. At longer wavelengths the signal rises with increasing depth because Raman scatter and phytoplankton fluorescence (Chlorophyll-*a* at 683 nm and for 120 & 150 m depths, possibly phycoerythrin at 589 nm) increase rapidly in the upwelling radiance signal relative to downwelling irradiance

was also 0–10 m), then a pattern of increasing $K_{d,UV}$ with depth to a peak at the DCM near 130 m. For broadband visible irradiance (PAR, 400–700 nm) the $K_{d,PAR}$ was higher above 15 m but paralleled the other sensors at depths below 60 m. The three instruments agreed closely for most depths over which the same wavelengths were measured, with the greatest variation occurring near the surface. Figure 5 shows the same type of pattern but in this case averaged for the period 1996–2002. There are parallel changes with depth in $K_{d,320}$, $K_{d,380}$, and $K_{d,blue}$ (summarized also in Table 1). Figure 6 shows that the ratio of K_{d320}:K_{d380} varied regularly with depth (K_{d320}:K_{d380} = –0.0031*depth + 2.32, r^2 = 0.68) for the 1996–2002 summer data. In

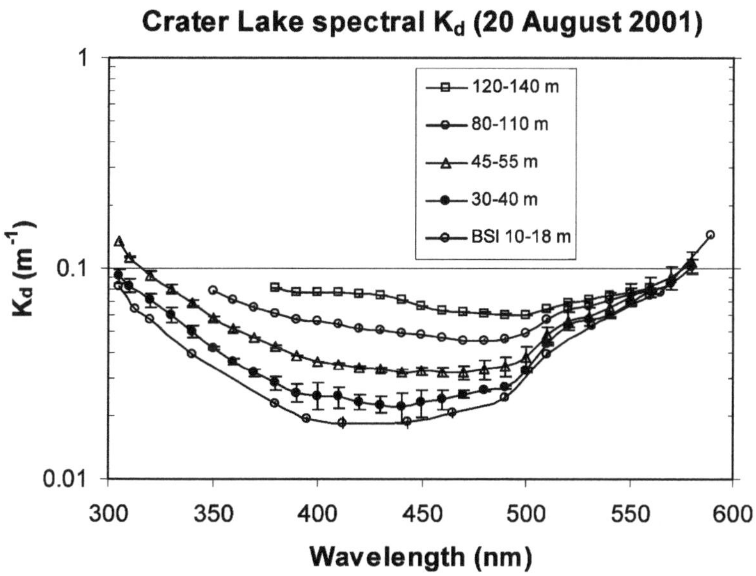

Fig. 3 Crater Lake spectral diffuse attenuation from spectral irradiance in Figure 1, plus an average for 10–18 m depth calculated from profiles by PRR-800 and PUV-2500 instruments from Biospherical Instruments, Inc. (BSI). LI-1800uw K_d was calculated for each depth interval and then averaged for a range of depths (error bars are ±S.E. of mean for depth ranges included in mean). BSI K_d was calculated for 10–18 m from polynomial regressions of $E_{d,Z}$ vs depth to reduce noise from surface reflections. Note that the minimum K_d was at 412 nm. Values for wavelengths longer than 560 nm are excluded below 40 m because of Raman scatter artifact described in Figure 2B. Shorter UV wavelengths are excluded at depths below 55 m because of instrument detection limits. The lowest $K_{d,320}$ value was 0.057 m^{-1} (PUV-2500, 10–18 m)

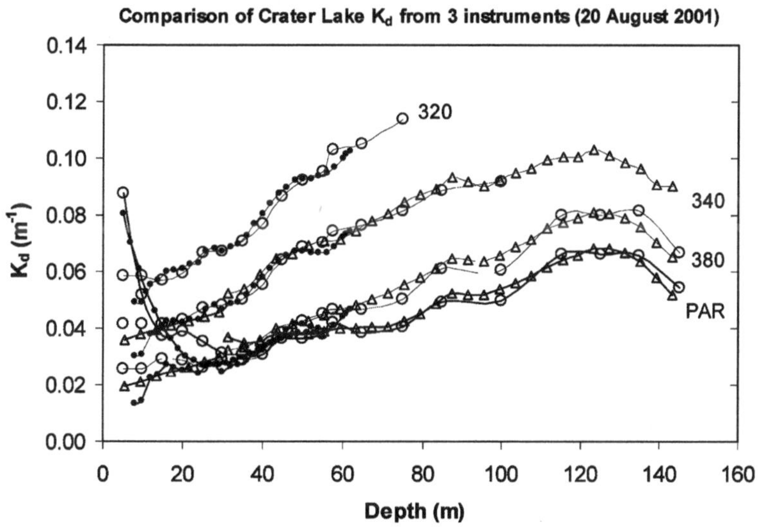

Fig. 4 K_d versus depth for selected wavelengths in Crater Lake (20 August 2001) showing minimum values at surface and increasing values with depth to the deep chlorophyll maximum (DCM). Four wavebands and three instruments are compared (smoothed with 8 m running averages above 60 m). Narrow-band signals show parallel changes with depth while the PAR waveband increases near the surface (where highly attenuated red wavelengths are detected). BSI instruments (PUV-2500, solid circles; PRR-800, solid triangles) give similar results to LI-COR LI-1800uw (open circles) except for greater noise in the latter near the surface because of long scan times. The 320 nm curve for LI-1800uw data was limited to about 75 m because of instrument sensitivity but the PUV-2500 was only lowered to 63 m. The PRR-800 radiometer was not equipped with a 320 nm sensor during this comparison and the PUV-2500 PAR signal is not included for greater clarity. The optically mixed depths correspond to the thermally mixed epilimnion (0–10 m, not shown). The lowest $K_{d,320}$ averages were 0.057 m^{-1} (PUV-2500, 10–18 m) and 0.058 m^{-1} (LI-1800uw, 0–20 m)

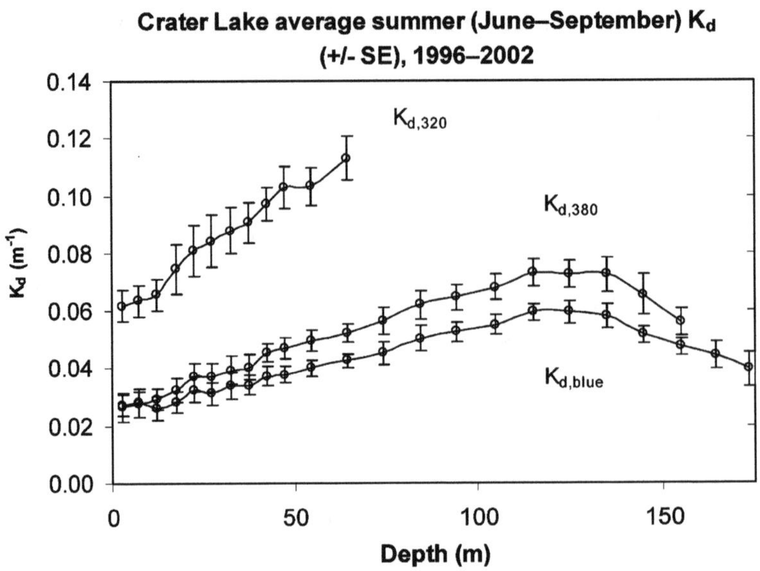

Fig. 5 Average K_d for three wavebands versus depth in Crater Lake (June–September 1996–2002, LI-1800 uw data, N = 15 dates, bars indicate + /– S.E.) showing the summer pattern with minimum K_d at the surface and maximum K_d at the deep chlorophyll maximum (DCM). $K_{d,blue}$ was calculated for the irradiance waveband 400–500 nm while $K_{d,320}$ and $K_{d,380}$ were calculated using the average from 2 nm above to 2 nm below the central wavelength. The average value for $K_{d,320}$ near the surface (0–5 m) was 0.062 m^{-1}

contrast, the ratios $K_{d,320}$:$K_{d,blue}$ and $K_{d,380}$:$K_{d,blue}$ were relatively constant for two depth ranges: 0–10 m and below 10 m. Related optical changes with depth based on particle absorption coefficients are shown in Fig. 12b (described below).

Table 1 Average diffuse attenuation coefficients and % attenuation of E_d (June–September, 1996–2002) by depth

K_d Depth (m)	E_d Depth (m)	$K_{d,320}$ (m^{-1})	2SE	$E_d\%_{320}$	$K_{d,380}$	2SE	$E_d\%_{380}$	$K_{d,blue}$	2SE	$E_d\%_{blue}$	K_{dPAR}	$E_d\%_{PAR}$
2.8	5.0	0.062	0.0054	73%	0.027	0.0049	88%	0.027	0.0036	87%	0.108	58%
7.3	10.0	0.064	0.0056	53%	0.028	0.0045	76%	0.028	0.0048	76%	0.066	42%
12.2	15.0	0.066	0.0055	38%	0.029	0.0036	66%	0.026	0.0040	67%	0.050	33%
17.5	20.0	0.074	0.0086	27%	0.032	0.0041	56%	0.028	0.0035	58%	0.048	26%
22.5	25.0	0.081	0.0089	18%	0.037	0.0050	47%	0.033	0.0045	49%	0.047	20%
27.5	30.0	0.084	0.0090	12%	0.037	0.0051	39%	0.031	0.0041	42%	0.044	16%
32.7	35.0	0.088	0.0081	7.5%	0.039	0.0048	32%	0.034	0.0048	36%	0.045	13%
37.7	40.0	0.091	0.0069	4.7%	0.040	0.0043	26%	0.034	0.0034	30%	0.045	10%
42.5	45.0	0.097	0.0058	2.9%	0.045	0.0033	21%	0.037	0.0035	25%	0.047	8.2%
47.5	50.0	0.103	0.0072	1.7%	0.047	0.0036	16%	0.038	0.0028	21%	0.048	6.5%
54.8	60.0	0.103	0.0065	0.62%	0.049	0.0037	10%	0.040	0.0028	14%	0.043	4.2%
64.5	70.0	0.113	0.0075	0.20%	0.052	0.0032	5.9%	0.043	0.0025	9.0%	0.045	2.7%
74.3	80.0	*0.122*		0.059%	0.056	0.0049	3.4%	0.045	0.0040	5.7%	0.047	1.7%
84.3	90.0	*0.134*		0.016%	0.062	0.0050	1.8%	0.050	0.0044	3.5%	0.051	1.0%
94.3	100.0				0.065	0.0045	1.0%	0.053	0.0035	2.0%	0.053	0.59%
104.7	110.0				0.068	0.0044	0.48%	0.055	0.0032	1.2%	0.055	0.34%
115.0	120.0				0.073	0.0046	0.23%	0.059	0.0029	0.65%	0.060	0.19%
125.0	130.0				0.073	0.0045	0.11%	0.059	0.0038	0.36%	0.059	0.10%
135.0	140.0				0.073	0.0059	0.054%	0.058	0.0040	0.201%	0.058	0.058%
145.0	150.0				0.065	0.0071	0.028%	0.052	0.0031	0.120%	0.052	0.034%
155.0	160.0				0.056	0.0047	0.016%	0.047	0.0027	0.075%	0.048	0.021%
164.6	170.0							0.044	0.0046	0.048%	0.044	0.014%
173.6	180.0							0.040	0.0061	0.032%	0.040	0.0092%

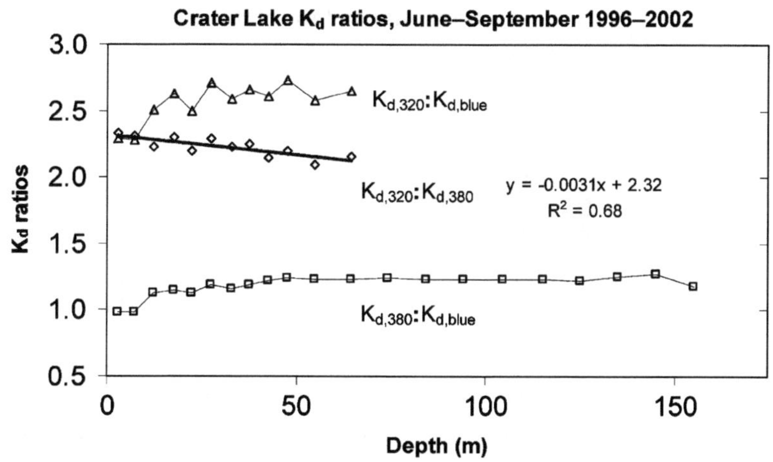

Fig. 6 Crater Lake K_d ratios from data in Figure 5 (June–September 1996–2002, LI-1800uw). Least squares regression for $K_{d,320}$:$K_{d,380}$ versus depth (m), $y = -0.0031x + 2.32$, $r^2 = 0.68$. Average 0–10 m ratios: $K_{d,320}$:$K_{d,blue}$, 2.28 (SE = 0.007), $K_{d,380}$:$K_{d,blue}$, 0.98 (SE = 0.002), $K_{d,320}$:$K_{d,380}$ = 2.32 (SE = 0.012). Average 0–40 m ratios: $K_{d,320}$:$K_{d,blue}$ = 2.52 (SE = 0.058); $K_{d,380}$:$K_{d,blue}$ = 1.11 (SE = 0.029); $K_{d,320}$:$K_{d,380}$ = 2.27 (SE = 0.016)

Seasonal and interannual changes in UV K_d for different depth ranges are summarized in Fig. 7. July and August were typically the months with the lowest $K_{d,320}$ averaged over 0–40 m, the depth range where maximal UVR impact on organisms was expected. If the single measurement in January is typical then winter values are much higher than mid-summer values for this depth range (Fig. 7a). Over the period from 1996 to 2002 the lowest summer average $K_{d,320}$ was observed in 2001; the highest were in 1998 and 1999. When July and August data are averaged over this period the greatest interannual variations in $K_{d,380}$ occurred in the depths from 20 m to 40 m and the least variations occurred from 100 m to 140 m. The range from 0 m to 20 m changed in parallel with 20–40 m except for 1999 (Fig. 7b).

Proxy measurements for UV attenuation versus depth; phytoplankton control of $K_{d,UV}$

Bio-optical signals provide several proxies for UV attenuation through the water column. The red beam transmissometer signal (from which c_{p660} is calculated) has been measured since 1987 in Crater Lake (Larson et al., 1996a) and provides an optical proxy for particulate organic carbon (POC) and in the upper 80 m it can serve as a proxy measurement for UV attenuation. The rationale for using cp660 as a proxy for $K_{d,UV}$ is described earlier. Depth profiles measured during 1999 for POC (calculated from c_{p660} based upon POC analysis of discrete samples), total organic carbon (TOC), and dissolved organic carbon (DOC, calculated as TOC-POC) are shown in Fig. 8A–C. The POC signal in the upper 130 m was much higher than at deeper depths and in this shallower range it increased slightly from July to September. TOC values showed similar trends but were more variable, especially the few measurements at deeper depths (not plotted below 150 m). The DOC values were of low accuracy because of combined errors in the POC and TOC measurements and difficulty of making low-level measurements, but they show a similar seasonal trend to that for POC and TOC and a consistent small peak near a depth of 80 m.

On some occasions when there has been an influx of suspended inorganic particles the c_{p660} signal does reflect the concentration of POC. Figure 9a shows a series of c_{p660} depth profiles during 1995, a summer that experienced a late snowmelt, and several large rain storms in June and July. Extra peaks are apparent in Fig. 9a above 25 m and below 200 m. The change in this turbidity signal over time is plotted in Fig. 9b for both the near-surface water (0–20 m) and representative deep water (>300 m) along with comparison data for the typical (dry) summer of 1994,

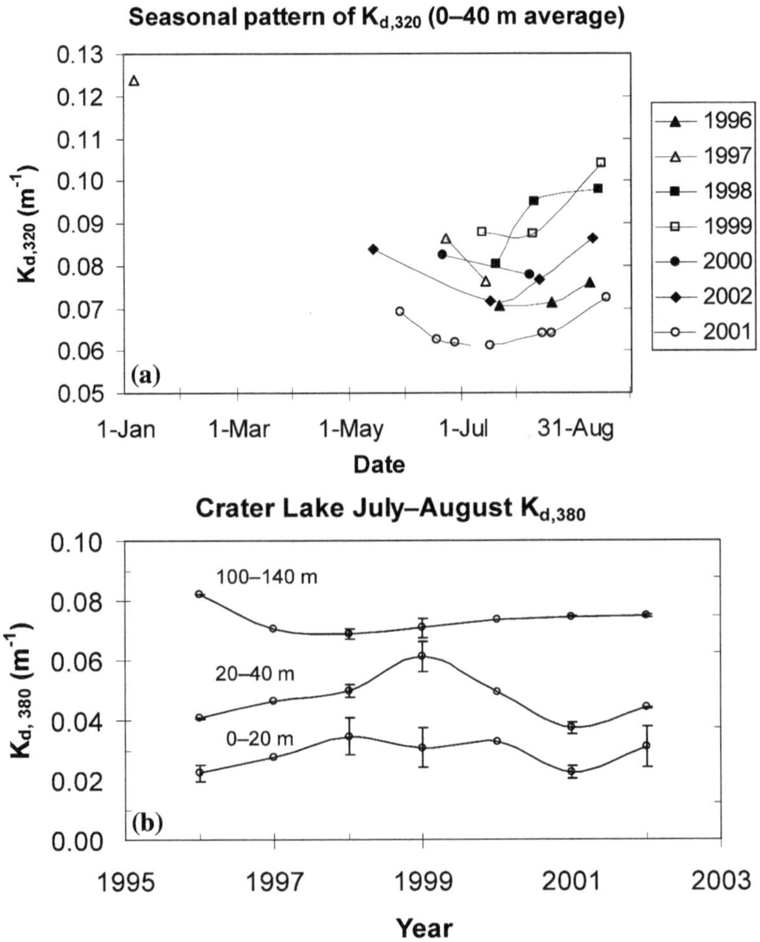

Fig. 7 (**a**) Seasonal pattern of $K_{d,320}$ in Crater Lake, 1996–2002 (averaged 0–40 m). Typically the minimum K_d occurred in July or August. The highest K_d values during this period were in 1998–1999, the lowest in 2001. (**b**) Interannual variation in $K_{d,380}$ for three depths, July–August average, 1996–2002. Little change occurred at the depth of the deep chlorophyll maximum (100–140 m). Larger variations occurred for depths 0–20 m and 20–40 m, which showed similar changes in 6 out of 7 years. Error bars (±1 S.E.) are shown for years with multiple measurements (N = 2, 1996, 1998, 1999, 2002; N = 3, 2001)

when the deep c_{p660} remained low throughout the summer.

Chlorophyll-*a* concentration can also serve as a proxy for UVR attenuation. Figure 10a shows the average concentrations from 1984 to 2002 for shallow (0–30 m) and deep (40–140 m) chlorophyll-*a* extracted from phytoplankton retained on a 0.45 micron filter. Also plotted is average monthly rain in July and August for this period. Except for 1986–1987 the deep and shallow concentrations tended to change roughly in parallel, with a maximum during the late 1980s and low values since 1996. Figure 10b shows the variations in $K_{d,320}$ (estimated by our Secchi depth proxy) plotted against the chlorophyll-*a* concentration in the 0–30 m depth range. The regression line for the dry months or months where precipitation occurred as snow or where runoff would have entered cold surface waters ($y = 015x + 0.08$, $r^2 = 0.44$) indicates that during periods with little runoff retained in the upper mixed layer about 44% of the variation in UV attenuation is explained by the absorption and scattering in UV wavelengths that co-varies with chlorophyll-*a* concentration.

Another bio-optical signal, chlorophyll-*a* fluorescence (F_{chl}), has been recorded in Crater Lake depth profiles since 1999 and provides a proxy for UV attenuation consistent with the dominant role of phytoplankton absorption and scattering. Figure 11a shows $K_{d,380}$ and F_{chl} for 20 August 2001 and the correspondence between the two signals is apparent. However at depths above about 80 m there are subtle differences between the two signals and for this upper part of the photic zone $K_{d,380}$ more closely resembles the c_{p660} signal (also plotted in Fig. 11a). A simple empirical model estimates

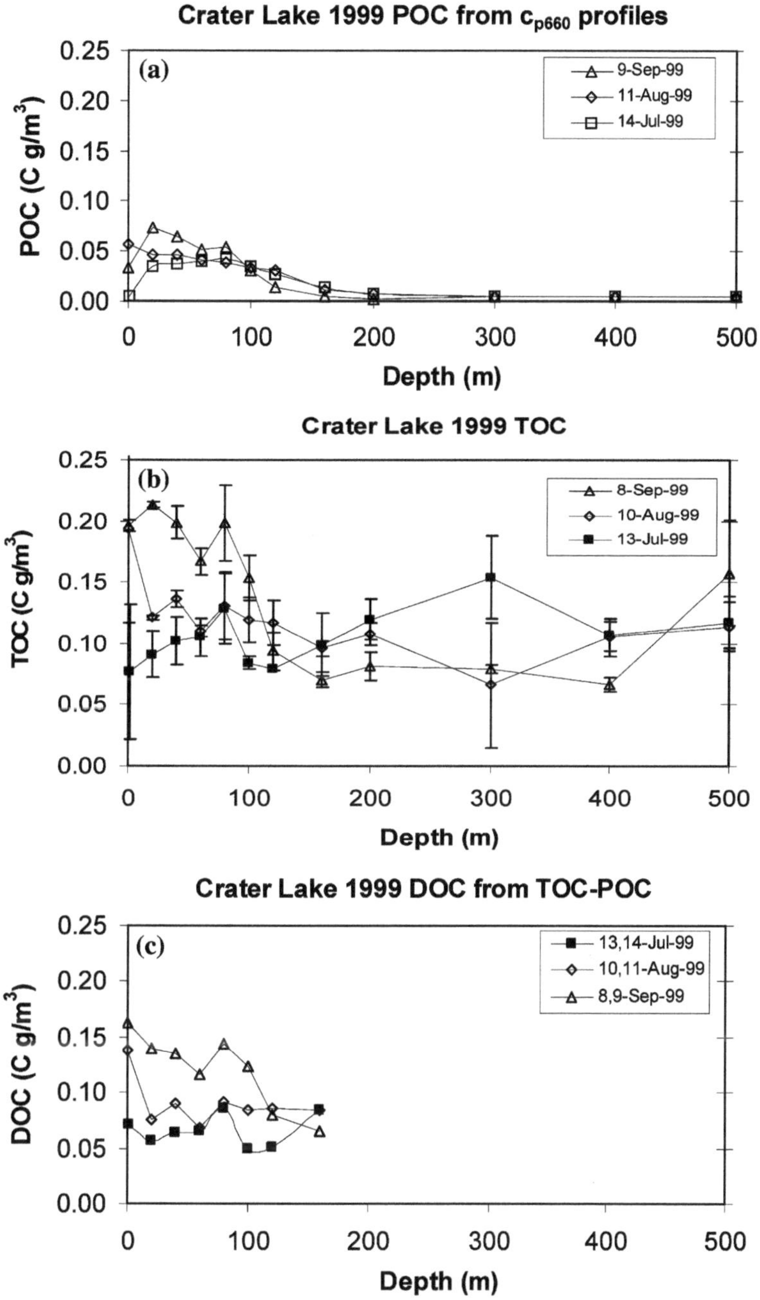

Fig. 8 (**a**) Crater Lake particulate organic carbon (POC, C g m^{-3}) versus depth, 1999. POC was derived from 2 m binned c_{p660} data (SeaTech transmissometer). Calculation: POC = (c_{p660} + 0.024-X)/0.032)/83.3 (proxy equation from Boss et al., 2007, but X was adjusted on each date to match POC over 300–500 m for all dates). Photic zone particulate carbon appears to increase over the course of the summer. (**b**) Crater Lake Total Organic Carbon (TOC) versus depth, 1999. TOC (carbon g m^{-3}) in whole water samples was determined from three analytical replicates with a Shimadzu TOC-5000 (error bars are ±S.E.). C. Crater Lake Dissolved Organic Carbon (DOC) versus depth, 1999. DOC calculated by difference (TOC-POC). Because no replicate samples were collected and analytical replicates for TOC were noisy, only the values above 200 m are plotted. DOC concentration appears to peak near 80 m, and to increase in the photic zone from July through September, consistent with a summer increase in CDOM observed in 2001 by Boss et al. (2007)

$K_{d,380}$ (heavy curve in Fig. 11a) from c_{p660} and F_{chl} (μg/l from fluorescence) by combining the two bio-optical signals: $K_{d,380} = 0.40{*}F_{Chl} + 0.36{*}c_{p660}$. The F_{chl} and c_{p660} signals are also combined in a ratio (F_{chl}/c_{p660}) in Figure 11b to show changes with depth and season for 2001 as a proxy for chlorophyll-*a*:carbon ratios. Figure 11b shows a distinct peak near the surface on 14 August (20–40 m) and a smaller surface peak on 20 August (40–60 m). Ratios on all dates increased with depth to a maximum corresponding to the DCM. Below the DCM the pattern varied with season, generally declining during the course of the summer.

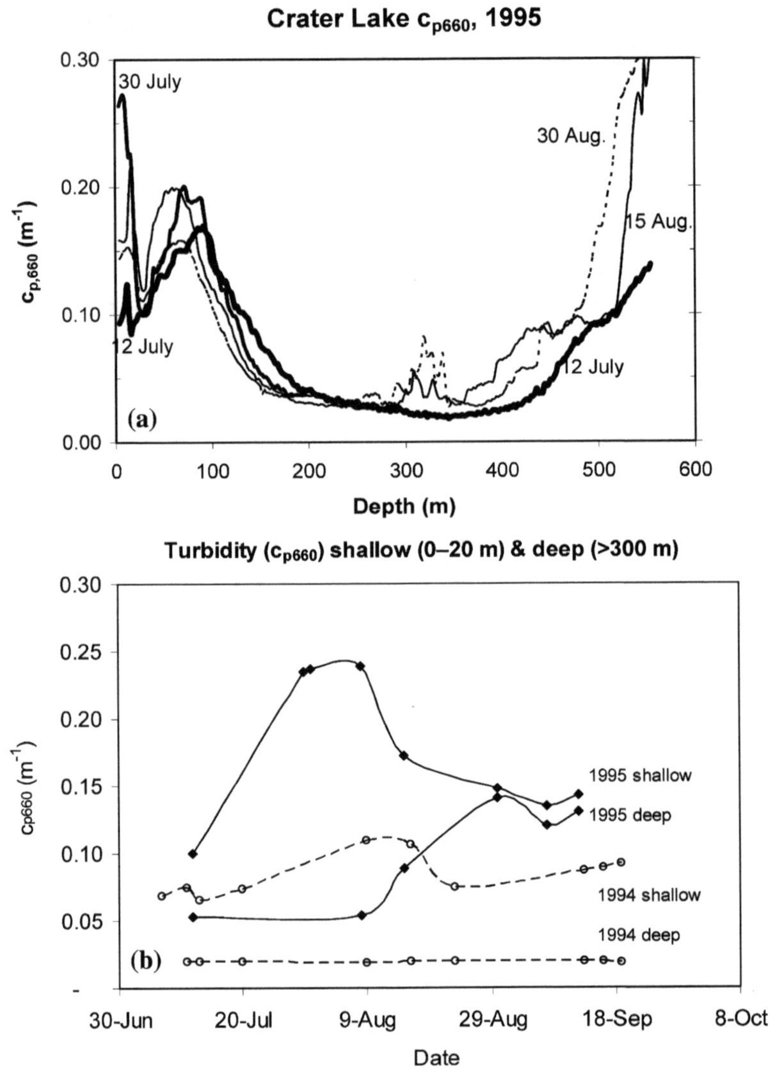

Fig. 9 (**a**). Crater Lake c_{p660} versus depth for 1995. Unusually high values near surface and deep in the water column are probably suspended mineral particles from storm runoff. The broad peak centered at 70 m is normal (assumed to be microbial biomass). Rain record at Crater Lake NPS Headquarters: 4–19 June = 4.7″, 6–12 July = 3.8″, 25 July = 0.9″. Snow pack also melted late this year (5 July 1995). (**b**) Crater Lake particles (c_{p660}) reached unusual levels during summer 1995 compared to 1994 (a typical year), coincident with high precipitation and runoff during Spring and Summer 1995 (see also Figure 10a). Both the surface and deep water (averages for 0–20 m and 300 m plotted) were elevated in early July and went much higher at the surface during the days after the 25 July thunderstorm with particles accumulating at the bottom 2–4 weeks after the storm

Possible UVR impacts on the Crater Lake ecosystem

Optical absorption by phytoplankton contributes directly to diffuse attenuation and absorption spectra can also reveal a protective response of phytoplankton to UVR exposure. Figure 12a shows spectral absorption measurements for five depths made 20 August 2001 on GF/F glass fiber filters. Distinct chlorophyll-*a* peaks are visible at 430 and 675 nm. Chlorophyll-*a* concentrations calculated from 675 nm peaks (0.038–0.38 mg m^{-3} using $ap^*_{675} = 0.040$) closely match the [Chl-*a*] calculated from data in Fig. 3 using the diffuse attenuation model of Morel and Maritorena 2001. Typical $ap^*_{675} = 0.035$ according to Sathyendranath et al. (1987).

The arrow in Fig. 12a indicates a peak at 325 nm that appears clearly at depths of 25 m and 50 m. This wavelength is consistent with UV-B protective compounds called micosporine-like amino acids or MAAs (Tartarotti et al., 2001) known to be produced by certain phytoplankton when exposed to UV-B radiation. Figure 12b shows depth trends for UV-B irradiance ($E_{d,320}$), particulate absorption ratios (a_{p330}:a_{p675} and a_{p440}:a_{p675}), and the F_{chl}:c_{p660} ratio (from the 20 August 2001 curve in Figure 11b). The ratio a_{p330}:a_{p675} follows a similar declining trend with depth as observed in

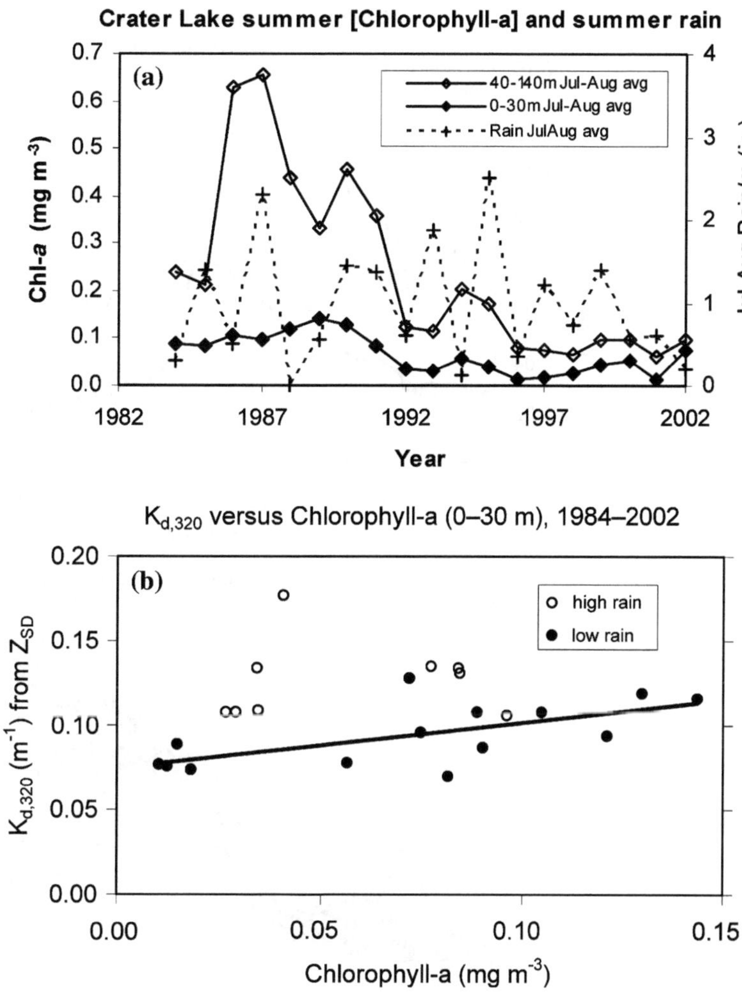

Fig. 10 (**a**) Crater Lake rain (inches per month at Crater Lake NPS Headquarters) and chlorophyll-*a* (0.45 um filter), averaged 0–30 m and 40–140 m (1984–2002 July–August). (**b**) Crater Lake $K_{d,320}$ (10–40 m) versus surface Chl-*a* (0–30 m, retained on 0.45 micron filter) for July and August 1984–2002. Only "dry" months (little rain or heavy but frozen precipitation) are included in the regression. Regression equation for dry months, $y = 0.15x + 0.08$ ($r^2 = 0.44$)

$K_{d,320}$:$K_{d,380}$ ratios (Fig. 6), suggesting that phytoplankton produce more MAAs relative to photosynthetic pigments when exposed to stronger UV-B irradiance.

One possible response to strong UVR in an aquatic ecosystem is inhibition of growth and survival for members of the plankton community. The low concentration of chlorophyll observed near the surface (Figs. 10a, 11, 12a) and scarcity of other organisms (Larson et al., 1996b) are consistent with this impact but other factors such as nutrient limitation could also be a dominant factor. A natural experiment with solar UV-B radiation has taken place over the past 20 years because of daily, seasonal, and interannual variations in stratospheric ozone above Crater Lake. Figure 13a shows the record of stratospheric ozone (averaged for several weeks before the typical mid-month sampling dates for phytoplankton). Also plotted are the chlorophyll-*a* concentrations from Fig. 10a. The large variations in phytoplankton apparent in Fig. 13a may be partly explained by atmospheric variations in ozone that lead to inverse variations in UV-B radiation reaching the lake surface. Figure 13b shows the positive regression relationship between chlorophyll-*a* and stratospheric ozone (Chl-*a*, 40–140 m: $y = 0.025x–7.4$, $r^2 = 0.49$; Chl-*a*, 0–40 m: $y = 0.004x–1.07$, $r^2 = 0.27$) consistent with UV-B irradiance accounting for 27–49% of the variation in phytoplankton pigment concentration either directly, or by inhibiting predators that feed on phytoplankton grazers.

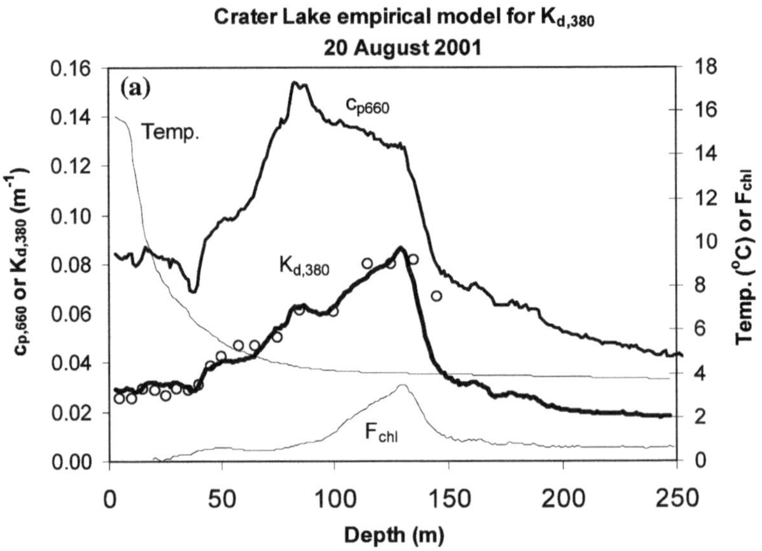

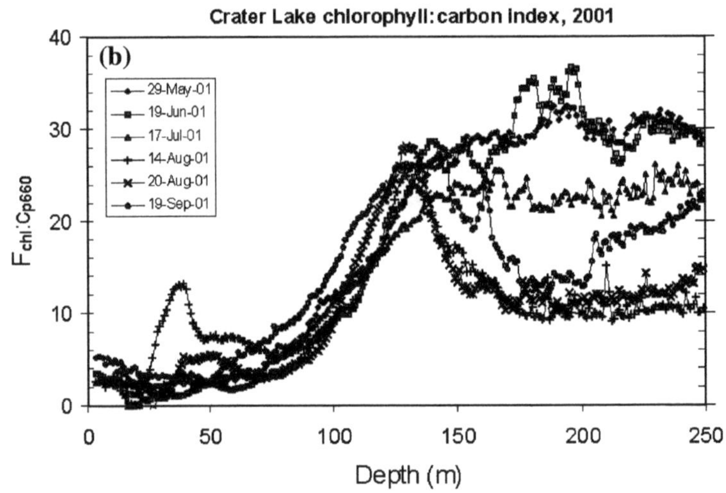

Fig. 11 (**a**) Crater Lake empirical model for $K_{d,380}$ versus depth based on bio-optical signals (20 August 2001). Temperature and depth from Seabird CTD; chlorophyll fluorescence (F_{chl}, relative units) from Wetlabs WetStar fluorometer; c_{p660} from SeaTech transmissometer (c_w = 0.411 m^{-1}); $K_{d,380}$ (open circles) from LI-1800uw scans. Model: $K_{d,380} = 0.40*F_{chl} + 0.36*c_{p660}$. (**b**) Crater Lake phytoplankton chlorophyll:carbon index for 2001 (from ratio of bio-optical signals, $F_{chl}:c_{p660}$)

Long-term proxies for $K_{d,UV}$ and decadal changes in UV transparency

Several proxies for UV attenuation can be used to look at long term changes in UV transparency because of the long time period covered by the primary measurements. The signal most similar to $K_{d,UV}$ is $K_{d,blue}$ measured by photometers equipped with changeable filters on deck and underwater sensors. LI-COR LI-1800uw radiometer data averaged over depths 0–40 m and for the period 1996–2002 were used to calculate K_d plotted in Fig. 14a. These were used to derive regression equations (see Fig. 14a caption) relating $K_{d,blue}$ (nominally 400–500 nm) to $K_{d,320}$ and $K_{d,380}$ (1996–2002). We then applied the equations to historic $K_{d,blue}$ data (gathered by photometers equipped with blue filters in 1940, 1969 and 1980–1991) to estimate $K_{d,320}$ and $K_{d,380}$. Figure 14b shows the time series for $K_{d,320}$ and $K_{d,380}$ derived from direct UV measurements (black symbols), from

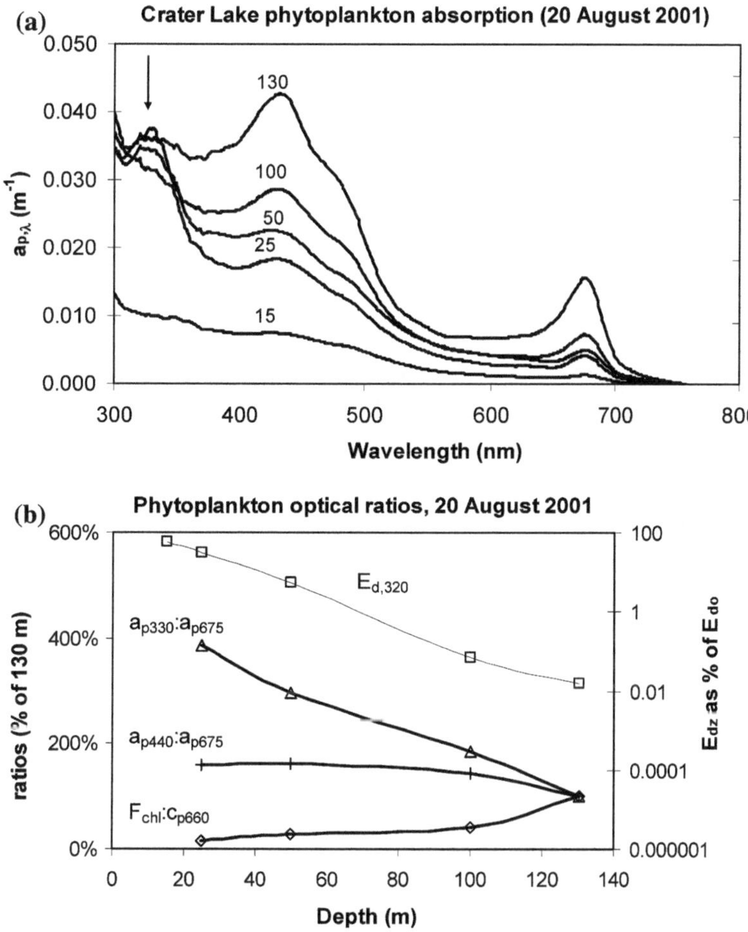

Fig. 12 (**a**) Crater Lake particulate absorption (a_p) spectra from 5 depths (20 August 2001). QFT method as modified by Lohrenz (2000) using GF/F filters (0.5–1.0 liters filtered); filter supported on quartz disc and scanned using a Shimadzu 160-UV spectrophotometer. Chlorophyll-*a* absorption peaks at 675 nm and 430 nm; apparent micosporine-like amino acid (MAA) peak at 325 nm (arrow) clearly visible in 25 m and 50 m samples. Chlorophyll-*a* concentrations calculated from 675 nm peaks (0.038–0.38 mg m^{-3} using $ap^*_{675} = 0.040$) closely match the [Chlorophyll-*a*] calculated from data in Figure 3 using the diffuse attenuation model of Morel and Maritorena 2001. Typical $ap^*_{675} = 0.035$ m^{-1} per mg m^{-3} according to Sathyendranath et al. (1987). (**b**) Absorption ratios consistent with photoprotection from UV-B in phytoplankton near the surface (MAA peak versus Chlorophyll-*a* red peak ratio follows attenuation trend for UV-B irradiance; Chlorophyll-*a* blue peak versus Chlorophyll-*a* red peak ratio remains constant with depth). Rising $F_{chl}:c_{p660}$ with depth suggests photoacclimation

$K_{d,blue}$ measurements with blue-filter photometers (light gray squares and triangles), and from radiometer measurements of only $K_{d,380}$ values (dark gray triangles) where $K_{d,320}$ was calculated using the relationship in Fig. 6. Based on this collection of direct and indirect calculations the lowest $K_{d,\ 320}$ (0.066 m^{-1} estimated from radiometer-measured $K_{d,380}$) occurred in 1966; $K_{d,\ 320}$ in 2001 was nearly identical (0.068 m^{-1}). The range among the annual summer averages for $K_{d,320}$ during this period (0.07–0.12 m^{-1}) represents approximately 35% variation above or below the mean.

The longest optical record in Crater Lake is from measurements of Secchi depth, beginning in 1896 (Larson et al., 1996a). The inverse of Secchi depth ($1/Z_{SD}$) should be correlated with other apparent and inherent optical properties as

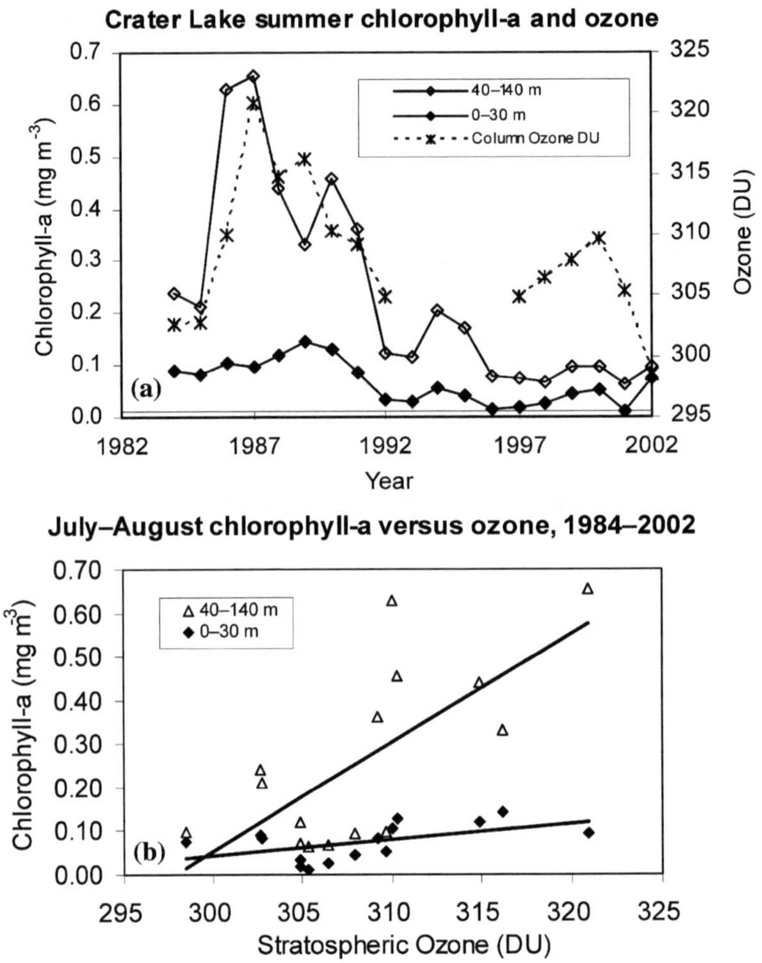

Fig. 13 (**a**) Crater Lake surface Chlorophyll-*a* (0–30 m average and 40–140 m average, retained on 0.45 micron filter) for July and August 1984–2002, and stratospheric ozone over Crater Lake during July and August (average of days 1–17 each month for latitude + 43, longitude –122, source Nimbus-7 TOMS sensor, 1978–1993, Earth Probe TOMS sensor, 1993–present, http://toms.gsfc.nasa.gov/teacher/ozone_overhead_archive.html). (**b**) Chlorophyll-*a* (1984–2002 averages for July & August, 0–30 m and 40–140 m retained on 0.45 micron filter) versus Stratospheric ozone (DU) averaged for first 17 days of the sample month (see Figure 13a for source). Regression equation for deep [Chlorophyll-*a*], $y = 0.025x–7.45$ ($r^2 = 0.49$); for shallow [Chlorophyll-*a*], $y = 0.0037x–1.07$ ($r^2 = 0.27$)

described earlier. When $K_{d,blue}$ is calculated from blue-filter photometer data and averaged from the surface to the Secchi depth measured on the same day, 13 years of summer averages (June–August) are well-correlated with $1/Z_{SD}$ (Fig. 15a, $K_{d,blue} = 2.08/Z_{SD}–0.026$; $r^2 = 0.54$, $N = 13$). Figure 15b compares $K_{d,blue}$ calculated from $1/Z_{SD}$ for the period 1937 through 2002 with $K_{d,blue}$ measured by photometer (averaged to Z_{SD} or 40 m) and by radiometer (averaged to 40 m). The agreement among $K_{d,blue}$ summer averages (July–August) from Photometer and $1/Z_{SD}$ data (regardless of whether data were collected on the same day) was within + 20% and –28% for the 12 years where both two types of data were available. Comparing $K_{d,blue}$ derived from radiometer and Z_{SD} the agreement was within + 7% and –35%. The largest difference between radiometer and $1/Z_{SD}$ averages occurred in 2000 when the summer averages were 0.046 m^{-1} from all $1/Z_{SD}$ data and 0.031 m^{-1} from the two-radiometer dates (June and August). The agreement improved when only $1/Z_{SD}$ data collected within 2 days of the radiometer profiles were compared

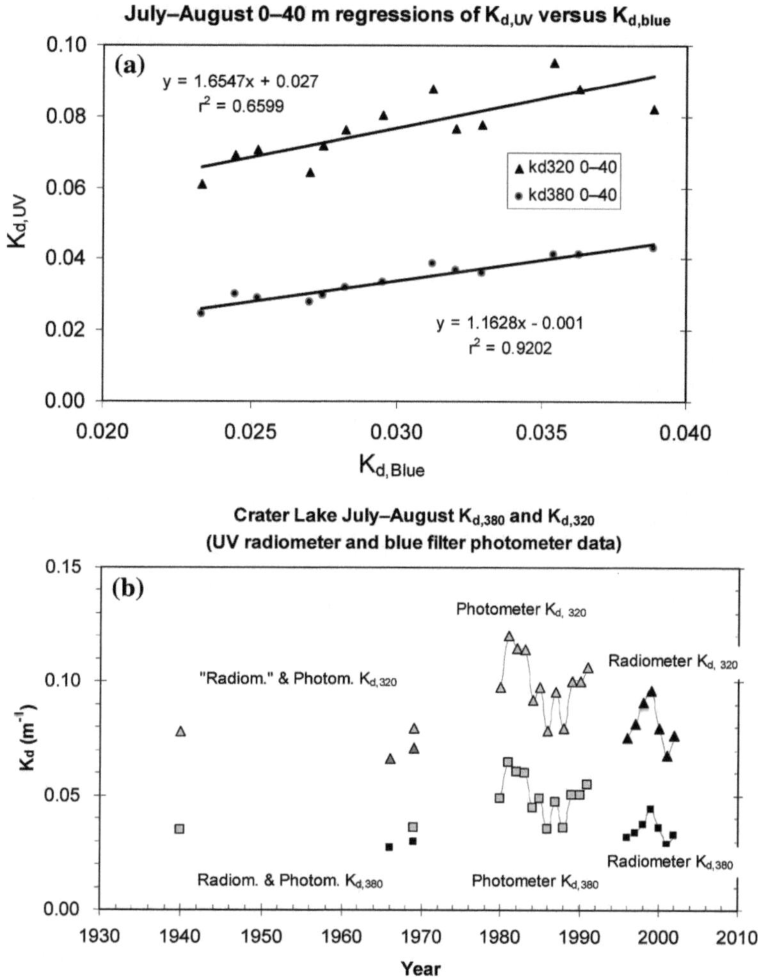

Fig. 14 (**a**) Regressions show that Crater Lake $K_{d,320}$ and $K_{d,380}$ can be calculated from $K_{d,blue}$ (400–500 nm) using data averaged over 0–40 m for July–August 1996–2002: $K_{d,320} = 1.65 * K_{d,blue} + 0.027$, $r^2 = 0.66$; $K_{d,380} = 1.16 * K_{d,blue} - 0.001$, $r^2 = 0.92$. (**b**) Time series for $K_{d,320}$ and $K_{d,380}$ (0–40 m) in Crater Lake using direct radiometer measurements (1966, 1969, 1996–2002) and proxy derived from $K_{d,blue}$ using blue-filter photometer data, (1940, 1969, 1980–1991). For 1966 $K_{d,380}$ was estimated from reported $E_{d,360–390\ nm}$ vs depth (5–25 m) yielding $K_{d,380} = 0.024$; after adjustment for 0–40 m depth range using recent patterns of K_d versus depth (Figures 5 and 6), $K_{d,380} = 0.027$. For 1966 and 1969 the $K_{d,320}$ radiometer data were estimated from measured radiometer $K_{d,380}$ using $K_{d,320} = K_{d,380} * = 2.27$

(agreement of summer averages was then within 1.5%). Differences can also be expected in part because instrument profiles covered 0–40 m while Secchi depth summer averages ranged from 23 m to 38 m during the period 1896–2002.

Using the relationships described above for estimating $K_{d,blue}$ from Secchi depth measurements and for estimating $K_{d,320}$ and $K_{d,380}$ from $K_{d,blue}$, we calculated $K_{d,320}$ and $K_{d,380}$ for the period 1896–2002. The resulting record of UV attenuation (averaged from the surface to depths ranging from 23 m to 44 m, depending on the Secchi depth) is shown in Fig. 16a from 1937 to 2002 along with radiometer-derived and photometer-derived values. In years when LI-1800uw radiometer or photometer records coincide with Secchi depth data there is a general correspondence between estimates except for a few years (early 1980s and 2000). The large spike in 1995 is based only on Z_{SD} measurements although c_{p660} data confirm the high attenuation near the surface (Fig. 9a and b). The range for $K_{d,320}$ based on Secchi depths is 0.08–0.19 m^{-1} (maximum 0.15 excluding 1995); based on the blue photometer data it is 0.08–0.15 m^{-1} (no measurements in 1995); based on the LI-1800uw it is 0.07–0.10 m^{-1} (1996–2002). From the detailed measurements starting in the 1978 there is a decadal pattern of peaks and valleys with amplitude of roughly ±50%, scattered with infrequent spikes of higher attenuation (e.g. 1995). By averaging all sources of data for $K_{d,320}$ for each year (Fig. 16b) the range and pattern from 1896 to 1969 was plotted. The values ranged from 0.07 to 0.14 m^{-1} from 1978 to 2002, and from 0.07 to 0.12 m^{-1} from 1896 to 1969.

Fig. 15 (**a**) Crater Lake regression of photometer $K_{d,blue}$ against inverse Secchi depth (units m^{-1}; annual averages for only same-day measurements, 1968–1991). Equation: $K_{d,blue} = 2.08/Z_{SD} - 0.026$ ($r^2 = 0.54$, $N = 13$ years). (**b**) Crater Lake time series of $K_{d,blue}$ from direct photometer (gray circles) and radiometer (black circles) measurements and proxy using Secchi depth (open circles). The spike in 1995 follows heavy summer rain in June and July (see Figure 9a). Many of the comparison points are for measurements made on different days in the same month

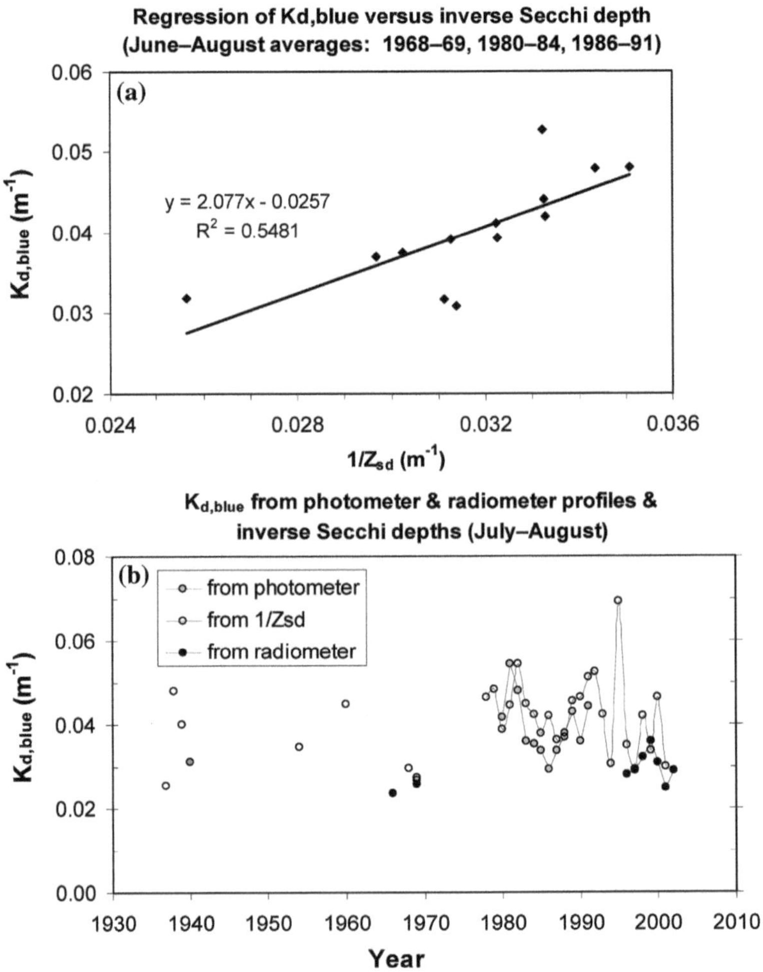

Discussion

Spectral measurements of incident and underwater solar radiation

Spectral measurements of downwelling irradiance (Fig. 1a), upwelling irradiance (Fig. 2a), spectral K_d (Fig. 3) in August 2001 were nearly identical to the 1960s data of Smith et al. (1973) except that the recent measurements included a greater range of wavelengths, responded to lower light levels, and in the case of E_u, resolved algal fluorescence distinctly at 683 nm. The LI-1800uw has been shown previously to record solar irradiance accurately within the constraints of its 8 nm bandwidth except at wavelengths below 308 nm (Kirk et al., 1994). Figure 1 demonstrates similar performance: matching within 96% on average and –13% and + 9% at all wavelengths from 308 nm to 400 nm in comparing a scan of incident irradiance with a spectrum generated by a high resolution radiative transfer model adjusted for local conditions and bandwidth.

Figures 2a and b show signs of "internal light sources" including chlorophyll-*a* fluorescence (683 nm), Raman scattering (impact at shorter wavelengths from 700 to 490 nm as depth increases), and possibly phycoerythrin fluorescence (589 nm, Hewes et al., 1998). The fluorescence signals indicate the presence of phytoplankton but cannot be directly used as indices of concentration because of photoacclimation and quenching near the surface. The detection of any internal source should serve as a warning because it invalidates attempts to measure diffuse attenuation at the wavelength and depth where it is detected.

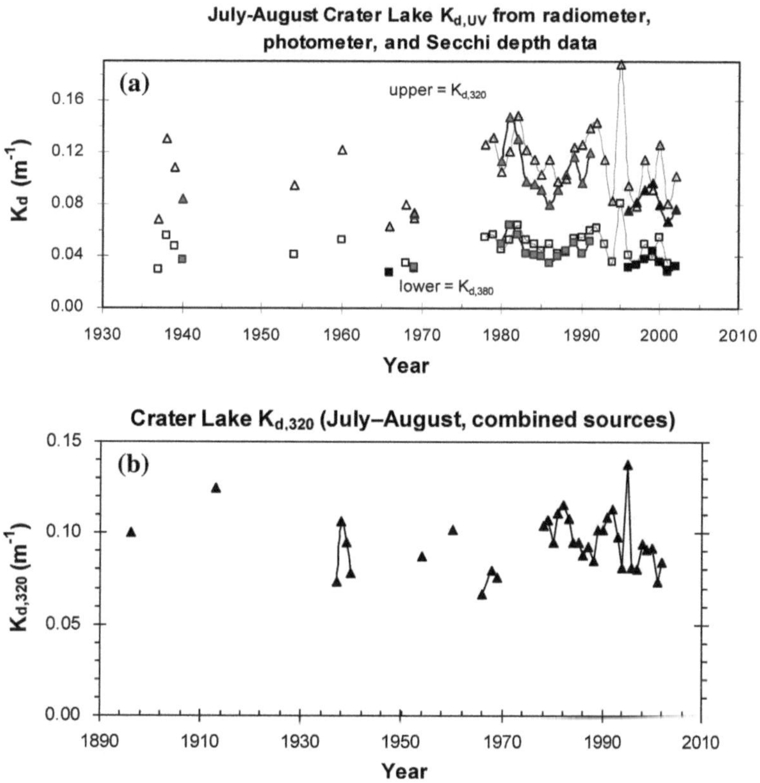

Fig. 16 (**a**) Crater Lake time series (1940–2002) of $K_{d,320}$ and $K_{d,380}$ using radiometer and photometer data from Figure 15B plus all valid Secchi disk data. Open symbols are from Secchi disk (triangles are $K_{d,320}$, squares are $K_{d,380}$); black symbols are direct from radiometer; light gray symbols are photometer data; dark gray triangles are $K_{d,320}$ derived from radiometer $K_{d,380}$. The spike in 1995 follows heavy rain in June and July. (**b**) Crater Lake times series of $K_{d,320}$ using average of all sources of data. Sample frequency prior to 1978 is inadequate to characterize cycles but maximum and minimum values from 1896 to 1969 (0.07–0.12) fall within the same range as the more frequent measurements since 1978 (0.07–0.14, or 0.07–0.12 without including the rainy summer of 1995)

While the conventional method of characterizing potential exposure to UVR is to measure diffuse attenuation, another factor which strongly influences exposure is incident irradiance. Our discovery of a positive correlation between phytoplankton chlorophyll-*a* and stratospheric ozone means that variations in incident UV-B irradiance could be playing an important role and thus UV-B irradiance should be monitored continuously on or near the lake.

The upwelling irradiance (Fig. 2a) and radiance spectra (not shown) help to explain the unusual color of Crater Lake. The low concentration of absorbing substances (e.g. phytoplankton and CDOM) near the surface clearly accounts for part of the phenomenon, thereby allowing for a greater optical role for water molecules to transmit and scatter light of certain wavelengths. Spectral reflectance models that predict the optical behavior of pure water (e.g. Morel & Prieur, 1977) indicate that while Crater Lake has extremely low levels of absorbing substances near the surface, it also contains particles that increase backscatter (for more on scattering, see Boss et al., 2007). The optical role of abundant glass-like suspended particles reported in the surface waters by Utterback et al. (1942) should be investigated.

Comparison of spectral measurements from 1969 in Crater Lake with similar measurements in Lake Tahoe (Smith et al., 1973) showed that Crater Lake had a greater proportion of short wavelengths in its upwelling spectrum and in the deep downwelling spectra. The differences between these two lakes appear to have increased since 1969 because of rising levels of nutrients, phytoplankton, and suspended mineral particles in L. Tahoe (Jassby et al., 1999).

Spectral diffuse attenuation, validation, and variations over space and time

In Crater Lake several intercomparisons have been carried out that have confirmed the ability to measure spectral diffuse attenuation accurately under extremely transparent conditions (e.g., Fig. 4) with the LI-1800uw. Its primary limitations in Crater Lake are the inability to accurately record wavelengths <308 nm in full sun, and the time required to complete a scan when irradiance is changing rapidly (e.g. within 15 m of the surface). Measurements of diffuse attenuation would not benefit from higher spectral resolution than its 8 nm bandwidth, in contrast to measurements of solar irradiance spectra, except in the case of the shortest UV-B wavelengths. This is because spectra of typical attenuating substances tend to vary gradually with wavelength, and because there are instrument performance tradeoffs so that increasing spectral resolution leads to decreasing sensitivity to low light levels underwater (Kirk et al., 1994). While the effective center wavelength of a broad irradiance waveband can change as a function of depth (Hargreaves, 2003), the LI-1800uw has been tested previously in lakes with substantially more CDOM than Crater Lake and proved capable of generating accurate K_d values there in comparison with BSI radiometers and a scanning radiometer that had a 2 nm bandwidth (Kirk et al., 1994).

The surface waters of Crater Lake are remarkably transparent to UV-B radiation. The relative attenuation of solar UV-B (320 nm) at a depth of 20 m in Crater Lake was 27% of surface irradiance on average (Table 1). Both the average $K_{d,320}$ of 0.062 m^{-1} (1996–2002) and minimum $K_{d,320}$ (August 2001, 10–18 m average) of 0.057 m^{-1}, for the surface of Crater Lake were lower than attenuation coefficients estimated for pure water ($K_{dw,320}$ = 0.09 m^{-1}, Smith & Baker, 1981). On only a few other occasions have such low $K_{d,UV\text{-}B}$ values been reported for natural waters. An ice covered lake in the dry valleys region of Antarctica (L. Vanda) had low levels of phytoplankton and DOC and $K_{d,320}$ = 0.055 m^{-1} near its surface (Vincent et al., 1998). An earlier study (Goldman et al., 1967) reported that the upper water column in L. Vanda was remarkably clear and similar to pure water based on blue photometer measurements ($K_{d,blue}$ = 0.031 on one occasion, identical to $K_{d,blue}$ average over 0–40 m for Crater Lake in July 1940, Utterback et al., 1942). In a region of the Gulf of Mexico away from the influence of Mississippi River runoff (flowing at the lowest rate in 52 years during July 1988 when measurements were made) Højerslev and Aarup (2002) reported $K_{d,310}$ = 0.071 m^{-1}. This value is equivalent to $K_{d,320}$ = 0.061 m^{-1} after wavelength conversion (using the exponential equation for CDOM absorption of Bricaud et al., 1981, but without the backscatter correction suggested by Markager & Vincent, 2000, using an exponent of –0.015 derived from regressing Crater Lake spectral K_d, averaged 0–40 m on 20 August 2001, against wavelengths from 305 nm to 380 nm). Morel and Maritorena (2001) also reported extremely low values ($K_{d,305}$ = 0.095, or $K_{d,320}$ = 0.076, converted as described above) for oligotrophic regions of the tropical Pacific Ocean where chlorophyll-a varied from 0.043 mg m^{-3} to 0.054 mg m^{-3} over depths 0–72 m. In each of these published reports of extremely low UV-B attenuation the authors noted the discrepancy between their measurements and the Smith and Baker (1981) estimates for K_w of pure water. Hargreaves (2003) estimated the UV-B attenuation of pure water from Crater Lake measurements; the new estimate ($K_{dw,320}$ = 0.045) represented a 50% reduction from the 1981 estimate. Other cases of extreme transparency probably exist where water is isolated from organic soils and algal productivity or other sources of CDOM by temperature, altitude, water currents, and strong solar UVR and where photobleaching can reduce absorption by any CDOM that is present.

From Figs. 4 and 5 one can see that the surface of Crater Lake typically has the greatest transparency but measurements here are problematic because of optical noise caused by waves and ripples reflecting and redirecting sunlight (Zaneveld et al., 2001). While a comparison of irradiance ($E_{d,0+}$) just above the surface with $E_{d,Z}$ at a depth of 15 or 20 m should remove much of the noise attributed to surface waves, this approach introduces uncertainty about the wavelength-specific transmittance through the air-water interface. With a profiling radiometer it is easier to compensate for this noise by lowering the instrument slowly in the upper 15 m so that multiple readings are available for averaging (of log-transformed irradiance) within 1 m, or larger "depth bins". Another method (we used this to generate the 10–18 m curve in Figure 3) involves fitting polynomial regressions to $E_{d,Z}$ versus Z for each wavelength of interest, taking care to use a sufficient number of significant figures for fitted parameters and to avoid extrapolation beyond the bounds of the fitted data. From the regression equations for each wavelength the specific values of $E_{d,Z}$ can then be calculated at several depths close to the surface and from these $K_{d,Z}$ can be calculated.

Morel and Prieur (1977) have used Crater Lake as an example of Case 1 natural water in their ocean classification scheme on the basis of autochthonous control of transparency. One indication of the oceanic nature of Crater Lake is the convergence of K_d spectra in Figure 3 and model K_d (not shown) calculated from the bio-optical model of Morel and Maritorena (2001). After increasing their values for K_{water} slightly (4.5%) to match the measured K_d at longer wavelengths not affected by phytoplankton, it was possible to fit each curve in Fig. 3 with a calculated K_d spectrum from 350 nm to 550 nm by adjusting a hypothetical chlorophyll-*a* value from which the model calculates K_d. On 20 August 2001 these fitted values for different depths ranged from 0.03 mg m^{-3} to 0.33 mg m^{-3} chlorophyll-*a* and compared well with chlorophyll-*a* concentrations estimated from particulate absorption at 675 nm (Fig. 12a), which ranged from 0.04 mg m^{-3} to 0.38 mg m^{-3} using a specific absorption coefficient at 675 nm of 0.04 m^{-1} per mg m^{-3}. For chlorophyll-*a* measurements from samples collected on 14 August 2001 and 11 September 2001 (chlorophyll-*a* extracted from 0.45 micron filters) the concentration of chlorophyll-*a* in the 0–30 m range of depths was 0.014 mg m^{-3} and the value near the DCM was 0.12 mg m^{-3}.

Seasonal and interannual variations in radiometer-derived $K_{d,320}$ and $K_{d,380}$ (Fig. 7A and B) were small but significant during the period 1996–2002 but few UV measurements are available from earlier years. The first Crater Lake UV measurement (August 1964) used an experimental instrument to characterize spectral horizontal radiance reflectance, 400–700 nm (Tyler, 1965). Although the spectrum was clearly related (inversely) to absorption in the upper meter of the water column, it is difficult to compare because of the unconventional approach. Measurements in August 1966 of upwelling and downwelling spectral irradiance to 25 m from 360 to 700 nm were more useful (Smith & Tyler, 1967; Tyler & Smith, 1970). When we recomputed K_d from reported E_d values at 0, 5, 15, and 25 m we discovered substantial noise in the blue and UV wavelengths but used an exponential model (Bricaud et al., 1981) to estimate K_d at 380 nm from wavelengths 360–390 nm (depth-averaged K_d using 5–15 and 15–25 m E_d data pairs) and then adjusted for shallow depths using the relationship in Figure 5 to compare with K_d averaged over 0–40 m (Fig. 12). After these adjustments the values were still historically at the low end of observed K_d suggesting that Crater Lake was extremely clear during the 1960s. The data collected in August 1969 included upwelling and downwelling spectral irradiance to 99 m over 360–700 nm (Smith et al., 1973) but no tables were provided; values were estimated from published graphs and showed that Crater Lake was similar in transparency to the average of the recent period (1996–2002). For both 1966 and 1969 the values for $K_{d,320}$ were not measured but were estimated from measured $K_{d,380}$ using the average relationship in Fig. 6.

Larson et al. (1996a) discussed determinants of visual clarity in Crater Lake, primarily phytoplankton and abiotic particles (storm-related suspended mineral sediments). Only a few other cases of phytoplankton influencing UV attenuation in clear lakes have been reported (e.g. Sommaruga, 2001). While we have argued above that phytoplankton control attenuation during dry periods (see Figs. 10b, 11a, 12a), there probably exists a contribution to K_d from suspended allochthonous particles (mineral particles and pollen) at all times (Morel & Prieur, 1977). This contribution may vary in response to wind, turbulence, and proximity to shore. Even without the uncertain contribution of suspended minerals, chlorophyll-*a* is not a perfect predictor of UV attenuation if phytoplankton vary in their composition of accessory pigments and MAAs. The additional contribution of CDOM produced by phytoplankton requires more investigation given the observation of Boss et al. (2007) that CDOM increases with depth and during the summer. The possible impact of fires on UV transparency has not been addressed but fires could influence $K_{d,UV}$ directly (smoke particles entering the water) and indirectly (e.g. influencing turbidity and nutrients in runoff).

Proxy measurements for UV attenuation versus depth

A number of indirect proxy measurements for UV attenuation have been utilized in our study. Proxies can be useful when conditions are not appropriate for measuring $E_{d,Z}$ directly (e.g. rapidly changing sky conditions, or low sun angle) or when a UV radiometer is not available. Many take advantage of some optical property of phytoplankton (e.g. scattering of the red beam of a transmissometer (e.g. 660 nm), fluorescence (F_{chl}, 683 nm), absorption at the red chlorophyll-*a* peak (675 nm) or over the blue part of the visible spectrum (400–500 nm). The strong correlation of F_{chl} from chlorophyll-*a* and $K_{d,380}$ (Fig. 11a) can be caused by the direct absorption of UVR by chlorophyll-*a*, but is also likely to vary with depth and UV wavelength because of the absorption of other molecules such as MAAs and because photoacclimation and photochemical quenching change the relationship between F_{chl} and [chlorophyll-*a*] (Boss et al., 2007). Variation in accessory and UV-B screening pigments is suggested by the changing ratios of K_{d320}:K_{d380} in Fig. 6, and of a_{p330}:a_{p675} in Fig. 12b. It is not uncommon in clear aquatic systems for the biomass peak to occur at a different depth from the chlorophyll-*a* maximum (Fennel & Boss, 2003).

The concentration of dissolved organic carbon (DOC) has frequently been used as a proxy for UV attenuation (reviewed in Hargreaves, 2003) but measurements in Crater Lake are limited to a few dates and low reproducibility (Fig. 8). Measurements of phytoplankton absorbance can be complicated by the tendency for MAAs to be released from phytoplankton during filtration (Laurion et al., 2003) as this will tend to cause elevated values in CDOM measured after filtration, and by the limits of detection using a 10 cm cuvette and laboratory spectrophotometer (but see Boss et al., 2007, for a long-pathlength in situ method). Published equations relating $K_{d,320}$ and DOC concentration (Hargreaves, 2003) vary because of the optical quality of DOC (photobleaching and source contribute to this variation). At low [DOC] the fit of this type of equation also is likely to depend on a correlation of [DOC] with [phytoplankton] because of the low absolute absorption by CDOM relative to phytoplankton. Within the uncertainty of the DOC measurements, Crater Lake is similar to the extremely transparent Lake Vanda, Antarctica. Comparing the DOC and UV attenuation data for L. Vanda and Crater Lake, Vincent et al. (1998) reported L. Vanda DOC = 0.3 g/m^3 and $K_{d,320}$ = 0.055, while our data for Crater Lake during summer 1999 (averaged over 0–40 m), show DOC = 0.1 g m^{-3} and $K_{d,320}$ = 0.09 m^{-1}. At this low level of DOC concentration the technique to account for instrument blanks is crucial (Sharp et al., 1993). The large difference in DOC-specific UV-B attenuation ($K_{d,320}$–$K_{dw,320}$):DOC = 0.03 for Lake Vanda and 0.5 for Crater Lake 0–40 m in 1999) suggests either a difference in DOC measurement technique or a difference in the optical contributions of particles and a_{CDOM}, for example by more mineral particles in Crater Lake.

To our knowledge particulate organic carbon (POC) has not been used as a proxy for $K_{d,UV}$. In Crater Lake we observed that UV attenuation can be characterized by a combination of c_{p660} and F_{chl}, the former more important near the surface and the latter more important at greater depths (Fig. 11a). This relationship seems reasonable in an oligotrophic lake with low levels of DOC because the c_{p660} detects scattering by cells when absorption by chlorophyll-*a* is suppressed at high light levels in the process of photoacclimation. The cells at shallower depths will still be attenuating UV wavelengths because of their inevitable content of other UV-absorbing molecules. Figure 11b clearly shows the consequence of photoacclimation in the phytoplankton community as the chlorophyll-*a* concentration per unit of carbon (and presumably per cell) is reduced at depths shallower than 50–75 m. Below this range of depths the chlorophyll-*a* concentration and the effective light absorbing properties of chlorophyll-*a* are increased as phytoplankton adapt to low light conditions. Along with the chlorophyll-*a* there are accessory pigments and macromolecules that absorb UVR. It is not clear if the pattern in Fig. 11b is modified by exposure to the high levels of UVR because one could propose a similar pattern as phytoplankton adapt to visible wavelengths. And wherever the microbial community develops in the water column, one can expect to find detrital particles and dissolved DOC co-varying with the living cells.

While depth profiles of c_{p660} typically show the pattern for biomass distribution in Crater Lake suggesting a biomass peak ranging from 50 m to 100 m (Boss et al., 2007) the c_{p660} signal also detects scattering caused by suspended mineral particles entering with runoff from heavy rain. The seasonal and depth patterns of c_{p660} for 1995, a year with unusually high optical attenuation and heavy summer rain, demonstrate the slow transport of sediments from the surface to the bottom. Depth profiles for 1995 (Fig. 9a) showed an unusual c_{p660} peak in turbidity close to the surface during July and August and peaks deeper in the water column (>250 m) that were not present in other years. A time course for the near-surface and near-bottom c_{p660} signal in Fig. 9b compares 1994, a more typical dry year, with 1995. During June and July of 1995 the Crater Lake weather records included several precipitation events and an unusually late date (5 July) for melting of the snow pack at the Park Headquarters weather station and one of the authors (MWB) recorded observations of downslope sediment transport into the water and turbidity in the lake surface in response to these storms. Water column peaks in the c_{p660} signal could be caused by mineral particles or phytoplankton. Records of [chlorophyll-*a*] averaged over 0–30 m do not show elevated values for July 1995 compared to other years (Fig. 15a). Turbidity traveling down the water column has appeared in transmissometer records from other years (e.g. 1991, 1996, 1997, 1998) but the entry of turbidity from runoff down the slopes and into the surface waters, followed by spreading across the lake by wind driven currents has only been reported for 1995.

Long-term proxies for $K_{d,UV}$

Development of the $K_{d,blue}$ signal as a proxy for $K_{d,UV}$ can be justified from empirical observations in Crater Lake. The attenuation spectra (e.g. Fig. 3) near the surface are relatively flat over the response waveband of the combined blue filter and Weston cell sensor using in underwater photometers (Crater Lake K_d averaged over 0–30 m varied spectrally by + 36% ($K_{d,500}$) and – 12% ($K_{d,400}$) compared to the mean $K_{d,400–500}$; the response bandwidth was 390–490 nm or nominally 400–500 nm). Also, the $K_{d,UV}$ values measured with the LI-1800uw radiometer gave nearly constant or predictable ratios for $K_{d,320}$:$K_{d,blue}$, $K_{d,380}$:$K_{d,blue}$, and $K_{d,320}$:$K_{d,380}$ in the upper water column (Fig. 6). While the LI-1800uw radiometer has not been directly compared with a Kahl blue-filter photometer, the agreement is excellent (Fig. 14b) between $K_{d,blue}$ from UV radiometer and photometer measurements in 1969 that occurred within 10 days of each other. $K_{d,UV}$ values from radiometer and photometer measurements made in the 1960s were similar to the data from 1996 to 2002 while on average the $K_{d,UV}$ values estimated by proxies from the 1980s were higher.

A combination of scattering and absorption contribute to Secchi depth signal and several theoretical and empirical papers have discussed the proportional relationship of $1/Z_{SD}$ and $K_d + c_p$ (Tyler, 1968; Preisendorfer, 1986). Our proxy relationship between $1/Z_{SD}$ and $K_{d,blue}$ is based on same-day measurements starting in 1968. While one cannot assume that c_p and K_d will change in parallel as conditions in the lake change, we have used the empirical pattern to establish an equation ($K_{d,blue} = 2.08/Z_{SD} - 0.026$; $r^2 = 0.54$, $N = 13$ years) and then convert $K_{d,blue}$ to $K_{d,320}$ or $K_{d,380}$ using the relationship described above. Although there are many possible sources of error in the proxy approach relating $K_{d,UV}$ and $1/Z_{SD}$, we estimate from regression residuals (Fig. 15a) that annual averages of $K_{d,320}$ estimated from $K_{d,blue}$ will fall within 12% of the values measured with a radiometer. This calculation assumes that conditions from 0 m to 40 m remain within the bounds that Crater Lake has experienced between the earliest paired measurements in 1968 and the present.

Decadal changes in UV transparency

Comparisons of our UVR attenuation measurements with those from radiometer data in the 1960s and from frequent proxy measurements since 1978 suggest that UV attenuation changes on decadal and longer cycles, perhaps in response to a combination of rain and stratospheric ozone influencing the phytoplankton community, with heavy summer rain occasionally influencing the level of suspended mineral particles. We also cannot rule out the impact of trophic interactions and winter precipitation or nutrient upwelling as additional factors influencing phytoplankton. Given the modest level of variation that our phytoplankton model accounts for, it is also possible that substantial variations in attenuation are caused by wind-driven changes in suspended mineral particles during dry periods. One possible explanation for lower average blue attenuation since 1992 (excluding storm-related peak in 1995) is the improvement to sewage disposal at the tourist facilities in 1991 (Larson, 2002). This hypothesis is considered unsupported by phytoplankton data by McIntire et al. (2007), and is also not supported by optical measurements once the pattern of rising attenuation in the early 1980s is placed into a longer time context (Fig. 16b). Climate variation is another likely factor to explain changes in blue and UV attenuation, possibly linked through periodic winter upwelling events that are thought to provide the majority of nutrients to phytoplankton each year (McManus et al., 1993; Larson et al., 1996a). Redmond (1999) reported a shift in prevailing winds in the mid-1970s over the region that was associated with the Pacific Decadal Oscillation (PDO, Mantua et al., 1997). The sparse attenuation estimates from proxy data since 1896 fall within the range of the better-characterized period since 1978, and from these we conclude that there has been no long-term change in UV transparency since 1896.

UVR impacts on the Crater Lake ecosystem

The incident spectrum of sunlight (upper curves in Fig. 1a) shows rapid attenuation at the shortest wavelengths because stratospheric ozone strongly absorbs UV-B wavelengths. Figure 13a also shows that summer stratospheric ozone can vary substantially (5–10%) from year to year. Thus exposure of planktonic organisms to UV-B will depend on atmospheric transparency as well as water column transparency. Because phytoplankton typically control the UVR penetration into the water column, if their growth rates or death rates are influenced by the increasing UV-B that accompanies a decline in ozone, the higher incident UV-B would create a water column that is more transparent to UVR and thus UV-B would penetrate deeper.

The observed correlation of chlorophyll-*a* with stratospheric ozone (Fig. 13b) suggests a direct and significant impact of UV-B on the phytoplankton community. More study of this phenomenon is needed because only two depth ranges (0–30 m and 40–140 m) and one ozone time period (the first 3 weeks of July and of August, averaged) have been investigated. Other controlling factors might have changed in parallel with ozone by coincidence or because the paths of the jet streams influence both ozone and weather. Rain has been suggested as a significant nutrient source for the nitrogen-depleted surface waters (McIntire et al., 2007). Of the three years since 1984 when the 0–30 m chlorophyll-*a* average for July–August dropped by more than 50% from the preceding year, two of those years (1992 and 2001) were also unusually dry (1992, 66% of average January–June precipitation; 2001, 60% of average).

The stronger correlation between ozone and deep chlorophyll-*a* compared with shallow chlorophyll-*a* (Fig. 13b) is curious, but the inferred presence of MAAs in near-surface phytoplankton (Fig. 12a, b), and signs of photoacclimation (Fig. 11b) provide the basis for several hypotheses. It is likely that species differ in a variety of ways that would impact their ability to thrive (survive) near the surface of Crater Lake. These include: their sensitivity both to UV-B and to high levels of PAR; their ability to produce MAAs; their cell size (small cells gain less benefit from intracellular compounds that absorb UV-B); and their ability to cope with scarce nutrients. We hypothesize that phytoplankton living near the surface of Crater Lake are more resistant to UV-B even when ozone levels are high and thus respond less to declines in ozone than phytoplankton at greater depths. Any cells that are growing near their UV-B limits when ozone is high may simply sink faster than their growth rate can replace them when ozone declines. Cells with less UV-B resistance that live deeper in the water column seem more likely to respond in this way. An alternative hypothesis is that cells living near the surface have lower chlorophyll-*a* concentration per cell (one possible interpretation of Fig. 11b), so that a similar percentage reduction in cell abundance in response to an ozone decline would involve a smaller change in chlorophyll-*a* in the surface waters compared to deeper in the water column.

Other signs of possible UVR impact on the Crater Lake ecosystem include the appearance of UV-resistant organisms and the scarcity or low diversity of organisms in the surface waters of the lake. A previous study reported that phytoplankton diversity and abundance were low in the upper 20 m and only one dominant species has been identified (*Nitzschia gracilis*) representing about 70% of net plankton and nanoplankton in this depth range from 1985 to 1988 (McIntire et al., 1996). According to McIntire et al. (2007), *Nitzschia gracilis* is dominant during the stratified summer period in the upper 20 m but the phytoplankton assemblage in the epilimnion also has higher densities of smaller species of cyanobacteria (*Aphanocapsa delicatissima* and *Synechocystis* sp.), and lower densities of dinoflagellates *Gymnodinium inversum* and *Peridinium inconspicuum*; the chrysophytes *Dinobryon sertularia* and *D. bavaricum* were good indicators of the lower epiliminon and upper metaliminion, although their mean relative abundance was below 7% of the total cell density.

We hypothesize that the microbial community will respond to UVR stress under the stable shallow-mixing conditions in Crater Lake with the appearance of highly adapted (and thus UVR resistant) species in the upper 30 m (average $Z_{10\%,320}$ = 32 m) and moderately resistant species down to 60 m (average $Z_{1\%,320}$ = 55 m). Urbach et al. (2001) identified two bacterioplankton taxa (CL120-10 verrucomicrobiales and ACK4 actinomycetes) from Crater Lake surface waters that are likely to be resistant to UVR. The deep chlorophyll maximum occurs between 120 m and 140 m (Larson et al., 1996a), close to the 0.1% depth for PAR (131 m summer average for 1996–2002). The biomass maximum (based on c_{p660}) occurs near the 1% depth for PAR (92 m summer average).

While the scarcity of zooplankton (Larson et al., 1996b) and phytoplankton (McIntire et al., 1996) at depths shallower than 40 m, where UV_{320} is more intense than 5–10% of incident irradiance during summer, is consistent with avoidance or poor survival because of exposure to UVR, other explanations for the phytoplankton distribution are possible. PAR irradiance is more intense than 10% of incident irradiance at depths above 44 m. The diatom *Nitzschia gracilis*, which frequently forms blooms near the surface (McIntire et al., 1996), is probably highly resistant to UVR. Some of our optical data are consistent with the presence of MAA photoprotective compounds in at least part of the phytoplankton community that has become adapted to conditions near the surface (ratio of $K_{d,320}$:$K_{d,380}$ in Fig. 6, phytoplankton spectral absorption at 330 nm in Fig. 12a, and the ratio of phytoplankton absorption at $a_{p,330}$:$a_{p,675}$ in Fig. 12b). Eisner et al., (2003) also found changed in vivo particulate absorption spectra ratios that corresponded to changes in phytoplankton pigments (photoprotective versus photosynthetic carotenoids) that were correlated with the light intensity in the water column. Tartarotti et al., (2001) observed a correlation between MAAs in zooplankton (derived from phytoplankton) and elevation in

Alpine lakes that they attributed to adaptations providing resistant to UVR exposure.

The absence of higher taxa near the surface of Crater Lake is also likely to be influenced directly or indirectly by UVR although lacking experimental evidence we cannot rule out other factors (grazing, predation, high levels of PAR). A benthic moss (*Drepanocladus aduncas*) has been reported in Crater Lake at depths 25–140 m with the greatest density 40–80 m (McIntire et al., 1994). Analysis of seasonal changes in morphology of this moss might reveal information about growth rate as a function of depth (Riis & Sand-Jensen, 1997) to explore the role of UVR in setting upper and lower depth limits for the population. Small zooplankton (rotifer taxa) have been reported at depths characteristic of moderate UVR but they are scarce above 40 m (Larson et al., 1996b). A factor that may influence the distribution of large zooplankton grazers is predator avoidance because the intensity of PAR irradiance at noon on a typical clear summer day exceeds 5 μmol m^{-2} s^{-1} at 100 m and 0.7 μmol m^{-2} s^{-1} at 150 m. Laboratory experiments (summarized in Kalff, 2002) have shown that the distance at which fish can perceive large zooplankton prey becomes limiting at PAR irradiance below 0.6 μmol m^{-2} s^{-1}; more illumination is required to capture smaller prey. Small planktivorous fish and large zooplankton might be forced to stay below an optimal feeding depth in Crater Lake during the day in order to avoid their respective visual predators and then migrate to the zone of maximal phytoplankton abundance at night or twilight to feed. In Crater Lake young Kokanee salmon (taken as prey by rainbow trout) are found near the surface at night but migrate down to 100 m during the day (Buktenica & Larson, 1996). Cladoceran zooplankton migrated diurnally as well, with larger species remaining deeper during the day and night than smaller species. By avoiding visual predators these zooplankton and small fish would also avoid exposure to strong UVR. Some fish are known to use UV-A wavelengths for vision and UV-A may penetrate deeper than visible wavelengths at dawn and dusk, or whenever the irradiance penetrating the lake surface is dominated by skylight rather than direct rays from the sun (Leech & Johnsen, 2003).

Future research on UVR in Crater Lake

Future work on UVR in the Crater Lake ecosystem should include experimental manipulations of UVR exposure of Crater Lake organisms (e.g. plankton and moss) to determine sensitivity of survival and productivity at different depths, continuous monitoring of incident UV-B and UV-A radiation, and better monitoring of the optical constituents of the water column that influence UVR penetration, especially in the upper 30 m where it is currently difficult to characterize UV attenuation. Crater Lake is a unique site for investigation of UVR effects on an aquatic ecosystem because it combines unusual UVR transparency with a wide range of depths below the mixed layer where summer exposure to UVR and PAR will be relatively stable from day to day except for the fluctuations in UV-B controlled by stratospheric ozone.

To explore both UV transparency and organic carbon dynamics, a CDOM fluorometer should be added to the suite of sensors in routine CTD profiles, and these data correlated periodically with measurements of particulate absorption spectra and particulate and dissolved organic carbon concentrations for different depths. The ecological control of the microbial community in different depth zones should be explored to clarify the relative importance of stratospheric ozone, nutrient sources (e.g. precipitation versus upwelling), and trophic interactions.

The response of Crater Lake to future climate change will likely involve its sensitivity to the timing and magnitude of changes in precipitation and stratospheric ozone. Precipitation appears to influence UVR transparency through influx of suspended sediments and phytoplankton nutrients. Ozone appears to protect the phytoplankton community from UV-B radiation, which may allow phytoplankton with moderate UV-B resistance in the upper water column to shield those deeper (and presumably more sensitive) organisms from UVR.

Conclusions

Crater Lake color and UVR transparency

Our recent UVR measurements show that Crater Lake is unusually transparent to UVR and visible wavelengths. In the DNA-damaging UV-B wavelengths diffuse attenuation ranges over values that are similar to, but in some cases more transparent than, other natural waters reported to date. Under average summer conditions (1996–2002) the intensity of a UV-B reference waveband (8-nm waveband centered at 320 nm) at 20 meters was 27% of surface irradiance, and at 40 m was 5% of surface irradiance (10% at 40 m under the especially clear conditions in 2001). The lowest UV-B diffuse attenuation coefficients ($K_{d,320}$) were somewhat lower than those reported for the clearest natural waters elsewhere (Lake Vanda, Antarctica, and several oligotrophic ocean regions), and were lower than the previously estimated attenuation by pure water (Smith & Baker, 1981).

Optical proxies and the impact of phytoplankton on UVR transparency

As in Case 1 marine waters, phytoplankton in Crater Lake normally are the dominant optical attenuator of UVR. During summer the UV attenuation rises from a minimum near the surface to a peak at the deep chlorophyll maximum (DCM) near 120 m. Measurements of $K_{d,UV}$ at different wavelengths parallel each other with increasing depth although only the longer UVR wavelengths can be detected deep in the water column. Optical proxies for UV attenuation include the scattering of light by phytoplankton in the upper part of the photic zone (c_{p660}, measured with a red beam transmissometer), and fluorescence of phytoplankton (F_{chl}, 683 nm) in the deeper regions (75–150 m). Other optical proxies for the depth range 0–40 m include inverse Secchi depth and diffuse attenuation in the broad blue waveband (400–500 nm), corresponding to historical underwater measurements with blue filter photometers. Only during rare heavy rains in summer does the optical dominance of phytoplankton give way to an optical signal from light scattering by suspended mineral particles and a likely breakdown of the proxy relationships. Excluding occasional wet summer months, $K_{d,UV}$ derived from Secchi depth was correlated with phytoplankton biomass as measured by chlorophyll-*a* concentration for the period 1984–2002.

UVR impact on the Crater Lake ecosystem

In the absence of UVR experiments the impact of UVR on lake biota can be inferred only from correlations. Since 1984 the summer average stratospheric ozone over Crater Lake was correlated with chlorophyll-*a* concentration in the photic zone (for depth ranges 0–30 m and 40–140 m), suggesting a direct or indirect impact of UV-B radiation on Crater Lake phytoplankton. Particulate absorption at a wavelength typical of UV-B protective pigments in phytoplankton varied with depth in proportion to UV-B exposure. Few planktonic organisms appear in the upper water column where daily exposure to UV-B irradiance is stronger than 5–10% of surface irradiance. Phytoplankton at depths where irradiance levels were stronger than 1% of incident UV-B and 15% of incident PAR developed maximal photoacclimation to strong sunlight. Visible light penetration at noon should become limiting at about 150 m for fish visual predation upon large zooplankton. Diurnal migration of small fish and large zooplankton sufficient to avoid visual predators would also prevent their exposure to strong UVR.

Long-term trends and possible impacts of future climate change

Our comparisons of nearly contiguous annual summer optical data since 1978 suggest that UV transparency changes on decadal and longer cycles, perhaps in response to a combination of rain and stratospheric ozone influencing the phytoplankton community, with heavy summer rain occasionally influencing the level of suspended mineral particles. We also cannot rule out the impact of trophic interactions and nutrient upwelling as additional factors influencing phytoplankton and UV transparency. The sparse attenuation estimates from proxy data between

1896 and 1978 fall within the range of the better-characterized period since 1978, and from these we infer that there has been no long-term change in UV transparency since 1896.

Acknowledgements We are grateful to John Morrow and Michael Holas of Biospherical Instruments, Inc. who provided profiling instruments for spectral radiance reflectance measurements in Figure 2 and for comparison measurements of UV irradiance and $K_{d,\lambda}$ in Figure 4. Valuable comments were also provided by Emmanuel Boss and two anonymous reviewers.

References

Ayoub, L. M., B. R. Hargreaves & D. P. Morris, 1996. UVR attenuation in lakes: Relative contribution of dissolved and particulate material. SPIE Ocean Optics XIII 2963: 338–343.

Bacon, C. R., J. V. Gardner, L. A. Mayer, M. W. Buktenica, P. Dartnell, D. W. Ramsey & J. E. Robinson, 2002. Morphology, volcanism, and mass wasting in Crater Lake, Oregon. Geological Society of America Bulletin 114(6): 675–692.

Baker, K. S. & R.C. Smith, 1979. Quasi-inherent characteristics of the diffuse attenuation coefficient for irradiance. Ocean Optics VI, SPIE 208: 60–63.

Barnard, G. P., 1938. The spectral sensitivity of selenium rectifier photoelectric cells. Proceedings of the Physical Society 51: 222–236.

Belzile, C., W. F. Vincent & M. Kumagai, 2002. Contribution of absorption and scattering to the attenuation of UV and photosynthetically available radiation in Lake Biwa. Limnology and Oceanography 47: 95–107.

Biospherical Instruments 1998. GUV Data Processing and Quality Control Procedures, C. Version 13, October 1998. Biospherical Instruments, Inc., San Diego, California, USA.

Bishop, J. K. B., 1986. The correction and suspended particulate matter calibration of SeaTech transmissometer data. Deep-Sea Research 33: 121–134.

Boss, E. S., R. W. Collier, G. L. Larson, K. Fennel & W. S. Pegau, 2007. Measurements of spectral optical properties and their relation to biogeochemical variables and processes in Crater Lake, Crater Lake National Park, OR. Hydrobiologia 574: 149–159.

Bricaud, A., A. Morel & L. Prieur, 1981. Absorption by dissolved organic matter in the sea (yellow substance) in the UV and visible domains. Limnology and Oceanography 26: 43–53.

Buktenica, M. W. & G. L. Larson, 1996. Ecology of Kokanee salmon and rainbow trout in Crater Lake, Oregon. Journal of Lake and Reservoir Management 12: 298–310.

Coblentz, W. W. & R. Stair, 1936. The evaluation of ultraviolet solar radiation of short wave-lengths. Proceedings of the National Academy of Science 22: 229–233.

Davies-Colley, R. J., 1988. Measuring water clarity with a black disk. Limnology and Oceanography 33: 616–623.

Dierssen, H. M. & R. C. Smith, 2000. Bio-optical properties and remote sensing ocean color algorithms for Antarctic Peninsula waters. Journal of Geophysical Research 105: 26301–26312.

Diller, J. S., 1897. Crater Lake, Oregon. National Geographic Magazine 8: 33–48 (cited in Larson et al., 1996a).

Eisner, L. B., M. S. Twardowski, T. J. Cowles & M. J. Perry, 2003. Resolving phytoplankton photoprotective:photosynthetic carotenoid ratios on fine scales using in situ special absorption measurements. Limnology and Oceanography 48: 632–646.

Fennel, K. & E. Boss, 2003. Subsurface maxima of phytoplankton and chlorophyll: Steady-state solutions from a simple model. Limnology and Oceanography 48: 1521–1534.

Goldman, C. R., D. T. Mason & J. E. Hobbie, 1967. Two antarctic desert lakes. Limnology and Oceanography 12: 295–310.

Gordon, H. R., 1989. Can the Lambert-Beer law be applied to the diffuse attenuation coefficient of ocean water? Limnology and Oceanography 34: 1389–1409.

Gordon, H. R., 1999. Contribution of Raman scattering to water-leaving radiance: A reexamination. Applied Optics 38: 3166–3174.

Haltrin, V. I., G. W. Kattawar & A. D. Weidman, 1997. Modeling of elastic and inelastic scattering effects in oceanic optics. SPIE 2963: 597–602.

Hargreaves, B. R., 2003. Water column optics and penetration of UVR. In Helbling E. W., & H. E. Zagarese (eds), UV Effects in Aquatic Organisms and Ecosystems, Comprehensive Series in Photochemical and Photobiological Sciences. Royal Society of Chemistry, Cambridge, UK: 59–105.

Helbling, E. W., V. Villafañe & O. Holm-Hansen, 1994. Effects of ultraviolet radiation on Antarctic marine phytoplankton photosynthesis with particular attention to the influence of mixing. In Weiler C. S. & P. A. Penhale (eds), Ultraviolet Radiation In Antarctica: Measurements and Biological Effects, Antarctic Research Series, Vol. 62. American Geophysical Union, Washington, DC: 207–227.

Hewes C. D., B. G. Mitchell, T. A.Moisan, M. Vernet & F. M. H. Reid, 1998.The phycobilin signatures of chloroplasts from three dinoflagellate species: A microanalytical study of Dinophysis caudata, D-fortii, and D-acuminata (Dinophysiales, Dinophyceae). Journal of Phycology 34: 945–951.

Højerslev, N. K., 1994. A history of early optical oceanographic instrument design in Scandinavia. In Spinrad R. W., K. L. Carder, M. J. Perry (eds), Ocean Optics. Oxford Univ. Press, New York: 118–147.

Højerslev, N. K. & T. Aarup, 2002. Optical measurements on the Louisiana shelf off the Mississippi River. Estuarine and Coastal Shelf Science 55: 599–611.

Jassby, A. D., C. R. Goldman, J. E. Reuter & R. C. Richards, 1999. Origins and scale dependence of temporal variability in the transparency of Lake Tahoe, California-Nevada. Limnology and Oceanography 44: 282–294.

Jerlov, N. G., 1950. Ultra-violet Radiation in the Sea. Nature 166: 111–112.

Johnson (=Jerlov), N. G., 1946. On anti-rachitic ultraviolet radiation in the sea. Medd. Oceanogr. Inst. Gothenburg. Ser. B. 3, 11 (cited in Højerslev 1994).

Kalff, J., 2002. Limnology: Inland Water Ecosystems. Prentice Hall: 592.

Kirk, J. T. O., 1994a. Optics of UV-B radiation in natural waters. Archiv für Hydrobiologie Beihefte Ergebnisse der Limnologie 43: 71–99.

Kirk, J. T. O., 1994b. Light & Photosynthesis in Aquatic Ecosystems, 2nd ed. Cambridge University Press, Cambridge.

Kirk, J. T. O., B. R. Hargreaves, D. P. Morris, R. Coffin, B. David, D. Frederickson, D. Karentz, D. Lean, M. Lesser, S. Madronich, J. H. Morrow, N. Nelson & N. Scully, 1994. Measurement of UV-B radiation in two freshwater lakes: an instrument intercomparison. Archiv für Hydrobiologie Beihefte Ergebnisse der Limnologie 43: 71–99.

Larson, D. W., 1972. Temperature, Transparency, and phytoplankton productivity in Crater Lake, Oregon. Limnology and Oceanography 17: 410–417.

Larson, D. W., 2002. Probing the Depths of Crater Lake. American Scientist 90: 64–71.

Larson, G. L., C. D. McIntire, M. Hurley & M. W. Bukenica, 1996a. Temperature, Water Chemistry, and Optical Properties of Crater Lake. Journal of Lake and Reservoir Management 12: 230–247.

Larson, G. L., C. D. McIntire, R. E. Truitt, M. W. Buktenica & E. Karnaugh-Thomas, 1996b. Zooplankton Assemblages in Crater Lake, Oregon, USA. Journal of Lake and Reservoir Management 12: 281–297.

Laurion, I., M. Ventura, J. Catalan, R. Psenner & R. Sommaruga, 2000. Attenuation of ultraviolet radiation in mountain lakes: factors controlling the among- and within-lake variability. Limnology and Oceanography 45: 1274–1288.

Laurion, I., F. Blouin, S. Roy, 2003. The quantitative filter technique for measuring phytoplankton absorption: Interference by MAAs in the UV waveband. Limnology and Oceanography: Methods 1: 1–9.

Leech, D. M., & S. Johnsen, 2003. Behavioral responses—UVR avoidance and vision. In Helbling E.W. & H. E. Zagarese (eds), UV Effects in Aquatic Organisms and Ecosystems, Comprehensive Series in Photochemical and Photobiological Sciences. Royal Society of Chemistry, Cambridge, UK: 455–484.

Lohrenz, S. E., 2000. A novel theoretical approach to correct for pathlength amplification and variable sampling loading in measurements of particulate spectral absorption by the quantitative filter technique. Journal of Plankton Research 22: 639–657.

MacIntyre, H. L., T. M. Kana, T. Anning & R. J. Geider, 2002. Photoacclimation of photosynthesis irradiance response curves and photosynthesis pigments in microalgae and cyanobacteria. Journal of Phycology 38: 17–38.

Madronich, S., 1993. UV radiation in the natural, perturbed atmosphere. In Tevini M. (eds), UVB Radiation and Ozone Depletion: Effects on Humans, Animals, Plants, Micro-organisms, and Materials. CRC Press, Boca Raton, Florida: 17–69.

Mantua, N. J., S. T. Hare, Y. Zhang, J. M. Wallace & R. C. Francis, 1997. A Pacific interdecadal climate oscillation with impacts on salmon production. Bulletin of the American Meteorological Society 78: 1069–1079.

Markager, S. & W. F. Vincent, 2000. Spectral light attenuation and the absorption of UV and blue light in natural waters. Limnology and Oceanography 45: 642–650.

McIntire, C. D., H. K. Phinney, G. L. Larson & M. Buktenica, 1994. Vertical distribution of a deep-water moss and associated epiphytes in Crater Lake, Oregon. Northwest Sciences 68: 11–21.

McIntire, C. D., G. L. Larson, R. E. Truitt & M. K Debacon, 1996. Taxonomic structure and productivity of phytoplankton assemblages in Crater Lake, Oregon. Lake and Reservoir Management 12: 259–280.

McIntire, C. D., G. L Larson & R. E. Truitt, 2007. Taxonomic composition and production dynamics of phytoplankton assemblages in Crater Lake, Oregon. Hydrobiologia 574: 179–204.

McManus, J., R. W. Collier & J. Dymond, 1993. Mixing processes in Crater Lake, Oregon. Journal of Geophysical Research 98: 18295–18307.

Mitchell, B.G., 1990. Algorithms for determining the absorption coefficient of aquatic particulates using the quantitative filter technique (QFT). Ocean Optics X, SPIE 1302: 137–148.

Morel, A., 1974. Optical Properties of pure water and pure sea water. In Jerlov N. G. & E. Steelmann Nielsen (eds), Optical Aspects of Oceanography. Academic Press, New York: 1–24.

Morel, A. & S. Maritorena, 2001. Bio-optical properties of oceanic waters: A reappraisal. Journal of Geophysical Research 106: 7163–7180.

Morel, A. & L. Prieur, 1977. Analysis of variations in ocean color. Limnology and Oceanography 22: 709–722.

Patterson, K. W., R. C. Smith & C. R. Booth, 1997. A method for removing a majority of the error in PUV attenuation coefficients due to spectral drift in response with depth in the water column. Ocean Optics XIII, SPIE 2963: 737–742.

Pegau, W. S., D. Gray & J. R. V. Zaneveld, 1997. Absorption and attenuation of visible and near-infrared light in water: dependence on temperature and salinity. Applied Optics 36: 6035–6046.

Pettit, E., 1936. On the color of Crater Lake water. Proceedings of the National Academy of Sciences of the United States of America 22: 139–146.

Pienitz, R. & W. F. Vincent, 2000. Effect of climate change relative to ozone depletion on UV exposure in subarctic lakes. Nature 404.6777: 484–487.

Pope, R. M. & E. S. Fry, 1997. Absorption spectrum (380–700 nm) of pure water. II. Integrating cavity measurements. Applied Optics 36: 8710–8723.

Preisendorfer, R. W., 1986. Secchi disk science: visual optics of natural waters. Limnology and Oceanography 31: 909–926.

Redmond, K. T., 1999. Crater Lake Evaporation and Climate Variability. Project report for subcontract No. B0023A-01.

Riis, T. & K. Sand-Jensen, 1997. Growth reconstruction and photosynthesis of aquatic mosses: influence of light, temperature and carbon dioxide at depth. Journal of Ecology 85: 359–372.

Sathyendranath, S, L. Lazzara & L. Prieur, 1987. Variations in the spectral values of specific absorption of phytoplankton. Limnology & Oceanography 32: 403–415.

Schindler, D. W. & P. J. Curtis, 1997. The role of DOC in protecting freshwaters subject to climatic warming and acidification from UV exposure. Biogeochemistry 36: 1–8.

Sharp, J. H., R. Benner, L. Bennett, C. A. Carson, R. Dow & S. E. Fitzwater, 1993. Re-evaluation of high temperature combustion and chemical oxidation measurements of dissolved organic carbon in seawater. Limnology and Oceanography 38: 1174–1782.

Smith, R. C. & J. E. Tyler, 1967. Optical Properties of Clear Natural Waters. Journal of the Optical Society of America 57: 589–595.

Smith, R. C., J. E. Tyler & C. R Goldman, 1973. Optical properties and color of Lake Tahoe and Crater Lake. Limnology and Oceanography 18: 189–199.

Smith, R. C., & J. E. Tyler, 1976. Transmission of solar radiation into natural waters. In Smith K. C. (eds), Photochemical and Photobiological Reviews, Vol. 1. Plenum Press, New York: 117–156.

Smith, R. C. & K. S. Baker, 1981. Optical Properties of the clearest natural waters (200–800 nm). Applied Optics 20: 177–184.

Sommaruga, R., 2001. The role of solar UV radiation in the ecology of alpine lakes. Journal of Photochemistry & Photobiology, B: Biology 62: 35–42.

Sosik, H. M., 1999. Storage of marine particulate samples for light-absorption measurements. Limnology and Oceanography 44: 1139–1141.

Tyler, J. E. & R. W Presiendorfer, 1962. Transmission of energy within the sea. In M.N. Hill (ed), The Sea, Vol. 1. Interscience, New York (cited in Tyler & Smith 1976): 397–451.

Tyler, J. E., 1965. In situ spectroscopy in ocean and lake waters. Journal of the Optical Society of America 55: 800–805.

Tyler, J. E., 1968. The Secchi Disc. Limnology and Oceanography 13: 1–6.

Tyler, J. E., & R. C. Smith, 1970. Measurements of Spectral Irradiance underwater. Gordon and Breach Science Publishers, New York.

Tartarotti, B., I. Laurion & R. Sommaruga, 2001. Large variability in the concentration of mycosporine-like amino acids among zooplankton from lakes located across an altitude gradient. Limnology and Oceanography 46: 1546–1552.

Urbach, E., K. L. Vergin, L. Young & A. Morse, 2001. Unusual bacterioplankton community structure in ultra-oligotrophic Crater Lake. Limnology and Oceanography 46: 557–572.

Utterback, C. L., L. D. Phifer & R. J. Robinson, 1942. Some Chemical, Planktonic, and Optical Characteristics of Crater Lake. Ecology 23: 97–103.

Vincent, W. F., R. Rae, I. Laurion, C. Howard-Williams & J. C. Priscu, 1998. Transparency of Antartctic ice-covered lakes to solar UV radiation. Limnology and Oceanography 43: 618–624.

Yentsch, C. S. & D.A. Phinney, 1989. A bridge between ocean optics and microbial ecology. Limnology and Oceanography 34: 1694–1705.

Zaneveld, J. R. & R. Bartz, 1984. Beam attenuation and absorption meters. SPIE Ocean Optics VII 486: 318–324.

Zaneveld, J. R. V. & W. S. Pegau, 2003. Robust underwater visibility parameter. Optics Express 11: 2997–3000.

Zaneveld, J. R. V., E. Boss, & A. Barnard, 2001. Influence of surface waves on measured and modeled irradiance profiles. Applied Optics 40: 1442–1449.

Hydrobiologia (2007) 574:141–148
DOI 10.1007/s10750-006-0349-z

CRATER LAKE, OREGON

Predicting Secchi disk depth from average beam attenuation in a deep, ultra-clear lake

Gary L. Larson · Robert L. Hoffman · Bruce R. Hargreaves · Robert W. Collier

© Springer Science+Business Media B.V. 2007

Abstract We addressed potential sources of error in estimating the water clarity of mountain lakes by investigating the use of beam transmissometer measurements to estimate Secchi disk depth. The optical properties Secchi disk depth (SD) and beam transmissometer attenuation (BA) were measured in Crater Lake (Crater Lake National Park, Oregon, USA) at a designated sampling station near the maximum depth of the lake. A standard 20 cm black and white disk was used to measure SD. The transmissometer light source had a nearly monochromatic wavelength of 660 nm and a path length of 25 cm. We created a SD prediction model by regression of the inverse SD of 13 measurements recorded on days when environmental conditions were acceptable for disk deployment with BA averaged over the same depth range as the measured SD. The relationship between inverse SD and averaged BA was significant and the average 95% confidence interval for predicted SD relative to the measured SD was ±1.6 m (range = –4.6 to 5.5 m) or ±5.0%. Eleven additional sample dates tested the accuracy of the predictive model. The average 95% confidence interval for these sample dates was ±0.7 m (range = –3.5 to 3.8 m) or ±2.2%. The 1996–2000 time-series means for measured and predicted SD varied by 0.1 m, and the medians varied by 0.5 m. The time-series mean annual measured and predicted SD's also varied little, with intra-annual differences between measured and predicted mean annual SD ranging from –2.1 to 0.1 m. The results demonstrated that this prediction model reliably estimated Secchi disk depths and can be used to significantly expand optical observations in an environment where the conditions for standardized SD deployments are limited.

Keywords Secchi disk depth · Beam attenuation · Transmissometer · Lakes

Guest Editors: Gary L. Larson, Robert Collier, and Mark W. Buktenica
Long-term Limnological Research and Monitoring at Crater Lake, Oregon

G. L. Larson · R. L. Hoffman (✉)
USGS Forest and Rangeland Ecosystem Science Center, 777 NW 9th Street, Suite 400, Corvallis, OR 97330, USA
e-mail: robert_hoffman@usgs.gov

B. R. Hargreaves
Department of Earth and Environmental Sciences, Lehigh University, 31 Williams Drive, Bethlehem, PA 18015, USA

R. W. Collier
College of Oceanic and Atmospheric Sciences, Oregon State University, 104 COAS Administration Bldg., Corvallis, OR 97331, USA

Introduction

Limnologists frequently use the Secchi disk to establish indices of lake water clarity (Preisendorfer, 1986). At Crater Lake (Crater Lake National Park, Oregon, USA), Secchi disk depth (SD) measurements have been an integral tool for measuring water clarity as part of a long-term program initiated in 1983 for monitoring water quality. Diller (1897) recorded the first Crater Lake water clarity measurement in 1896 using a white dinner plate. Subsequent Secchi disk depths were measured in 1913, 1937–1939, 1954, 1960, 1968–1969, and 1978–2004. SD measurements have been performed primarily June–September, but measurements have also been completed in March (once), January–April–October (twice each), and May (four times).

Clarity measurements obtained using a Secchi disk typically are the averages of descending and ascending disk readings. Although the Secchi disk is relatively easy to use, several sources of variation can influence Secchi disk depth measurements (Preisendorfer, 1986). The visual acuity of the individual observing the Secchi disk can be a primary source of error when measuring Secchi disk depth. In lakes with clarity measurements exceeding 20 m, for example, disk size becomes an important source of error because a standard 20 cm disk has an apparent diameter of 1 cm at a depth of 20 m and only 0.5 cm at 40 m (Larson & Buktenica, 1998). Even under ideal environmental conditions, the small size of the disk at depths greater than 25 m makes it difficult to determine the depth at which the disk disappears and, especially, the ascending depth of reappearance. Lake surface condition is another source of variation for Secchi disk clarity measurements. Larson & Buktenica (1998) provide evidence that rippled lake surface conditions reduce Secchi disk clarity readings by several meters in Crater Lake. Of even more concern, however, is that mountain lakes like Crater Lake can have extended periods of time when environmental conditions are unacceptable for Secchi disk deployment. Secchi disk transparency measurements, however, have been used to reliably predict the trophic state of natural lakes (Carlson, 1977; Canfield & Bachmann, 1981; Burns et al., 1999), and for determining, along with other water quality variables (e.g., primary productivity, nitrate, total phosphorus, and chlorophyll-*a*), potential change in lake trophic status (Goldman, 1988; Burns, 2001).

A beam transmissometer also has measured Crater Lake water clarity (Smith et al., 1973; Larson et al., 1996). Measurements related to the Crater Lake long-term water quality monitoring program began in 1987. The beam transmissometer measures light (or beam) attenuation in water due to absorption and scattering (Davies-Cooley et al., 1993). Attenuation can be related to the presence and/or concentrations in the water column of phytoplankton, chlorophyll, colored dissolved organic matter (CDOM), particulate organic carbon, and other types of organic and inorganic particulate matter (Topliss et al., 1989; Hall et al., 1999; Binder & DuRand, 2002; Hodges & Rudnick, 2004; Gallegos, 2005). Researchers also have demonstrated that beam attenuation as measured by a beam transmissometer can be used to estimate physiological variables associated with primary productivity (Behrenfeld & Boss, 2003), for measuring turbidity (Udy et al., 2005), and for elucidating cycles of pico- and ultraphytoplankton cellular abundance (Binder & DuRand, 2002). A beam transmissometer can be deployed at any time of the day or night and when lake surface conditions are less than ideal for Secchi disk deployment. Also, beam transmissometer clarity measurements are not biased because: (1) the instrument can be regularly calibrated for accuracy using a standard instrument calibration protocol, and (2) light transmission measurements do not require a subjective human observer.

We address, in this article, the possibility of reducing sources of error in measuring lake water clarity using a Secchi disk by investigating the use of beam transmissometer readings for estimating Secchi disk clarity. The objectives of this study are to: (1) investigate how Secchi disk and beam transmissometer water clarity measurements for the Crater Lake long-term water quality monitoring dataset compare, and (2) create a prediction model for using beam transmissometer water clarity measurements for estimating Secchi disk depth at Crater Lake and other montane lakes.

Methods

Crater Lake (Crater Lake National Park) is located along the crest of the Cascade Range in southern Oregon. The lake covers the floor of a caldera that formed about 6,850 years ago (Bacon et al., 2002). At a reference elevation of 1,882 m, the lake has a surface area of 53.2 km^2 and a maximum depth of 594 m (Byrne, 1965; Phillips & Van Denburg, 1968; Bacon et al., 2002). The lake has exceptionally blue waters, with Secchi disk clarity readings averaging 31.3 ± 5.5 m, June–September, 1978–2002 (unpublished data).

Secchi disk depth (SD) measurements and light transmission profiles for 13 sample dates (1989–2000; Table 1) were used to develop and calibrate a SD prediction model for Crater Lake. The measurements were made at Station 13 near the maximum depth of the lake on days when conditions were relatively optimal for Secchi disk deployment (i.e., between 1,000 and 1,400 h (Pacific Standard Time) under clear or high-haze sky and calm to slightly rippled (< 1.5 cm) lake surface conditions (Table 1)). A standard 20-cm black and white disk attached to a metered cable measured SD, calculated as the average of two to three measurements of descending and ascending SD recorded for each sample date. SD measurements were recorded at tenth-meter intervals (see Table 1). Although Larson & Buktenica (1998) determined that the average variation in SD measurements among trained observers was limited (i.e., < 1 m or < 5%), we decided to use the measurements of one observer (i.e., Mark Buktenica) for consistency. The average difference between the minimum and maximum measured SD for the 13 sample dates was 1.6 m, and individual sample date differences ranged from 0.2 to 4.5 m. A Seatech beam transmissometer mounted vertically and powered by a Seabird Seacat Profiler (model SBE19) generated light transmission profiles on the same day as SD measurements. The transmissometer light source had a nearly monochromatic wavelength of 660 nm with a path length of 25 cm. The unit, attached to a metered cable, was lowered into the lake at a rate of approximately 0.5 m s^{-1}, and two measurements of percent light transmission were recorded per second. Percent light transmission measurements in the upper 1.5 m of the water column were typically unreliable, so all light transmission profiles began at a depth of 2 m. Light transmission profiles were recorded during midday, typically beginning between 1,300 and 1,400 h (Pacific Standard Time).

Light transmission data for each of the 13-sample dates included percent transmittance and beam attenuation coefficient (c660). Seasoft Version 4.234 software (Seabird Electronics, Inc., 1808 136th Place NE, Bellevue, WA 98005) calculated these data for each sample date and the Seasoft Binavg Program averaged c660 at

Table 1 Data for 13 sampling dates used to calibrate the Secchi disk depth (SD) prediction model

Date	SD_{meas}	SD_{round}	$1/SD_{round}$	Zbins	cp660avg	SD_{pred}	Diff	SC
5/12/1989	33.1	33.0	0.0303	63	0.0725	30.8	–2.3	SR
7/06/1989	34.4	34.5	0.0290	66	0.0484	35.2	+0.8	SR
7/16/1990	29.0	29.0	0.0345	65	0.0516	34.6	+5.6	Calm
9/16/1991	26.1	26.0	0.0385	49	0.0948	27.7	+1.6	Calm
9/26/1991	29.5	29.5	0.0339	56	0.0845	29.0	–0.5	Calm
6/03/1992	33.7	33.5	0.0299	64	0.0839	29.1	–4.6	Calm
9/09/1993	30.9	31.0	0.0323	59	0.0559	33.7	+2.8	Calm
9/12/1995	24.7	24.5	0.0408	46	0.1013	26.9	+2.2	Calm
6/25/1997	42.2	42.0	0.0238	81	0.0369	37.8	–4.4	SR
8/04/1997	35.3	35.5	0.0282	68	0.0619	32.6	–2.7	SR
7/21/1998	36.6	36.5	0.0274	70	0.0461	35.7	–0.9	SR
7/18/2000	30.4	30.5	0.0328	58	0.0723	30.9	+0.5	Calm
8/15/2000	28.4	28.5	0.0351	54	0.1012	26.9	–1.5	Calm

SD_{meas} = measured SD; SD_{round} = SD_{meas} rounded to nearest 0.5 m; Zbins = number of half-meter bins from 2 m to SD_{round}; cp660avg = particulate red beam attenuation coefficient averaged over the depth range of SD_{round}; SD_{pred} = predicted SD; Diff = $SD_{pred} - SD_{meas}$; SC = lake surface conditions; SR = slight ripple (< 1.5 cm) half-meter intervals. SD measurements were rounded to the nearest 0.5 m (SD_{round}; see Table 1) to conform to the average c660 half-meter depth intervals. A particulate c660 (i.e., cp660avg) was then calculated at each half-meter interval by subtracting from each c660 the constants for absorption (i.e., aw660) and scattering ($bw660_0$) coefficients for pure water. The constants were derived from laboratory measurements (i.e., aw660 = 0.410 (Pope & Fry, 1997); bw660 = 0.0007 (Morel, 1974)), and aw660 was adjusted for temperature dependence (i.e., aw660 = 0.410 – 0.002 = 0.408; Pegau et al., 1997).

The SD prediction model was calibrated using steps 1–6 in Table 2. Predicted Secchi disk depths (SD_{pred}) for the 13 model calibration sample dates were calculated as part of the regression analysis of the relationship between $1/SD_{round}$ and average cp660 (i.e., cp660avg) to $1/SD_{round}$ (Table 1). Once the prediction model was calibrated, 11 additional sample dates (Table 3) were used to evaluate the relative accuracy of the prediction model. Seven of the 11 measurements were performed when lake surface conditions were less than optimal for Secchi disk deployment (i.e., surface conditions ranged from 2.5–5 cm ripples with pollen and dust on the lake water surface to 15 cm tour-boat-generated rolling waves (Table 3)). Measurements for these sample dates were used as part of the model validation to evaluate how well the prediction model would predict SD on days when lake surface condition was less than ideal for measuring water clarity using a Secchi disk. The same observer (i.e., Mark Buktenica) also made the 11 measurements performed at Station 13. The average difference between the minimum and maximum SD for these measurements was 2.2 m, and individual sample date differences ranged from 0.4 to 8.4 m. SD_{pred} was calculated using Steps 7–9 in Table 2. The SD depth range of 2–31.5 m was used because the 31.5 m maximum depth was the rounded summer (i.e., June–September) mean depth for SD measurements in the Crater Lake database, 1978–2002. The values for slope and intercept used in the SD_{pred} calculation formula (Step 9, Table 2) were derived from the model calibration regression analysis (Fig. 1). It should be noted that the maximum depth and the values for slope and intercept will be different for each lake for which SD is to be predicted. SD measurements and predicted SD for light transmission data collected 1996–2000 were used to examine the time-series pattern of measured SD and SD_{pred}, and mean annual (i.e., June–September, 1996–2000) measured and predicted SD were calculated using these data.

Table 2 Steps for calibrating the Secchi disk depth prediction model, and for calculating predicted Secchi disk depth

Step	Procedure for each sample date
Model calibration	
1.	Using a bin-averaging program (e.g., Seasoft Version 4.234) calculate a beam attenuation coefficient (c660) at half-meter intervals from 2 m to the nearest 0.5 m of the measured Secchi disk depth (i.e., this depth is identified as SD_{round} in Table 1, column 3);
2.	Calculate a particulate beam attenuation coefficient (cp660) at each half-meter interval by subtracting the constants aw660 = 0.408 and bw660 = 0.0007 from each half-meter c660;
3.	Calculate the average cp660 (cp660avg; Table 1, column 6) over the depth range of SD_{round} by dividing the cumulative cp660 by the total number of half-meter depth bins (Table 1, column 5) from 2 m to SD_{round};
4.	Calculate inverse SD_{round} (Table 1, column 4) as $1/SD_{round}$;
5.	Perform a simple linear regression (e.g., NCSS 2000; Hintze 1998) of $1/SD_{round}$ (dependent variable) and cp660avg (independent variable) for model calibration sample dates (Fig. 1).
6.	Convert the predicted $1/SD_{round}$ value calculated for each sample date during the regression analysis to predicted SD and subtract from the measured SD.
Calculation of predicted Secchi disk depth (SD_{pred})	
7.	Calculate a c660 and cp660 at half-meter intervals from 2 to 31.5 m;
8.	Calculate an average cp660 over this depth range;
9.	Calculate SD_{pred} using the formula: $SD_{pred} = 1/(\text{cp660avg} \times \text{slope} + \text{intercept})$, with values for slope and intercept derived from the model calibration regression.

Table 3 Data for 11 sampling dates used to evaluate the accuracy of the Secchi disk depth (SD) prediction model

Sample date	cp660avg	SD_{pred}	SD_{meas}	Diff	SC
7/11/1994	0.0299	39.5	36.7	+2.8	SR
7/13/1994	0.0387	37.4	36.0	+1.4	R, PD
8/16/1994	0.0607	32.8	36.3	–3.5	W
8/23/1994	0.0299	39.5	41.3	–1.8	R
9/16/1994	0.0340	38.5	35.9	+2.6	R
8/30/1995	0.0933	27.9	24.3	+3.6	R
7/08/1996	0.0538	34.1	34.4	+0.3	R
7/13/1999	0.0436	36.2	37.6	–1.4	W
8/03/1999	0.0767	30.2	33.6	–3.4	Calm
9/08/1999	0.1750	20.2	21.6	–1.4	SR
6/20/2000	0.0730	30.8	32.4	–1.6	SR

There were 60 half-meter depth bins per sample date (i.e., 2–31.5 m). Cp660avg = average red beam attenuation coefficient; SD_{pred} = predicted SD; SD_{meas} = measured SD; Diff = $SD_{pred} - SD_{meas}$; SC = lake surface conditions; SR = slight ripple (<1 5 cm); R = ripple (2.5–5.0 cm); W = boat generated waves (15 cm); PD = pollen and dust

Results

The relationship between inverse Secchi disk depth ($1/SD_{round}$) and average particulate beam attenuation coefficient (cp660avg) to SD_{round} for the 13 model calibration sample dates (Table 1, Fig. 1) was significant ($P = 0.0015$). The r^2 was 0.62, the slope was 0.1671, the intercept was 0.0203, and the Pearson correlation was 0.79 (Fig. 1). The predicted SD (SD_{pred}) calculated as part of the regression analysis for each of the 13 model calibration sample dates was compared to the measured SD (SD_{meas}) for each date (Table 1). The range of the differences between SD_{pred} and SD_{meas} was –4.6–5.5 m, and SD_{pred} was estimated with an average 95% confidence interval of ±1.6 m or ±5.0%. The SD_{pred} lower 95% confidence intervals ranged from –1.7 to –4.3 m below SD_{pred} and the SD_{pred} upper 95% confidence intervals ranged from 1.9 to 5.6 m above SD_{pred}.

The relationship of $1/SD_{round}$ with cp660avg indicated that cp660avg could be a useful proxy for predicting SD in Crater Lake. When the accuracy of the prediction model was evaluated using the 11 additional sample dates (Table 3),

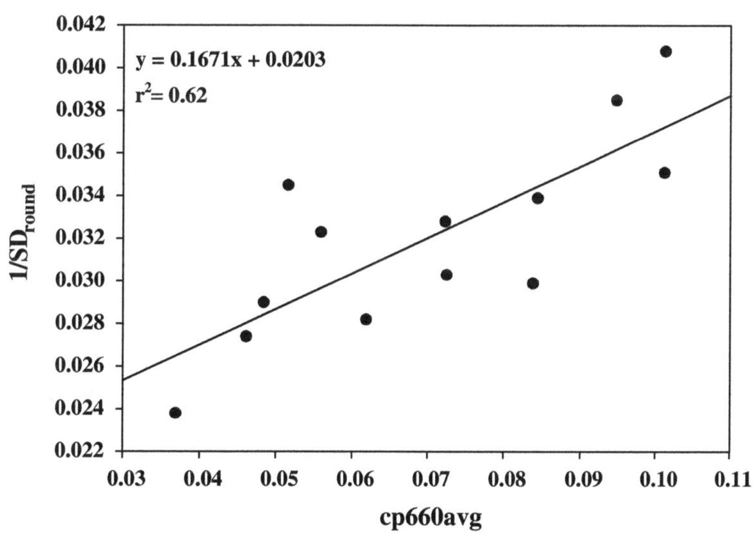

Fig. 1 Linear regression of $1/SD_{round}$ and cp660avg for the 13 sample dates used to calibrate the prediction model

SD_{pred} as compared to SD_{meas} was estimated with an average 95% confidence interval of ±0.7 m or ±2.2%. The range of the differences between SD_{pred} and SD_{meas} was –3.5–3.6 m. Four Secchi disk depths were over-estimated and seven were under-estimated (Table 3, Fig. 2).

The 1996–2000 time-series of SD_{meas} and SD_{pred} (Fig. 3) showed that the variance among the SD_{meas} and SD_{pred} points in the time-series was small. The time-series means for measured and predicted SD varied by 0.1 m and the medians varied by 0.5 m (SD_{meas}, n = 25, mean =

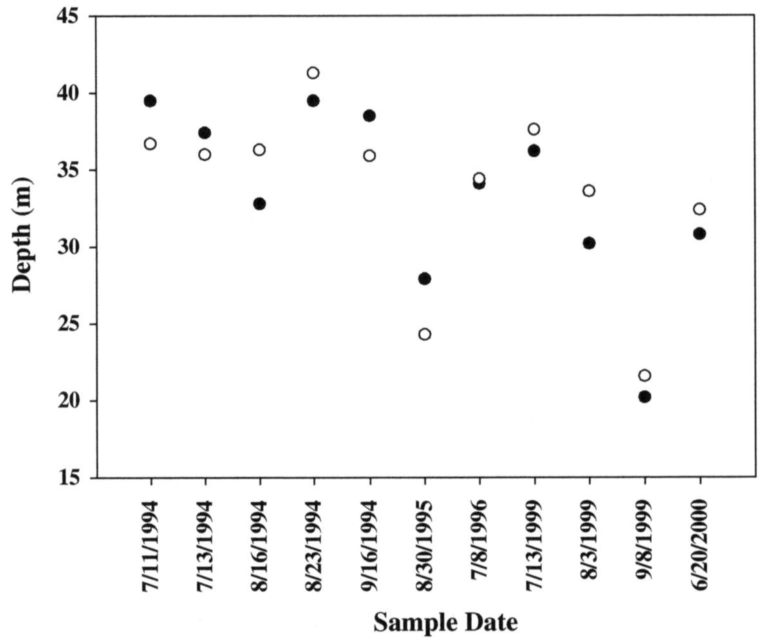

Fig. 2 Measured (○) and predicted (•) Secchi disk depths for the 11 sample dates used to evaluate the prediction model

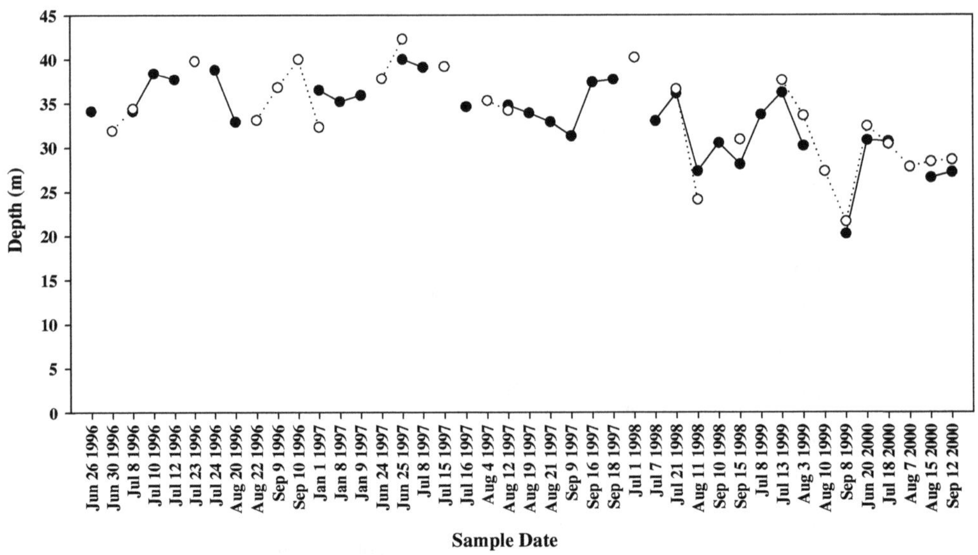

Fig. 3 Measured (○) and predicted (•) Secchi disk depths, 1996–2000. The time-series shows that the variance among the measured and predicted depths is small, and demonstrates the utility of using average beam attenuation to predict Secchi disk depth

Table 4 Mean annual measured and predicted SD for the months June–September, 1996–2000

Year	SD-type	*n*	Mean	Stdev	Range
1996	Measured	6	36.0	3.4	31.9–40.0
	Predicted	6	36.0	2.6	32.9–38.8
1997	Measured	5	37.8	3.2	34.2–42.3
	Predicted	9	35.7	2.9	31.3–40.0
1998	Measured	4	32.9	7.0	24.1–40.2
	Predicted	5	31.0	3.6	27.3–36.1
1999	Measured	4	30.0	7.0	21.6–37.6
	Predicted	4	30.1	7.0	20.2–36.2
2000	Measured	5	29.5	1.9	27.8–32.4
	Predicted	4	28.8	2.2	26.6–30.8

Stdev = standard deviation. The unit for Mean, Stdev and Range is meters

33.5 ± 5.3 m, median = 33.6 m, range = 21.6–42.3 m; SD_{pred}, $n = 31$, mean = 33.4 ± 4.4 m, median = 34.1 m, range = 20.2–40.0 m). The time-series mean annual measured and predicted SD's (Table 4) also varied little. The range of the differences in mean annual SD_{meas} and SD_{pred} was –2.1 to 0.1 m.

Discussion

These results demonstrate that a commercial transmissometer that emits a red (660 nm) beam attenuation signal can be used to estimate SD when environmental conditions are unacceptable for Secchi disk deployment. The signal is responsive to water column factors that also can partly affect SD as measured by human observation (Wetzel, 1983; Baker & Lavelle, 1984; Preisendorfer, 1986; Spinard et al., 1989). These factors include: (1) the scattering (which obscures an image) and absorption (which dims an image) of incident light caused primarily by the distribution and concentration of abiotic and biotic particles; (2) dissolved substances (e.g., CDOM) suspended in the water column; and (3) the absorptive capacity of water molecules.

Environmental conditions and human and instrument error can account for discrepancies between SD measured by human observation and SD predicted using averaged beam attenuation. Rough lake surface conditions, incorrect sun-angle, time-of-day, as well as the difficulty of observing a 20-cm Secchi disk at the water clarity depths that occur in Crater Lake (i.e., 21.6–42.2 m, this study) affect the ability of a human observer to identify accurately the depths of disk disappearance and reappearance. Instrument drift from acceptable calibration values can cause inaccurate beam attenuation measurements. Unlike SD, observer bias or unacceptable above water-surface environmental conditions that affect human estimation of SD do not affect the beam attenuation signal. However, the Secchi disk and transmissometer should be deployed at the same time since small changes in the types and densities of particles in the water column can affect the measurement of water clarity (Pilgram, 1984; Hojerslev, 1986) using either technique, and decrease their relative comparability.

The SD prediction model described in this article offers aquatic resource managers and investigators the option of using electronic measurement of water clarity instead of human measurement using a Secchi disk. This option would remove the element of human bias from water clarity depth estimates. It would also allow for the measurement of water clarity depth at any time during day and night, and make possible the automated measurement of water clarity depth from a mooring (i.e., if the instrument can be effectively and routinely calibrated and fouling of the mooring can be kept to a minimum). We hypothesize that this SD prediction model could be used at any lentic site. However, since the model was developed using SD and beam transmissometer measurements from ultraoligotrophic Crater Lake, it is probably most applicable to other oligotrophic and ultraoligotrophic montane lakes. Managers and investigators interested in using this SD prediction model will need to create for the lake or lakes of interest, a database of historical SD and beam transmissometer measurements obtained during environmental conditions appropriate for measuring SD. The relationship between $1/SD_{round}$ and cp660avg and a SD prediction model can be determined and developed using these data and the steps outlined in Table 2. SD can be estimated from measurements of cp660avg once this has been completed.

Acknowledgements We thank Mark Buktenica and Scott Girdner for their assistance in the field and reviewing the manuscript. Special thanks to C. David McIntire and Emmanuel Boss for discussing various aspects of the project and reviewing the manuscript. The project was funded by Crater Lake National Park and the USGS Forest and Rangeland Ecosystem Science Center.

References

Bacon, C. R., J. V. Gardner, L. A. Mayer, M. W. Buktenica, P. Dartnell, D. W. Ramsey & J. E. Robinson, 2002. Morphology, volcanism, and mass wasting in Crater Lake, Oregon. Geological Society of America Bulletin 114: 675–692.

Baker, E. T. & J. W. Lavelle, 1984. The effect of particle size on the light attenuation coefficient of natural suspensions. Journal of Geophysical Research 89: 8197–8203.

Behrenfeld, M. J. & E. Boss, 2003. The beam attenuation to chlorophyll ratio: an optical index of phytoplankton physiology in the surface ocean? Deep Sea Research Part I: Oceanographic Research Papers 50: 1537–1549.

Binder, B. J. & M. D. DuRand, 2002. Diel cycles in surface waters of the equatorial Pacific. Deep Sea Research Part II: Topical Studies in Oceanography 49: 2601–2617.

Burns, N. M., 2001. Trends in the ecosystem structure and water quality of Crater Lake, Oregon, from 1985 to 2000. Preliminary Report prepared for the United States Geological Survey. Lake Consultancy Report 2001/1, 31 p.

Burns, N. M., J. C. Rutherford & J. S. Clayton, 1999. A monitoring and classification system for New Zealand lakes and reservoirs. Lake and Reservoir Management 15: 255–271.

Byrne, J. V., 1965. Morphology of Crater lake. Limnology and Oceanography 10: 462–465.

Canfield, D. E. & R. W. Bachmann, 1981. Prediction of total phosphorus concentrations, chlorophyll *a*, and Secchi depths in natural and artificial lakes. Canadian Journal of Fisheries and Aquatic Sciences 38: 414–423.

Carlson, R. E., 1977. A trophic state index for lakes. Limnology and Oceanography 22: 361–369.

Davies-Colley, R. J., W. N. Vant & D. G. Smith, 1993. Colour and Clarity of Natural Waters, science and management of optical water quality. Ellis Horwood, New York, 310 p.

Diller, J. S., 1897. Crater Lake, Oregon. National Geographic Magazine 8: 33–48.

Gallegos, C. L., 2005. Optical water quality of a blackwater river estuary: the Lower St. Johns River, Florida, USA. Estuarine, Coastal and Shelf Science 63: 57–72.

Goldman, C. R., 1988. Primary productivity, nutrients, and transparency during the early onset of eutrophication in ultra-oligotrophic Lake Tahoe, California-Nevada. Limnology and Oceanography 33: 1321–1333.

Hall I. R., S. Schmidt, I. N. McCave & J. L. Reyss, 1999. Particulate matter distribution and $^{234}Th/^{238}U$ disequilibrium along the Northern Iberian Margin: implications for particulate organic carbon export. Deep Sea Research Part I: Oceanographic Research Papers 47: 557–582.

Hintze, J. L., 1998. NCSS 2000 Statistical System for Windows. Kaysville, Utah.

Hodges, B. A. & D. L. Rudnick, 2004. Simple models of steady deep maxima in chlorophyll and biomass. Deep Sea Research Part I 51: 999–1015.

Højerslev, N. K., 1986. Visibility of the sea with special reference to the Secchi disc. SPIE Ocean Optics VIII 637: 294–305.

Larson, G. L. & M. W. Buktenica, 1998. Variability of Secchi disk readings in an exceptionally clear and deep caldera lake. Archiv fur Hydrobiologie 141: 377–388.

Larson, G. L., C. D. McIntire, C. D. Hurley & M. W. Buktenica, 1996. Temperature, water chemistry, and optical properties of Crater Lake. Lake and Reservoir Management 12: 230–247.

Morel, A., 1974. Optical properties of pure water and pure sea water. In Jerlov N. G. & E. S. Nielsen (eds), Optical Aspects of Oceanography, Academic Press, New York, 1–24.

Pegau, W. S., D. Gray & J. R. V. Zaneveld, 1997. Absorption and attenuation of visible and near-infrared light in water: dependence on temperature and salinity. Applied Optics 36: 6035–6046.

Phillips, K. N. & A. S. Van Denburg, 1968. Hydrology of Crater, East, and Davis Lakes, Oregon. U.S. Geological Survey, Water Supply Paper 1859-E.

Pilgram, D. A., 1984. The Secchi disk in principle and in use. The Hydrograph Journal 33: 25–30.

Pope, R. M. & E. S. Fry, 1997. Absorption spectrum (380–700 nm) of pure water. II. Integrating cavity measurements. Applied Optics 36: 8710–8723.

Preisendorfer, R. W., 1986. Secchi disk science: visual optics of natural waters. Limnology and Oceanography 31: 909–926.

Smith, R. C., J. E. Tyler & C. R. Goldman, 1973. Optical properties and color of Lake Tahoe and Crater Lake. Limnology and Oceanography 18: 189–199.

Spinard, R. W., C. M. Yentsch, J. Brown, Q. Dortch, E. Haugen, N. Revelante & L. Shapiro, 1989. The response of beam attenuation to heterotrophic growth in a natural population of plankton. Limnology and Oceanography 34: 1601–1605.

Topliss, B. J., J. R. Miller & E. P. W. Horne, 1989. Ocean optical measurements- II. Statistical analysis of data from Canadian eastern Arctic waters. Continental Shelf Research 9: 133–152.

Udy, J., M. Gall, B. Longstaff, K. Moore, C. Roelfsema, D. R. Spooner & S. Albert, 2005. Water quality: a combined approach to investigate gradients of change in the Great Barrier Reef, Australia. Marine Pollution Bulletin 51: 224–238.

Wetzel, R. G., 1983. Limnology. 2nd edn. Saunders College Publishing, PA, 767 p.

Hydrobiologia (2007) 574:149–159
DOI 10.1007/s10750-006-2609-3

Measurements of spectral optical properties and their relation to biogeochemical variables and processes in Crater Lake, Crater Lake National Park, OR

Emmanuel S. Boss · Robert Collier · Gary Larson · Katja Fennel · W. S. Pegau

© Springer Science+Business Media B.V. 2007

Abstract Spectral inherent optical properties (IOPs) have been measured at Crater Lake, OR, an extremely clear sub-alpine lake. Indeed Pure water IOPs are major contributors to the total IOPs, and thus to the color of the lake. Variations in the spatial distribution of IOPs were observed in June and September 2001, and reflect biogeochemical processes in the lake. Absorption by colored dissolved organic material increases with depth and between June and September in the upper 300 m. This pattern is consistent with a net release of dissolved organic materials from primary and secondary production through the summer and its photo-oxidation near the surface. Waters fed by a tributary near the lake's rim exhibited low levels of absorption by dissolved organic materials. Scattering is mostly dominated by organic particulate material, though inorganic material is found to enter the lake from the rim following a rain storm. Several similarities to oceanic oligotrophic regions are observed: (a) The Beam attenuation correlates well with particulate organic material (POM) and the relationship is similar to that observed in the open ocean. (b) The specific absorption of colored dissolved organic material has a value similar to that of open ocean humic material. (c) The distribution of chlorophyll with depth does not follow the distribution of particulate organic material due to photo-acclimation resulting in a subsurface pigment maximum located about 50 m below the POM maximum.

Keywords Crater Lake · Optics · Biogeochemistry · Backscattering coefficient

Guest Editors: Gray L. Larson, Robert Collier, and Mark W. Buktenica
Long-term Limnological Research and Monitoring at Crater Lake, Oregon.

E. S. Boss (✉)
University of Maine, 5741 Libby Hall, Orono, ME 04469, USA
e-mail: emmanuel.boss@maine.edu

R. Collier · W. S. Pegau
Oregon State University, COAS, 104 Ocean. Admin. Bldg., Corvallis, OR 97331, USA

G. Larson
USGS Forest and Rangeland Ecosystem Science Center, 3200 Jefferson Way, Corvallis, OR 97331, USA

K. Fennel
IMCS, Rutgers University, 71 Dudley Road, New Brunswick, NJ 08901, USA

Introduction

Novel commercial in-situ spectral optical instrumentation, developed in the past decade, has opened oceans and lakes to exploration of biogeochemical processes at sub-1 m scales, scales, which have not been accessible previously. From a data poor field, with only a handful of absorption spectra measured per water column, hydrological optics has become a data rich field with spectra generated at sub-Hz and thus at sub-1 m resolution. Until the recent development of a commercial in-situ spectral absorption and attenuation sensor (Moore et al., 1997) and backscattering sensors (Maffione & Dana, 1997; Zaneveld et al., 2003), absorption and backscattering were not measured in high resolution, routinely, and in-situ.

The absorption and backscattering coefficients are important inherent optical properties, in particular since they are tightly linked to the color of a water body (e.g., Gordon et al., 1988). The link between the color of a water body and its inherent optical properties (properties not affected by the illumination conditions such as absorption, scattering, and attenuation, hereafter IOP) is provided by the radiative-transfer equation (e.g., Mobley, 1994); knowledge of the IOP and the illumination conditions is sufficient to predict the color of the lake that will be perceived by an observer.

IOP are routinely inverted to provide information about the concentration and nature of the particulates and dissolved substances in aquatic environments. Inversions of IOP have been used to estimate the volume of total suspended matter from beam attenuation (Spinrad & Zaneveld, 1982), particulate organic matter concentration from beam attenuation (Bishop, 1999), dissolved organic matter concentration from absorption by dissolved material (e.g., review by Blough & Green, 1995), chlorophyll concentration from fluorescence or particulate absorption, particulate size distribution from spectral particulate beam attenuation (Boss et al., 2001a), and bulk particulate composition from the particulate backscattering ratio (Twardowski et al., 2001; Boss et al., 2004).

Measurements of IOP in lakes have mostly been limited to single-beam transmissometers and land-based spectrophotometry (e.g., review by Bukata et al., 1995). In this paper we present data of spectral IOP measured in-situ in Crater Lake, OR.

Crater Lake is a sub-alpine lake (Altitude 1,883 m), 589 m deep (deepest in the US), and enclosed by a volcano's caldera with a radius of nearly 3 km at the lake's surface (Klimasauskas et al., 2002). Crater Lake has been a natural laboratory for hydrological optics for nearly 70 years. Pettit (1936) performed measurements of spectral backscattering there and found them to be comparable to distilled water. Smith & Baker (1981) used it as part of their dataset to determine the absorption of the clearest natural waters. In this paper we present the first *spectral* IOP measurements in Crater Lake. The clarity of Crater Lake provides a challenge to the novel spectral IOP instrumentation, since the signal is very weak. If this instrumentation is found to work satisfactorily at Crater Lake, it is likely to perform well in most natural conditions (assuming that the path-length is reduced in very turbid waters).

Material and methods

Sampling

Data were collected during two sampling periods. In June 2001 we sampled following a rain storm. We sampled at two stations in the lake. Sta. 13, over the deepest area of the lake (42°56′ N, 122°06′ W), was sampled on two consecutive days (June 28–29) down to 300 m depth (a subset of these data was analyzed in Fennel & Boss, 2003). On June 28 we also sampled a shallow (6 m) station where a turbid whitish stream of water flowed into the lake from the caldera walls. We returned to Sta. 13 on September 19th and sampled to within 20 m of the lake's bottom (596 m).

Data collection and method of analysis

Spectral absorption and attenuation, hydrographic properties and spectral backscattering were measured on a winched package with a single WETLabs' ac-9, a SeaBird SBE25 and a HobiLabs Hydroscat-6 (HS-6), respectively. Additionally, a WETLabs' Eco-VSF was used to measure backscattering in September.

Chlorophyll fluorescence was measured with a WETLabs WetStar. At each station two repeated casts were taken, one of which had 0.2 μm filters (Gelman Suporcap 100) one attached to each intakes of the absorption, and attenuation tubes of the ac-9 for the measurements of CDM absorption (Colored Dissolved Material, operationally material that goes through a 0.2 μm filter). In September we also used a 1 μm filter (Whatman Capsule) at each intake of the ac-9. New filters were used for each field campaign.

Data were binned to 1 m bins by computing the median of the properties in that bin. Particulate properties were computed by subtracting the binned measurement in a vertical profile of spectral absorption and attenuation with the 0.2 μm pre-filter from a profile performed without a filter. Some errors may be caused by the temporal departure between the two measurements (approximately 30 min), given the presence of internal waves confirmed in temperature profiles. However, these departures are likely to result in small uncertainties in the derived particulate absorption and attenuation due to the slowly varying profile of CDM with depth (see below). Note that differences between two successive measurements, with and without filter, result in a particulate profile that is independent of the instrument calibration (assuming we sample the same waters). Thus the likely errors of the measurements of absorption and attenuation are limited by the precision of the instrument ($\pm 0.001\ m^{-1}$) and the environmental variability and not the accuracy of the calibration (performed with pure water before and after the cruise at the author's lab following the method of Twardowski et al., 1999). Note that an additional uncertainty may be caused by changes in filter efficiency with use. Consistency of CDM values between consecutive profiles (not shown) suggested that this was not a significant problem during the deployments.

Backscattering by particles with the HS-6 (measures scattering at one angle, centered about a scattering angle of 140°) was computed using the method outlined in Boss & Pegau (2001). The volume scattering function measured (of water and particles) was multiplied by a factor of 1.12 to account for a correction in the calibration factor due to Spectralon reflectivity that has been detected in Dec. 2002 (D. Dana, personal communication, 2004). Application of this factor resulted in an improved agreement between the different ways of estimating the backscattering coefficient than achieved in Boss et al. (2004) (see below). The Eco-VSF (measures scattering at three angles in the back direction, centered at 100, 125, and 150°) was processed following Zaneveld et al. (2003). We had problems with two wavelengths of the HS-6; the 555-channel was observed to shift values within casts in a manner uncorrelated with the other wavelengths and to have the largest standard deviation during calibration. The 676 nm channel was highly correlated with chlorophyll concentration (probably due to its recording of some chlorophyll fluorescence excited at 676 and emitted at 681 nm), unlike the remaining four channels. We therefore present only four channels (440, 488, 530, 620 nm) of the HS-6 data (Fig. 1).

It is interesting to note that the contribution of molecular backscattering by pure water to the total backscattering coefficient in our measurements (the sum of water and particulate backscattering) varies from nearly 60% to 30%

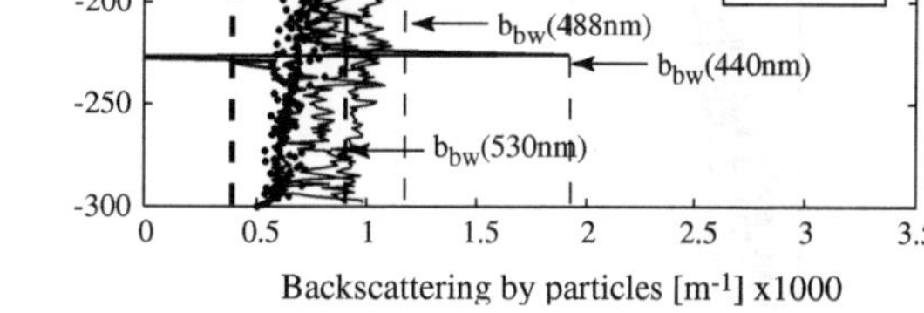

Fig. 1 Vertical distribution of the particulate backscattering coefficient obtained at four wavelengths (HS-6) and at a single wavelength (Eco-VSF) on Sep. 19, 2001 at St. 13. The straight lines, denotes the values of backscattering by pure water based on Morel (1974). The regression between the backscattering coefficients at 530 nm between the measurements by the two backscattering instruments is $b_{bp}(530, \text{EcoVSF}) = 0.97 b_{bp}(530, \text{HS6})$, and the correlation coefficient between them is 0.97. The spike in the HS-6 (530 nm) could not be explained

between blue and red wavelengths at depth (Fig. 1). In the calculation of particulate backscattering the contribution from water was subtracted out of the total backscattering assuming the backscattering by water to be constant and equal to the values of Morel (1974). The remaining backscattering is attributed to particles as it is customary, to assume a negligible contribution of CDM to scattering and backscattering (e.g., Mobley, 1994).

Chlorophyll concentration was estimated in three ways:

1. Chlorophyll fluorescence (WETLabs WetStar) calibrated against chlorophyll estimates from discrete samples analyzed with HPLC.
2. The line-height of the chlorophyll absorption peak in the red: ([chl] = $\{a_p(676) - (39/65 \cdot a_p(650) + 26/65 \cdot a_p(715))\}/0.014$), e.g., Davis et al. (1997), where a_p denotes particulate absorption, and the wavelength is given in brackets. The value of the chlorophyll-specific absorption, $a^*(676) = 0.014\ m^2\ g\ chl^{-1}$ used here is consistent with published literature values for the oceanic assemblages of phytoplankton (e.g., Sosik & Mitchell, 1995) and was chosen here based on regression of HPLC determination of [chl] versus absorption based [chl]. $a^*(676)$ is likely to vary with depth, as packaging within cells changes its value. Published values of $a^*(676)$ vary from $0.008\ m^2\ g\ chl^{-1}$ to $0.023\ m^2\ g\ chl^{-1}$ (Bricaud et al., 1995; Sosik & Mitchell, 1995).
3. Discrete chlorophyll concentration obtained with HPLC.

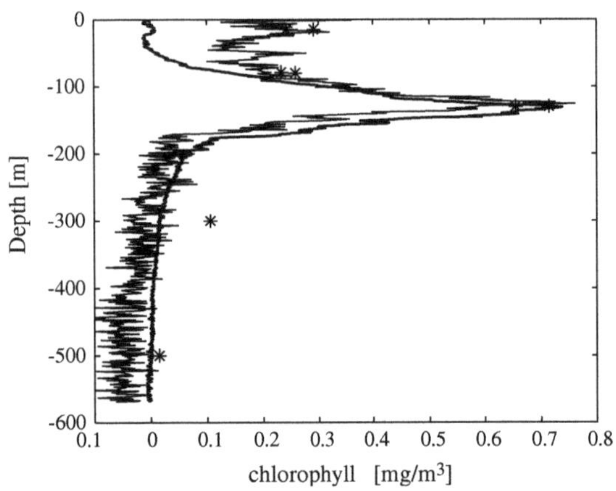

Fig. 2 Comparison between three estimates of chlorophyll; WetStar fluorometer (bold black line), chlorophyll estimated from particulate absorption line height (black), and HPLC (asterisks) based on measurements at St. 13, Sep. 19, 2001. Note that, unlike fluorescence, the absorption estimate of chlorophyll concentration does not suffer from non-photochemical quenching near the surface

The squared correlation coefficient between the chlorophyll fluorescence and absorption is $R^2 = 0.83$ (Fig. 2, $N = 570$). The absorption estimated [chl] has the advantage of not suffering from near-surface non-photochemical quenching which reduces the fluorescence-yield of phytoplankton exposed to high light near the surface. The uncertainty for the absorption-estimated [chl] is $\pm 0.2\ \mu g\ l^{-1}$ ($\sim mg\ m^{-3}$) based on the uncertainty in ac-9 measurements (see below, and not taking into account the possible bias in a^* which is likely to add and uncertainty of ±30%). Negative values ($\sim -0.05\ \mu g\ l^{-1}$) in absorption based chlorophyll estimates observed in Fig. 2 are indicative of too big a removal of the detrital absorption from the particulate absorption in the calculation of phytoplankton absorption at 676 nm. Note that these values are well below the uncertainty in the absorption-based chlorophyll.

Slopes of the particulate size distribution (the "PSD slope") were estimated from the attenuation spectrum by fitting a hyperbolic function to the particulate attenuation spectrum (e.g., Boss et al., 2001a). This method provides an estimate for the broad PSD slope assuming a hyperbolic particulate PSD (see Boss et al., 2001b, for a critical analysis of the assumptions). This method provides an estimate of the spatial changes in the mean particle size. We fitted a similar function to the particulate backscattering spectrum to assess whether there is any relation between this spectrum and that of particulate beam attenuation, and found no significant relation between the two (not shown). If the particles were non-absorbing and the measurements were error free we would have expected the slopes to be similar based on Mie theory (Morel, 1973).

Bulk particulate index of refraction was estimated using the method of Twardowski et al.

(2001). The bulk index of refraction provides an insight into the particulate composition; phytoplankton and detritus tend to have a low index of refraction (n = 1.02–1.1) due to a large water fraction while inorganic particles have a larger index of refraction (n = 1.12–1.24) (see Twardowski et al., 2001; Boss et al., 2004, for detailed discussion). Twardowski et al. (2001) developed a method to obtain information on the bulk index of refraction from knowledge of the particulate size distribution and the particulate backscattering ratio (the ratio of particulate backscattering to particulate scattering). Assuming the particulate size distribution (PSD) to be hyperbolic, a relationship has been found between the hyperbolic slope of the PSD and the spectral slope of particulate beam attenuation (e.g., Boss et al., 2001a, and references therein).

Discrete samples for chlorophyll and particulate organic carbon (POC) were collected in September and processed with the same methods as in Urbach et al. (2001). In addition, cells were collected in September for Flow-cytometric analysis.

Uncertainties in measurements

While with calibration, the error of the ac-9 are assumed to be O(0.005 m^{-1}) (Twardowski et al., 1999) we estimate the particulate absorption and attenuation measurement presented here to be accurate to within O(0.002 m^{-1}). This is because the values for the particulate attenuation and absorption were measured using a single instrument deployed with and without a pre-filter. Assuming the in-water properties to stay constant between the two consecutive profiles (30 min) the uncertainty is only limited by the instrument precision (0.001 m^{-1}) and stability of the instrument measuring the same waters O(0.002 m^{-1}). Note that changes in filter efficiency with time would increase these uncertainties. As noted above, we have not noticed changes in CDM concentration at depth between consecutive profiles, indicating no noticeable change in filter performance.

Note, however, that to this date there is a large uncertainty in the particulate absorption in the blue region of the spectrum due to issues associated with the scattering correction of particulate absorption. Different methods of the "Scattering correction" (e.g., Zaneveld et al., 1994) deviate by up to 20% in their estimate of absorption in the blue, while converging towards the red part of the spectrum. The scattering correction method used here was method 3 of Zaneveld et al. (1994), in which a fixed portion of scattering is removed from the measured absorption. The proportion of scattering that is removed equals the measured ratio of absorption and scattering at 715 nm, consistent with assuming zero absorption by particles at 715 nm. Uncertainty in colored dissolved material absorption (CDM) is 0.005 m^{-1}, a sizable uncertainty relative to the absolute CDM absorption measured in Crater Lake. To reduce this uncertainty the two independent measurements of CDM absorption obtained by the a- and c-sides of the ac-9 were averaged, reducing the uncertainty to 0.0035 m^{-1}.

Backscattering errors are assumed to be smaller than 15% of the signal based on uncertainties in converting light backscattered at 140 to the backscattering coefficient (Maffione & Dana, 1997; Boss et al., 2004). An additional uncertainty based on variability in the instruments' calibration history is ±0.0002 m^{-1} (D. Dana, 2004, personal communication). Comparison of particulate backscattering estimates at 530 nm based on the two backscattering instruments reveals that the two are highly correlated (R^2 = 0.94, N = 298), with the Eco-VSF being on average 0.97 times the HS-6 and with an offset that is not significantly different from zero (Fig. 1). This is an excellent agreement, in particular given that the two instruments are calibrated differently and that the backscattering coefficient is computed differently from each measurement (Zaneveld et al., 2003). It also indicates that the uncertainties quoted above are probably too conservative.

Uncertainties in the particulate backscattering ratio, the ratio of backscattering to total scattering, are likely to be less than 20%, based on propagating the errors of scattering and backscattering. Uncertainties in the discrete measurements were estimated from replicate samples and are 0.05 mg l^{-1} for chlorophyll and nearly 30% for POC.

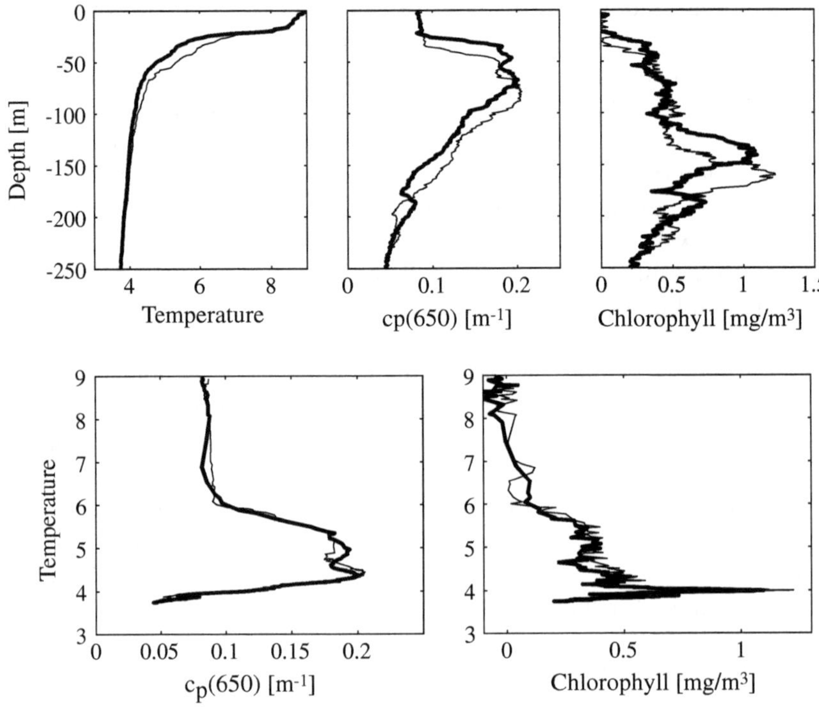

Fig. 3 Measurements of temperature, particulate attenuation at 650 nm, and chlorophyll (absorption based) as function of depth (top row) and of particulate attenuation at 650 nm, and chlorophyll as function of temperature (bottom row). Measurements were taken at St. 13 on June 28 (thin) and June 29 (bold), 2001

Results

Spatial and temporal variability in particulate properties

The profiles of particulate properties in June vary strongly between the 2 days of sampling (Figs. 3, 4). This variability is mostly due to the activity of large internal waves (amplitude >20 m) set up by a storm that passed by a day prior to sampling. When plotted against temperature, both chlorophyll and beam attenuation are very similar for both sampling days, indicating the (mostly) conservative behavior of these properties in the presence of internal waves over a single day.

Particulate properties vary in the water column as function of depth and differ between the June and September sampling periods (Figs. 4, 5). There does not seem to be a relation between particulate properties and stratification except for the elevated phytoplankton biomass near the surface in September. The vertical distribution of the backscattering coefficient is similar to the beam attenuation in September but not in June. The resulting change in the backscattering ratio with depth in June, suggests a change in the bulk particulate composition with depth.

Beam-c and POC

When regressing the discrete particulate organic carbon measurements (POC, mmol C m^{-3}) and the particulate beam attenuation measurements (c_p, m^{-1}, extrapolated to 660 from measurements at 650 and 676) we find:

$$c_p(660) = POC^{*}0.032 - 0.024 (R^2 = 0.996).$$

This regression is similar to that found by Gunderson et al. (1998) for the Arabian Sea (0.031 and – 0.007, for slope and intercept, respectively). While the correlation coefficient is high, the confidence in the slope and intercept is low due to the 30% difference between replicate measurements of POC (we used the mean of the replicates to derive the regression). Given that large uncertainty, the regression is also consistent with data from Station Aloha near Hawaii (Fennel & Boss, 2003) and the Southern ocean (Gardner

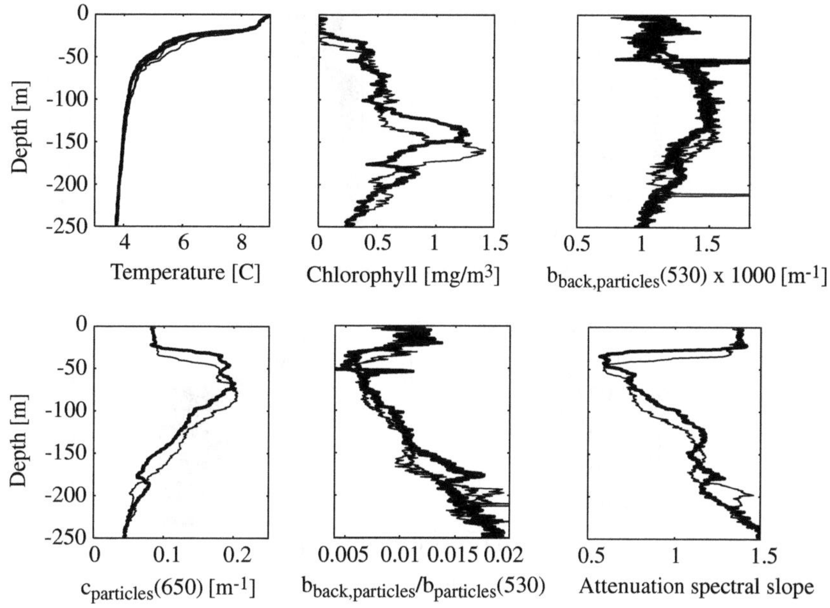

Fig. 4 Measurements of temperature (4 casts), chlorophyll (absorption based), particulate backscattering, particulate beam attenuation, backscattering ratio, and the spectral slope of beam attenuation at St. 13 on June 28 (thin) and June 29 (bold), 2001. Two of the four temperature casts and the chlorophyll and beam attenuation data are the same as those in Fig. 3

et al., 2000) where the slopes were found to be $O(0.02[m^{-1} (mmol\ C\ m^{-3})^{-1}])$.

Chlorophyll versus biomass

In both sampling periods we observed chlorophyll and particulate attenuation to be decoupled. Similar decoupling between phytoplankton bio-volume and chlorophyll has been observed in past measurements in Crater Lake (McIntire et al., 1996: Fig. 5). The decoupling is due to phyto-plankton photo-acclimation, which is characterized by a typical nearly exponential increase in the chlorophyll/c_p ratio and the ratio of chlorophyll/phytoplankton-bio-volume from the surface down to the chlorophyll maximum (e.g., Kitchen & Zaneveld, 1990; Fennel & Boss, 2003: Fig. 7). This relationship is discussed in more depth in Fennel & Boss (2003), and illustrates the problem associated with using chlorophyll as a proxy phytoplankton biomass when analyzing vertical profiles.

Contribution of sub-micron particles

Sub-micron particles are found to contribute nearly 30% of chlorophyll at the chlorophyll maximum in September (Fig. 5), though their contribution to absorption or attenuation is less than 20%. Flow-cytometric analysis of samples from the lake (Sherr & Sherr, personal communication, 2002) indicates that the phytoplankton sampled on the same day are neither *Synechococcus* nor *Prochlorococcus*, the dominating sub-micron nano-phytoplankton in the oceans (based on fluorescence and scattering characteristics). We currently do not know the taxonomy of these sub-micron particles that are observed to contribute significantly to the optical properties of Crate Lake. McIntire et al. (1996) observed the presence of several species of picoplankton with diameters on the order of 1 μm but could not identify microscopically smaller cells.

Size distribution

Changes in the slope of the beam attenuation between June and September are consistent with, in general, smaller particle size in September than in June, except right next to the surface. This is consistent with the surface waters at Crater Lake being regularly dominated by the relatively large and elongated (about 70 μm long and a few micron wide) diatom *Nitzschia gracilis* in August and September (McIntire et al., 1996). In both sampling seasons, the chlorophyll maximum region is dominated by particles with a similar size distribution (c_p slope ~1).

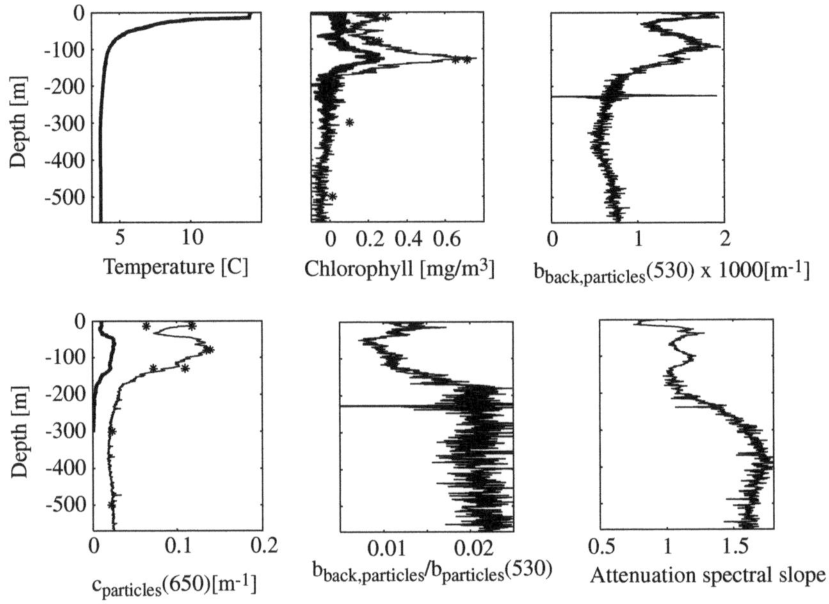

Fig. 5 Measurements of temperature, chlorophyll (absorption based), particulate backscattering, particulate beam attenuation, particulate backscattering ratio, and the spectral slope of particulate beam attenuation at St. 13 on Sep 19, 2001. The thin line represents total particulate measurements while the bold line represents the fraction smaller than 1 μm. Stars represent discrete chlorophyll-*a* (HPLC) and POC measurements. HydroScat-6 backscattering is denoted by a black line and extends only to 300 m. Eco-VSF backscattering is denoted in gray

Particulate composition and their bulk index of refraction

Sampling in June near the caldera's wall, in waters fed to the lake by a tributary, reveals a level of particulate attenuation nearly five times higher than that found near the surface at St. 13. These waters are dominated by particulate material (Fig. 6) with a ratio of particulate scattering to particulate attenuation greater than 0.93. This material has a backscattering ratio greater than 0.017 and a beam attenuation slope of nearly 1.2. Based on the analysis of Twardowski et al. (2001; Fig. 7 here) this is likely to be inorganic material, characteristic of clay minerals. This material is likely contributing to the relatively high backscattering ratio observed for the surface waters and the waters below the chlorophyll maxima at St. 13 (Fig. 7). At 50 m, however, the water of St. 13 has a backscattering ratio characteristic of organic material such as phytoplankton (0.005).

The bulk index of refraction at St. 13 in September is found to be low (<1.1) consistent with organic, phytoplankton, and detritus (Fig. 7). The particles with the highest index of refraction are located below 200 m. This may be due to a mix of detrital particles (e.g., inorganic phytoplankton shells, small inorganic particles) and heterotrophic organisms. The variability in estimated PSD slope and index of refraction the full water column of Crater Lake is indicative of the different assemblages of particles that occupy different depths of the lake.

Colored dissolved material and DOM

Absorption by dissolved material at 440 nm increases monotonically with depth in June while having a subsurface maximum at 130 m in September (Fig. 8). Between June and September it increases mostly between 100 m and 300 m, below the layer of maximum attenuation (maximum POC and phytoplankton biomass), and in

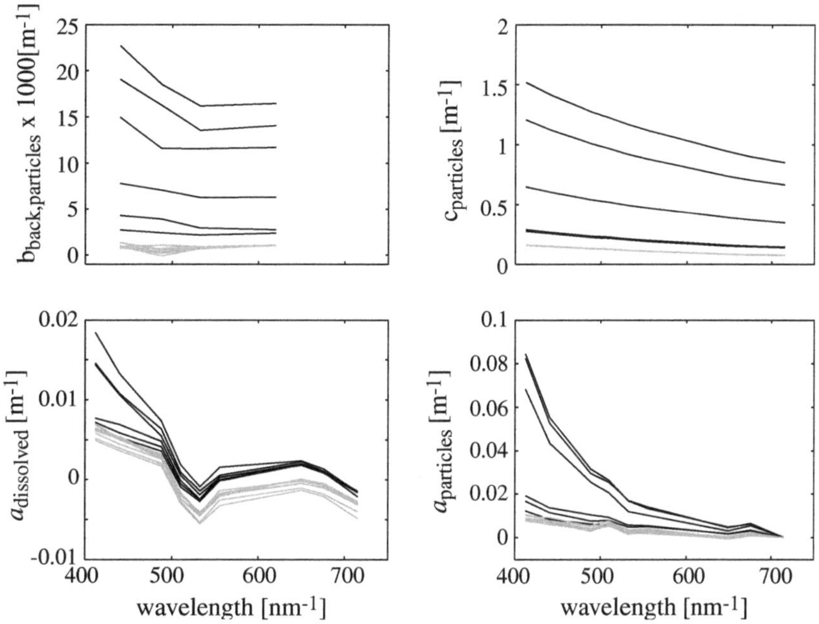

Fig. 6 Spectra of particulate attenuation (c_p), particulate scattering (b_p), particulate absorption (a_p), dissolved absorption (a_{CDM}) and particulate backscattering (b_{bp}, multiplied by 50) averaged over 1 m at a station near the edge of the caldera where a stream entered into the lake on June 28, 2001 (black lines). Highest values were measured closest to the bottom and decreased monotonically towards the surface. Gray lines denote the same properties in the upper 10 m at the center of the lake (st. 13)

the area of maximum particulate gradient. The observed accumulation is consistent with a release of dissolved organic material as a by-product of primary and secondary production within the water column and near the bottom (where a slight increase in CDM is also observed). The dissolved absorption measured near the tributary in June is similar in magnitude to that at the center of the lake (Fig. 6), implying that no significant colored dissolved material (and by conjunction no significant amount of dissolved material) is input into the lake from the caldera's rim at that location. Unfortunately, large uncertainties in CDM values (± 0.0035 m^{-1}) did not allow us to compute a reliable CDM spectral slope (e.g., spectra in Fig. 6).

Past measurements of total organic carbon (TOC) at Crater Lake suggest a background value of about 0.1 g C m^{-3}, dominated by dissolved organic carbon (DOC) (Urbach et al., 2001; Hargreaves et al., 2007). This magnitude of DOC is at least a factor of five lower than open ocean values (compared to values found in the Bermuda Atlantic and the Hawaiian Ocean time series). For an absorption value of 0.01 ± 0.003 m^{-1} at 440 nm (Fig. 8) it implies a DOC specific absorption coefficient of 0.001 ± 0.0005 m^2 (g C)$^{-1}$. This value of the CDM specific absorption is very low and similar to values measured for marine humic acid (Blough & Green, 1995).

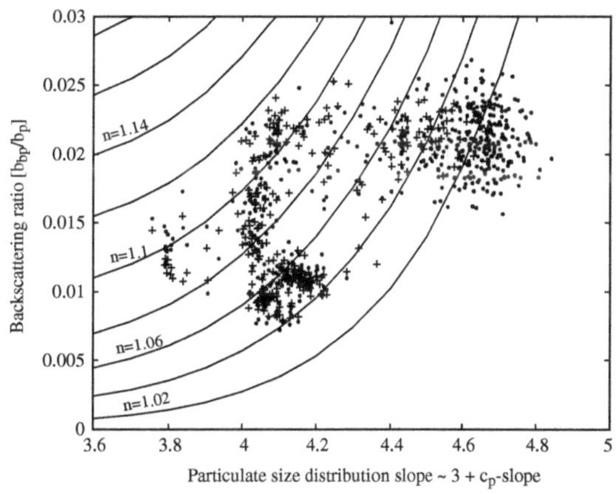

Fig. 7 Diagram depicting the backscattering-ratio against the estimated slope of the particulate size distribution (computed as 3 + spectral slope of c_p) for profile data collected at St. 13, Sep. 19 2001. Contours represent lines of equal index of refraction (n) based on Mie theory (see Twardowski et al., 2001; Boss et al., 2004). The +-sign denotes the backscattering ratio based on measurements with the HS-6 for the upper 300 m, while the dots denote the backscattering ratio based on the Eco-VSF, both at 530 nm. Low values of the bulk index of refraction (1.02–1.1) imply dominance by organic material

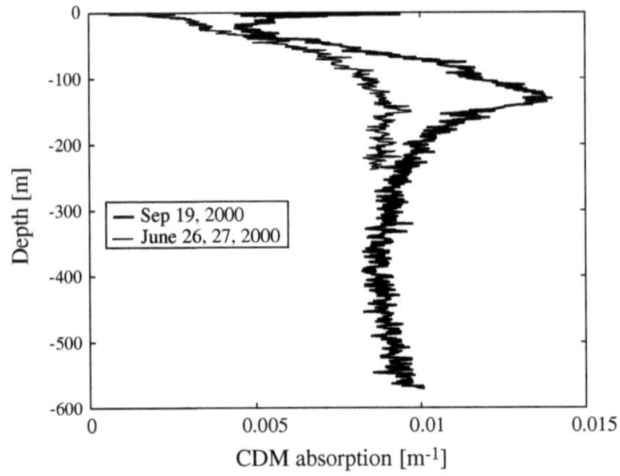

Fig. 8 Distribution of absorption by colored dissolved material at 440 nm in June (thin) and September (bold) 2001. The June profile is an average of the profiles measured on both days (28 and 29 June). Note that the uncertainty in the CDM measurement is 0.0035 m^{-1}, which is large compared to measurement (see Methods section)

Discussion

We find modern optical sensors to perform well despite the clarity of Crater Lake. These sensors provide us with high vertical resolution measurements of optical proxies of biogeochemical variables, which cannot be achieved using standard techniques.

The picture that emerges from the measurements is that the waters of Carter Lake are highly stratified optically in June and September at depths where density stratification is weak. Large amplitude internal waves were observed and may provide an important mechanism for bringing nutrients closer to the surface. Photo-acclimation in phytoplankton causes a decorrelation of the particulate beam attenuation (and also POC and phytoplankton bio-volume) and chlorophyll concentration; Chlorophyll exhibits a maximum 40–50 m deeper than that of POC.

Interestingly, optical estimates and vertical distribution of POC, and chlorophyll compare well with the relationships and distributions observed in open ocean environments. In addition, the specific absorption of CDM is found comparable to that of oceanic humic materials. These similarities suggest that biogeochemical lake studies may be relevant for understanding oceanic processes.

Sub-micron particles are found to contribute significantly to the chlorophyll concentration. To this date we do not know what are the organisms dominating this fraction in Crater Lake (McIntire, 2002, personal communication). CDM absorption, a proxy for dissolved organic material, is found to increase from spring to fall, most likely a by-product of primary and secondary production.

The instruments we have used for this study lend themselves to autonomous deployment on moorings. We strongly encourage such deployment in Crater Lake and similar environments to obtain measurements of biogeochemical proxies with high temporal resolution. Such deployments will allow the study of the lake response to forcing varying from episodic events to climate change. Real-time broadcast of the data can be used for adaptive sampling, where field sampling is driven by observed changes in the lake hydrographic and optical properties.

Acknowledgements We thank F. Baratange for assistance in the field and lab. Thanks to L. Eisner, F. Prahl, E. and B. Sherr for the analysis of the discrete samples. Comments by T. Swift and two anonymous reviewers are gratefully acknowledged. This work as been funded by the United States Geological Survey. The instrumentation used was purchased with funding by ONR and NASA to S. Pegau and E. Boss.

References

Bishop, J. K. B., 1999. Transmissometer measurement of POC. Deep-Sea Research I 46: 353–369.

Blough, N. V. & S. A. Green, 1995. Spectroscopic characterization and remote sensing of non living organic matter. In Zepp R. G. & C. Sonntag (eds), The Role of Non-living Organic Matter in the Earth's Carbon Cycle. Willey, Chichester: 23–45.

Boss, E. & W. S. Pegau, 2001. The relationship of light scattering at an angle in the backward direction to the backscattering coefficient. Applied Optics 40: 5503–5507.

Boss, E., W. S. Pegau, W. D. Gardner, J. R. V. Zaneveld, A. H. Barnard, M. S. Twardowski, G. C. Chang & T. D. Dickey, 2001a. The spectral particulate attenuation and particle size distribution in the bottom boundary layer of a continental shelf. Journal of Geophysical Research 106: 9509–9516.

Boss, E., M. S. Twardowski & S. Herring, 2001b. The shape of the particulate beam attenuation spectrum and its relaton to the size distribution of oceanic particles. Applied Optics 40: 4885–4893.

Boss, E., W. S. Pegau, M. Lee, M. S. Twardowski, E. Shybanov, G. Korotaev & F. Baratange, 2004. The particulate backscattering ratio at LEO 15 and its use to study particles composition and distribution. Journal of Geophysical Research 109, C0101410.1029/2002JC001514.

Bricaud, A., M. Babin, A. Morel & H. Claustre, 1995. Variability in the chlorophyll-specific absorption coefficients of natural phytoplankton: analysis and parameterization. Journal of Geophysical Research (C – Oceans) 100: 13,321–13,332.

Bukata, R. P., J. H. Jerome, K. Y. Kondratyev & D. V. Pozdnyakov, 1995. Optical Properties and Remote Sensing of Inland and Coastal Waters. CRC Press, Boca Raton, Florida.

Davis, R. F., C. C. Moore, J. R. V. Zaneveld & J. M. Napp, 1997. Reducing the effects of fouling on chlorophyll estimates derived from long-term deployments of optical instruments. Journal of Geophysical Research 102: 5851–5855.

Fennel, K. & E. Boss, 2003. Subsurface maxima of phytoplankton and chlorophyll—steady state solutions from a simple model. Limnology and Oceanography 48: 1521–1534.

Gardner, W. D., M. J. Richardson & W. O. Smith, 2000. Seasonal patterns of water column particulate organic carbon and fluxes in the Ross Sea, Antarctica. Deep-Sea Research II 47: 3423–3449.

Gordon, H. R., O. B. Brown, R. E. Evans, J. W. Brown, R. C. Smith, K. C. Baker & D. C. Clark, 1988. A semianalytic model of ocean color. Journal of Geophysical Research 96: 10909–10924.

Gunderson, J. S., W. D. Gardner, M. J. Richardson & I. D. Walsh, 1998. Effects of monsoons on the seasonal and spatial distribution of POC and chlorophyll in the Arabian Sea. Deep-Sea Research II 45: 2103–2132.

Hargreaves, B. R., S. F. Gardiner, M. W. Buktenica, R. W. Collier, E. Urbach & G. L. Larson, 2007. Ultraviolet radiation and bio-optics in Crater Lake, Oregon. Hydrobiologia 574: 107–140.

Kitchen, J. C. & J. R. Zaneveld, 1990. On the noncorrelation of the vertical structure of light scattering and chlorophyll *a* in case I waters. Journal of Geophysical Research 95: 20237–20246.

Klimasauskas, E., C. Bacon & J. Alexander, 2002. Mount Mazama and Crater Lake: Growth and Destruction of a Cascade Volcano U.S. Geological Survey Fact Sheet 092-02.

Maffione, R. A. & D. R. Dana, 1997. Instruments and methods for measuring the backward-scattering coefficient of ocean waters. Applied Optics 36: 6057–6067.

McIntire, C. D., G. L. Larson, R. E. Truitt & M. K. Debacon, 1996. Taxonomic structure and productivity of phytoplankton assemblages in Crater Lake, Oregon. Journal of Lake and Reservoir Management 12: 259–280.

Mobley, C. D., 1994. Light and Water: Radiative Transfer in Natural Waters. Academic Press, San Diego.

Moore, C., E. J. Bruce, W. S. Pegau & A. D. Weideman, 1997. WET Labs ac-9: field calibration protocol, deployment techniques, data processing, and design improvements. "Ocean Optics XII", Proceedings of the Society of Photo Optics Instrumentation Engineers 2963: 725–730.

Morel, A., 1973. Diffusion de la lumiere par les eaux de mer. Resultat experimentaux et approch theorique. In Agard Lecture Series 61 on Optics of the Sea~ Advisory Group for Aerospace Research and Development; NATO, London: 3.1.1–76.

Morel, A., 1974. Optical properties of pure water and pure sea water. In Jerlov, N. G. & E. S. Nielsen (eds), Optical Aspects of Oceanography. Academic Press, NewYork.

Pettit, E., 1936. On the color of Crate Lake water. Proceedings, National Academy of Science 22: 139–146.

Smith, R. C. & K. S. Baker, 1981. Optical properties of the clearest natural waters (200–800 nm). Applied Optics 20: 177–184.

Sosik, H. M. & B. G. Mitchell, 1995. Light absorption by phytoplankton, photosynthetic pigments and detritus in the California Current System. Deep Sea Research 42: 1717–1728.

Spinrad, R. W. & J. R. V. Zaneveld, 1982. An analysis of the optical features of the near-bottom and bottom nepholoid layers in the area of the Scotian Rise. Journal of Geophysical Research 87: 9553–9561.

Twardowski, M. S., J. M. Sullivan, P. L. Donaghay & J. R. V. Zaneveld, 1999. Microscale quantification of the absorption by dissolved and particulate material in coastal waters with an ac-9. Journal of Atmospheric Oceanic Technology 16: 691–707.

Twardowski, M. S., E. Boss, J. B. Macdonald, W. S. Pegau, A. H. Barnard & J. R. V. Zaneveld, 2001. A model for estimating bulk refractive index from the optical backscattering ratio and the implications for understanding particle composition in Case I and Case II waters. Journal of Geophysical Research 106: 14129–14142.

Urbach, E., K. L. Vergin, L. Young, A. Morse, G. L. Larson & S. J. Giovannoni, 2001. Unusual bacterioplankton community structure in ultra-oligotrophic Crater Lake. Limnology and Oceanography 46: 557–572.

Zaneveld, J. R. V., J. C. Kitchen & C. C. Moore, 1994. Scattering error correction of reflecting tube absorption meters. In Ackleson, S. (ed.), Ocean Optics XII, Proc. SPIE 2258: 44–55.

Zaneveld, J. R. V., S. Pegau & J. L. Mueller, 2003. Volume scattering function and backscattering coefficients: instruments, characterization, field measurements and data analysis protocols. In Mueller, J. L., G. S. Fargion & C. R. McClain (eds.), Ocean Optics Protocols for Satellite Ocean Color Sensor Validation, Revision 4, Volume IV: Inherent Optical Properties: Instruments, Characterizations, Field Measurements and Data Analysis Protocols, NASA Tech. Memo., 2003-211621/Rev4-Vol.IV. Greenbelt: NASA Goddard Space Flight Center: 65–76.

Hydrobiologia (2007) 574:161–177
DOI 10.1007/s10750-006-0351-5

Bacterioplankton communities of Crater Lake, OR: dynamic changes with euphotic zone food web structure and stable deep water populations

Ena Urbach · Kevin L. Vergin · Gary L. Larson · Stephen J. Giovannoni

© Springer Science+Business Media B.V. 2007

Abstract The distribution of bacterial and archaeal species in Crater Lake plankton varies dramatically over depth and with time, as assessed by hybridization of group-specific oligonucleotides to RNA extracted from lakewater. Nonmetric, multidimensional scaling (MDS) analysis of relative bacterial phylotype densities revealed complex relationships among assemblages sampled from depth profiles in July, August and September of 1997 through 1999. CL500-11 green nonsulfur bacteria (Phylum Chloroflexi) and marine Group I crenarchaeota are consistently dominant groups in the oxygenated deep waters at 300 and 500 m. Other phylotypes found in the deep waters are similar to surface and mid-depth populations and vary with time. Euphotic zone assemblages are dominated either by β-proteobacteria or CL120-10 verrucomicrobia, and ACK4 actinomycetes. MDS analyses of euphotic zone populations in relation to environmental variables and phytoplankton and zooplankton population structures reveal apparent links between *Daphnia pulicaria* zooplankton population densities and microbial community structure. These patterns may reflect food web interactions that link kokanee salmon population densities to community structure of the bacterioplankton, via fish predation on *Daphnia* with cascading consequences to *Daphnia* bacterivory and predation on bacterivorous protists. These results demonstrate a stable bottom-water microbial community. They also extend previous observations of food web-driven changes in euphotic zone bacterioplankton community structure to an oligotrophic setting.

Keywords Crater Lake · Bacterioplankton community structure · Multidimensional scaling · Green nonsulfur bacteria · Marine Group I crenarchaeota · *Daphnia* predation

Guest Editors: Gary L. Larson, Robert Collier, and Mark W. Buktenica
Long-term Limnological Research and Monitoring at Crater Lake, Oregon

E. Urbach · K. L. Vergin · S. J. Giovannoni
Department of Microbiology, Oregon State University, Corvallis, OR, USA

G. L. Larson
USGS Forest and Rangeland Ecosystem Science Center, Corvallis, OR, USA

Present Address:
E. Urbach (✉)
eMetagen Corporation, 3591 Anderson St., Suite 207, Madison, WI 53704, USA
e-mail: ena.urbach@emetagen.com

Introduction

Crater Lake is a high-altitude, ultraoligotrophic lake in the Cascade Mountains of the Northwestern United States. Formed by the catastrophic eruption and collapse of a volcanic mountain approximately 6500 years ago, Crater Lake has been the subject of a wide range of studies characterizing the physics, chemistry, biology and ecology of the system (Larson, 1996). An unusual community of planktonic bacteria and archaea was identified in the lake, based on phylogenetic analysis of SSU rRNA genes sampled during August of 1997 (Urbach et al., 2001). At that time, the upper water column was dominated by two groups of currently uncultivated bacteria: the CL120-10 evolutionary cluster in the *Verrucomicrobiales*, and the ACK4 cluster in the *Actinomycetales*. At the same time, the lower water column contained a novel group of green nonsulfur (GNS) bacteria (Chloroflexi), CL500-11, and archaea in the marine Group I crenarchaeota clade. The relative importance of these groups inferred from DNA analysis was confirmed by oligonucleotide hybridization to RNA from the same lakewater samples. Microbial populations in other aerobic lakes of many trophic levels and elevations, from the Alps to the Netherlands, are consistently dominated by β-proteobacteria, with significant populations of ACK4 actinomycetes and cytophaga/flavobacteria and other, minor groups (Bahr et al., 1996; Hiorns et al., 1997; Semenova & Kuznedelov, 1998; Methé & Zehr, 1999; Glöckner et al., 2000; Zwart et al., 2002b; Zwisler et al., 2003). Comparison of the physical, chemical and biological characteristics of Crater Lake to 13 other lakes characterized by rRNA gene analysis suggested that Crater Lake's microbial communities were likely structured by low concentrations of available trace metals and dissolved organic matter, chemistry of infiltrating hydrothermal waters or high levels of ultraviolet light (Urbach et al., 2001).

Crater Lake is an extraordinarily oligotrophic, deep lake with deep penetration of ultraviolet light, an oxygenated hypolimnion, low concentrations of metals and dissolved organic carbon, and influx of hydrothermal fluids (R.W. Collier, pers. comm.; Collier et al., 1991, 1993; McManus et al., 1996; Nelson et al., 1996; Crawford & Collier, 1997; Hargreaves et al., this issue). At all depths, dissolved N/P molar ratios fall below the Redfield ratio of 16, suggesting that autotrophic growth may be N-limited (Redfield, 1958). Bioassay studies also suggest that phytoplankton productivity may be co-limited by N and trace metals, with different size fractions showing differences in growth stimulation by N and metal mixtures (Lane & Goldman, 1984; Groeger & Teitjen, 1993; Groeger, 2007). Crater lake shares a number of features with oligotrophic, open ocean waters, including oxygenated deep waters, the possibility of combined N and trace metal limitation, a deep chlorophyll maximum, and exceptionally high light transmittance (Urbach et al., 2001).

Organisms that live in Crater Lake must contend with the extreme and unusual environmental conditions: high fluxes of ultraviolet light, low concentrations of dissolved organic carbon, hydrothermal influx at the bottom of the lake and deep hydrography, combined with N and potentially trace metal limitations. Available data do not show much variation in these factors (with the possible exception of NH_4^+ concentrations), and they are likely to be responsible for the unusual bacterioplankton taxa found in the lake. CL120-10 verrucomicrobia and CL0-1 OP10 bacteria have been found in low numbers near the mouth of the Columbia River and CL120-10 in a Siberian reservoir, but, at present, CL500-11 GNS and CL500-3 planctomycetes are unique to Crater Lake, and marine Group I archaea have not been found in other freshwater planktonic populations, though they have been reported to occur in sediments (MacGregor et al., 1997; Crump et al., 1999; Urbach et al., 2001; Trusova & Gladyshev, 2002). Our earlier study presented a snapshot of Crater Lake prokaryotic plankton and identified correlates with the unusual organisms found there.

Here we present analysis of Crater Lake bacterioplankton populations during July through September of 1997, 1998 and 1999, with comparison to time series measurements of physical, chemical and biological factors. These analyses suggest that depth and food web structure are major factors influencing Crater Lake bacterioplankton populations.

Methods

Sampling and sample processing

Lakewater samples were collected from the deck of the R/V Neuston at Station 13 (42°56′ N, 122°06′ W) over the deepest part of Crater Lake during July through September of 1997–1999 (7/15/97, 8/19/97, 9/16/97, 7/21/98, 8/11/98, 9/15/98, 7/13/99, 8/10/99 and 9/7/99). Samples for bacterioplankton population structure and biomass analyses were collected from 12 depths (0, 20, 40, 60, 80, 100, 120, 140, 160, 180, 300 and 500 m), except during July 1997 and July 1999, when subsets of these depths were sampled (0, 20, 40, 60, 80, 100, 120, 140, 160, 180, and 0, 20, 80, 120, 500, respectively). Procedures for collection and processing of these samples were as previously reported (Urbach et al., 2001). Sample collection for chemical, phytoplankton and zooplankton analyses, and measurements of temperature, transmissivity, light and primary productivity profiles occurred within 2 days of bacterioplankton sampling (methods cited in Urbach et al., 2001).

Oligonucleotide probe hybridization

rRNAs from microbial phylogenetic groups were quantified by hybridization of group-specific oligonucleotide probes to Crater Lake RNA samples arrayed onto nylon membranes (Giovannoni et al., 1990b, 1996). Probes were designed to hybridize specifically to phylogenetic groups identified in three Crater Lake SSU rDNA libraries made from water collected during August, 1997, as well as to freshwater phylotypes identified in other clone libraries and broad, kingdom-specific probes (Urbach et al., 2001). An additional probe hybridizing to SAR202 GNS bacteria (including the Crater Lake sublineage, CGTAGGAGTGGGGGCCG, stringent wash temperature 45°C) was included in these analyses. Quantification standards included RNAs extracted from cultured organisms (when these were available) and synthetic rRNAs transcribed in vitro from SSU rDNA clones (Giovannoni et al., 1990a; Polz & Cavanaugh, 1997). Hybridizations were normalized to sample-specific totals of bacteria + archaea + eukarya (for kingdoms) or to totals of bacterial group-specific probes minus plastid probes (for bacterial phylotypes).

MDS community analysis

Nonmetric, multidimensional scaling (MDS) and other routines implemented in the computer program PRIMER were used to compare bacterial phylotype, phytoplankton and zooplankton assemblages in Crater Lake samples (Kruskal, 1964; Clarke & Gorley, 2001). Bacterioplankton populations were compared by Bray–Curtis distance calculations using untransformed population data (fraction of nonplastid bacterial rRNA for each phylotype) in PRIMER, and the resulting distance matrix used to infer 2- or 3-dimensional MDS plots. MDS plots were constructed for individual samples and for averaged samples from the same depth or samples from the same date averaged over all depths in the euphotic zone (0–180 m). In some plots, values of environmental variables or population data for phytoplankton or zooplankton taxa or bacterioplankton phylotypes were superimposed on MDS plots as bubbles of proportionate size by specifying options in the graph properties menu of PRIMER. Phytoplankton and zooplankton populations were compared by Bray–Curtis distance calculations using square-root transformed population data consisting of integrated biomass densities for species from the surface to 200 m. The quality of an MDS plot as a representation of relationships among populations is reported in the PRIMER output as a "stress" value. Stress values for MDS analyses were compared to 1% cutoff values for random distributions (Sturrock & Rocha, 2000).

Statistical differences between clusters identified in MDS plots were investigated by a randomization method. Distances among member populations in clusters in an MDS plot were compared to a distribution of such distances calculated from the same plot with labels randomly reassigned in as many ways as possible, or 1000 times, whichever is less (ANOSIM) (Clarke & Green, 1988). Contributions of individual species to differences between clusters were calculated with the SIMPER routine (Clarke 1993).

Results

Temperature, nutrients and primary productivity depth profiles

Primary productivity varied among sampling dates, even though temperature and nutrient distributions were essentially the same during all sampling periods (Fig. 1). Primary productivity, measured as ^{14}C-bicarbonate incorporation into plankton retained by 0.45 μm pore-size filters, varied from an euphotic zone integrated value of 17.8 mg C m^{-2} h^{-1} during July, 1997 to 80.7 mg C m^{-2} h^{-1} during July, 1999. Relatively low productivity (below 0.4 mg C m^{-3} h^{-1} at each depth) prevailed during 1997 and 1998 and generally higher productivity (exceeding 0.8 mg C m^{-3} h^{-1} at one or more depths) on every sampling date during 1999. Thermal structure of the water column showed relatively little variation during these periods: thermocline depth ranged from 30 to 60 m, with surface temperatures between 14 and 17°C. Nutrient concentration depth profiles were similarly uniform, with NO_3^--N below the level of detection (~0.07 μM) in the euphotic zone (above 200 m), rising with depth to 0.86–1.07 μM in bottom waters. NH_4^+-N was generally below the level of detection (~0.07–0.14 μM) throughout the water column, occasionally rising above the baseline (but not exceeding 0.5 μM). Concentrations of PO_4^{-3}-P (~0.3 μM) and total P (~0.8 μM) were nearly uniform throughout the water column, with the euphotic zone depleted relative to the hypolimnion by roughly 0.1 μM dissolved P (not shown) (Dymond et al., 1996). The thermal, nutrient and primary productivity depth profiles were within the limits of normal variation observed during 20 years of regular sampling at Crater Lake (Collier et al., 1990; Larson et al., 1996; McIntire et al., 1996; Boss et al., 2007; Groeger, 2007; Larson et al., (2007a).

Bacterioplankton population structure

Oligonucleotide hybridization to RNA extracted from lakewater revealed dynamic euphotic zone and stable deep water microbial populations (Figs. 2, 3). Above 200 m depth, CL120-10 verrucomicrobia accounted for the largest proportion of nonplastid bacterial rRNA during August, 1997, July and September, 1998 and August and September, 1999; β-proteobacteria dominated this zone during July, 1997 and August, 1998 and ACK4 actinomycetes during September, 1997 and July, 1999 (Fig. 2). On many of these dates, β-proteobacteria were most numerous at the surface (0 and 20 m), giving way to other microbial groups deeper in the euphotic zone. The minor phylotypes C111 actinomycetes, CL0-1 candidate division OP10 bacteria, CL500-3 planctomycetes, SAR202 GNS and SAR11 α-proteobacteria were generally present at or below 5% of the nonplastid bacterial rRNA, though most minor groups exhibited occasional population peaks, e.g., CL0-1 during September, 1998 and SAR202 during September, 1997. Bottom waters at 300 and 500 m were largely populated by CL500-11 GNS, which consistently accounted for about 50% of the nonplastid bacterial rRNA at these depths. Other bacterial phylotypes found in the euphotic zone generally were also present at lower densities near the bottom of the lake. Variation in euphotic zone bacterial populations presented a complex pattern with a consistent cast of dominant microbes, CL120-10, ACK4 and β-proteobacteria, while bottom water bacterial populations were consistently dominated by CL500-11.

Kingdom-specific oligonucleotide probes revealed a pattern in which bacteria and eukarya dominated the euphotic zone, with bacteria rising to 50–80% in deep waters, accompanied by archaea, contributing ~10%, and eukarya making up the remainder of the total rRNA (Fig. 3). In the euphotic zone, the ratio of bacterial to eukaryal rRNA varied. During 1997 and July, 1998, eukaryal and bacterial sequences accounted for roughly 75 and 25% of total rRNA, respectively, while during August and September, 1998 and all three months during 1999, bacteria and eukarya contributed nearly equally. Variation in these ratios could not be accounted for by simple biomass comparisons for eukaryotes and prokaryotes in the euphotic zone (Table 1). Variations in rRNA per unit biomass is likely to vary dramatically among eukaryotic phytoplankton taxa, and variations in phytoplankton population structure may account for most of the variation in relative hybridization signals in the upper 200 m (see below). In deep, hypolimnetic waters, the eukarya constituted ~20% of total rRNA, which is likely the result of eukaryotes sinking from the surface, with some contribution by eukaryotic microzooplankters indigenous to the bottom waters. Removing the eukaryotes from rRNA calculations, archaea consistently contributed 7–22% to total prokaryotic (bacterial plus archaeal) rRNA in bottom waters (Table 2).

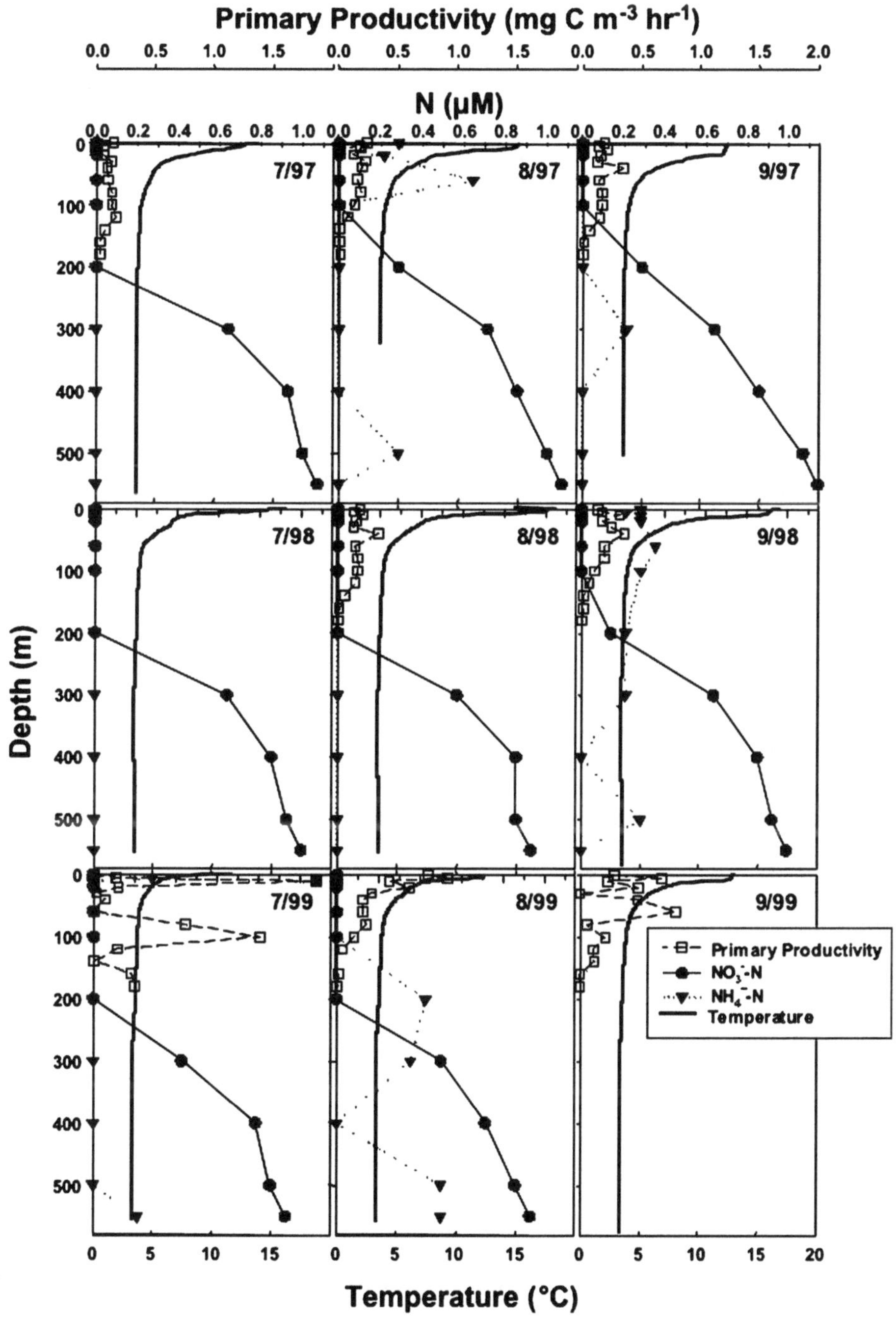

Fig. 1 Primary productivity, nitrogenous nutrient concentrations and temperature depth profiles for Crater Lake during 1997–1999. Measurements and samples were collected within 2 days of sampling for microbial population structure. No primary productivity data were collected during July, 1998

Prokaryotic cell densities

Prokaryotic cell densities assessed by counts of DAPI-stained cells, varied from approximately 2.0×10^5 to 1.8×10^6 cells ml^{-1}, and generally did not vary significantly with depth (Fig. 4). In the euphotic zone, highest cell densities were recorded during 1998 and lowest during 1999. There was only a single sampling date for bacterial cell numbers during 1997, but average cell numbers and inferred biomass were significantly higher during 1998 than 1999 (mean and SD $5.6 \times 10^5 \pm 3.4 \times 10^4$ and $3.7 \times 10^5 \pm 2.8 \times 10^4$, respectively).

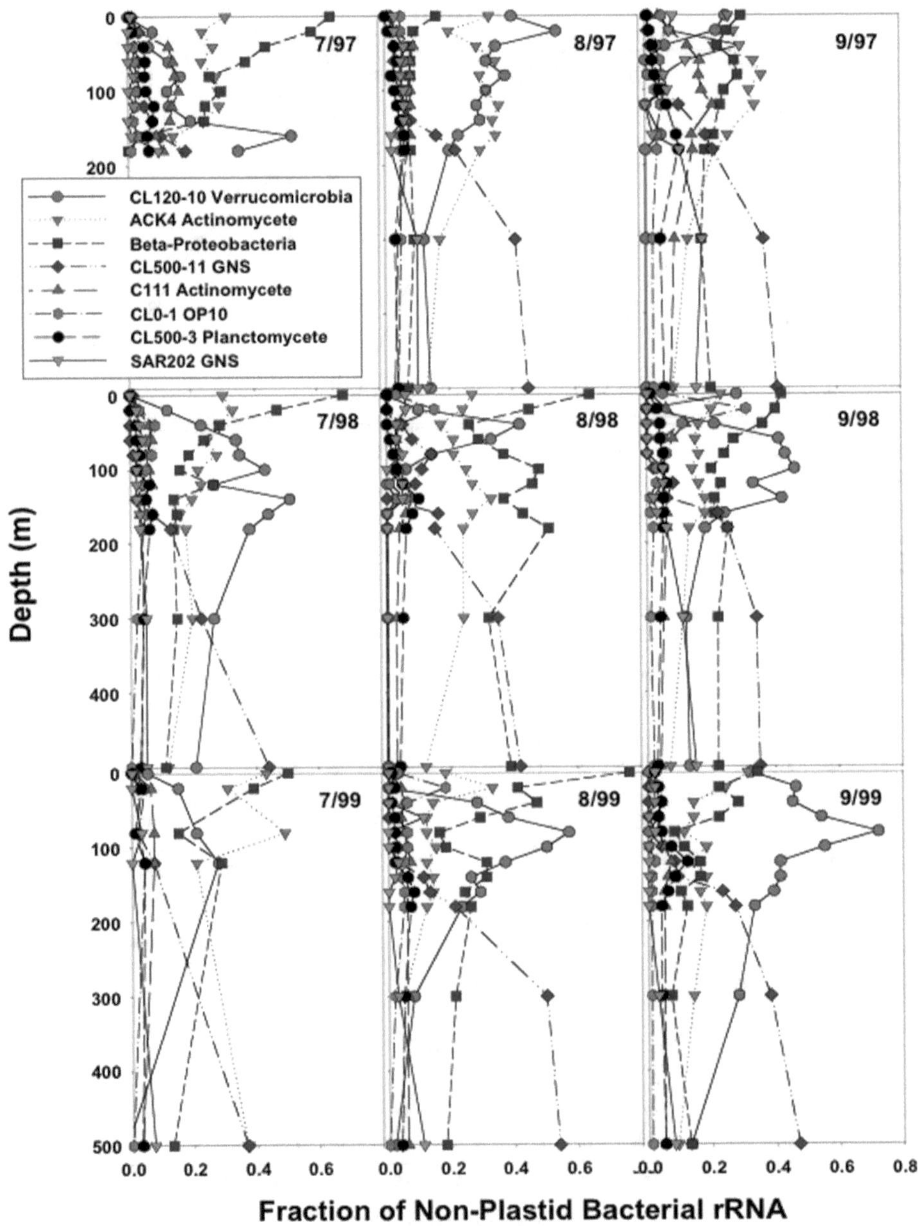

Fig. 2 Depth distributions of bacterial phylogenetic groups in Crater Lake as determined by relative hybridization of group-specific oligonucleotide probes to lake water RNA. Population densities of the freshwater SAR11 clade of α-proteobacteria were below 0.05 in all samples and are omitted from the figure for clarity

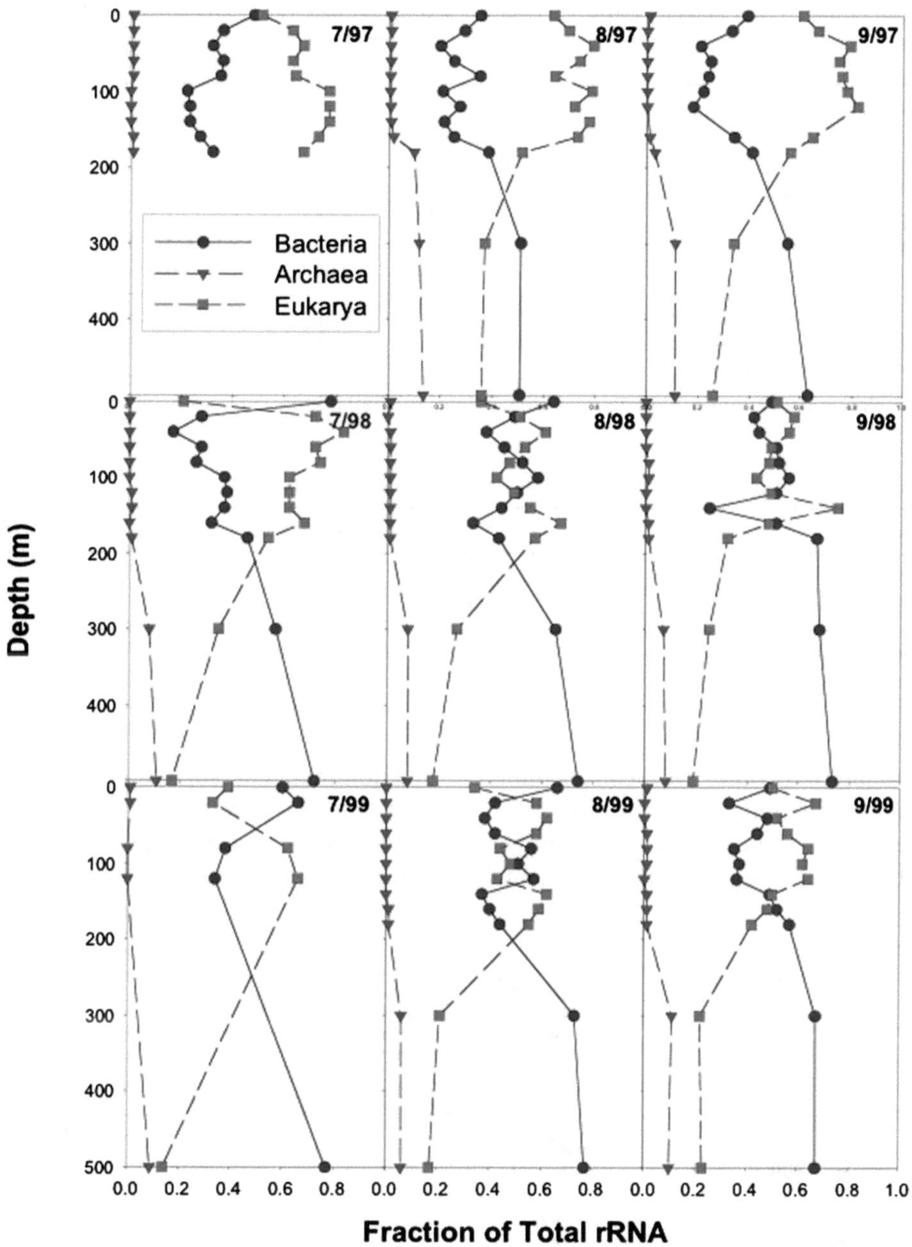

Fig. 3 Depth distributions of rRNA sequences for the kingdoms bacteria, archaea and eukarya as determined by relative hybridization of group-specific oligonucleotide probes to Crater Lake lake water RNA

Phytoplankton population structure

Phytoplankton populations showed high variability (McIntire et al., this issue). On a biomass basis, bacillariophytes often dominated the upper 20 m of the water column, with chlorophytes, chrysophytes, bacillariophytes and pyrrophytes (dinoflagellates) rising to dominance on different days and different depths. The most common surface-dominant bacillariophyte was *Nitzschia gracilis*, with *Nitzschia vermicularis* replacing it during July, 1997. In deeper waters, *Stephanodiscus hantzchii* contributed the most bacillariophyte biomass. Pyrrophytes were most often *Gymnodi-*

Table 1 Integrated biomass* and biomass ratios for Crater Lake planktonic food web elements

	Jul-97	Aug-97	Sep-97	Jul-98	Aug-98	Sep-98	Jul-99	Aug-99	Sep-99
Bacterioplankton (Whole Water Column)	ND	9.0	ND	11.5	12.7	10.2	6.9	8.7	8.1
Bacterioplankton (Euphotic Zone)	ND	4.8	ND	5.6	4.8	4.4	2.5	3.5	3.6
Phytoplankton (Euphotic Zone)	78.7	101.0	65.8	17.2	65.3	31.1	75.8	60.5	65.9
Zooplankton (Euphotic Zone)	2.7	8.6	0.4	0.7	1.1	0.8	0.7	2.0	3.7
Euks/Proks (Euphotic Zone)	ND	22.71	ND	3.18	13.77	7.21	30.70	17.67	19.45
Phyto/Proks (Euphotic Zone)	ND	20.93	ND	3.06	13.55	7.03	30.40	17.11	18.41
Zoo/Phyto	0.03	0.09	0.01	0.04	0.02	0.03	0.01	0.03	0.06

* Biomass (g m^{-2}) is integrated over the upper 200 m (euphotic zone) or upper 550 m (whole water column). Prokaryotic biomass is calculated from cell density measurements on the basis of 20 fg C/cell (Lee and Fuhrman, 1987), assuming that carbon comprises 50% of the biomass

Euks: eukaryotes; Proks: prokaryotes; Zoo: zooplankton; Phyto: phytoplankton; ND: no data

Table 2 Archaeal fraction of prokaryotic rRNA in abyssal waters of Crater Lake*

Depth	Aug-97	Sep-97	Jul-98	Aug-98	Sep-98	Jul-99	Aug-99	Sep-99
300 m	0.19	0.16	0.13	0.11	0.09	ND	0.08	0.15
500 m	0.22	0.15	0.13	0.10	0.10	0.10	0.07	0.13

* Calculated from oligonucleotide hybridization data as archaea/bacteria+archaea

ND: no data

nium inversum and/or *Peridinium inconspicuum*, with *Gymnodinium fuscum* appearing during 1997 and September, 1999. The dominant chlorophyte on a biomass basis was *Mougeotia laetevirens*, and the dominant chrysophytes were *Tribonema affine* and *Dinobryon sertularia*. Cyanobacteria (cyanophytes) were present, but did not contribute significantly to phytoplankton biomass.

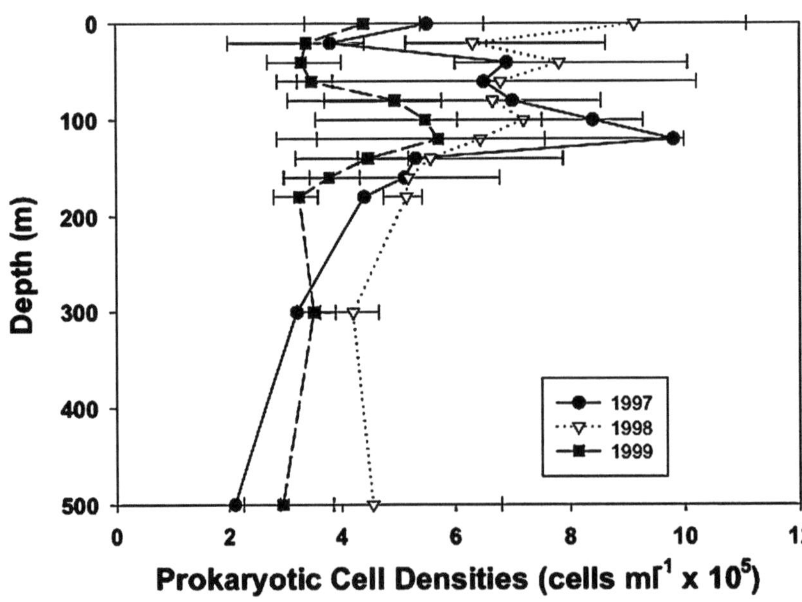

Fig. 4 Depth distributions of average prokaryotic cell densities for 1997–1999, determined by microscope counts of cells stained with DAPI. Error bars indicate the range of average cell densities on individual dates. 1997 data are for August only

Zooplankton population structure

Zooplankton population structure varied mostly on an annual basis, as assessed by biomass (Larson et al., 2007b). During 1997, the rotifer *Kellicottia longispina* was dominant, with *Synchaeta oblonga* appearing around 50 m during September. During 1998, the cladoceran *Daphnia pulicaria* and rotifer *S. oblonga* contributed the most biomass at different depths, with *K. longispina* equaling these species during September. During 1999, *D. pulicaria* was again dominant, accompanied by the rotifer *Filinia terminalis*.

Biomass comparisons

Integrated biomass for phytoplankton, zooplankton and bacterioplankton illustrate changing relationships among these food web components (Table 1). Bacterioplankton biomass was least variable, with euphotic zone totals ranging from 2.5 to 5.6 g m^{-2}. Phytoplankton biomass varied almost 6-fold, from 17.2 to 101.0 g m^{-2}, and zooplankton varied 21-fold, from 0.4 to 8.4 g m^{-2}. Changes in food web element ratios reflect dynamic variations in sources and sinks, producers and consumers of biomass viewed from the perspective of gross trophic levels.

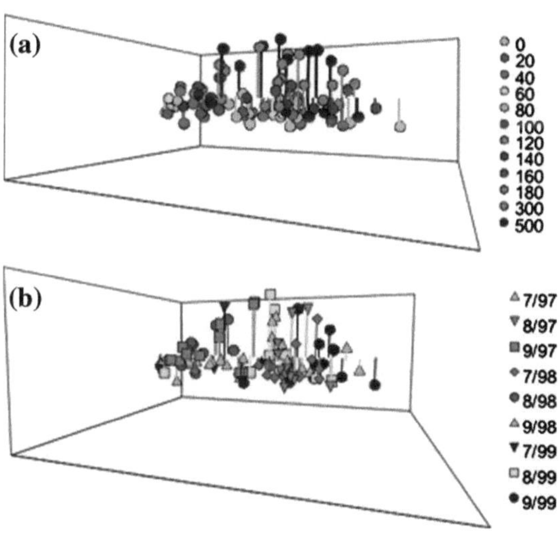

Fig. 5 Three-dimensional MDS plot illustrating rank similarity relationships among all Crater Lake populations sampled. Both panels illustrate the same plot, with symbols keyed according to (**a**) sampling depth and (**b**) sampling date. Lines are drawn between symbols and the $z = 0$ plane to aid visualization. Stress for the plot is 0.08 ($P < 0.01$)

MDS analyses

MDS analysis revealed patterns in bacterial population structure at different depths and times. Euphotic zone and deepwater bacterial populations appear distinct in a three-dimensional MDS plot illustrating relationships among populations in all 98 bacterioplankton samples collected for this study, color-coded according to depth (Fig. 5a). Nonparametric permutation analysis of the rank similarity matrix underpinning the MDS plot confirms that euphotic zone (0–160 m) and deepwater (300–500 m) populations are distinct ($P = 0.001$, one-way ANOSIM). Dissimilarity between the euphotic zone and deep water populations was principally attributable to GNS, characterizing the deep populations (35.1% average contribution to Bray–Curtis dissimilarity in comparisons of euphotic and deep water populations) and CL120-10 verrucomicrobia, β-proteobacteria and ACK4 actinomycetes, characterizing the euphotic zone populations (20.5, 15.5 and 12.7%, respectively).

Re-coloring of the whole-dataset MDS plot to distinguish sampling dates offers a different perspective on the bacterial population data, revealing additional, superimposed patterns (Fig. 5b). This plot reveals that similarity among populations sampled on the same date contributes to the overall pattern of population structure. The graphical effect of this phenomenon is a closer association of same-date symbols in the deep and shallow water clusters, e.g., on 8/97, contrasting with the pattern of more typical water columns, e.g., 8/99 (Fig. 5b).

Averaged over time, bacterial populations at different depths form a continuum from 0 to 500 m in an MDS plot, with an arched distribution characteristic of populations arranged according to a single, strong gradient (Fig. 6; Field et al., 1982). Coding each population according to densities of particular phylotypes illustrates stratification of phylogenetic groups to different depths in the water column. On average, β-proteobacteria are their largest fraction of bacterial populations at 0 m, with a gradient of

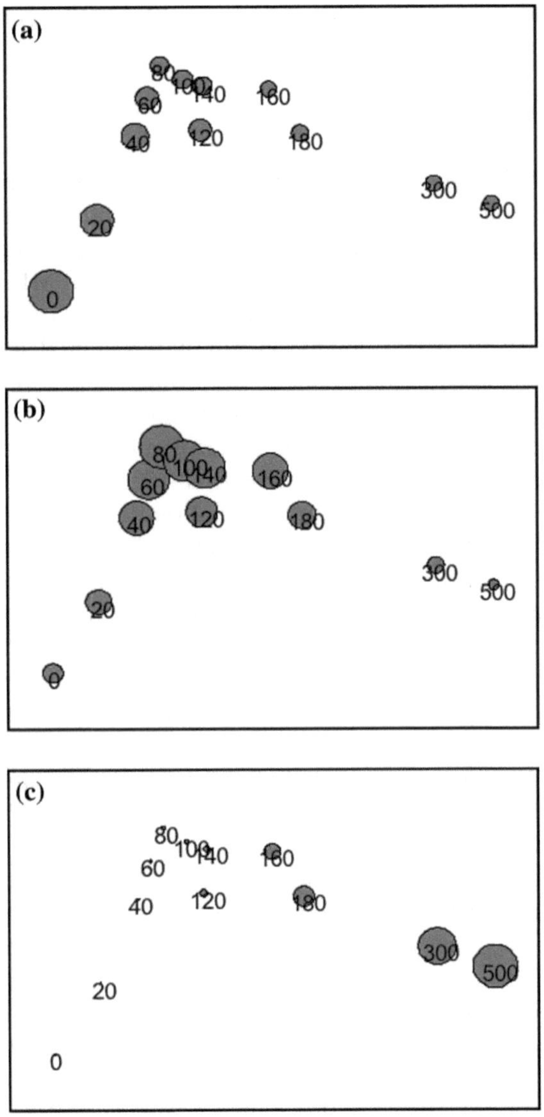

Fig. 6 Two-dimensional MDS plot illustrating rank similarity relationships among samples at different depths, averaged over all dates. The three panels show the same plot with symbols sized in proportion to date-averaged rRNA fractions for a different bacterial phylogenetic group in each panel. Panel (**a**) β-proteobacteria (0.18–0.49), (**b**) CL120-10 verrucomicrobia (0.08–0.33) and (**c**) CL500-11 GNS (0.01–0.43). Stress for the plot is 0.01 ($P < 0.01$)

decreasing representation in successively deeper waters (Fig. 6a). In contrast, CL120-10 verrucomicrobia are their largest fraction of the population at mid depths, between 40 and 160 m, and CL500-11 GNS form a gradient with higest representation at 500 m, decreasing to low proportions at mid depths (Fig. 6b, c). Stratification of individual phylotypes is the basis for the distinction between euphotic zone and deep water populations.

Analysis of averaged euphotic zone bacterial populations on different dates yields an MDS plot that, again, forms an arched, continuous distribution (Fig. 7). In this plot, populations are not arranged according to consecutive dates and do

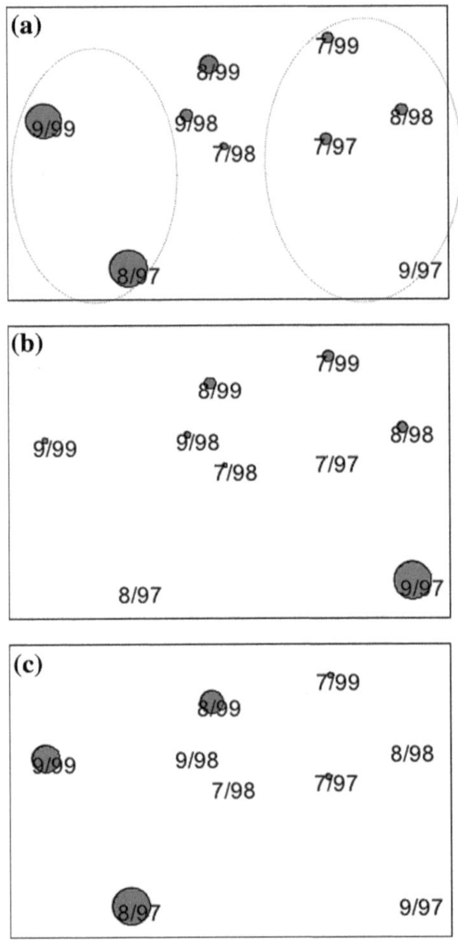

Fig. 7 Two-dimensional MDS plot illustrating rank similarity relationships among bacterioplankton populations in the euphotic zone on different dates (averaged over 0–180 m). Symbols are sized in proportion to integrated biomass densities for (**a**) *Daphnia pulicaria* zooplankton (0–1.4 × 10^3 mg m^{-2}), (**b**) *Ochromonas* spp. phytoplankton (0–7.1 × 10^2 mg m^{-2}) and (**c**) *Polyarthra dolichoptera* zooplankton (0–2.1 × 10^2 mg m^{-2}). Dotted ellipses in panel (**a**) encircle "high *Daphnia*-associated" (left) and "low *Daphnia*-associated" (right) population clusters. Stress for the plot is 0.06 ($P < 0.01$)

not cluster by year or months in different years. Instead, the distribution is arranged according to gradients of integrated population densities of particular zooplankton and phytoplankton taxa: *D. pulicaria*, summed species of the phytoplankton genus *Ochromonas* and the zooplankton species *Polyarthra dolichoptera* (Fig. 7a–c). Physical and chemical variables, including ultraviolet and photosynthetically available light, light scattering by particles, temperature, P and N nutrients, chlorophyll and dissolved organic carbon failed to show correlations with the surface bacterioplankton community MDS plot; similarly uninformative comparisons were made with the phytoplankton and zooplankton community compositions and primary productivity (not shown). Considered as clusters, populations at the extreme left and extreme right of the MDS plot (September, 1999 and August, 1997 versus July and September, 1997, August, 1998 and July, 1999, respectively, Fig. 7a) were different ($P = 0.067$, the lowest value obtainable for permutations among the two groups, one-way ANOSIM). CL120-10 (high on the two dates to the left of the plot) contributed 39.6% to the difference between the two groups and β-proteobacteria and ACK4 actinomycetes (high on the four dates to the right of the plot) contributed 29.9 and 12.0%, respectively. The two groups of averaged euphotic zone bacterial populations represent high *Daphnia*-associated populations on the left and low *Daphnia*-associated on the right of the plot.

Depth-integrated phytoplankton and zooplankton populations fall into patterns which can be grouped according to year when analyzed as MDS plots (Fig. 8a, b). Permutation analyses confirm that groupings according to year are significant for both phytoplankton and zooplankton ($P = 0.004$ and 0.011, respectively, one-way ANOSIM). Factors that do not vary on an annual basis also contribute to relationships among both sets of populations. Eukaryotic to prokaryotic biomass ratios in the euphotic zone form an apparent gradient superimposed upon the interannual variation in the phytoplankton MDS plot (Fig. 8a; Table 1). Similarly, *D. pulicaria* biomass contributes to zooplankton population structure, but not enough to disrupt significant clustering of these populations by year (Fig. 8b). Although multiple factors contribute to relationships among Crater Lake phytoplankton and zooplankton populations, they fall into patterns that differ from relationships among bacterioplankton populations.

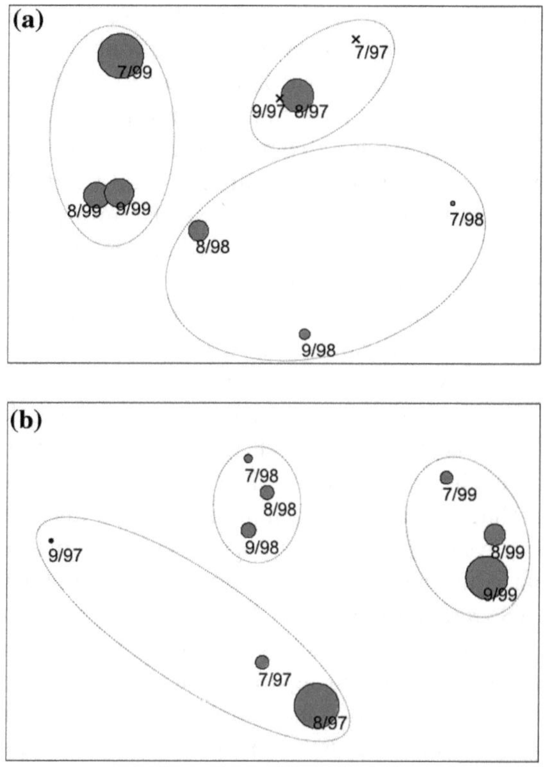

Fig. 8 Two-dimensional MDS plots illustrating rank similarity relationships among depth-integrated populations on different dates. Dotted ellipses encircle populations grouped by year. Panel (**a**) phytoplankton; symbols are sized in proportion to eukaryotic/prokaryotic integrated biomass ratios (Table 1), crosses indicate phytoplankton populations from July and September, 1997, for which biomass ratios were not available, and stress for the plot is 0.10 ($P < 0.01$). Panel (**b**) zooplankton; symbols are sized in proportion to depth-integrated *D. pulicaria* biomass (0–1.4×10^3 mg m^{-2}), and stress for the plot is 0.03 ($P < 0.01$)

Discussion

Bacterioplankton in Crater Lake are revealed as a combination of actively changing euphotic zone populations with relatively stable deep-water inhabitants. Surface bacterioplankton can change dramatically on a monthly timescale, with sequential months showing fluctuations along a continuum from verrucomicrobia- to β-proteobacteria-dominated bacterioplankton and back, with irregular timing (Fig. 2). Deep waters were consistently dominated by archaea and CL500-11 GNS, with changing populations of minor phylotypes (Figs. 2, 3; Table 2). Ordination analyses further characterize Crater Lake bacterioplankton and correlate population transitions with biotic and environmental variables, identifying potentially important interactions.

The environment and identities of Crater Lake deep water taxa suggest hypotheses for their biogeochemical roles. The GNS and crenarchaeota divisions each include a wide diversity of physiological types, suggesting physiologies that could also pertain to uncultivated CL500-11 GNS and marine group I crenarchaeota in the deep waters of Crater Lake. Crater Lake's waters are oxygenated throughout, but H_2 and other reduced compounds may enter bottom waters in hydrothermal fluids (concentrations have not been measured, R. Collier, pers. comm.). Oxygen is therefore the likely electron acceptor, with reduced organic and inorganic electron donors possible. For example, either of the two taxa could be an aerobic heterotroph similar to the genus *Herpetosiphon* GNS bacteria. Alternatively, the uncultivated organisms could be chemoautotrophs like the crenarchaeota *Methanococcus* and *Pyrococcus* (Adrian et al., 2000; He et al., 2003). In Lake Soyang, CL500-11 bacteria have been identified near an oxic/anoxic boundary, suggesting that this organism may acquire reduced compounds from the anoxic zone (Choi et al., 2001; Zwart et al., 2002a). The possibility of carbon fixation by Crater Lake GNS is also consistent with photoautotrophy by the GNS genus *Chloroflexus*, though actual GNS chemoautotrophy has not been described. A final hypothesis provides roles for both deep water phylotypes: oceanic marine group I crenarchaeota may be autotrophs, as evidenced by dark incorporation of ^{13}C bicarbonate into biphytanyl-membrane lipids in the North Sea (Wuchter et al., 2003). In the broad GNS sublineage that includes CL500-11, there is only a single cultured isolate. This organism ferments sucrose, glucose and arabinose to produce H_2, acetate and CO_2, and its growth is improved by anaerobic co-cultivation with an H_2-utilizing methanogen (Sekiguchi et al., 2001). Thus, a mutualistic association of similar, though oxygen-tolerant, GNS bacteria and crenarchaeota would be consistent with the little that is known about physiology and biogeochemistry of their relatives. Future work will shed light on the potentially unusual physiologies of Crater Lake deep water microorganisms.

Relationships among complex communities and correlations with environmental and biotic variables are best interpreted by graphical ordination methods such as nonmetric multidimensional scaling (MDS) (Kruskal, 1964; Schiffman et al., 1981). MDS is a graphical ordination technique that makes few model assumptions about data or relationships among samples, and thus is appropriate for analysis of ecological population data (Field et al., 1982; Clarke, 1993). MDS produces two or three-dimensional "best fit" plots, representing rank similarity matrices, to compare communities using species composition data. MDS plots illustrate the best approximations to relationships among populations, which are arranged arbitrarily in space, without orientation to any axes. Similar populations are placed close to each other in MDS plots, and the statistical significance of apparent groupings can be assessed by permutation procedures that are not dependent upon normality or metricity of the data (Clarke & Green, 1988). Calculations based on the Bray–Curtis population similarity index can be used to compare contributions of individual species to the differences between groups and can also be used to identify discriminating species (Clarke, 1993). Patterns of environmental and biotic variables can be superimposed upon MDS plots to identify factors correlated with complex patterns of variation in population structure (Field et al., 1982). We applied MDS to evaluate relationships among Crater Lake bacterioplankton populations and physical and chemical variables as well as phytoplankton and zooplankton populations.

For Crater Lake data, MDS illustrates a significant distinction between euphotic zone and deep water bacterial population groups. This bipartite pattern is consistent with observations in other deep, oxic lakes and in open ocean waters

(Figs. 7a, 8; Bel'kova et al., 1996; Giovannoni et al., 1996; Field et al., 1997; Semenova & Kuznedelov, 1998; Karner et al., 2001).

Prokaryotic populations can be structured by predation, in addition to competition with other microorganisms and the physical, chemical and nutritional constraints of their environment. Large-bodied *Daphnia* are keystone predators in lakewater communities, feeding on phytoplankton, protists and bacterioplankton (Güde, 1988; Jürgens, 1994). Effects of *Daphnia* on bacterioplankton are complex: they remove nanoflagellate microzooplankton (heterotrophic and mixotrophic), that are otherwise the major bacteriovores, and they are also efficient feeders on all but the smallest bacterioplankton (Peterson et al., 1978; Nygaard & Tobiesen, 1993; Langenheder & Jurgens, 2001; Jürgens & Matz, 2002; Sherr & Sherr, 2002). In the absence of *Daphnia*, nanoflagellates deplete bacteria in the 0.15 to 2.0 μm^3 size range, selecting for dominant bacteria that are larger and/or smaller than their range of prey sizes (Gonzalez et al., 1990; Hahn & Höfle, 1999). When flagellates are replaced by *Daphnia*, the bacterioplankton size spectrum may expand to fill the middle size classes, or shrink to include only tiny cells resistant to *Daphnia* predation (Jürgens et al., 1994; Jürgens et al., 1999). Although some bacterial species are pleiotrophic and can adapt their sizes for survival, this is not true of all, and the genetic structure of bacterioplankton populations can change in concert with cell size-selection by predators (Hahn & Höfle, 2001). It has been hypothesized that control of bacterioplankton community structure by predation may be characteristic of eutrophic and mesotrophic lakes, and that oligotrophic lakes should be controlled by bottom-up, environmental factors (Hahn & Höfle, 2001).

In Crater Lake, variations in euphotic zone bacterial populations show a striking relationship to *D. pulicaria* densities, but not to "bottom-up" variables such as primary productivity (Figs. 1, 2, 7a). High *Daphnia*-associated and low *Daphnia*-associated populations form distinct groups in MDS analysis, although the small number of possible permutations precludes statistical significance (Fig. 7a). Populations of nanoflagellates and ciliates have been identified in Crater Lake, and it is likely they are the dominant bacteriovores when *Daphnia* densities are low (L. Eisner, pers. comm.). The high *Daphnia*-associated populations are characterized by high proportions of CL120-10 verrucomicrobia (accounting for 39.6% of the average dissimilarity between members of the groups). The size distribution of prokaryotic cells was not determined during this investigation. However, cultured members of the verrucomicrobia are generally small in size, and the CL120-10 cluster is specifically related to cultured ultramicrobacteria (Janssen et al., 1997; Urbach et al., 2001). This suggests that CL120-10 cells may be small and dominate during *Daphnia* predation due to predator selection against all but the tiniest organisms. Our results extend observations of predator-associated changes in bacterioplankton community structure to an oligotrophic setting.

Experimental studies characterizing other lake systems by using chemostats and filtered lakewater microcosms identified taxonomic groups selected by different predators. These studies demonstrated decreased proportions of β-proteobacteria and increased α-proteobacteria (and, in one case, cytophaga/flavobacteria) under conditions of nanoflagellate grazing, relative to predator-free conditions (Pernthaler et al., 1997; Langenheder & Jurgens, 2001; Simek et al., 2001). These taxonomic shifts were sometimes accompanied by formation of grazing-resistant bacterial filaments, identified as β-proteobacteria, α-proteobacteria or cytophaga/flavobacteria in different experiments (Pernthaler et al., 1997; Simek et al., 1999). Addition of *Daphnia* caused a β-proteobacteria and cytophaga/flavobacteria-dominated population to remain apparently unchanged, although a related experiment demonstrated latent changes revealed by PCR/DGGE analysis (Langenheder & Jurgens, 2001; Degans et al., 2002). Differences in bacterioplankton populations under nanoflagellate or *Daphnia* predation are in broad agreement with data from Crater Lake, although details of these results are in apparent conflict.

Major differences in Crater Lake bacterioplankton populations under presumed nanoflagellate or *Daphnia* predation regimes include a shift from β-proteobacteria and ACK4 actinomycetes-dominated euphotic zone populations when *Daphnia* are rare to CL120-10-dominated populations when *Daphnia* are plentiful. Data for Crater Lake also suggest that α-proteobacteria in the freshwater SAR11 clade never rise above 5% of the population (other a-proteobacteria were not assessed, as they were not found among 237 bacterial clones in our Crater Lake clone libraries) (Urbach et al., 2001). As α-proteobacteria and cytophaga/flavobacteria were not significantly detected in Crater Lake, their growth may be hindered by the oligotrophic, potentially N and metal limited, UV-irradiated, low DOC environment, a potential example of bottom up control. β-proteobacteria and ACK4 actinomycetes consequently dominate under conditions of top-down, flagellate control. The taxonomic shift observed upon transition to a *Daphnia* predation regime is more distinct in Crater Lake than seen in the experimental systems. Thus, in Crater Lake, dominance of β-proteobacteria and ACK4 actinomycetes is consistent with flagellate control of bacterioplankton community structure in the absence of vigorous α-proteobacterial and cytophaga/flavobacterial populations, and the transition to CL120-10 verrucomicrobial dominance is consistent with a rise in *Daphnia* predation.

In Crater Lake, changes in phytoplankton and metazooplankton population structures follow different temporal patterns from the euphotic zone bacterioplankton (Figs. 7, 8). The arrangement of averaged euphotic zone bacterioplankton populations in an MDS plot shows no simple relationship to the time line, while phytoplankton and zooplankton integrated populations are significantly clustered by year. Although thermal profiles and total N and P do not vary much from year to year during the sampling periods, significant inter-annual variation in winter mixing depth recorded by moored thermisters imply annually variable upwelling of nutrient-enriched, hypolimnetic waters, leading to the establishment of distinct phytoplankton and zooplankton populations and primary productivity rates by the time of first sampling (Fig. 1; McManus et al., 1993, 1996; Crawford & Collier, 1997). Thus, it is likely that environmental conditions of the previous winter set the stage for phytoplankton and zooplanktonic succession each year, and determine rates of productivity that persist throughout the succeeding months of thermal stratification.

In limnological studies, it can be difficult to differentiate selective effects of phytoplankton and zooplankton on lake bacterioplankton, as the temporal pattern of phytoplankton and bacterioplankton population transitions can sometimes coincide, perhaps due to Daphnia predation on both phytoplankton and bacterioplankton (Kent et al., 2004). In Crater Lake, this common confounding pattern is not seen; the gross population structure of the phytoplankton community changes on an annual basis and is not correlated with changes in bacterioplankton populations or *Daphnia* biomass densities (Figs. 7, 8a). Evidence of *Daphnia* predation is seen in densities of the minor zooplankton taxon *P. dolichoptera* and the mixotrophic phytoplankter *Ochromonas* (summed species), which are positively and negatively correlated with *Daphnia*, respectively (Fig. 7b, c). Like Daphnia, *P. dolichoptera* rotifers feed on protozoans. Their positive correlation with *Daphnia* populations suggest they prey upon species that are not effectively cropped by *Daphnia*, or their growth may be promoted by similar factors (Buikema et al., 1978; Arndt, 1993; Jürgens et al., 1996). Chrysophytes, such as *Ochromonas*, are prey for *Daphnia* and feed on prokarytes in competition with *Daphnia*, both consistent with a negative correlation between *Ochromonas* and the cladoceran (Daley et al., 1973). On a biomass basis, *Ochromonas* and *P. dolichoptera* are minor taxa, their population densities are likely to be consequences of variations in *Daphnia* densities, and they are not likely to be driving shifts in bacterioplankton community structure. Thus, in this study, the structure of the phytoplankton population does not appear to be a major influence on bacterioplankton populations, suggesting that euphotic zone bacterioplankton population structure is determined by a combination of *Daphnia* predation and environmental constraints.

Crater Lake is a deep, extraordinarily oligotrophic lake with deep penetration of ultraviolet light, low concentrations of trace metals and dissolved organic carbon, and an influx of hydrothermal fluids. These characteristics form the basic, unchanging constraints that select for adapted microbes, some of which have not been found anywhere else. In bottom waters, nearly stable year-round environmental conditions promote stable populations dominated by unusual microorganisms. The patterns of variation in euphotic zone bacterioplankton, phytoplankton and zooplankton communities imply an additional structuring factor: dynamically changing bacteriovore populations may cause variation in the surviving bacterioplankton assemblage. *Daphnia* populations rise and fall in accord with seasonal life history constraints and predation by planktivorous fish (vulnerable, in turn, to piscivorous fish), giving rise to large variations in intra- and interannual population densities (Jürgens, 1994). Correlations between boom and bust cycles of kokanee salmon and *D. pulicaria* in Crater Lake suggest that, in this system, *Daphnia* are largely controlled by these highly variable fish populations (Buktenica & Girdner, this issue). These results suggest that cascading, large scale food web effects from fish through zooplankton may impact bacterioplankton population structure in an oligotrophic lake. Future inquiries into the Crater Lake microbial ecosystem will address questions and hypotheses raised by these findings and exploit this extraordinary site to further deconstruct the workings of complex ecological systems. (Lee & Fuhrman, 1987).

Acknowledgements We are grateful to M. Buktenica and S. Girdner of the National Park Service, and L. Young, A. Morse, R. Hoffman and R. Truitt of OSU for their invaluable help with sampling and laboratory analyses. We thank T. Kratz and A. Yannarell of University of Wisconsin-Madison for helpful comments and E. Triplett for generous hospitality. This work was supported by NSF Grant DEB-9709012 to S.J.G. and E.U.

References

Adrian, L., U. Szewzyk, J. Wecke & H. Görisch, 2000. Bacterial dehalorespiration with chlorinated benzenes. Nature 408: 580–583.

Arndt, H., 1993. Rotifers as predators on components of the microbial web (bacteria, heterotrophic flagellates, ciliates)—a review. Hydrobiologia 255: 231–246.

Bahr, M., J. E. Hobbie & M. L. Sogin, 1996. Bacterial diversity in an arctic lake: a freshwater SAR11 cluster. Aquatic Microbial Ecology 11: 271–277.

Bel'kova, N. L., L. Y. Denisova, E. N. Manakova, E. F. Zaichikov & M. A. Grachev, 1996. Species diversity of deep-water microorganisms of Lake Baikal: analysis of 16S rRNA sequences. Doklady Biological Sciences 348: 692–695.

Boss, E., R. W. Collier & G. L. Larson, 2007. Measurements of spectral optical properties and their relation to biogeochemical variables and processes in Crater Lake, Crater Lake National Park, OR.

Buikema, A. L. Jr., J. D. Miller & W. H. Yongue Jr., 1978. Effects of algae and protozoans on the dynamics of *Polyarthra vulgaris*. Verhandlungen - Internationale Vereinigung für theoretische und angewandte Limnologie 20: 2395–2399.

Buktenica, M. & S. Girdner, this issue. Life history studies of kokanee salmon and rainbow trout in a deep, ultraoligotrophic caldera lake.

Choi, K., B. Kim & U. -H. Lee, 2001. Characteristics of dissolved organic carbon in three layers of a deep reservoir, Lake Soyang, Korea. International Review of Hydrobiology 86: 63–76.

Clarke, K. R., 1993. Non-parametric multivariate analyses of changes in community structure. Australian Journal of Ecology 18: 117–143.

Clarke, K. R. & R. N. Gorley, 2001. PRIMER. *in*. PRIMER-E Ltd., Plymouth, UK.

Clarke, K. R. & R. H. Green, 1988. Statistical design and analysis for a 'biological effects' study. Marine Ecology Progress Series 46: 213–226.

Collier, R. W., J. Dymond, J. McManus & J. Lupton, 1990. Chemical and physical properties of the water column at Crater Lake, OR. In Drake, E., G. L. Larson, J. Dymond & R. W. Collier (eds), Crater Lake: An Ecosystem Study. AAAS: 69–80.

Collier, R., J. Dymond & J. McManus, 1991. Studies of hydrothermal processes in Crater Lake, OR. Cooperative Agreement No. CA 9000-3-0003. Subagreement No 7., National Park Service, PNW Region, Seattle, WA.

Collier, R., J. Dymond & J. McManus, 1993. Studies of hydrothermal processes. NPS/PNROSU/NRTR-93/03, U.S. Department of the Interior, National Park Service, Pacific Northwest Region, Seattle, WA.

Crawford, G. B. & R. W. Collier, 1997. Observations of a deep-mixing event in Crater Lake, Oregon. Limnology and Oceanography 42: 299–306.

Crump, B. C., E. V. Armbrust & J. A. Baross, 1999. Phylogenetic analysis of particle-attached and free-living bacterial communities in the Columbia River, its estuary, and the adjacent coastal ocean. Applied and Environmental Microbiology 65: 3192–3204.

Daley, R. J., G. P. Morris & S. R. Brown, 1973. Phagotrophic ingestion of a blue-green-alga by *Ochromonas*. Journal of Protozoology 20: 58–61.

Degans, H., E. Zollner, K. Van der Gucht, L. De Meester & K. Jurgens, 2002. Rapid *Daphnia*-mediated changes in microbial community structure: an experimental study. FEMS Microbiology Ecology 42: 137–149.

Dymond, J., R. Collier, J. McMannus & G. L. Larson, 1996. Unbalanced particle flux budget in Crater Lake, Oregon: implications for edge effects and sediment focusing in lakes. Limnology and Oceanography 41: 732–743.

Field, J. G., K. R. Clarke & R. M. Warwick, 1982. A practical strategy for analysing multispecies distribution patterns. Marine Ecology Progress Series 8: 37–52.

Field, K. G., D. Gordon, M. Rappé, E. Urbach, K. Vergin & S. J. Giovannoni, 1997. Diversity and depth-specific distribution of SAR11 cluster rRNA genes from marine planktonic bacteria. Applied and Environmental Microbiology 63: 63–76.

Giovannoni, S. J., T. B. Britschgi, C. L. Moyer & K. G. Field, 1990a. Genetic diversity in Sargasso Sea bacterioplankton. Nature 345: 60–62.

Giovannoni, S. J., E. F. DeLong, T. M. Schmidt & N. R. Pace, 1990b. Tangential flow filtration and preliminary phylogenetic analysis of marine picoplankton. Applied and Environmental Microbiology 56: 2572–2575.

Giovannoni, S. J., M. S. Rappé, K. L. Vergin & N. L. Adair, 1996. 16S rRNA genes reveal stratified open ocean bacterioplankton populations related to the Green Non-Sulfur bacteria. Proceedings of the National Academy of Sciences of the United States of America 93: 7979–7984.

Glöckner, F. O., E. Zaichikov, N. Belkova, L. Denissova, J. Pernthaler, A. Pernthaler & R. Amann, 2000. Comparative 16S rRNA analysis of lake bacterioplankton reveals globally distributed phylogenetic clusters including an abundant group of Actinobacteria. Applied and Environmental Microbiology 66: 5053–5065.

Gonzalez, J., E. B. Sherr & B. F. Sherr, 1990. Size-selective grazing on bacteria by natural assemblages of estuarine flagellates and ciliates. Applied and Environmental Microbiology 56: 583–589.

Groeger, A., 2007. Nutrient limitation in Crater Lake, Oregon.

Groeger, A. W. & T. C. Teitjen, 1993. Physiological responses of nutrient-limited phytoplankton to nutrient addition. Verhandlungen – Internationale Vereinigung für theoretische und angewandte Limnologie 25: 370–372.

Güde, H., 1988. Direct and indirect influences of crustacean zooplankton on bacterioplankton of Lake Constance. Hydrobiologia 159: 63–73.

Hahn, M. W. & M. G. Höfle, 1999. Flagellate predation on a bacterial model community: interplay of size-selective grazing, specific bacterial cell size, and bacterial community composition. Applied and Environmental Microbiology 65: 4863–4872.

Hahn, M. W. & M. G. Höfle, 2001. Grazing of protozoa and its effect on populations of aquatic bacteria. FEMS Microbiology Ecology 35: 113–121.

Hargreaves, B. R., S. Girdner, M. Buktenica, R. W. Collier, E. Urbach & G. L. Larson, this issue. Ultraviolet radiation and bio-optics in Crater Lake, Oregon.

He, J., K. M. Ritalahti, M. R. Aiello & F. E. Loffler, 2003. Complete detoxification of vinyl chloride by an anaerobic enrichment culture and identification of the reductively dechlorinating population as a *Dehalococcoides* species. Applied and Environmental Microbiology 69: 996–1003.

Hiorns, W. D., B. A. Methé, S. A. Nierzwicki-Bauer & J. P. Zehr, 1997. Bacterial diversity in Adirondack mountain lakes as revealed by 16S rRNA gene sequences. Applied and Environmental Microbiology 63: 2957–2960.

Janssen, P. H., A. Schuhmann, E. Mörschel & F. A. Rainey, 1997. Novel anaerobic ultramicrobacteria belonging to the *Verrucomicrobiales* lineage of bacterial descent isolated by dilution culture from anoxic rice paddy soil. Applied and Environmental Microbiology 63: 1382–1388.

Jürgens, K., 1994. Impact of *Daphnia* on planktonic microbial food webs—a review. Marine Microbial Food Webs 8: 295–324.

Jürgens, K. & C. Matz, 2002. Predation as a shaping force for the phenotypic and genotypic comosition of planktonic bacteria. Antonie van Leeuwenhoek 81: 413–434.

Jürgens, K., H. Arndt & K. O. Rothhaupt, 1994. Zooplankton-mediated changes of bacterial community structure. Microbial Ecology 27: 27–42.

Jürgens, K., S. A. Wickham, K. O. Rothhaupt & B. Santer, 1996. Feeding rates of macro- and microzooplankton on heterotrophic nanoflagellates. Limnology and Oceanography,: 1833–1839.

Jürgens, K., J. Pernthaler, S. Schalla & R. Amann, 1999. Morphological and compositional changes in a planktonic bacterial community in response to enhanced protozoan grazing. Applied and Environmental Microbiology 65: 1241–1250.

Karner, M. B., E. F. DeLong & D. M. Karl, 2001. Archaeal dominance in the mesopelagic zone of the Pacific Ocean. Nature 409: 507–510.

Kent, A., S. Jones, A. Yannarell, J. Graham, G. Lauster, T. Kratz & E. Triplett, 2004. Annual patterns in bacterioplankton community variability in a humic lake. Microbial Ecology 48: 550–560.

Kruskal, J. B., 1964. Multidimensional scaling by optimizing goodness of fit to a nonmetric hypothesis. Psychometrika 29: 1–27.

Lane, J. L. & C. R. Goldman, 1984. Size-fractionation of natural phytoplankton communities in nutrient bioassay studies. Hydrobiologia 118: 219–223.

Langenheder, S. & K. Jurgens, 2001. Regulation of bacterial biomass and community structure by metazoan and protozoan predation. Limnology and Oceanography 46: 121–134.

Larson, G. L., 1996. Development of a 10-year limnological study of Crater Lake, Crater Lake National Park, Oregon, USA. Lake and Reservoir Management 12: 221–229.

Larson, G. L., C. D. McIntire, R. E. Truitt, M. W. Buktenica & E. Darnaugh-Thomas, 1996. Zooplankton Assemblages in Crater Lake, Oregon, USA. Lake and Reservoir Management 12: 281–297.

Larson, G. L., C. D. McIntire, M. Buktenica, S. Girdner & R. Hoffman, 2007a. Water quality and optical properties of Crater Lake, Oregon. .

Larson, G. L., C. D. McIntire, M. Buktenica, S. Girdner & R. Truitt, 2007b. Distribution and abundance of zooplankton populations in Crater Lake, Oregon.

Lee, S. & J. A. Fuhrman, 1987. Relationships between biovolume and biomass of naturally derived marine bacterioplankton. Applied and Environmental Microbiology 53: 1298–1303.

MacGregor, B. J., D. P. Moser, E. W. Alm, K. H. Nealson & D. A. Stahl, 1997. Crenarchaeota in Lake Michigan sediment. Applied and Environmental Microbiology 63: 1178–1181.

McIntire, C. D., G. L. Larson, R. E. Truitt & M. K. Debacon, 1996. Taxonomic structure and productivity of phytoplankton assemblages in Crater Lake, Oregon. Lake and Reservoir Management 12: 259–280.

McIntire, C. D., G. L. Larson & R. Truitt, this issue. Taxonomic composition and production dynamics of phytoplankton assemblages in Crater Lake, Oregon.

McManus, J., R. W. Collier & J. Dymond, 1993. Mixing processes in Crater Lake, Oregon. Journal of Geophysical Research 98: 295–307.

McManus, J., R. Collier, J. Dymond, C. G. Wheat & G. L. Larson, 1996. Spatial and temporal distribution of dissolved oxygen in Crater Lake, Oregon. Limnology and Oceanography 41: 722–731.

Methé, B. A. & J. P. Zehr, 1999. Diversity of bacterial communities in Adirondack lakes: do species assemblages reflect lake water chemistry? Hydrobiologia 401: 77–96.

Nelson, P. O., J. F. Reilly & G. L. Larson, 1996. Chemical solute mass balance of Crater Lake, Oregon. Lake and Reservoir Management 12: 248–258.

Nygaard, K. & A. Tobiesen, 1993. Bacterivory in algae—a survival strategy during nutrient limitation. Limnology and Oceanography 38: 273–279.

Pernthaler, J., T. Posch, K. Simek, J. Vrba, R. Amann & R. Psenner, 1997. Contrasting bacterial strategies to coexist with a flagellate predator in an experimental microbial assemblage. Applied and Environmental Microbiology 63: 596–601.

Peterson, B. J., J. E. Hobbie & J. F. Haney, 1978. *Daphnia* grazing on natural bacteria. Limnology and Oceanography 23: 1039–1044.

Polz, M. F. & C. M. Cavanaugh, 1997. A simple method for quantification of uncultured microorganisms in the environment based on in vitro transcription of 16S rRNA. Applied and Environmental Microbiology 63: 1028–1033.

Redfield, A. C., 1958. The biological control of chemical factors in the environment. American Scientist 46: 205–221.

Schiffman, S. S., M. L. Reynolds & F. W. Young, 1981. Introduction to multi-dimensional scaling. Theory, methods and applications. Academic Press, London.

Sekiguchi, Y., H. Takahashi, Y. Kamagata, A. Ohashi & H. Harada, 2001. In situ detection, isolation, and physiological properties of a thin filamentous microorganism abundant in methanogenic granular sludges: a novel isolate affiliated with a clone cluster, the Green Non-sulfur Bacteria, Subdivision I. Applied and Environmental Microbiology 67: 5740–5749.

Semenova, E. A. & K. D. Kuznedelov, 1998. A study of the biodiversity of Baikal picoplankton by comparative analysis of 16S rRNA gene 5′-terminal regions. Molecular Biology 32: 754–760.

Sherr, E. B. & B. F. Sherr, 2002. Significance of predation by protists in aquatic microbial food webs. Antonie van Leeuwenhoek 81: 293–308.

Simek, K., P. Kojecka, J. Nedoma, P. Hartman, J. Vrba & J. R. Dolan, 1999. Shifts in bacterial community composition associated with different microzooplankton size fractions in a eutrophic reservoir. Limnology and Oceanography 44: 1634–1644.

Simek, K., J. Pernthaler, M. G. Weinbauer, K. Hornak, J. R. Dolan, J. Nedoma, M. Masin & R. Amann, 2001. Changes in bacterial community composition and dynamics and viral mortality rates associated with enhanced flagellate grazing in a mesoeutrophic reservoir. Applied and Environmental Microbiology 67: 2723–2733.

Sturrock, K. & J. Rocha, 2000. A multidimensional scaling stress evaluation table. Field Methods 12: 49–60.

Trusova, M. Y. & M. I. Gladyshev, 2002. Phylogenetic diversity of winter bacterioplankton of eutrophic Siberian reservoirs as revealed by 16S rRNA gene sequences. Microbial Ecology 44: 252–259.

Urbach, E., K. L. Vergin, L. Young, A. Morse, G. L. Larson & S. J. Giovannoni, 2001. Unusual bacterioplankton community structure in ultra-oligotrophic Crater Lake. Limnology and Oceanography 46: 557–572.

Wuchter, C., S. Schouten, H. T. S. Boschker & J. S. S. Damste, 2003. Bicarbonate uptake by marine Crenarchaeota. FEMS Microbiology Letters 219: 203–207.

Zwart, G., B. C. Crump, M. P. K. V. Agterveld, F. Hagen & S. K. Han, 2002a. Typical freshwater bacteria: an analysis of available 16S rRNA gene sequences from plankton of lakes and rivers. Aquatic Microbial Ecology 28: 141–155.

Zwart, G., B. C. Crump, M. P. K. V. Agterveld, F. Hagen & S. K. Han, 2002b. Typical freshwater bacteria: an analysis of available 16S rRNA gene sequences from plankton of lakes and rivers. Aquatic Microbial Ecology 28: 141–155.

Zwisler, W., N. Selje & M. Simon, 2003. Seasonal patterns of the bacterioplankton community composition in a large mesotrophic lake. Aquatic Microbial Ecology 31: 211–225.

Hydrobiologia (2007) 574:179–204
DOI 10.1007/s10750-006-0352-4

CRATER LAKE, OREGON

Seasonal and interannual variability in the taxonomic composition and production dynamics of phytoplankton assemblages in Crater Lake, Oregon

C. David McIntire · Gary L. Larson · Robert E. Truitt

© Springer Science+Business Media B.V. 2007

Abstract Taxonomic composition and production dynamics of phytoplankton assemblages in Crater Lake, Oregon, were examined during time periods between 1984 and 2000. The objectives of the study were (1) to investigate spatial and temporal patterns in species composition, chlorophyll concentration, and primary productivity relative to seasonal patterns of water circulation; (2) to explore relationships between water column chemistry and the taxonomic composition of the phytoplankton; and (3) to determine effects of primary and secondary consumers on the phytoplankton assemblage. An analysis of 690 samples obtained on 50 sampling dates from 14 depths in the water column found a total of 163 phytoplankton taxa, 134 of which were identified to genus and 101 were identified to the species or variety level of classification. Dominant species by density or biovolume included *Nitzschia gracilis, Stephanodiscus hantzschii, Ankistrodesmus spiralis, Mougeotia parvula, Dinobryon sertularia, Tribonema affine, Aphanocapsa delicatissima, Synechocystis* sp., *Gymnodinium inversum*, and *Peridinium inconspicuum*. When the lake was thermally stratified in late summer, some of these species exhibited a stratified vertical distribution in the water column. A cluster analysis of these data also revealed a vertical stratification of the flora from the middle of the summer through the early fall. Multivariate test statistics indicated that there was a significant relationship between the species composition of the phytoplankton and a corresponding set of chemical variables measured for samples from the water column. In this case, concentrations of total phosphorus, ammonia, total Kjeldahl nitrogen, and alkalinity were associated with interannual changes in the flora; whereas pH and concentrations of dissolved oxygen, orthophosphate, nitrate, and silicon were more closely related to spatial variation and thermal stratification. The maximum chlorophyll concentration when the lake was thermally stratified in August and September was usually between depths of 100 m and 120 m. In comparison, the depth of maximum primary production ranged from 60 m to 80 m at this time of year.

Guest Editors: Gary L. Larson, Robert Collier, and Mark W. Buktenica
Long-term Limnological Research and Monitoring at Crater Lake, Oregon

C. D. McIntire (✉)
Department of Botany and Plant Pathology, Oregon State University, Corvallis, OR 97331, USA
e-mail: saxojazz@comcast.net

G. L. Larson
U. S. Geological Service, Forest and Rangeland Ecosystem Science Center, 3200 SW Jefferson Way, Corvallis, OR 97330, USA

R. E. Truitt
National Park Service, 1512 East Main Street, Ashland, OR 97520, USA

Regression analysis detected a weak negative relationship between chlorophyll concentration and Secchi disk depth, a measure of lake transparency. However, interannual changes in chlorophyll concentration and the species composition of the phytoplankton could not be explained by the removal of the septic field near Rim Village or by patterns of upwelling from the deep lake. An alternative trophic hypothesis proposes that the productivity of Crater Lake is controlled primarily by long-term patterns of climatic change that regulate the supply of allochthonous nutrients.

Keywords Crater Lake · Phytoplankton · Chlorophyll · Primary productivity · Species composition

Introduction

In 1982 the United States Congress authorized an intensive 10-year study of the Crater Lake ecosystem during which a large number of physical, chemical, and biological variables were investigated by modern limnological methods. This study included a preliminary investigation of species composition of the phytoplankton at two study sites from 1982 to 1984, and the subsequent initiation of a sampling program that was designed to reveal long-term spatial and temporal patterns in the taxonomic structure of the flora at Station 13, a location representative of the deepest basin in the lake. In addition, measurements of chlorophyll concentration and carbon-14 primary production were made between 1984 and 1990 and between 1986 and 1990, respectively. Results of the phytoplankton work during the 10-year study (Debacon & McIntire, 1991; McIntire et al., 1996) indicated that spatial and temporal distributions of primary productivity, chlorophyll, and algal populations were closely associated with patterns of circulation in the water column of the lake. When the lake was thermally stratified in late summer and early fall, phytoplankton populations exhibited a stratified distribution, and the maximum concentration of chlorophyll was usually found at depths between 100 m and 140 m. At this time of year, phytoplankton assemblages had lower species diversity and higher dominance in the epilimnion than in the metalimnion and upper part of the hypolimnion. In contrast, in winter and spring, when the lake water was mixing to a depth of between 200 m and 250 m (McManus et al., 1993), there was a more uniform distribution of chlorophyll and phytoplankton species in the euphotic zone (0–120 m) of the water column.

At the conclusion of the 10-year study, the phytoplankton sampling program was continued through the fall of 2000 as a part of a long-term monitoring program. In this paper, we extend the analysis of the patterns in chlorophyll concentration, primary productivity, and species composition in the water column of Crater Lake to include data obtained during the monitoring program. Specifically, data sets included in the analysis represented time periods from 1984 to 1999 (chlorophyll), from 1986 to 2000 (carbon-14 primary production), and from 1988 to 2000 (species composition). The objectives of the extended analysis were to describe long-term patterns in the structural and functional attributes of the phytoplankton; to explore relationships between water column chemistry and the species composition of the phytoplankton; to determine effects of primary and secondary consumers on the phytoplankton assemblage; and to examine spatial and temporal patterns in species composition, chlorophyll concentration, and primary productivity in relation to seasonal patterns of water circulation. Hypotheses examined in relation to these objectives were (1) the taxonomic structure of phytoplankton assemblages in the water column is associated with the dynamics of chemically defined water masses in the lake; (2) chlorophyll concentration and phytoplankton cell biovolume in the upper 40 m of the lake has a negative relationship with lake transparency; (3) seasonal changes in spatial distribution of phytoplankton populations are closely related to the process of thermal stratification; (4) phytoplankton biomass is limited and regulated by the processes of herbivory and predation by primary and secondary consumers; and (5) autotrophic processes in the water column of the lake are controlled by physical and chemical processes that generate tradeoffs between nutrient and light energy limitation.

Study site

Crater Lake is located in southern Oregon (USA) at a latitude of 42°56′ N and longitude of 122°06′ W. The lake is almost circular in shape with a diameter that ranges from 8 km to 10 km. The surface area of the lake is 53.2 km^2, and the maximum and mean depths are 589 m and 325 m, respectively. The lake was formed about 6,850 years ago after a climactic eruption and subsequent collapse of the top of Mount Mazama. Therefore, the catchment area of the lake is confined to the adjacent caldera wall, an area of only 15 km^2. Sources of water and nutrients for the lake include direct precipitation and atmospheric fallout, surface run-off and snowmelt from the surrounding catchment area, inputs from permanent and ephemeral streams and springs within the caldera, and inputs from underwater hydrothermal springs. Crater Lake is classified as an ultra-oligotrophic system, as it exhibits a high Secchi Disk transparency, usually between 28 m and 38 m, and very low concentrations of nutrients essential for plant growth. Additional details of the geological history, morphometry, and the chemical and physical properties of Crater Lake were described in publications by Byrne (1965), Phillips & Van Denburgh (1968), Bacon (1983), Bacon & Lanphere (1990), Barber & Nelson (1990), Nathenson & Thompson (1990), McManus (1992), and G. L. Larson et al. (1996).

Methods

Sampling methods

Preliminary sampling of phytoplankton in Crater Lake between 1982 and 1984 indicated that the floras in the water column of the two deepest basins of the lake (Stations 13 and 23) were similar (McIntire et al., 1996: Fig. 1). Beginning in 1985, all phytoplankton samples were obtained with 4-l Van Dorn bottles at Station 13, an intensive study site located in the deepest basin of the lake, approximately 3 km south of Cleetwood Cove (McIntire et al., 1996: Fig. 1). Data included in the analysis of species composition represented 690 samples obtained on 50 sampling dates from 14 depths in the water column (water surface and 5, 10, 20, 30, 40, 60, 80, 100, 120, 140, 160, 180, and 200 m below the surface). The time period corresponding to this set of samples extended from January 1988 through September 2000, a period when the flora was identified and counted by the same technician (Robert E. Truitt). There were 10 missing observations in the data set, 6 of which occurred in October 1989 when weather conditions restricted field activities.

Samples for the analysis of chlorophyll concentration in Crater Lake were obtained between June 1984 and September 1999 from 18 depths in the water column (water surface and depths of 5, 10, 20, 30, 40, 60, 80, 100, 120, 140, 160, 180, 200, 225, 250, 275, and 300 m below the surface). All samples were collected at Station 13 with 4-l Van Dorn bottles on the same day that samples were taken for the analysis of species composition. Water was transferred from the Van Dorn bottles to 500-ml opaque polyethylene bottles, and a few drops of $MgCO_3$ solution were added to each sample. All samples were stored and refrigerated in a dark container before transport to the laboratory for pigment analysis.

Water samples for measurements of carbon-14 primary production were obtained at Station 13 with 4-l Van Dorn bottles from the water surface and from depths of 5, 10, 20, 30, 40, 60, 80, 100, 120, 140, 160, and 180 m below the surface. A subsample of water from each bottle was transferred to a 125-ml transparent glass bottle and a 125-ml bottle that was covered with black tape. Water in the glass bottles was inoculated with 1 ml of radioactive carbon-14 $NaHCO_3$ solution (specific activity from 1.35 to 4.00 μCi ml^{-1}), and resuspended in situ at corresponding depths for approximately 4 h, usually between 1,000 h and 1,400 h. After the 4-h incubation period, the bottles were retrieved, refrigerated in a dark container, and transported to the laboratory for processing. Data reported in the results section of the paper represented samples obtained between 1986 and 1990, and between 1996 and 2000.

Chemical data reported in this paper were derived from samples obtained at Station 13 on 65 sampling dates between May 1986 and September 2000. These samples were collected with 4-l Van Dorn bottles from the lake surface, and from

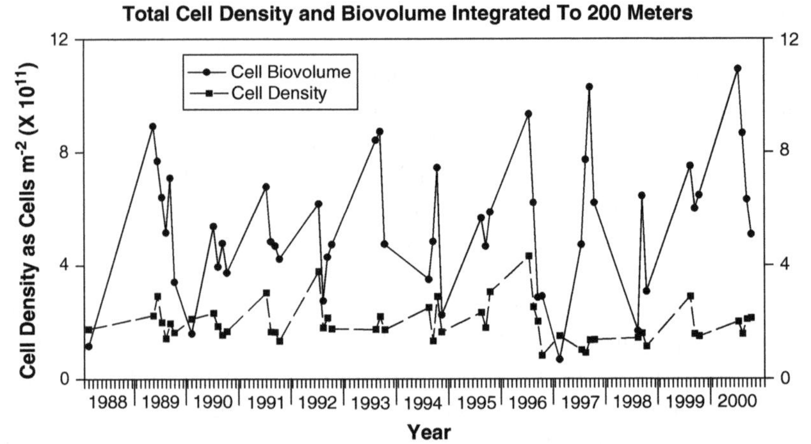

Fig. 1 Total phytoplankton cell density and biovolume integrated to a depth of 200 m in Crater Lake between January 1988 and September 2000

depths of 5, 10, 20, 60, 100, 200, 300, 400, 500, and 550 m below the surface. Chemical variables included in the statistical analysis were concentrations of dissolved oxygen, total Kjeldahl nitrogen, orthophosphate, total phosphorus, nitrate, ammonia, and silicon, as well as measurements of pH, alkalinity, and conductivity.

Laboratory methods

G. L. Larson et al. (1996) described laboratory methods for the measurement of chemical variables. Laboratory procedures for the quantitative analysis of phytoplankton species composition, chlorophyll concentration, and carbon-14 primary productivity were reported in detail by McIntire et al. (1996).

Water samples for the analysis of species composition were fixed in Lugol's solution and concentrated by allowing the cells to settle for 72 h. Subsamples of 50 ml were transferred to plastic chambers, and 300 algal cells were identified and counted at a magnification of 1,750×. This procedure allowed the enumeration of cells down to a minimum diameter of 1 μm. McIntire & Overton (1971) determined that a sample size of 300 cells was sufficient for the estimation of community composition parameters (e.g., heterogeneity and redundancy) and the relative abundance of dominant taxa.

Chlorophyll concentration in the water samples was determined by a fluorometric procedure (APHA 1985). Samples were filtered through a 0.45 μm Millipore filter, and pigments were extracted by placing the filters in 15-ml glass tubes to which 10 ml of 90% aqueous acetone was added. After centrifugation, the samples were processed in a Turner Designs Model 10 Fluorometer using standards provided by the Environmental Protection Agency. Beginning in 1990, samples for chlorophyll analysis also were fractionated into two components, material retained on a 2 μm Millipore filter, and material retained on a 0.22 μm filter after the sample had passed through the 2-μm filter. The purpose of this procedure was to separate picoplankton chlorophyll (cells < 2 μm in diameter) from chlorophyll produced by nannoplankton (cells 2–20 μm) and netplankton (cells > 20 μm).

Phytoplankton primary productivity was determined by the carbon-14 method described by Goldman (1963). All samples were filtered through a 0.45 μm Millipore filter and counted in a Beckman Liquid Scintilation Counter (Model LS9000).

Statistical methods

For the analysis of taxonomic structure, cell densities estimated from 690 samples obtained on 50 sampling dates and at 14 different depths in the water column (surface to a depth of 200 m) were organized into a data matrix. The samples were classified into a 4-cluster structure by Two Way Indicator Species Analysis (TWINSPAN). Briefly, the TWINSPAN program classifies samples (sites) and species simultaneously and constructs an ordered two-way table from a sample-

Table 1 List of 40 phytoplankton taxa used in the TWINSPAN analysis of 690 samples obtained from the Crater Lake water column between January 1988 and September 2000

Division	Taxon	Density	Biovolume
Bacillariophyta	*Asterionella formosa* Hass	4,274.52	3,932,556.00
	Nitzschia gracilis Hantzsch	39,965.83	18,903,835.80
	Stephanodiscus hantzschii Grun.	17,108.20	3,866,452.10
	Synedra delicatissima W. Smith	741.44	3,707,184.40
	Synedra radians Kutz.	594.14	718,909.40
Chlorophyta	*Ankistrodesmus falcatus* (Corda) Ralfs.	1,307.76	231,473.50
	Ankistrodesmus falcatus v. acicularis (Braun) West	1,276.82	140,449.90
	Ankistrodesmus spiralis (Turner) Lemm.	10,831.52	324,945.60
	Mougeotia parvula Hassel	19,174.90	8,130,159.50
	Oocystis pusilla Hansgrig	2,325.96	760,590.30
Chrysophyta	*Chromulina* sp. 1	2,466.77	95,217.30
	Chromulina minor Pasch.	744.1	21,578.90
	Chromulina parvula Conr.	8,965.64	154,208.90
	Chrysophyta sp. 1	13,819.42	237,694.10
	Dinobryon bavaricum Imhoff.	6,605.73	508,641.00
	Dinobryon sertularia Ehr.	23,098.24	9,216,196.40
	Ochromonas sp.4	1,177.14	161,268.50
	Ochromonas elegans Dolf.	3,636.92	822,672.00
	Ochromonas miniscula Conrad	906.13	30,355.40
	Ochromonas ovalis Dolf.	1,170.47	117,632.60
	Ochromonas verrucosa Skuja	1,208.09	35,034.60
	Ophiocytium capitatum Wolle	1,679.40	329,666.20
	Ophiocytium cochleare A. Br.	1,342.03	263,440.50
	Ophiocytium parvulum (Perty) A.Br.	8,065.20	1,583,198.60
	Pseudokephyrion sp. 2	678.79	46,836.60
	Pseudopedinella sp. 1	5,886.75	5,439,355.50
	Tribonema affine G.S. West	14,655.92	27,157,425.00
Cryptophyta	*Chroomonas acuta* Uter.	7,677.57	264,108.40
	Crytomonas erosa Ehrenberg	1,191.11	403,668.80
	Rhodomonas lacustris Pascher et Ruttner	2,026.31	2,111,415.00
	Rhodomonas minuta Skuja	860.82	708,454.90
	Rhodomonas minuta var. nannoplantica Skuja	841.54	213,750.20
Cyanobacteria	*Aphanocapsa delicatissima* West & West	335,884.40	604,591.90
	Chroococcus sp. 1	9,773.26	17,591.90
	Dactylococcopsis fascicularis Lemm.	858.2	9,869.30
	Lyngbya limnetica Lemm.	8,283.69	33,963.10
	Synechocystis sp. 1	100,621.49	10,213,081.70
Pyrrophyta	*Amphidinium luteum* Skuja	1,330.19	460,777.40
	Gymnodinium inversum Nygaard	2,058.43	11,197,865.70
	Peridinium inconspicuum Lemm.	2,126.37	14,353,026.70

Total density and biovolume of each taxon is the summation of all density and biovolume values over all samples, where the density and biovolume of each taxon in each sample are expressed as cells ml^{-1} and μm^3 ml^{-1}, respectively

by-species matrix (Hill et al., 1975; Jongman et al., 1987). The data file used for this analysis included 40 taxa representing 6 major divisions of the plant kingdom (Table 1). Rare taxa were eliminated from the data file to facilitate pattern recognition. The 40 taxa used in the analysis represented 97.4 and 98.7% of the total cell density and total cell biovolume recorded during the study, respectively. Results of the TWINSPAN analysis were reported as a four-cluster diagram and a corresponding table of mean densities of each taxon for each cluster.

A holistic approach to the examination of the relationship between the phytoplankton flora and water column chemistry involved the combination of classification and discriminant analysis. In this case, a matrix of chemical data obtained from the water column was reordered so that the samples were classified according to the 4-cluster structure derived from the TWINSPAN analysis of the phytoplankton data. Subsequently, the matrix of chemical data was analyzed by a discriminant analysis (Pimentel, 1979) to determine the degree to which patterns in the chemical data conformed to the cluster structure of the phytoplankton data. Results of this analysis were reported as a plot of the sample points for each cluster in discriminant space after the discriminant factors were interpreted in terms of the individual chemical variables.

Results

Cell density and biovolume

Debacon & McIntire (1990) published a list of 88 taxa found in phytoplankton samples from the water column of Crater Lake during the period from June 1985 through April 1987. By the year 2000, the list of taxonomic entities identified to the division level had increased to 163 taxa, 134 of which were identified to genus and 101 were identified to either the species or variety level of classification. The mean biovolume of these taxa ranged from 1.8 μm^3 $cell^{-1}$ (*Aphanocapsa delicatissima* and *Chroococcus* sp.) to 42,000 μm^3 $cell^{-1}$ (*Nitzschia tryblionella*). All but four taxa had a mean biovolume of less than 10,000 μm^3 $cell^{-1}$, whereas the mean biovolume of 133 taxa was less than 1,000 μm^3 $cell^{-1}$. Of the 163 taxa identified to the division level, there were 56 diatoms (Bacillariophyta), 24 chlorophytes (Chlorophyta), 53 chrysophytes (Chrysophyta), 5 cryptomonads (Cryptophyta), 12 cyanobacteria, 12 dinoflagellates (Pyrrhophyta), and 1 euglenoid (Euglenophyta).

Total density and biovolume of 40 dominant phytoplankton taxa are listed in Table 1. These totals were calculated by summing the densities of each taxon over the 690 samples used in the TWINSPAN analysis, where the density and biovolume in each sample were expressed as cells ml^{-1} and μm^3 ml^{-1}, respectively. Numerically, the most abundant taxa (total density > 10,000 cells) were *Nitzschia gracilis, Stephanodiscus hantzschii, Ankistrodesmus spiralis, Mougeotia parvula*, an unidentified chrysophyte, *Dinobryon sertularia, Tribonema affine, Aphanocapsa delicatissima*, and *Synechocystis* sp. Of these taxa, a relatively small cyanobacterium (*A. delicatissima*) had the highest total density in the data set. However, this species was not among the taxa with the highest total cell biovolume because of its small cell size (<3 μm in diameter). Taxa with the highest total cell biovolume (>10,000,000 μm^3) were *Nitzschia gracilis, Tribonema affine, Synechocystis* sp., *Gymnodinium inversum*, and *Peridinium inconspicuum*. The latter two species, the dinoflagellates, were usually present in small numbers, but because of their relatively large cell sizes, they made significant contributions to the cell biovolume of the flora when they were present in the water column.

Temporal plots of total phytoplankton cell density and biovolume integrated to a depth of 200 m in the water column are presented in Fig. 1. The lack of correspondence between the two graphs indicated that there was a periodic change in the average cell size of the flora. Correlation coefficients expressing the covariance between cell density and chlorophyll concentration and between cell biovolume and chlorophyll concentration were low, $r = 0.20$ and 0.07, respectively. Likewise, the corresponding correlation coefficient between carbon-14 primary production and either cell density or biovolume also was low ($r = 0.18$ and 0.02).

Seasonal patterns in the vertical distribution of phytoplankton in the upper 200 m of the water column were determined by plotting mean cell density and biovolume against depth for 7 months of the year (Figs. 2 and 3). Data were not obtained in the months of February, March, November, and December. The distribution of cells and biovolume was more uniform throughout the water column in January than in the other months represented by the data. Patterns in the spring data were less certain because of the small sample size (only one sample from April and May). Vertical distributions of cell density and biovolume in June were similar with maxima at a depth of 80 m below the water surface. The greatest discrepancy between patterns in cell density and biovolume were found for the month of August after the lake was thermally stratified. At this time of year, maximum mean cell biovolume was in the upper 10 m of the water column, whereas mean cell

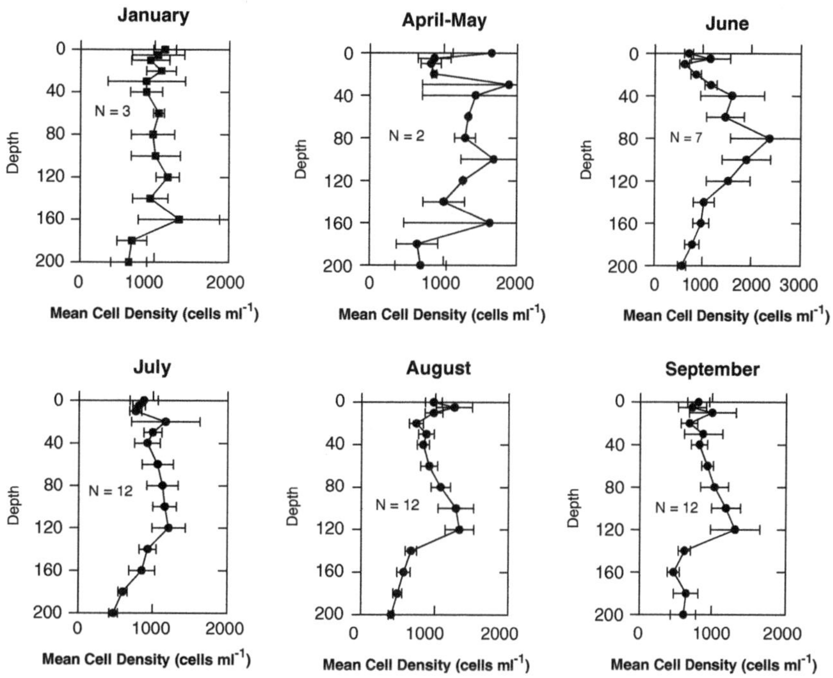

Fig. 2 Mean vertical distribution of total phytoplankton cell density in Crater Lake between January 1988 and September 2000. *N* values are the number of samples, and the horizontal bars represent the standard error of the mean

density maxima were at depths of 5 m and between 100 m and 120 m. In September, thermal stratification continued throughout the month, and maximum mean cell density also was at a depth of 120 m. In contrast, mean cell biovolume in September was greatest and more uniformly distributed in the water column between the water surface and a depth 100 m.

When the lake was thermally stratified in August, some of the dominant phytoplankton taxa

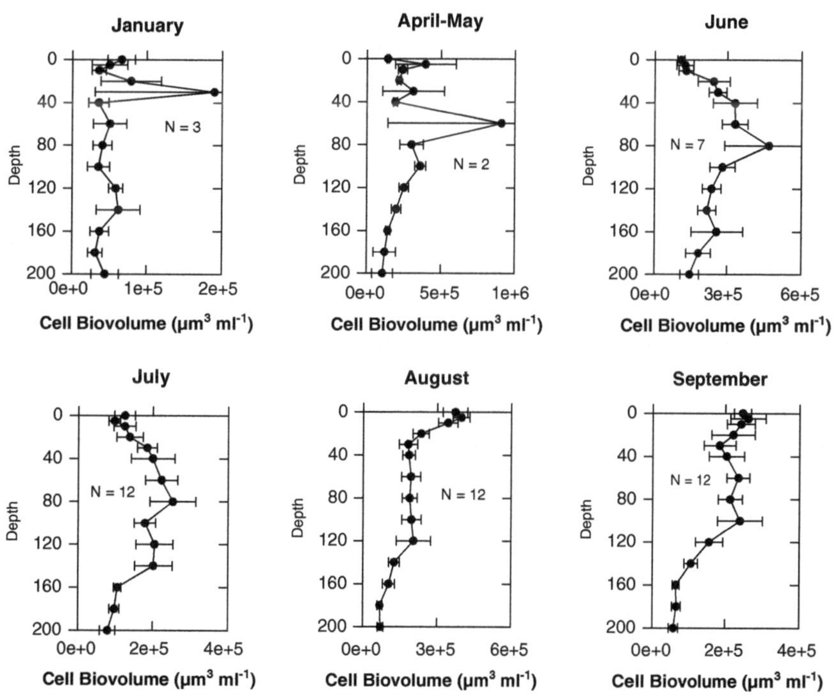

Fig. 3 Mean vertical distribution of total phytoplankton cell biovolume in Crater Lake between January 1988 and September 2000. *N* values are the number of samples, and the horizontal bars represent the standard error of the mean

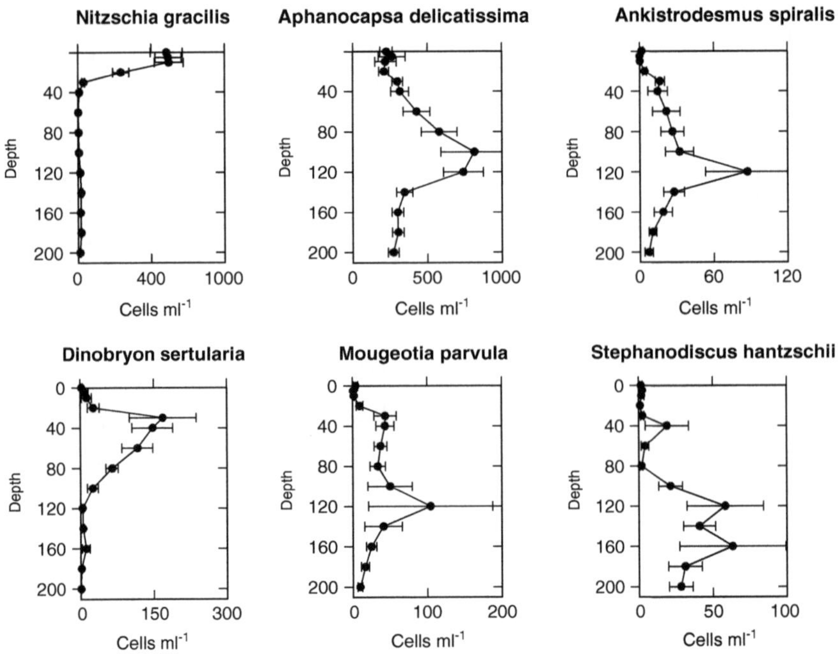

Fig. 4 Mean vertical distribution of selected phytoplankton species during August between 1989 and 2000. Horizonal bars represent the standard error of the mean

exhibited a stratified vertical distribution in the water column (Fig. 4). For example, *Nitzschia gracilis* was dominant in the epilimnion from the lake surface to a depth of 20 m, whereas *Dinobryon sertularia* usually reached its maximum density in the metalimnion between depths of 40 m and 60 m. Examples of taxa with deep-water maxima (100–160 m below the surface) were *Aphanocapsa delicatissima*, *Ankistrodesmus spiralis*, and *Stephanodiscus hantzschii.*

Cell size

Seasonal, and spatial patterns of change in cell size were examined by calculating the mean ratio of cell biovolume to cell density. In this case, mean size was expressed as μm^3 cell^{-1}. Mean ratios were calculated for 14 depths (surface to a depth of 200 m) and five months of the year (January, June, July, August, and September) from data obtained from 1988 through 2000. Annual variation in mean cell size for this period of time also was examined by calculating the mean biovolume to density ratio throughout the entire water column (down to 200 m) for all phytoplankton samples obtained each year during the month of August, when the lake exhibited thermal stratification and physical properties of the water column were similar from year to year.

In January, mean cell size was relatively small and exhibited little variation between the lake surface and a depth of 200 m (Fig. 5). The average cell size increased between January and early summer, and was slightly less in the near-surface water than at depths below 10 m in June and July. However, during thermal stratification (August and September), mean cell size was considerably greater in the epilimnion (>300 μm^3 cell^{-1}) than at depths greater than 20 m below the lake surface. Below a depth of 20 m, mean cell size in August and September was similar to values calculated for June and July. Mean cell size in August was relatively low in 1989, 1990, and 1991, as compared to the years from 1992 through 2000 (Fig. 6). The year 1992 apparently was a period of change in the pattern of cell size, as there was considerable variation in cell size between the water surface and a depth of 200 m.

Classification of phytoplankton samples

The TWINSPAN classification of 690 phytoplankton samples obtained from January 1988 to

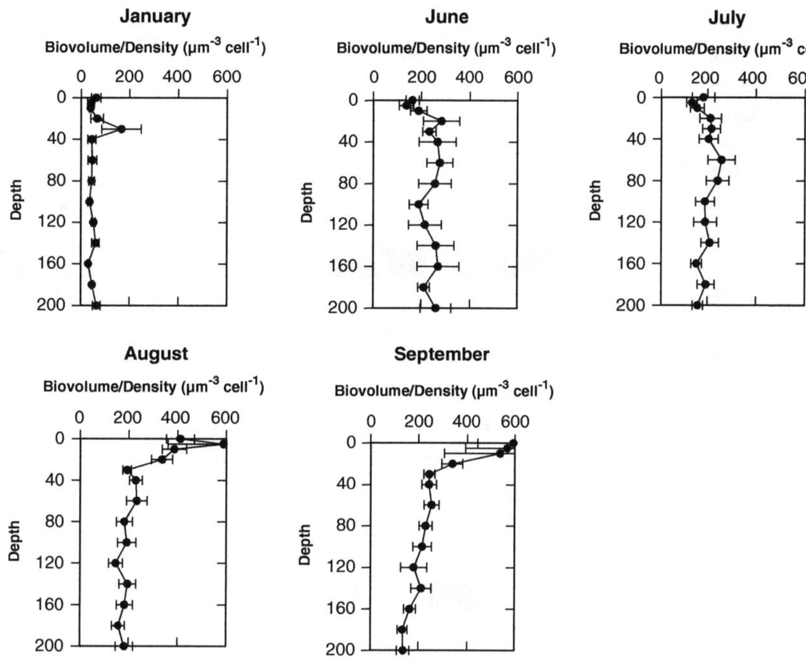

Fig. 5 Mean depth profiles of the ratio of the phytoplankton cell biovolume to cell density in Crater Lake between January 1988 and September 2000. Horizonal bars represent the standard error of the mean

September 2000 at depths between the water surface and a depth of 200 m identified four groups of samples that represented spatial and temporal patterns in the flora (Fig. 7). In general, the cluster diagram illustrates a vertical stratification of the flora from the middle of the summer through early fall, a pattern related to the onset and establishment of thermal stratification in the water column. Moreover, the analysis indicated that the species composition of the phytoplankton usually was uniform throughout the water column in January, in the spring, and sometimes in early summer. This pattern corresponded to periods of mixing in the upper 200–250 m of the lake. The classification analysis also indicated that the relative abundance and species composition of the flora in 1988, 1989, and 1990, represented mostly by clusters 3 and 4, were different from the flora that was present between 1992 and 2000 (clusters 1 and 2). Therefore, years 1991 and 1992

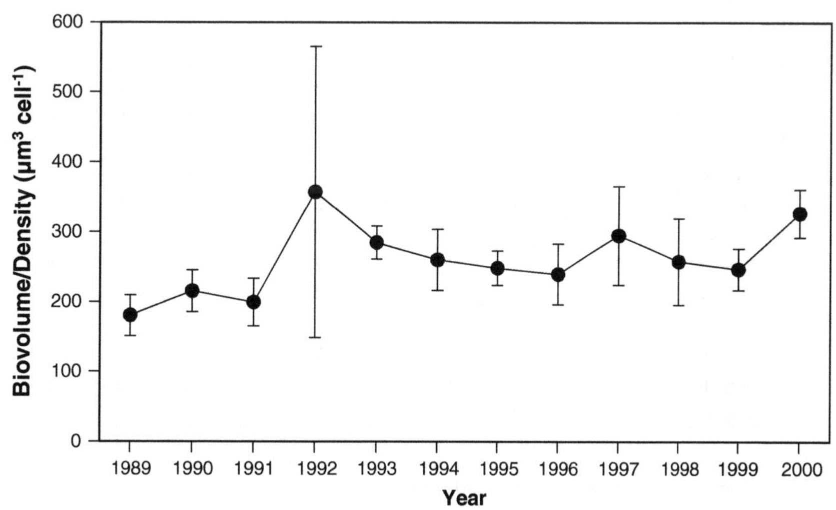

Fig. 6 Mean ratio of the phytoplankton cell biovolume to cell density for all samples collected each year during the month of August from 1989 to 2000. Vertical bars represent the standard error of the mean for samples collected from the water surface to a depth of 200 m

Depth	0	5	10	20	30	40	60	80	100	120	140	160	180	200
Jan-88	3	3	3	3		3	3	3	3	3	3	3	3	3
Apr-89	3	3	3	3	3	3	3	3	3	3	3	3	3	3
May-89	3	3	3	3	3	3	3	3	3	3	3	3	3	3
Jun-89	3	3	3	3	3	3	3	3	3	3	3	3	3	3
Jul-89	4	4	4	4	3	3	3	3	3	3	3	3	3	3
Aug-89	4	4	4	4	3	3	3	3	4	3	3	3	3	3
Sep-89	4	4	4	4	4	4	3	4	4	3	3	3	3	3
Oct-89	4	4		4			4	3	3	4				3
Jan-90	3	3	3	3	3	3	3	3	3	3	3	3	3	3
Jun-90	4	4	4	4	4	4	3	3	3	3	3	3	3	3
Jul-90	4	4	4	4	4	4	3	3	3	3	3	3	3	3
Aug-90	4	1	4	4	4	3	3	3	3	3	3	3	3	3
Sep-90	4	1	1	4	4	3	3	2	3	3	3	3	4	2
Jun-91	2	2	2	2	1	2	2	1	2	2	2	2	2	2
Jul-91	1	1	1	1	2	2	2	3	2	2	3	2	2	2
Aug-91	4	4	1	1	1	2	2	2	2	2	2	2	2	2
Sep-91	4	4	4	1	1	2	2	2	2	2	2	2	2	2
Jun-92	4	1	1	1	1	1	2	2	2	2	2	2	2	2
Jul-92	1	1	1		1	2	1	2	1	2	2	2	2	2
Aug-92	1	1	1	1	1	1	1	2	2	2	2	2	1	2
Sep-92	1	1	1	1	1	1	1	1	2	2	2	2	2	1
Jul-93	1	2	2	1	2	2	2	2	2	2	2	2	2	2
Aug-93	1	1	1	1	1	2	2	2	2	2	2	2	2	2
Sep-93	1	1	1	1	1	1	2	2	2	2	2	2	2	2
Jul-94	1	2	1	1	2	2	2	2	2	2	2	2	2	2
Aug-94	1	1	1	1	1	1		2	2		2	2	2	2
Sep-94	1	1	1	1	1	1	2	2	2	2	2	2	2	1
Oct-94	1	1	1	1	1	1	2	2	2	1	2	2	2	2
Jul-95	4	1	1	2	1	2	1	2	2	2	2	2	2	2
Aug-95	1	1	1	1	1	2	2	2	2	2	2	2	2	2
Sep-95	1	1	1	1	1	1	2	2	2	2	2	2	2	2
Jun-96	2	2	1	2	2	2	2	2	2	2	2	2	2	2
Jul-96	1	1	1	1	2	2	2	2	2	2	2	2	2	2
Aug-96	1	1	1	1	2	2	2	2	2	2	2	2	2	2
Sep-96	1	1	1	1	2	2	2	2	2	2	2	2	2	2
Jan-97	2	2	2	2	2	2	2	2	2	2	2	2	2	2
Jun-97	2	2	2	2	2	2	2	2	2	2	2	2	2	2
Jul-97	1	1	1	1	2	2	2	2	2	2	2	2	2	2
Aug-97	1	1	1	1	2	2	2	2	2	2	2	2	2	2
Sep-97	1	1	1	1	1	2	2	2	2	2	2	2	2	2
Jul-98	1	1	1	1	1	2	2	2	2	2	2	2	2	2
Aug-98	1	1	1	1	1	2	2	2	2	2	2	2	2	2
Sep-98	1	1	1	1	1	2	2	2	2	2	2	2	2	2
Jul-99	2	2	2	2	2	2	2	2	2	2	2	2	2	2
Aug-99	1	1	1	1	1	2	2	2	2	2	2	2	2	2
Sep-99	1	1	1	1	1	2	2	2	2	2	2	2	2	2
Jun-00	2	2	2	2	2	2	2	2	2	2	2	2	2	2
Jul-00	1	1	1	2	2	2	2	2	2	2	2	2	2	2
Aug-00	1	4	1	1	2	2	2	2	2	2	2	2	2	2
Sep-00	1	1	1	1	1	2	2	2	2	2	2	2	2	2

Fig. 7 Four-cluster structure of Crater Lake phytoplankton data for the period of January 1988 to September 2000. The analysis was based on cell densities of 40 dominant taxa

represented a transition period that brought about detectable changes in variables associated with the dynamics of the phytoplankton assemblage.

Mean percentage abundances of 40 phytoplankton taxa in samples for each of the clusters illustrated in Fig. 7 are listed in Table 2. Cluster 1 was composed of samples mostly from the upper 40 m of the water column between July 1992 and September 2000 during periods when the lake was thermally stratified. Dominant taxa in these samples were *Nitzschia gracilis*, *Synechocystis* sp., and *Aphanocapsa delicatissima*. Other less abundant taxa that were indicators for cluster 1 included *Pseudopedinella* sp., *Gymnodinium inversum*, *Peridinium inconspicuum*, and *Dinobryon bavaricum*. Cluster 2 was represented by samples obtained between June 1991 and September 2000, mostly at depths between 40 m and 200 m during thermal stratification, and between

Table 2 Mean percentage abundance of 40 phytoplankton taxa found in 690 samples partitioned into four clusters by the classification program TWINSPAN

Taxon	Cluster 1	Cluster 2	Cluster 3	Cluster 4
Aphanocapsa delicatissima West & West	34.56	50.53	59.31	39.82
Synechocystis sp.	16.23	14.12	10.96	20.95
Nitzschia gracilis Hantzsch	20.62	2.09	1.21	14.48
Dinobryon sertularia Ehr.	3.20	3.37	2.39	6.66
Chrysophyta, unidentified	3.56	2.71	1.84	2.10
Mougeotia parvula Hassel	1.83	4.67	2.46	0.76
Stephanodiscus hantzschii Grun.	0.28	4.86	3.19	0.17
Dinobryon bavaricum Imhoff.	5.64	0.07	0.25	1.34
Chromulina parvula Conr.	1.51	0.64	2.56	1.62
Ophiocytium parvulum (Perty) A.Br.	0.01	0.14	4.42	1.34
Tribonema affine G.S. West	0.19	4.51	0.61	0.06
Ankistrodesmus spiralis (Turner) Lemm.	1.00	2.64	0.75	0.08
Pseudopedinella sp.	2.03	0.52	0.58	1.24
Chroococcus sp.	1.91	2.09	0.00	0.02
Chromulina sp.	0.10	0.20	0.49	2.78
Peridinium inconspicuum Lemm.	1.65	0.14	0.45	0.80
Chroomonas acuta Uter.	0.11	1.50	1.08	0.07
Rhodomonas lacustris Pascher *et* Ruttner	0.00	0.03	0.89	1.79
Gymnodinium inversum Nygaard	1.22	0.30	0.15	0.26
Ochromonas elegans Dolf.	1.18	0.43	0.22	0.03
Asterionella formosa Hass	0.13	1.10	0.63	0.00
Ophiocytium capitatum Wolle	0.00	0.05	0.85	0.65
Lyngbya limnetica Lemm.	0.05	1.28	0.05	0.02
Oocystis pusilla Hansgrig	0.93	0.36	0.02	0.02
Ankistrodesmus falcatus (Corda) Ralfs.	0.00	0.01	0.83	0.29
Amphidinium luteum Skuja	0.52	0.13	0.10	0.35
Ophiocytium cochleare A. Br.	0.00	0.03	0.88	0.12
Ochromonas miniscula Conrad	0.07	0.01	0.28	0.64
Ochromonas verrucosa Skuja	0.00	0.00	0.85	0.12
Ochromonas ovalis Dolf.	0.41	0.09	0.20	0.24
Chromulina minor Pasch.	0.24	0.02	0.07	0.40
Synedra delicatissima W. Smith	0.01	0.03	0.30	0.34
Ochromonas sp.	0.33	0.28	0.02	0.00
Dactylococcopsis fascicularis Lemm.	0.00	0.00	0.41	0.19
Ankistrodesmus falcatus v. *acicularis* (Braun) West	0.19	0.34	0.02	0.04
Synedra radians Kutz.	0.00	0.01	0.33	0.12
Rhodomonas minuta Skuja	0.00	0.12	0.20	0.08
Pseudokephyrion sp.	0.22	0.11	0.05	0.00
Crytomonas erosa Ehrenberg	0.00	0.33	0.03	0.00
Rhodomonas minuta var. *nannoplantica* Skuja	0.08	0.14	0.07	0.03

The cluster numbers correspond to the numbers in the cluster diagram (Fig. 8)

the water surface and a depth of 200 m when the lake was not thermally stratified. Although *Synechocystis* sp. and *A. delicatissima* also were very abundant in the samples of cluster 2, better indicator taxa for this cluster were *Tribonema affine*, *Mougeotia parvula*, *Stephanodiscus hantzschii*, and *Ankistrodesmus spiralis*.

Phytoplankton samples collected between January 1988 and September 1990 were represented by cluster 3 and cluster 4 (Fig. 7). Samples in cluster 3 and cluster 4 had a similar spatial distribution as samples in cluster 2 and cluster 1, respectively. In other words, samples in cluster 3 were from deeper water (>20 or 40 m) during thermal stratification, and throughout the water column when the lake was not stratified; and samples in cluster 4 were typically found in the epilimnion. Samples in both of these clusters also had relatively high mean proportions of cyanobacteria (*Synechocystis* sp. and *Aphanocapsa delicatissima*). *Nitzschia gracilis* was a good indicator for cluster 4 (Table 2), but its relative abundance in this group of samples was less than in the samples from cluster 1. Other indicators of cluster 4 were *Chromulina* sp. and *Dinobryon sertularia*; and some good indicators of cluster 3 included *Chromulina parvula*, *Ophiocytium parvulum*, and the highest mean relative abundance of *A. delicatissima*. As mentioned in the previous section, mean cell size of organisms found in 1989, 1990, and 1991, corresponding to clusters 3 and 4, was less than the mean cell size found in samples obtained after 1991, i.e., samples mostly in clusters 1 and 2 (Figs. 6 and 7).

Relationships between phytoplankton and chemical variables

Chemical variables in the water column were related to the entire phytoplankton assemblage by grouping the water column chemical data according to the 4-cluster structure found for the phytoplankton data (Fig. 7). For this analysis, the data were reduced to 148 sets of samples, because it was necessary to eliminate phytoplankton samples for which there was no matching chemical data. Mean values for ten chemical variables corresponding to the 4-cluster structure of the phytoplankton (Fig. 7) are listed in Table 3. Phytoplankton samples from the epilimnion (clusters 1 and 4) tended to be associated with water with slightly lower concentrations of dissolved oxygen and orthophosphate than samples from other the parts of the lake. Phytoplankton samples collected from 1989 through 1990 (clusters 3 and 4) were found in water with lower concentrations of TKN and higher concentrations of ammonia than samples collected after 1992, indicating that the concentration of organic nitrogen increased between 1991 and 1992. Moreover, samples obtained from the hypolimnion, and at times during the year when the water column was mixing (clusters 2 and 3), were collected from water with higher concentrations of nitrate than water from the near surface layer during thermal stratification.

The distribution of the chemical samples in relation to the 4-cluster structure of the phytoplankton also was examined by a canonical analysis of discriminance. In this case, chemical data again were grouped according to the 4-cluster structure of the phytoplankton, and linear combinations of the variables were calculated so that the among-group variance was maximized. A plot of the first two canonical factors (Fig. 8) indicated that there was considerable overlap among the four groups of samples. However, there was a tendency for the samples in each group to be concentrated in a specific area of the ordination: left and upper left (cluster 1); left and lower left (cluster 2); right and lower right (cluster 3); and upper and upper right (cluster 4). Correlation coefficients between the first two factors and individual chemical variables indicated that Factor 1 represented a contrast between total phosphorus/ammonia and TKN/alkalinity, whereas Factor 2 displayed a contrast between silicon/pH and dissolved oxygen/orthophosphate/nitrate (Table 4). Therefore, the ordination graph is consistent with the conclusion that samples in clusters 1 and 2 tended to come from water with a relatively high alkalinity and concentration of TKN in comparison with samples in the other clusters, whereas samples in clusters 3 and 4 were associated with higher concentrations of ammonia and total phosphorus. Chemical differences displayed by Factor 1 related more to temporal variation, in this case 1989–1991 compared with

Table 3 Mean values for chemical variables corresponding to the four-cluster structure of phytoplankton data from the water column of Crater Lake 1989–2000

Variables	Cluster 1	Cluster 2	Cluster 3	Cluster 4
Dissolved Oxygen (mg l^{-1})	9.2820	10.3522	10.5160	9.0793
TKN (mg l^{-1})	0.0226	0.0225	0.0167	0.0184
Orthophosphate (mg l^{-1})	0.0096	0.0103	0.0103	0.0095
Total Phosphorus (mg l^{-1})	0.0225	0.0227	0.0293	0.0237
Nitrate (mg l^{-1})	0.0002	0.0012	0.0007	0.0001
Ammonia (mg l^{-1})	0.0012	0.0016	0.0051	0.0052
Silicon (mg l^{-1})	8.3988	8.3959	8.3590	8.5056
pH	7.7600	7.7313	7.6953	7.7607
Alkalinity (mg l^{-1})	7.1366	7.1428	7.0179	7.0698
Conductivity (μmhos cm^{-1})	112.1462	114.7906	114.1698	115.4025

1992–2000, than to spatial variation. In contrast, Factor 2 was more closely related to spatial variation during thermal stratification, particularly with respect to gradients of nitrate and dissolved oxygen from the epilimnion to the hypolimnion during thermal stratification. For example, most of the samples in cluster 1 and cluster 4 tend to be located towards the upper end of the Factor 2 axis. These samples were obtained mostly from the epilimnion where the dissolved oxygen and nitrate were relatively low during the summer and early fall.

Multivariate test statistics (Wilks' lamda and the corresponding F-value) indicated that there was a significant difference ($P < 0.001$) among the cluster centroids for the clusters displayed in Fig. 8. Univariate F tests for the individual chemical variables listed in Tables 3 and 4 also

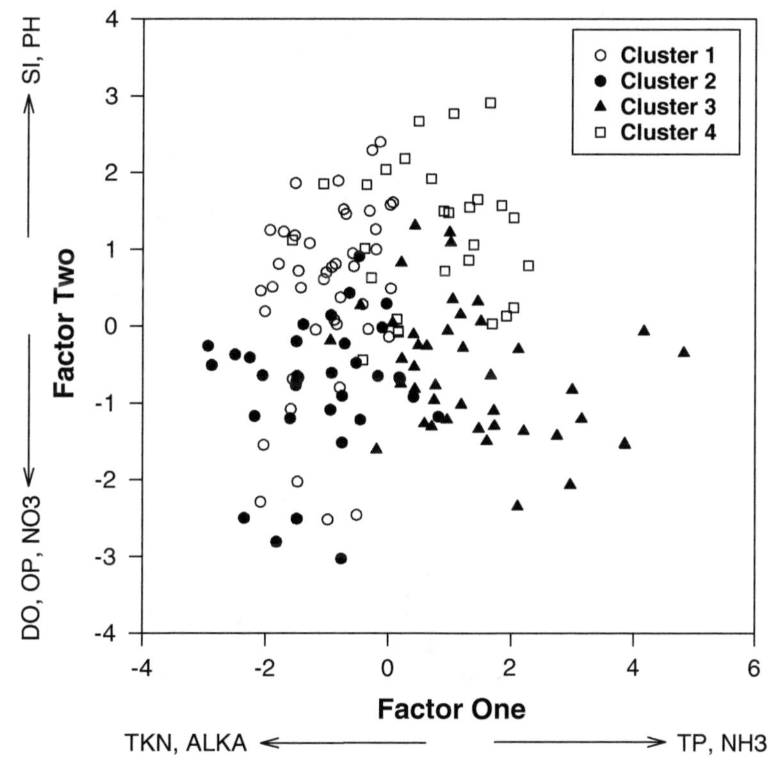

Fig. 8 Discriminant analysis of Crater Lake water column chemical data relative to the four-cluster structure of the phytoplankton data illustrated in Fig. 8. Axis interpretation is based on the correlation between each chemical variable and the first two discriminant factors (see Table 4)

Table 4 Correlation coefficients expressing the covariance between ten chemical variables and the first two factors derived from a discriminant analysis of water column chemical data relative to the 4-cluster structure of the phytoplankton

Variable	Factor 1	Factor 2
Dissolved Oxygen	0.212	**–0.790**
TKN	**–0.522**	0.022
Orthophosphate	0.117	**–0.667**
Total Phosphorus	**0.490**	–0.290
Nitrate	–0.024	**–0.421**
Ammonia	**0.700**	0.120
Silicon	–0.004	**0.503**
pH	–0.313	**0.486**
Alkalinity	**–0.743**	0.094
Conductivity	0.082	–0.014

Factor loadings greater than 0.4 are emphasized in bold type and correspond to the directional arrows in Fig. 11

found significant differences ($P < 0.01$, d.f. = 3, 144) among the cluster means for all variables except conductivity ($P = 0.484$). Although the four clusters exhibited overlapping distributions, the discriminant function classified 70.3% of the samples into their assigned clusters.

Chlorophyll and primary production

The concentration of chlorophyll in the upper 200 m of the water column of Crater Lake was relatively high from 1986 to 1991 (Fig. 9). The chlorophyll maximum in 1987 corresponded to the highest rate of primary production measured between 1986 and 2000 (Fig. 10). Data obtained between 1984 and 1999 indicated that chlorophyll

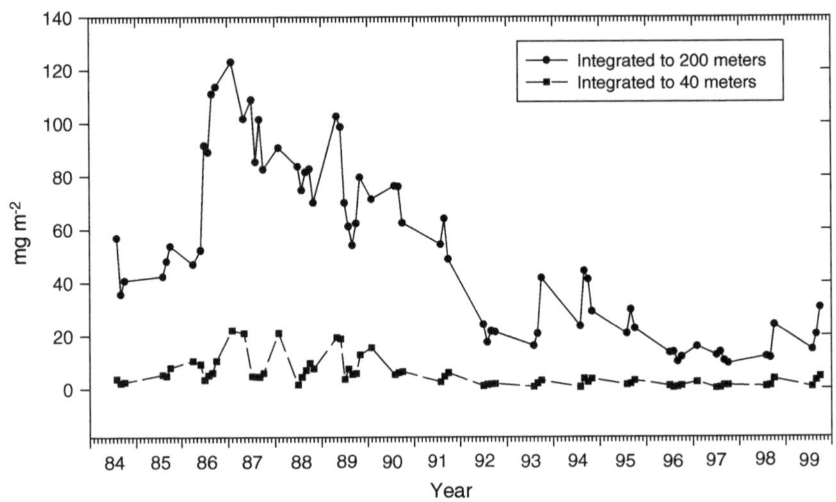

Fig. 9 Chlorophyll concentration in the water column of Crater Lake integrated to the depths 40 m and 200 m. Data were obtained between 1984 and 1999

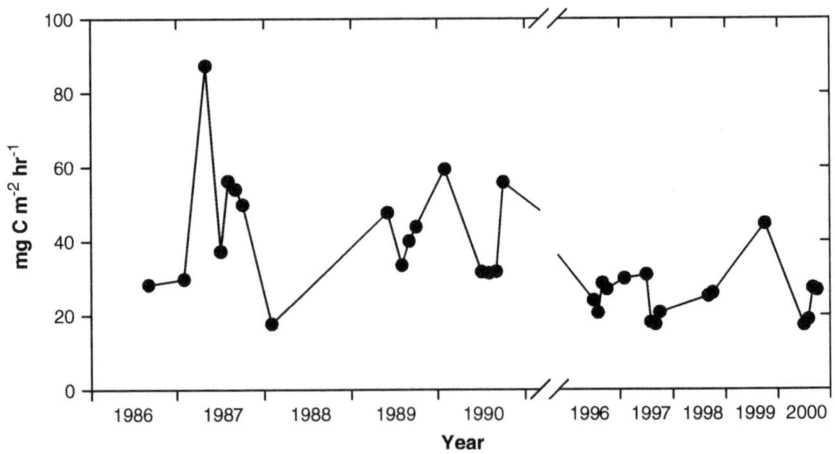

Fig. 10 Carbon-14 primary productivity in the water column of Crater Lake integrated to the depth of 180 m. Data were obtained between 1986 and 1990, and from 1996 to 2000

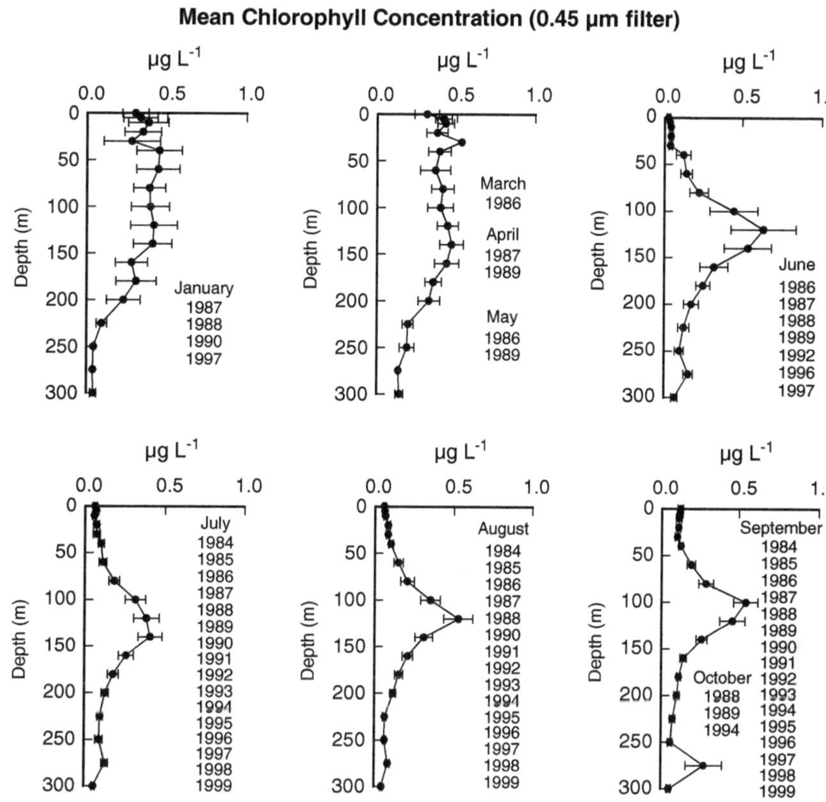

Fig. 11 Vertical depth profiles of chlorophyll concentration in the water column of Crater Lake from 1984 to 1999. Horizontal bars represent the standard error of the mean

was uniformly distributed in the water column between the lake surface and a depth of approximately 150 m in the winter and spring (Fig. 11). During the summer months and early fall, chlorophyll reached mean maximum concentrations at depths between 100 m and 140 m. The mean maximum concentration when the lake was thermally stratified in August and September was usually between depths of 100 m and 120 m, whereas the mean depth of maximum primary production varied between 60 m and 80 m (Fig. 12).

Vertical distribution of chlorophyll in the lake at Station 13 also was partitioned into two size fractions: the fraction associated with particles retained by a 2.0 μm filter (nannoplankton and net plankton) and a fraction retained by a 0.22 μm filter, after it had passed through a 2.0 μm filter (picoplankton). Results of this analysis are presented as a comparison of one set of samples obtained in January, when the lake was circulating to a depth of at least 200 m, and 7 sets of samples obtained in August, when the lake was thermally stratified (Fig. 13). Patterns for the months of July and September were similar to the mean pattern found for the month of August. In January, the ratio of the 0.22 μm fraction to the 2.0 μm fraction ranged from 0.85 (water surface) to 0.15 (300 m below the surface) with a mean ratio for all depths of 0.47. This relatively high mean ratio corresponded to a small mean cell size of organisms collected in three sets of samples obtained during the month of January (Fig. 5). However, when thermal stratification was present, picoplankton chlorophyll represented a smaller fraction of the total chlorophyll in the water column. Mean ratios of the 0.22 μm fraction to the 2.0 μm fraction for the months of June, July, August, September, and October were 0.17, 0.14, 0.18, 0.20, and 0.19, respectively.

Correlation coefficients expressing relationships between chlorophyll concentration at various depths and corresponding values for 8 chemical variables were relatively low: 0.195 (dissolved oxygen), 0.346 (pH), –0.036 (alkalin-

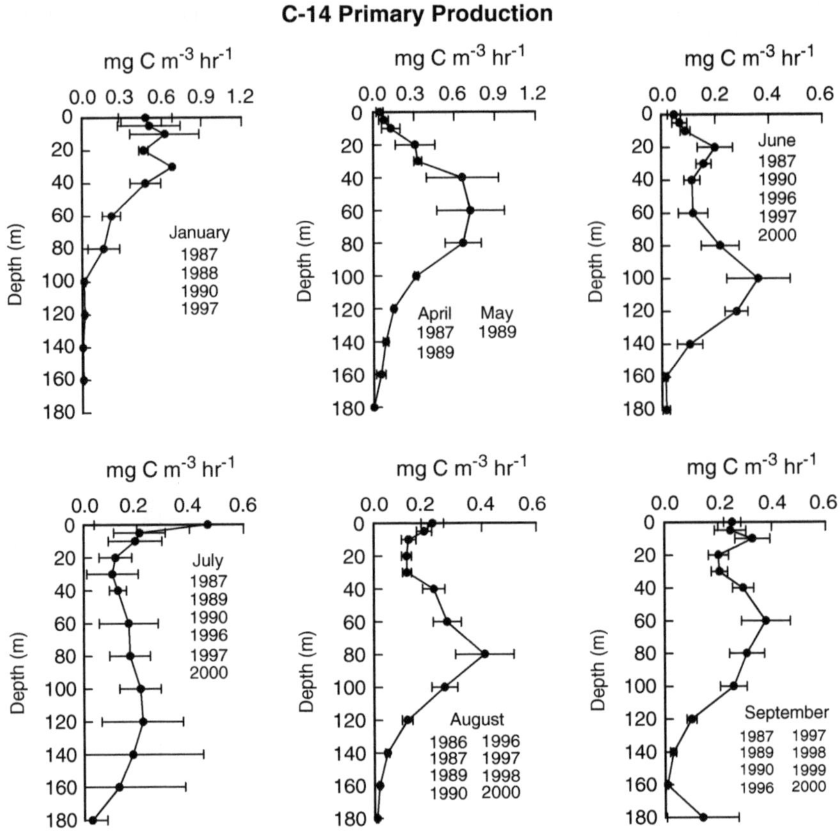

Fig. 12 Vertical depth profiles of carbon-14 primary productivity in the water column of Crater Lake from 1986 to 2000. Horizontal bars represent the standard error of the mean

ity), –0.041 (conductivity), 0.102 (total phosphorus), 0.006 (orthophosphate), 0.094 (total Kjeldahl nitrogen), and –0.252 (nitrate). Because of the large number of samples (261), the coefficients for pH and nitrate were significant at $P < 0.01$.

Vertical depth profiles of carbon-14 primary production were summarized for all observations obtained between 1986 and 2000 (Fig. 12). In January, the maximum mean rate of primary production (0.68 mg C m^{-3} h^{-1}) occurred in the upper 40 m of the water column, whereas, during the spring of the year, maximum mean rates (0.66–0.72 mg C m^{-3} h^{-1}) were distributed between depths of 40 m and 80 m. In some years the lake began to stratify earlier than in other years, and the onset of thermal stratification usually occurred in either June or July. Consequently, the month of July exhibited more variation in the vertical pattern of primary production than the other months. During thermal stratification in August and September, the depth of the mean maximum rate of primary production gradually decreased from 80 m (0.41 mg C m^{-3} h^{-1}) to 60 m (0.38 mg C m^{-3} h^{-1}), a pattern that coincided with a shift in the chlorophyll maximum from a depth of 120 m to 100 m (Fig. 11).

The temporal pattern of carbon-14 primary production was hard to interpret, because of the failure to obtain satisfactory measurements for the 5-year period from 1991 to 1995 (Fig. 10). The highest value integrated to 200 m (87.35 mg C m^{-2} h^{-1}) was obtained in April 1987, during a period of relatively high chlorophyll concentrations (Fig. 9). Primary production was relatively low in January 1988 (17.7 mg C m^{-2} h^{-1}) and in all months sampled during 1996, 1997, 1998, and 2000 (17.5–30.9 mg C m^{-2} h^{-1}). The correlation coefficient expressing the covariance between chlorophyll concentration integrated to a depth of 200 m and carbon-14 primary production integrated to a depth of 180 m was 0.528 ($N = 28$, $P < 0.01$).

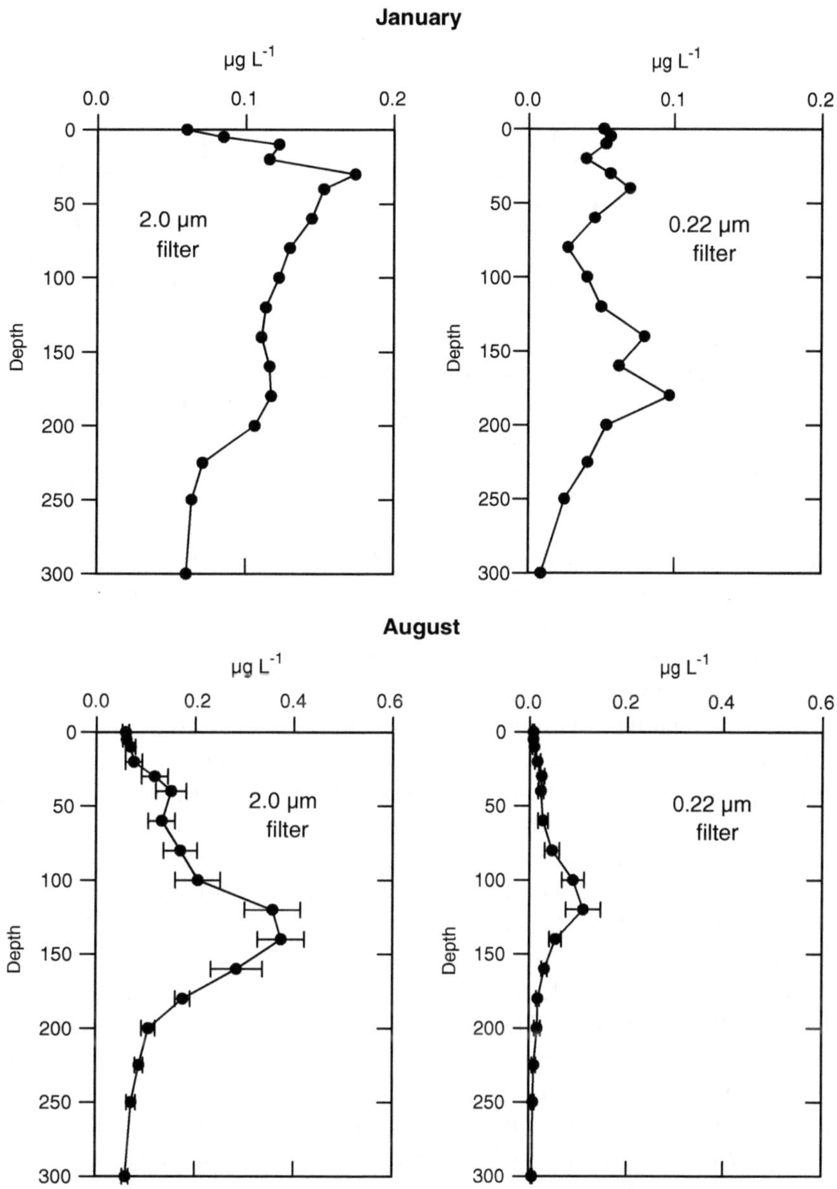

Fig. 13 Vertical profiles of chlorophyll concentration in the water column of Crater Lake partitioned into >2.0 μm and >0.22 to <2.0 μm fractions for the months of January and August. Only one set of observations was available for January. Horizontal bars for August represent the standard error of the mean

Relationships between phytoplankton and lake transparency

Regression analysis detected a weak negative relationship between chlorophyll concentration integrated to a depth of 40 m and Secchi disk depth, a measure of lake transparency (Fig. 14). In this case, the model was fitted for a plot of the natural logarithm of the integrated chlorophyll concentration against the Secchi disk depth:

$$\hat{Y} = 35.64 - 4.419X,$$

where $\hat{Y}$ was the predicted Secchi disk depth in meters and X was the natural logarithm (plus 1) of the chlorophyll concentration integrated to 40 m. This relationship was derived from 29 pairs of measurements that could be matched up for the same time period. Although the regression was significant ($P = 0.011$), the correlation coefficient was relatively low (–0.464).

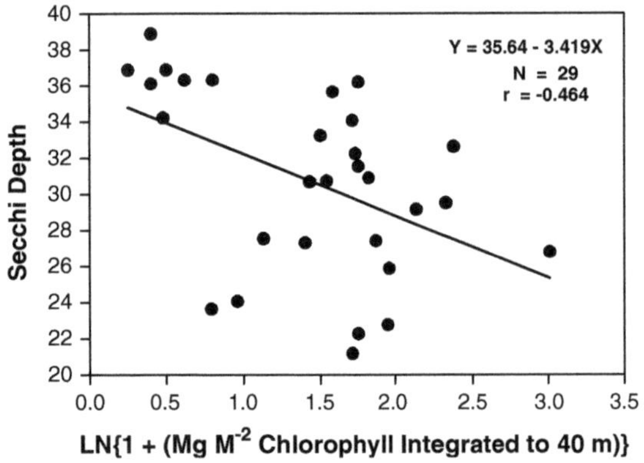

Fig. 14 Relationship between chlorophyll concentration integrated to a depth of 40 m and Secchi disk transparency in the water column of Crater Lake

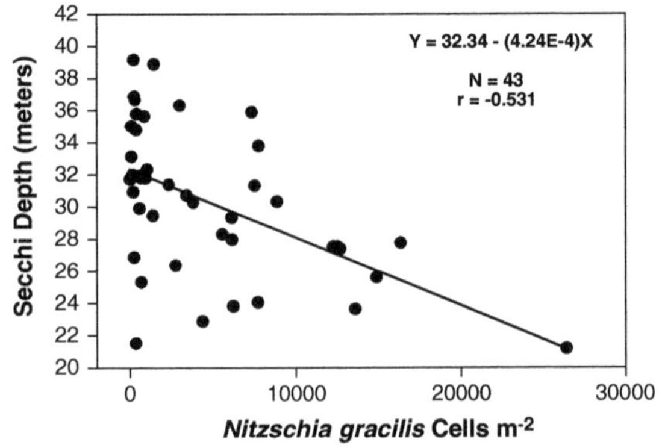

Fig. 15 Relationship between the density of *Nitzschia gracilis* integrated to a depth of 40 m and Secchi disk transparency in the water column of Crater Lake in August when the lake was thermally stratified

Based on data obtained in 1982, 1983, and 1984, D. W. Larson et al. (1990) proposed that Secchi disk transparency is inversely related to the density of *Nitzschia gracilis* in the upper 40 m of the water column. This hypothesis was examined by investigating relationships between: (1) the density of *N. gracilis* in the upper 40 m of the water column during the month of August from 1989 to 2000, and corresponding mean Secchi disk depths; and (2) the density of *N. gracilis* in the upper 40 m of the water column and the corresponding mean Secchi disk depths for all matching sampling dates from 1989 to 2000. Relationship (2) also included effects seasonal changes in these variables. The correlation coefficient expressing covariance between variables for relationship (1) was low (–0.176, $P > 0.05$), indicating that annual fluctuations in the density of *N. gracilis* during the period of thermal stratification between 1988 and 2000 were not significantly related to annual variations in lake transparency. However, there was a weak, but significant negative correlation ($r = -0.531$, $P < 0.01$) between *N. gracilis* density integrated to 40 m and Secchi disk transparency when data from all matching sampling dates were used in the analysis (Fig. 15). Therefore, results of these analyses indicated that seasonal fluctuations in the density of *N. gracilis* between 1988 and 2000 were more closely related to lake transparency than annual variation in the density of this species.

Temporal trajectories of *Nitzschia gracilis* cell density integrated to 200 m, the nitrate concentration in Spring 42, and Secchi disk depth are compared in Fig. 16. The mean concentration of nitrate in Spring 42 (0.238 mg l^{-1}), a spring located below the tourist center at Rim Village, was almost seven times greater than mean concentrations calculated for nine other springs (0.035 mg l^{-1}) that were studied intensively between 1983 and 2000. Moreover, there were concerns that sewage contamination of ground water associated with human activities at Rim Village was responsible, in part, for the high concentrations of nitrate in Spring 42 and a possible deterioration of the optical properties of the lake (D. W. Larson et al., 1990). In response to these concerns, the use of the septic field in the vicinity of Rim Village was discontinued in 1991, and sewage discharge was routed off the rim of the caldera. Patterns illustrated in Fig. 16 failed to exhibit any obvious annual trends in the nitrate concentration in Spring 42 between 1983 and 2000, and in the Secchi disk depth and cell density of *N. gracilis* between 1988 and 2000. However, there was considerable seasonal variation in values for these variables.

Discussion

Based on data obtained between 1982 and 2000, it is clear that the species composition of the phy-

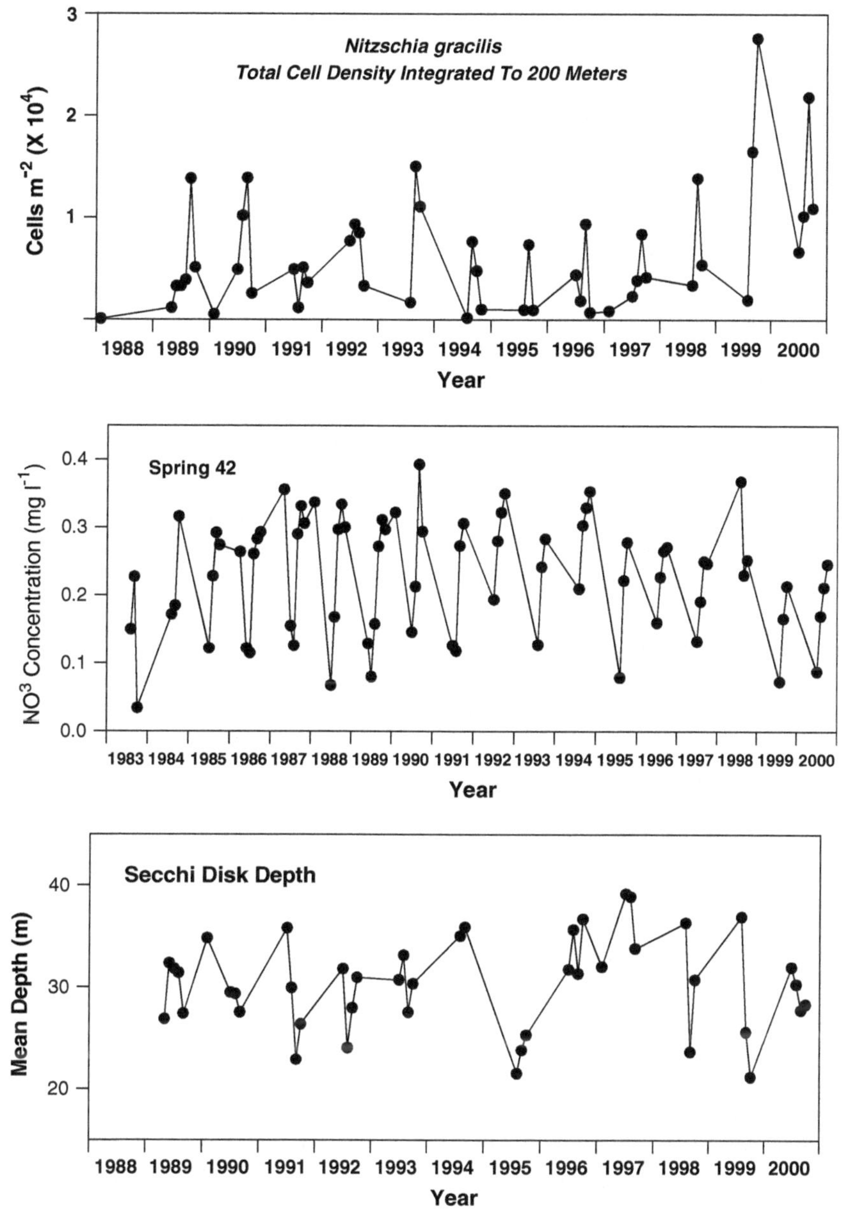

Fig. 16 Temporal trajectories of the density of *Nitzschia gracilis* integrated to a depth of 200 m, the concentration of nitrate in Spring 42, and Secchi disk depth in Crater Lake from 1988 to 2000 (*Nitzschia gracilis* and Secchi disk depth) and from 1983 to 2000 (nitrate in Spring 42)

toplankton assemblage in Crater Lake is considerably more diverse than the community structure reported from the results of earlier studies (Kemmerer et al., 1924; Brode, 1938; Utterback et al., 1942; D. W. Larson et al., 1987). A synthesis of data obtained between 1985 and 1990 (McIntire et al., 1996) and data analyzed in this paper (1988–2000) revealed a complex assemblage of microalgae and cyanobacteria that exhibited continuous spatial and temporal changes in cell density, cell biovolume, species composition, and species richness and dominance. Although seasonal patterns related to the process of thermal stratification were evident, and to some degree, predictable from year to year, the addition of data obtained between 1991 through

2000 also allowed the investigation of annual changes in the flora over a period of 16 years.

Debacon & McIntire (1991) and McIntire et al. (1996) identified a taxonomic discontinuity in the phytoplankton of Crater Lake that was related to the onset and establishment of thermal stratification. This discontinuity, which occurred at depths between 20 m and 40 m, also was evident in the 1991–2000 data, and corresponded to a transition zone between the epilimnion and metalimnion. Although the water temperature near the lake surface began to increase in June, the process of thermal stratification usually was not complete until August, and normally persisted into the latter part of September or early October. When the lake was not thermally stratified, the species composition of the flora was more uniformly distributed between the lake surface and a depth of 200 m.

The most conspicuous indicator species for the flora in the epilimnion was *Nitzschia gracilis*, the same taxon that was dominant in the upper 20 m of the water column in the study by D. W. Larson et al. (1987). Because of the lack of taxonomic information before 1970, it was impossible to estimate when *N. gracilis* became a prominent member of the phytoplankton assemblage. Utterback et al. (1942) reported the presence of the genus *Nitzschia* in the lake, but apparently did not identify specimens to the species level. Compared to other taxa that were abundant numerically in the epilimnion, *N. gracilis* had a relatively high cell biovolume and was easy to identify under a light microscope. However, the phytoplankton assemblage in the epilimnion also exhibited high densities of smaller species of cyanobacteria, namely *Aphanocapsa delicatissima* and *Synechocystis* sp. Based on the cluster analysis of the cell density data, mean relative abundances of *N. gracilis*, *A. delicatissima*, and *Synechocystis* sp. in samples from the epiliminion ranged from 14 to 21%, from 34 to 40%, and from 16 to 21%, respectively. The dinoflagellates *Gymnodinium inversum* and *Peridinium inconspicuum* were present in low densities in the epilimnion between 1992 and 2000, and made a significant contribution to the total cell biovolume because of their large cell size. Moreover, *Dinobryon sertularia* and *D. bavaricum* also were good indicators of the lower epiliminon and upper metaliminion, although their mean relative abundance was below 7% of the total cell density.

The phytoplankton in the epiliminion of Crater Lake must tolerate an environment that receives relatively high inputs of UV radiation, and is essentially without nutrient inputs from the hypolimnion for a period of 2–3 months each year. Also, nutrient inputs from the atmosphere and caldera wall in August and September, when the lake is thermally stratified, are minimal, and are usually restricted to small inputs from adjacent springs. Consequently, mean concentrations of essential plant nutrients in the upper 20 m of the water column during the month of August are very low: (1) no detectable nitrate; (2) orthophosphate concentrations between 0.009 mg l^{-1} and 0.010 mg l^{-1}; and total phosphorus and total Kjeldahl nitrogen concentrations between 0.023 mg l^{-1} and 0.025 mg l^{-1} and between 0.021 mg l^{-1} and 0.024 mg l^{-1}, respectively. These data indicate that all available nitrogen is rapidly converted into organic materials. Unfortunately, the physiological characteristics of the Crater Lake populations of *Nitzschia gracilis, Aphanocapsa delicatissima, Synechocystis* sp., the dinoflagellates, and the two species of *Dinobryon* have not been investigated in isolation. All of these taxa apparently can survive extreme nitrogen impoverishment, as the ratio of total nitrogen to total phosphorus was usually one or less in the euphotic zone. Groeger & Tietjen (1993) conducted a series of tests with Crater Lake water and found that test organisms grew faster with the introduction of nitrogen and iron than with nitrogen alone.

The systematics of the dominant taxa in the epiliminion is not clearly understood. A search of distributional records of diatoms at the Philadelphia Academy of Natural Sciences (Charles Reimer, pers. comm.) revealed that morphological entities identified as *N. gracilis* have been reported to occur in a wide range of habitats (e.g., in lakes, ponds, bogs, streams, and rivers). Also, organisms recorded as *Synechocystis* sp. from Crater Lake data were originally identified as *Chlorella* sp., a small chlorophyte (McIntire et al., 1996). In spite of the uncertainties over taxonomic position, the flora of the epilimnion consisted of recognizable entities that can be investigated in isolation by standard culture methods and newer molecular techniques. For example, one interesting question for future research is whether or not coccoid cyanobacteria in the water column have the physiological capacity for nitrogen fixation.

McIntire et al. (1996) found that phytoplankton sampled from 1985 through 1988 was more diverse, i.e., with more species and lower dominance, in the hypolimnion than in the epilimnion. However, this analysis excluded organisms below 5 μm in their longest dimension. The TWINSPAN analysis of the 1989–2000 data indicated that, based on cell density, the flora in the hypolimnion was dominated by picoplankton, in this case *Aphanocapsa delicatissima*. This species also reached relatively high densities in the epilimnion, but in the hypolimnion, it usually represented greater that 50% of the total cell density. However, because of its small cell size, its contribution to the total cell biovolume in the hypolimnion was less than some of the larger species (e.g., *Tribonema affine, Mougeotia parvula, Stephanodiscus hantzschii, Asterionella formosa*, and *Synechocystis* sp.).

Reynolds et al. (2002) presented a classification system for freshwater phytoplankton in which 31 functional associations were identified in relation to environmental tolerances and sensitivities. Taxonomic assemblages identified in samples from Crater Lake (clusters 1–4) were similar to some of the functional groups proposed by this system. Collectively, groups T, Z, E, and possibly L_o of the classification system roughly correspond to functional groups of taxa defined by the cluster analysis. For example, group Z, an assemblage characterized by prokaryotic picoplankton, associated with clear mixed water layers and low nutrients, occurred in the water column of Crater Lake throughout the year, although the relative abundance of typical species varied seasonally and from year to year. Moreover, typical representatives of groups T (*Mougeotia* and *Tribonema*), E (*Dinobryon*), and L_o (*Peridinium*) often occurred in Crater Lake, but were usually associated with picoplankton (group Z). According to the classification system (Reynolds et al., 2002), taxa associated with groups E and L_o occur in small, oligotrophic, base poor lakes, or the epilimnia of mesotrophic lakes, and are tolerant of low nutrient concentrations; whereas taxa in group T are tolerant of light deficiency and are found in deep, well-mixed epilimnia. In Crater Lake, the most obvious taxonomic discontinuity occurred in the water column when the lake was stratified, and when *Nitzschia gracilis* was a dominant species in the epilimnion.

The vertical distribution of chlorophyll and seasonal patterns of primary productivity in Crater Lake are consistent with the oligotrophic pattern described by the trophic state hypothesis proposed by Moll & Stoermer (1982): one primary production maximum in the spring each year and a deep-water chlorophyll maximum below the metalimnion. The first quantitative investigations of chlorophyll concentration and primary productivity in the water column of Crater Lake were initiated in 1967 (Donaldson & Hoffman, 1968; D. W. Larson, 1972). Results of these studies indicated that the depth of maximum carbon-14 primary productivity during July and August usually occurred between 70 m and 100 m below the lake surface, and that the highest concentration of chlorophyll was found at a depth of 110 m. In studies conducted between 1978 and 1983, D. W. Larson et al. (1987) reported primary production maxima at depths similar to those found between 1967 and 1969, and chlorophyll maxima located between depths of 120 m and 140 m, a range well below the depth of 1% of the irradiance at the lake surface (D. W. Larson & Hurley, 1993).

An analysis of all chlorophyll data obtained between 1984 and 2000 clearly indicated that the vertical distribution of this variable in the water column varied seasonally. The mean concentration of chlorophyll was uniformly distributed in the water column in January and during the spring months when the lake was circulating to a depth of between 200 and 250 m, whereas chlorophyll exhibited a distinct mean maximum concentration at depths between 100 and 140 m with the onset and development of thermal stratification during the summer and early fall. A deep-water chlorophyll maximum also has been reported for other oligotrophic lakes in the western United States, e.g., Lake Tahoe (Kiefer et al., 1972; Abbott et al., 1984; Coon et al., 1987),

Waldo Lake (Salinas & Larson, 2000), and Mowich Lake (G. L. Larson, 2000); in Lake Michigan (Brooks & Torke, 1977; Moll & Stoermer, 1982); and in the Pacific Ocean (Anderson, 1969; Venrick et al., 1973). A comparison of autotrophic properties in Crater Lake and Lake Tahoe indicated that the depths of the chlorophyll maximum in the two lakes were similar during thermal stratification, usually between 100 m and 120 m below the lake surface. Abbott et al. (1984) concluded that the formation of the deep-water chlorophyll maximum in Lake Tahoe was controlled by turbulent diffusion, nutrient supply, and the availability of light energy, and that shade adaptations and cell sinking were relatively unimportant processes. However, Coon et al. (1987) presented evidence that supported the hypothesis that after the deep-water chlorophyll maximum in Lake Tahoe is established, it is maintained primarily by in situ growth of algal cells, and secondarily by the sinking of cells from depths above the maximum.

The chlorophyll maximum in Lake Tahoe was closely associated with a nitracline (Coon et al., 1987). In Crater Lake there was no detectable nitrate down to a depth of 100 m during thermal stratification in August and September, and the mean concentrations of nitrate at depths of 200 m and 550 m were only 0.003 mg l^{-1} and 0.015 mg l^{-1}, respectively. Therefore, the chlorophyll maximum in Crater Lake may be located at a depth near the upper threshold of detectable nitrate. Chemical data also indicated that the nitrate concentration in the upper 200 m of the water column gradually decreased from a mean of 0.002 mg l^{-1} in January to no detectable nitrate in June, presumably in response to a seasonal maximum in phytoplankton productivity during the spring months. The mean nitrate, chlorophyll, and primary productivity depth profiles for July were irregular and inconsistent with profiles for the other months of the year, patterns probably related to physical processes associated with the onset of thermal stratification.

Maximum carbon-14 primary productivity in Crater Lake was in the upper 30 m of the water column in January, between depths of 40 m and 80 m in the spring, and approximately 40 m above the chlorophyll maximum in August and September, when the lake was thermally stratified. Possible mechanisms that accounted for the 40-m depth difference between chlorophyll and primary production maxima included: (1) sinking of living cells (Coon et al., 1987) and the accumulation of detrital chlorophyll, (2) physiological adaptations to low light intensities (Tilzer & Goldman, 1978), (3) a switch to heterotrophic nutrition below the compensation depth (Vincent & Goldman, 1980; Sandgren, 1988), and (4) selective grazing by zooplankton above the chlorophyll maximum (Venrick et al., 1973). Although the positions of the chlorophyll maximum in the water column of Crater Lake, Waldo Lake, and Lake Tahoe were similar, the primary production maximum in Waldo Lake and Lake Tahoe was usually in the epilimnion (Salinas & Larson, 2000; Goldman & Amezaga, 1975), whereas the corresponding maximum in Crater Lake was between 60 m and 80 m below the lake surface. Therefore, Waldo Lake and Lake Tahoe probably receive a greater input of allochthonous nutrients from the surrounding watershed than the inputs to Crater Lake from the adjacent caldera wall and direct atmospheric deposition. Dodds et al. (1991) found that heterotrophic regeneration by organisms less than 3 μm in diameter supplied most of the inorganic nitrogen and phosphorous in the epilimnion of Flathead Lake during thermal statification, when allochthonous inputs of these nutrients were less than 5% of biotic regeneration. Moreover, it is now established that some non-heterocystis unicellular cyanobacteria (e.g., *Synechococcus*) can fix nitrogen (Rippka et al., 1971) when other essential nutrients are available, a process that may help support populations of other algal taxa during periods of severe nitrogen limitation. Similar mechanisms may be active in Crater Lake, because the epilimnion is virtually sealed off from nutrient inputs from the hypolimnion in late summer and early fall when allochthonous inputs from the caldera and atmosphere are minimal.

An examination of all phytoplankton data obtained from Crater Lake between 1984 and 2000 revealed a discontinuity in the temporal trajectories of chlorophyll concentration, mean cell size, and the taxonomic structure of the phytoplankton. Between 1986 and 1991, total chlorophyll in the water column was relatively high, mean cell size was relatively low, and the species composition of the flora was slightly different than the taxonomic structure observed between 1992 and 2000 (Figs. 6, 7, and 9). Whether or not a similar discontinuity was present in the temporal pattern of carbon-14 primary production was less clear, because of the lack of data for the period from 1991 to 1995. However, the highest values for total primary production in the water column were recorded before 1991, i.e., in 1987, 1989, and 1990 (Fig. 10). Temporal patterns of total cell density and biovolume in the water column exhibited seasonal fluctuations, but were inconsistent with the discontinuity in species composition, cell size, primary production, and chlorophyll observed between 1991 and 1992. A possible explanation for this discrepancy may be related to problems of detecting and enumerating picoplanktonic organisms less than 2 μm in diameter under the light microscope. In any case, the collection of variables related to water column autotrophy (chlorophyll concentration, mean cell size, and species composition) clearly indicated that a change in the structure and dynamics of the flora occurred during a 2-year period (1991–1992), whereas these variables exhibited relatively little change in behavior between 1992 and 2000.

Results of the discriminant analysis indicated that the temporal discontinuity in the phytoplankton assemblage was related, in part, to differences in the chemical properties of the water column. Possible mechanisms that accounted for changes in the chemical properties of the lake were (1) variations in the pattern of lake circulation and the depth of mixing; (2) fluctuations in allochthonous nutrient inputs as indicated by changes in lake level and patterns of precipitation; and (3) the termination of the septic field and changes in the sewage system at the Rim Village area in 1991.

Dymond et al. (1996) estimated that upward mixing of the deep-water nitrate pool (below the depth of 200 m) accounted for more 85% of the new nitrogen transported to the euphotic zone of Crater Lake. An example of a deep-mixing event was recorded in 1993 by Crawford & Collier (1997). Such events undoubtedly have pronounced effects on the seasonal and interannual dynamics of autotrophic processes in the lake. However, an examination of available physical data revealed that temporal changes in lake circulation and patterns of upwelling from the deep lake did not explain the 1991–1992 discontinuity in variables related to the phytoplankton assemblage (Bob Collier, pers. com.).

If it is assumed that lake level is an indicator of allochthonous nutrient inputs, the decline in lake level between 1986 and 1992 may be indirectly related to the decrease in total chlorophyll in the upper 200 m of the water column. However, a subsequent increase in lake level between 1995 and 2000 was not associated with a concurrent increase in chlorophyll. Total zooplankton biomass was relatively high in 1995, 1996, and 1997, and populations of *Bosmina longirostris* and *Daphnia pulicaria* reached maxima in 1994 and 1995, and in 1997 and 1999, respectively, patterns that could account for the failure of the chlorophyll concentration to increase with the corresponding increase in lake level after 1994. Total zooplankton biomass was low between 1986 and 1991 when the chlorophyll concentration was relatively high, indicating that zooplankton populations probably were not food limited during that period of time. However, a large increase in the salmon population occurred in 1989 and 1990 and persisted for several years (Buktenica et al., this issue), suggesting that fish predation on zooplankton may have accounted, in part, for the corresponding low zooplankton biomass and relatively high chlorophyll concentration.

In summary, patterns in the phytoplankton, zooplankton, and kokanee populations suggested that annual changes in the productive capacity of Crater Lake were driven by fluctuations in the supply of nutrients from allochthonous sources, and that state variables associated with these populations varied in response to both "bottom-up" and "top-down" processes. When productive capacity was high between 1986 and 1989, phytoplankton biomass was high and zooplankton biomass was low. However, as productive capacity increased again between 1995 and 1997, fish biomass was relatively low, and the increase in productive capacity was manifested through a corresponding increase in zooplankton biomass. Beginning in 1996, there was a switch in dominant crustaceans in the zooplankton from the relatively small species of *Bosmina* to the larger species of *Daphnia*, a change that corresponded to an increase in fish biomass and the proportion of larger fish in the kokanee population. As the biomass of both the zooplankton and fish increased after 1993, phytoplankton biomass remained relatively low, indicating a concurrent decrease in turnover time and a corresponding increase in energy transfer between phytoplankton populations and primary and secondary consumers. *Daphnia* disappeared from the zooplankton assemblage in 2000 after the kokanee population reached a maximum biomass in 1999.

If long-term patterns of the productive capacity of Crater Lake are determined by allochthonous inputs of nutrients, the source and magnitude of such inputs must be identified in order to track and predict the trajectories of state variables most closely related to management objectives. Obviously, precipitation represents a major source of nutrients, either as direct inputs from the atmosphere or indirect inputs from adjacent springs and runoff from the annual cycle of snowmelt within the caldera. These inputs also are strongly affected by short-term and long-term changes in air quality. In addition, there was considerable concern that human activities were gradually becoming a significant source of allochthonous nutrients, a conclusion based, in part, on (1) a comparison of Secchi disk readings in 1978 and the 1980's with limited historical data obtained before 1978 (D. W. Larson, 1984), (2) the relatively high nitrate concentration in Spring 42, a spring located below a major tourist area at Rim Village (D. W. Larson et al., 1990), and (3) a seasonal increase in the density of *Nitzschia gracilis* in August each year when the lake is thermally stratified (D. W. Larson et al., 1990). These observations and concerns about possible anthropogenic effects on lake clarity provided the impetus for the establishment of a 10-year research program in 1982, a subsequent monitoring program, and the removal of the septic field near Rim Village in 1991.

Whether or not human activities contributed significantly to changes in optical properties of the lake between 1968 and 2000 is still controversial. D. W. Larson (2002) presented two alternative hypotheses, one that strongly implicates the effects of the sewage system at Rim Village, and another view that emphasizes the influence of upwelling events and hydrothermal activities in the deep basins of the lake. Data presented in this paper suggested a third hypothesis. This hypothesis states that the productive capacity of Crater Lake is controlled primarily by long-term patterns of climatic change that regulate the supply of allochthonous nutrients. Evidence in support of this hypothesis includes: (1) the patterns of upwelling events did not correspond to the pronounced decrease in total chlorophyll and changes in taxonomic composition of the phytoplankton observed between 1990 and 1992; (2) there were no obvious changes in density of *Nitzschia gracilis*, the nitrate concentration in Spring 42, or the Secchi disk depth after the sewage system was altered in 1991 (Fig. 16); (3) of the 10 springs receiving intensive study, Spring 42 had the lowest mean concentration of total Kjeldahl nitrogen; and (4) the relatively low total chlorophyll concentration observed after the elimination of the septic field can be explained by trophic interactions with primary and secondary consumers, i.e., an increase in productive capacity between 1994 and 1999 that corresponded to concurrent increases in total zooplankton biomass and the biomass of kokanee salmon, while phytoplankton biomass remained at a relatively low level. In reality, the clarity of Crater Lake over the past 30 years may have been determined by a complex set of variables that reacted to changes consistent with aspects of all three of the proposed hypotheses.

References

Abbott, M. R., K. L. Denman, T. M. Powell, P. J. Richerson, R. C. Richards & C. R. Goldman, 1984. Mixing and the dynamics of the deep chlorophyll maximum in Lake Tahoe. Limnology and Oceanography 29: 862–878.

Anderson, G. C., 1969. Subsurface chlorophyll maximum in the Northeast Pacific Ocean. Limnology and Oceanography 14: 386–391.

Bacon, C. R., 1983. Eruptive history of Mount Mazama and Crater Lake caldera, Cascade Range, USA. Journal of Volcanology and Geothermal Research 18: 57–115.

Bacon, C. R. & M. A. Lanphere, 1990. The geologic setting of Crater Lake, Oregon. In Drake, E. T., G. L. Larson, J. Dymond & R. Collier (eds), Crater Lake: An Ecosytem Study. Pacific Division. American Association for Advancement of Science, San Francisco, CA, 19–27.

Barber, J. H. Jr. & C. H. Nelson, 1990. Sedimentary history of Crater Lake caldera, Oregon. In Drake, E. T., G. L. Larson, J. Dymond & R. Collier (eds), Crater Lake: An Ecosytem Study. Pacific Division. American Association for Advancement of Science, San Francisco, CA, 29–39.

Brode, J. S., 1938. The denizens of Crater Lake. Northwest Science 12: 298–310.

Brooks, A. S. & B. G. Torke, 1977. Vertical and seasonal distribution of chlorophyll *a* in Lake Michigan. Journal of the Fisheries Research Board of Canada 34: 2280–2287.

Buktenica, M. W., S. F. Girdner, G. L. Larson & C. D. McIntire, this issue. Variability of kokanee and rainbow trout food habits, distribution, and population dynamics, in an ultraoligotrophic lake with no manipulative management.

Byrne, J. V., 1965. Morphometry of Crater Lake, Oregon. Limnology and Oceanography 10: 462–465.

Coon, T. G., M. Lopez, P. J. Richerson, T. M. Powell & C. R. Goldman, 1987. Summer dynamics of the deep chlorophyll maximum in Lake Tahoe. Journal of Plankton Research 9: 327–344.

Crawford, G. B. & R. W. Collier, 1997. Observations of a deep-mixing event in Crater Lake, Oregon. Limnology and Oceanography 42: 299–306.

Debacon, M. K. & C. D. McIntire, 1990. Spatial and temporal patterns in the phytoplankton of Crater Lake (1985–1987). In Drake, E. T., G. L. Larson, J. Dymond & R. Collier (eds), Crater Lake: An Ecosytem Study. Pacific Division, American Association for Advancement of Science, San Francisco, CA, 167–175.

Debacon, M. K. & C. D. McIntire, 1991. Taxonomic structure of phytoplankton assemblages in Crater Lake, Oregon, USA. Freshwater Biology 25: 95–104.

Dodds, W. K., J. C. Priscu & B. K. Ellis, 1991. Seasonal uptake and regeneration of inorganic nitrogen and phosphorus in a large oligotrophic lake: size-fractionation and antibiotic treatment. Journal of Plankton Research 13: 1339–1358.

Donaldson, J. R. & F. O. Hoffman, 1968. Zooplankton population dynamics, Crater Lake. Crater Lake Report 2. National Park Service, Crater Lake, OR.

Dymond, J., R. Collier & J. McManus, 1996. Unbalanced particle flux budgets in Crater Lake, Oregon: implications for edge effects and sediment focusing in lakes. Limnology and Oceanography 41: 732–743.

Goldman, C. R., 1963. The measurement of primary productivity and limiting factors in freshwater with carbon-14. In Doty, M. S. (ed.), Primary Productivity Measurement, Marine and Freshwater. U. S. Atomic Energy Commission TID-7633: 103–113.

Goldman, C. R. & E. de Amezaga, 1975. Spatial and temporal changes in the primary productivity of Lake Tahoe, California-Nevada, between 1959 and 1971. Verhandlungen Internationale Vereinigung für Theoretische und Angewandte Limnologie 19: 812–825.

Groeger, A. W. & T. E. Tietjen, 1993. Physiological responses of nutrient-limited phytoplankton to nutrient addition. Verhandlungen Internationale Vereinigung für Theoretische und Angewandte Limnologie 25: 370–272.

Hill, M. O., R. G. H. Bunce & M. W. Shaw, 1975. Indicator species analysis, a divisive polythetic method of classification, and its application to a survey of native pinewoods in Scotland. Journal of Ecology 63: 597–613.

Jongman, R. H. G., C. J. F. ter Braak & O. F. R. van Tongeren, 1987. Data Analysis in Community and Landscape Ecology. Centre for Agricultural Publishing and Documentation (Pudoc), Wageningen.

Kemmerer, G., F. Bovard & W. R. Boorman, 1924. Northwestern lakes of the United States: biological and chemical studies with reference to possibilities in production of fish. Bulletin of the U. S. Bureau of Fisheries 39: 51–140.

Kiefer, D. A., O. Holm-Hansen, C. R. Goldman, R. Richards & T. Berman, 1972. Phytoplankton in Lake Tahoe: deep-living populations. Limnology and Oceanography 17: 418–422.

Larson, D. W., 1972. Temperature, transparency, and phytoplankton productivity in Crater Lake, Oregon. Limnology and Oceanography 17: 410–417.

Larson, D. W., 1984. The Crater Lake study: detection of possible optical deterioration of a rare, unusually deep caldera lake in Oregon, USA. Verhandlungen Internationale Vereinigung für Theoretische und Angewandte Limnologie 22: 513–517.

Larson, D. W., 2002. Probing the depths of Crater Lake. American Scientist 90: 64–71.

Larson, D. W., C. N. Dahm & N. S. Geiger, 1987. Vertical partitioning of the phytoplankton assemblage in ultraoligotrophic Crater Lake, Oregon, USA. Freshwater Biology 18: 429–442.

Larson, D. W., C. N. Dahm & N. S. Geiger, 1990. Limnological response of Crater Lake to possible long-term sewage influx. In Drake, E. T., G. L. Larson, J. Dymond & R. Collier (eds), Crater Lake: An Ecosytem Study. Pacific Division, American Association for Advancement of Science, San Francisco, CA, 197–212.

Larson, G. L., 2000. Chlorophyll maxima in mountain ponds and lakes, Mount Rainier National Park, Washington State, USA. Lake and Reservoir Management 16: 333–339.

Larson, G. L. & M. Hurley, 1993. Photometer. In Larson, G. L., C. D. McIntire & R. Jacobs (eds), Crater Lake Limnological Studies Final Report. National Park Service Technical Report NPS/PNROSU/NRTR-93/03: 317–329.

Larson, G. L., C. D. McIntire, M. Hurley & M. W. Buktenica, 1996. Temperature, water chemistry, and optical properties of Crater Lake. Lake and Reservoir Management 12: 230–247.

McManus, J., 1992. On chemical and physical limnology of Crater Lake, Oregon. PhD. thesis. Oregon State University, Corvallis.

McManus , J., R. W. Collier & J. Dymond, 1993. Mixing processes in Crater Lake, Oregon. Journal of Geophysics Research 98: 18,295–18,307.

McIntire, C. D. & W. S. Overton, 1971. Distributional patterns in assemblages of attached diatoms from Yaquina Estuary, Oregon. Ecology 52: 758–777.

McIntire, C. D., G. L. Larson, R. E. Truitt & M. K. Debacon, 1996. Taxonomic structure and productivity of phytoplankton assemblages in Crater Lake, Oregon. Lake and Reservoir Management 12: 259–280.

Moll, R. A. & E. F. Stoermer, 1982. A hypothesis relating trophic status and subsurface chlorophyll maxima of lakes. Archiv für Hydrobiologie 94: 425–440.

Nathenson, M. & J. M. Thompson, 1990. Chemistry of Crater Lake, Oregon, and nearby springs in relation to weathering. In Drake, E. T., G. L. Larson, J. Dymond & R. Collier (eds), Crater Lake: An Ecosytem Study. Pacific Division, American Association for Advancement of Science, San Francisco, CA, 115–126.

Phillips, K. N. & A. S. Van Denburgh, 1968. Hydrology of Crater, East, and Davis lakes, Oregon. U. S. Geological Survey Water-Supply Paper 1859-E.

Pimentel, R. A., 1979. Morphometrics, The Multivariate Analysis of Biological Data. Kendall-Hunt, Dubuque, Iowa.

Reynolds, C. S., V. Huszar, C. Kruk, L. Naselli-Flores & S. Melo, 2002. Towards a functional classification of freshwater phytoplankton. Journal of Plankton Research 24: 417–428.

Rippka, R., A. Neilson, R. Kunisawa & G. Cohen-Bazire, 1971. Nitrogen fixation by unicellular blue-green algae. Archives of Microbiology 76: 341–348.

Salinas, J. & D. W. Larson, 2000. Phytoplankton primary production and light in Waldo Lake, Oregon. Lake and Reservoir Management 16: 71–84.

Sandgren, C. D., 1988. The ecology of chrysophyte flagellates: their growth and perennation strategies as freshwater phytoplankton. In Sandgren, C. D. (ed.), Growth and Reproductive Strategies of Freshwater Phytoplankton. Cambridge University Press, Cambridge, UK, 9–104.

Tilzer, M. M. & C. R. Goldman, 1978. Importance of mixing, thermal stratification and light adaptation for phytoplankton productivity in Lake Tahoe (California-Nevada). Ecology 59: 810–821.

Utterback, C. L., L. D. Phifer & R. J. Robinson, 1942. Some chemical, planktonic and optical characteristics of Crater Lake. Ecology 23: 97–103.

Venrick, E. L., J. A. McGowan & A. W. Mantla, 1973. Deep maxima of photosynthetic chlorophyll in the Pacific Ocean. Fishery Bulletin 71: 41–52.

Vincent, W. F. & C. R. Goldman, 1980. Evidence for algal heterotrophy in Lake Tahoe, California-Nevada. Limnology and Oceanography 25: 89–99.

Hydrobiologia (2007) 574:205–216
DOI 10.1007/s10750-006-0353-3

Nutrient limitation in Crater Lake, Oregon

Alan W. Groeger

© Springer Science+Business Media B.V. 2007

Abstract Experiments were carried out to determine what nutrient (or nutrients) was primarily responsible for limiting phytoplankton productivity in ultraoligotrophic Crater Lake. The experiments included in situ and laboratory nutrient addition bioassays utilizing the natural phytoplankton community, *Selenastrum capricornutum* bottle assays, photosynthetic responses, photosynthetic carbon metabolism, and response of dark uptake of $^{14}CO_2$ with the addition of NH_4^+. The results suggested that a trace metal(s) or its availability was the primary factor limiting the epilimnetic phytoplankton productivity. Nitrogen was extremely low, and quickly became limiting with the addition of trace metals and a chelator. Iron is the most likely candidate as the limiting nutrient. Trace metals and nitrogen are also both important in limiting phytoplankton at 100 m, a depth where biologically mediated turnover of nutrients seems to be more important.

Guest Editors: Gary L. Larson, Robert Collier, and Mark W. Buktenica
Long-term Limnological Research and Monitoring at Crater Lake, Oregon

A. W. Groeger (✉)
Aquatic Station, Department of Biology, Texas State University—San Marcos, San Marcos, TX 78666, USA
e-mail: ag11@txstate.edu

Keywords Crater Lake · Phytoplankton · Nutrient limitation · Trace metals · Nitrogen

Introduction

Crater Lake, in southwestern Oregon in the United States, has been long known internationally for its clear water and ultraoligotrophic nature (Hutchinson, 1957; Larson, 1988). Crater Lake is an exceptional lake, and is commonly found in the outer fringes of frequency distributions of limnological variables among lakes. The formation of the caldera has resulted in a very deep basin with an extreme relative depth (Hutchinson, 1957), a long water residence time (approximately 150 year), and an "extreme" example of a small catchment area: lake surface area (0.2). Together these characteristics suggest an ecosystem in which nutrient inputs are dominated by atmospheric deposition (Kalff, 2002). The combination of high light availability and extremely low productivity within the euphotic zone is normally indicative of extremely low concentrations of nutrients essential for the algal growth and very low concentrations of "yellow organic acids" (Kirk, 1983) that characterize inputs of allochthonous dissolved organic carbon from the drainage basin. In the upper 100 m of the Crater Lake water column dissolved inorganic nitrogen concentrations are commonly $\leq 0.08\ \mu M\ l^{-1}$ (Larson et al., 1996a), dissolved organic carbon (DOC) is very low (<1 μM) (Dymond et al., 1996), and summer chlorophyll *a* concentrations commonly range from 0.005 to 0.3 $\mu g\ l^{-1}$ (Larson, 1998). Collier et al. (1990) reported that Crater Lake trace metals are "extremely" low, some being less than concentrations found in open ocean waters.

An important question, from both a management perspective as well as basic limnology, is which nutrient (or nutrients) is primarily responsible for limiting phytoplankton growth and biomass in Crater Lake. A few studies in the past have addressed nutrient limitation within Crater Lake. Lane & Goldman (1984) carried out size-fractionated nutrient enrichment bioassays on a single sampling date in August 1981 with water from 0 m and 100 m depths and simulated lake conditions. They concluded that nitrogen and trace metals were limiting to Crater Lake phytoplankton at both depths, though different size classes responded to enrichment differently. Experiments on near-surface waters from the growing season of 1987 (Groeger & Tietjen, 1993) suggested that a certain metal (a mixture of six essential metals was used) plus a chelator was necessary, sometimes with a nitrogen source, to stimulate certain physiological processes in epilimnetic Crater Lake phytoplankton. The purpose of the present study was to apply a number of different methods for determining which nutrients are primarily responsible for controlling the productivity of this unique ecosystem.

Methods

Experiments were carried out in the summer of 1987 and 1997. All samples were collected at Station 13 (except where otherwise noted) following the procedures and using equipment described by Larson et al. (1996a). Station 13 was located over the deepest part of the lake, and was the primary sampling site for much of the monitoring efforts on Crater Lake (Larson, 1988). In 1987 water was collected at 5 m during the regular monitoring program in June and July, and 10 m in August and September. The samples were transported back to the laboratory in Corvallis, Oregon on ice within 30 h of their collection. All the procedures described in 1987 used aliquots from these samples. In 1997, water was collected on July 28 (long-term in situ nutrient addition assays) and July 29 (all other experiments). In 1997 all work was done with in situ incubations and laboratory work was done either at the boathouse on Wizard Island or the water laboratory at the Crater Lake National Park headquarters.

Statistical analyses for the experiments described below were carried out using SPSS 8.0 for Windows based systems. If a significant difference ($p < 0.05$) was found between treatments using the ANOVA, a pairwise comparison of the individual treatment means and range tests (homogeneous subset analysis) were performed with Tukey's HSD (honestly significant difference) test.

Long-term laboratory bioassays (1987)

Two hundred milliliters of near-surface water were apportioned into acid-washed 500 ml glass flasks and various nutrient additions were made (Table 1). The flasks were capped with foam plugs, and gently shaken on a shaker platform under 24 h light (≈100 $\mu E\ m^{-2}\ s^{-1}$) at 21°C for 7–10 days. All treatments were carried out in duplicate, and the response was measured daily as in vivo fluorescence.

Long-term in situ bioassays (1997)

Water for these assays was collected at station 13 on July 28 from two different depths. Representing the epilimnion, a series of 14 vertically oriented 4 l bottles were collected simultaneously from depths of 3, 4, 5, 6, 7, and 8 m and pooled in 30 l carboys. For the deep samples, 12 vertically oriented 4 l bottles were collected simultaneously from depths of 98, 99, 100, 101, 102, and 103 m and pooled in 30 l carboys. Deep samples were shaded from direct sunlight during the procedure. For each depth, 40 1 l polyethylene acid-washed cubitainers (representing five replicates of eight different treatments, Table 1) were rinsed with either epilimnetic or deep water, and filled in the shaded cabin of the research boat. Cubitainers were set up to avoid any head space, were placed in a double set of wide mesh bags, and were hung

Table 1 The various treatments and corresponding concentrations utilized in the1987 and 1997 long-term bioassays

Treatment	1987 Long-term bioassays	1997 Long-term bioassays
Control	No additions	No additions
+N	71 μM NH_4^+	2.86 μM NH_4^+
$+NO_3$	71 μM NO_3^-	
+P	16 μM PO_4^-	0.19 μM PO_4^-
+E	0.89 μM Na_2EDTA	
+ME	0.89 μM Na_2EDTA plus 0.33 μM B, 0.48 μM Mn, 0.024 μM Zn, 0.006 μM Co, 0.00006 μM Cu, 0.03 μM Mo and 2.87 μM Fe	1.12 μM Na_2EDTA plus 0.09 μM Mn, 0.008 μM Zn, 0.004 μM Co, 0.004 μM Cu, 0.003 μM Mo and 1.17 μM Fe
+FeEN	0.89 μM Na_2EDTA plus 2.87 μM Fe and +N at conc. from above	
+MEN	+ME and +N at conc. from above	+ME and +N at conc. from above
+MEP	+ME and +P at conc. from above	+ME and +P at conc. from above
+NP		+N and +P at conc. from above
+MENP	+ME, +N and +P at conc. from above	+ME, +N and +P at conc. from above
+EN	+E and +N at conc. from above	
+FeENP	+FeEN and +P at conc. from above	
+ENP	+E, +N and +P at conc. from above	

at 10 and 82 m, for the "epilimnion" samples and "deep" samples, respectively.

Cubitainers were recovered after 7 days on August 4, placed in coolers, and transported back to the Park laboratory. The total volume of each cubitainer was measured in a graduated cylinder and the complete volume was filtered through a 47-mm GF/F glass fiber filter and frozen. Chlorophyll *a* concentrations were determined fluorometrically following the techniques of Burnison (1980).

Enhancement of dark carbon uptake

The concept behind this method was that under nitrogen limitation, an algal community or species would increase its uptake of CO_2 in the dark when exposed to elevated NH_4^+. The tests were carried out in August 1987 and in 1997 from near-surface samples and closely followed the method of Yentsch et al. (1977). In 1987, triplicate controls and $+NH_4^+$ treatment (71 μM NH_4^+ final concentration) were set up in 130 ml dark bottles and incubated for 10 h at 20°C. On July 29, 1997 quintuplicate controls, $+NH_4$ treatment (4.6 μM NH_4^+ final concentration), and $+NH_4$ + ME treatment (4.6 μM NH_4^+ final concentration plus the +ME treatment given in Table 1 for 1997), in 250 ml bottles. These bottles were placed in a cooler filled with lake water in complete darkness, and filtered after 16 h. In 1987 2.22×10^6 Bq of $^{14}C–HCO_3^-$ were added to each bottle, and in 1997 13.32×10^6 Bq were added. After the incubation, the samples were filtered onto GF/F filters and ^{14}C on the filters was quantified through liquid scintillation counting.

Response of photosynthesis and photosynthetic carbon metabolism in response to nutrient additions

In 1987 and 1997 five experiments measuring photosynthetic response to nutrient additions and photosynthetic carbon metabolism (allocation of carbon into various molecular fractions) were carried out on near-surface waters from Station 13. For photosynthetic carbon metabolism, the filtered samples were chemically fractionated (complete description of the method in Li et al., 1980 and Groeger & Kimmel, 1988) into four biochemical pools: the macromolecules protein, polysaccharide (which includes nucleic acids), lipids, and a pool of smaller molecules referred to as soluble metabolites after an incubation with $^{14}C–HCO_3^-$. The ^{14}C in each pool was quantified by liquid scintillation counting.

Experiment 1

In July 1987 200 ml were apportioned into duplicate glass 500 ml flasks with the following treatments: control; $+NH_4$, final concentration of

7.7 μM NH_4^+; $+NO_3$, final concentration of 7.7 μM NO_3^-; +P, final concentration of 0.3 μM P L; and +ME, the EDTA-metals solution in Table 1. To each flask 2.22×10^6 Bq of $^{14}C–HCO_3^-$ were added, and the flasks were placed on a lighted shaker table, and 20 ml subsamples were collected five times over an 8 h incubation.

Experiment 2

In July 1987 a photosynthetic carbon metabolism experiment was set up in duplicate flasks as in Experiment 1, with the treatments being a control, a $+NH_4$, and a $+NO_3$ at the same concentrations used in Experiment 1. To each flask 3.33×10^6 Bq of $^{14}C–HCO_3^-$ were added and incubated for 5 h.

Experiment 3

In August 1987, the photosynthetic carbon metabolism experiment was repeated with 200 ml of water apportioned into triplicate flasks with the following treatments: control; $+NH_4$; $+NO_3$; +P; and the +ME; all at the same concentrations used in Experiment 1, and the incubation in light was 10 h.

In 1997 two experiments to support the long-term bioassay in the epilimnion were carried out on near surface water collected at station 13 on July 29. The first (Experiment 4) was a delayed photosynthetic response to nutrient enrichment with 130 ml glass bottles set up as: three Control, two +N, two +P, two +ME, two +MEN, and two +NP replicate bottles. Final concentrations in these bottles were the same as the cubitainers in the long-term in situ bioassays described above. The bottles were hooked to a harness and hung at a depth of about 7 m in an embayment of Wizard Island. These bottles were retrieved after two days, 11.1×10^6 Bq of $^{14}C–HCO_3^-$ was added per bottle, and the bottles were resuspended in the lake for 8 h of daylight after which they were filtered to measure photosynthesis.

In the second experiment from 1997 (Experiment 5) six 1 l clear polyethylene bottles were filled with 1 l from the surface waters at station 13. Three duplicated treatments were set up: Control, +N, and +MEN. The concentrations of nutrients added were the same as those used in the long-term in situ bioassay. To each bottle, 35.5×10^6 Bq $^{14}C–HCO_3^-$ were added, and these bottles were incubated in the lake at a depth of 7 m. These bottles were removed at intervals over the 66 h incubation at which time 200 ml aliquots were collected on 47 mm GF/F filters. After the fourth sampling period, the bottles were wrapped in black plastic (to ensure total darkness) placed in a cooler with lake water to maintain their in situ incubation temperature, transported back to the laboratory and filtered after 17 h in the dark.

Selenastrum Algal Assay Bottle Test

In June and July 1987, water from various depths (depths and specific treatments listed in Table 4) at Station 13 and two springs was collected to carry out the *Selenastrum capricornutum* Algal Assay Bottle Test (Miller et al., 1978) using the bioassay facilities at the U.S. EPA lab in Corvallis, Oregon. Lake water was filtered through a 0.45 µm pore size filter, placed in flasks which were inoculated with a final concentration of 1000 cells ml^{-1} of *S. capricornutum*, and amended with various nutrient additions at the same concentrations shown in Table 1, except all treatments with EDTA had a final concentration of 2.97 µM EDTA, and the +ME-Fe treatment, which is EDTA plus all the metals in Table 1 with iron excluded. Triplicate flasks for each treatment were sampled on day 4, 5, 7, 10, 12, and 14, and biomass response was determined with a Coulter counter (Miller et al., 1978).

Results

The water column

In both 1987 and 1997 there was virtually no detectable NO_3^- (<0.08 µM) above a depth of 300 m in Crater Lake, and summer NH_4^+ was usually below detection levels (<0.8 µM) in the epilimnion. In both years soluble reactive phosphorus ranged from 0.3–0.6 µM in the upper 200 m of the water column, and silica concentrations were ≥0.28 mM (Larson, 1988, 1998). In 1987, the sampling dates for this study covered the period of stratification. In 1997 experiments were carried out at a time when there was a distinct vertical zonation of temperature and distribution of chlorophyll *a* within the water column (Fig. 1).

Long-term laboratory bioassays (1987)

In the July bioassay, there were no significant differences in in vivo fluorescence between the treatments (Fig. 2). There was a non-significant pattern that occurred in all three 1987 bioassay experiments though, in which after three to four days the metals and EDTA addition (+ME) showed an increase of at least 20% over the individual nutrient additions (NH_4^+, NO_3^+, and P), and control. No combination treatments were used in July. In August and September a minimal combination of either +ME with NH_4^+ (+MEN) or EDTA alone with NH_4^+ (+EN) was needed to show a significant stimulation over the other single or combination nutrient treatments.

Long-term in situ bioassays (1997)

From the initial pooled samples collected on July 28, initial chlorophyll *a* concentrations were 0.02 μg/l and 0.06 μg/l in the epilimnetic and deep waters, respectively. There was a significant difference between the means of the treatments in the epilimnetic experiment (Table 2, Fig. 3), with chlorophyll *a* concentrations significantly higher in all treatments that included a nitrogen addition (+N, +MEN, +NP, and +MENP; Fig. 3) relative to the control. The +N chlorophyll *a*, however, was significantly less of an increase than the other three combination treatments. In the deep samples there was also a significant difference between the means of the treatments (Table 2), and chlorophyll *a* concentrations were significantly enhanced in all the combination treatments, including +MEN, +MEP, +NP, and +MENP (Fig. 3). The Tukey post hoc HSD analysis indicated that +MENP > +MEN and +MEP > +NP.

Enhancement of dark carbon uptake

In 1987 and 1997 the addition of $+NH_4^+$ or combination addition of ammonium and metals (1997 only) did not significantly stimulate the dark uptake of inorganic carbon in the epilimnetic samples (Table 3).

Response of photosynthesis and photosynthetic carbon metabolism in response to nutrient additions

Experiments 1 & 2

There was no significant difference between photosynthetic uptake of carbon in the five treatments in Experiment 1 in July 1987 (Fig. 4).

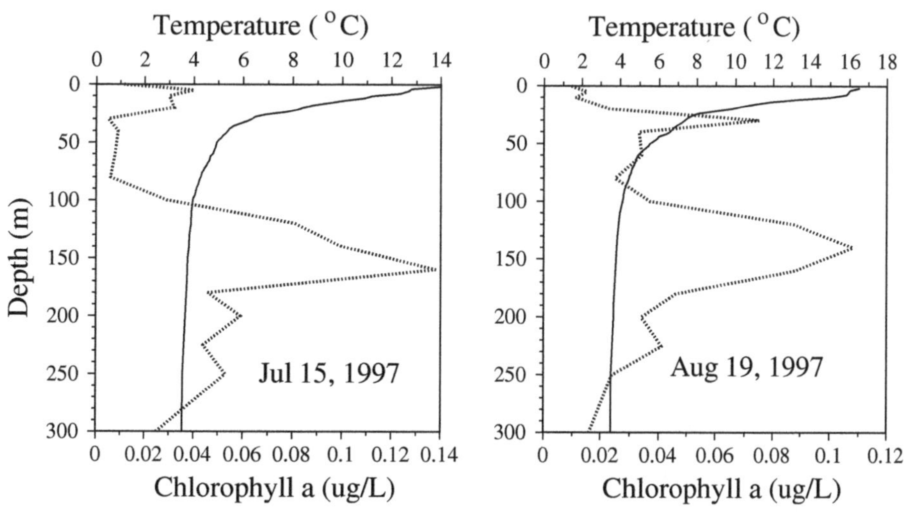

Fig. 1 Temperature (solid lines) and chlorophyll *a* concentrations (dotted line) in Crater Lake, July and August 1997

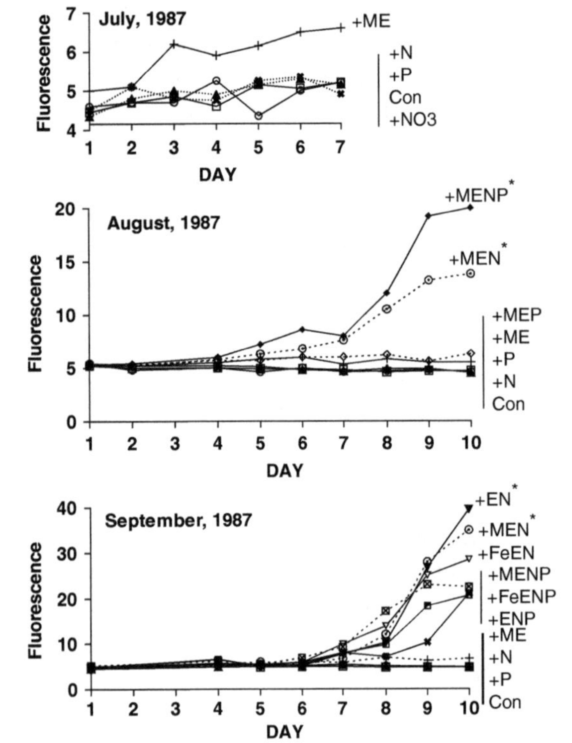

Fig. 2 In vivo chlorophyll fluorescence response in the 1987 long-term laboratory bioassays using near-surface Crater Lake phytoplankton. Each point represents the mean of duplicate flasks. Treatments explained in Table 1, (* indicates difference from the control, $p < 0.05$)

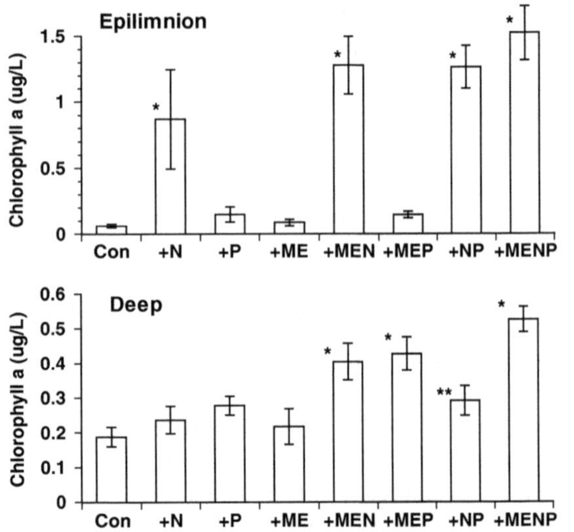

Fig. 3 Final chlorophyll *a* concentrations in the 1997 in situ bioassays in the epilimnion and deep samples. Error bars represent ±1 standard deviation. (* indicates significantly different from the control (Con) at $p < 0.0005$, ** indicates $p < 0.05$)

At the same time there was no significant response to the nutrient additions (NO_3^- and NH_4^+) relative to the control in patterns of photosynthetic carbon metabolism in Experiment 2, with all treatments being within 10% of each other in ^{14}C uptake. Protein and polysaccharide synthesis accounted for approximately 32% and 48% of the accumulated carbon, respectively, with the lipid and soluble metabolites making up about 10% each in this experiment (Fig. 5).

Experiment 3

In August 1987, nutrient additions had no significant effect on the photosynthetic rate (data not shown) in any of the treatments relative to the control. The metals–EDTA treatment did significantly stimulate protein synthesis rates over that of the control (Fig. 6).

Experiments 4 & 5

There appeared to be an enhancement of photosynthesis in treatments that included an addition of metals and EDTA (Fig. 7) in Experiment 4, with the other additions having no effect, though the small number of replicates did not show a statistical difference. When the treatments were pooled into a metals and EDTA addition group (the +ME and +MEN treatments) or other (control, +N, +P, and +NP), the group that had the

Table 2 ANOVA results for the effects of nutrient additions on final chlorophyll *a* concentrations in the epilimnion and deep from the Crater Lake in situ bioassays in 1997

		Sum of squares	df	Mean square	F	Significance
Epilimnion	Between groups	13.340	7	1.906	57.69	<0.0005
	Within groups	1.024	31	0.03304		
	Total	14.364	38			
Deep Samples	Between groups	0.464	7	0.0631	37.72	<0.0005
	Within groups	0.0527	30	0.00176		
	Total	0.517	37			

Table 3 Enhancement of dark carbon uptake in Crater Lake

	Control	$+NH_4^+$	$+NH_4^+$ + ME
August, 1987	12,505 (858)	13,303 (690)	
July, 1997	356,016 (23,279)	382,044 (33,060)	382,080 (28,560)

Values represent Bq of ^{14}C retained on the filter. Standard deviations in parentheses, $n = 3$ in 1987 and 5 in 1997

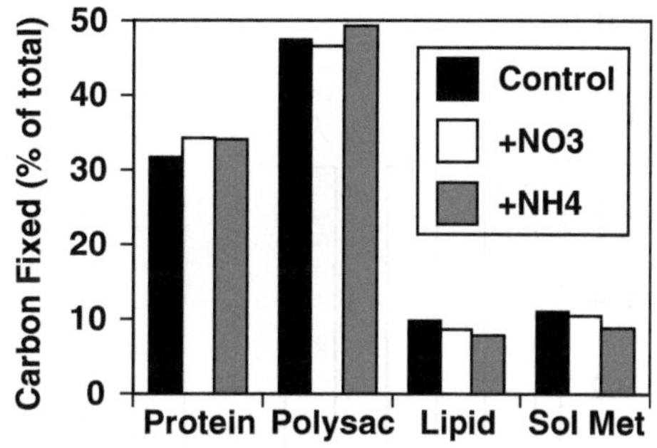

Fig. 5 Response of photosynthetic carbon metabolism to the addition of NO_3^- or NH_4^+ in July, 1987. Each treatment is the mean of two replicates

metal and EDTA mixture added had a significantly higher rate of photosynthesis ($p < 0.0005$, *t*-test) than did bottles with no metals and EDTA added.

Within the first incubation period (20.75 h) in Experiment 5, the +MEN treatment had accumulated more carbon through photosynthetic processes than did the control and +N treatments, and this trend continued to increase through out the 66 h of the experiment (Fig. 8). The control and +N treatments appeared to respond similarly, over the time span of the experiment. Total carbon increase in the +MEN treatment was 11.4 mg C m^{-3}, and was 9.2 and 8.5 mg C m^{-3} in the control and +N treatment, respectively. The difference in carbon accumulation was mostly due to increased accumulation of the protein within the phytoplankton (Fig. 8). The percent of total photosynthate allocated to protein was also highest in the +MEN treatment, with protein accounting for 71% of the carbon accumulated, which was significantly higher than the protein in control and +N, which were 58% and 66%, respectively.

Selenastrum Algal Assay Bottle Test

The *S. capricornutum* assays showed that an addition of EDTA (whether it included any trace metals or not) significantly stimulated growth at all depths and times (Table 4). In the near-surface waters (collected from 10 m) in both months there was no significant response to +N or +P, though these treatments stimulated growth at some of the lower depths. Generally, the pattern of *S. capricornutum* response at Station 13 was +ENP > +EN > +E and +EP > +NP. The +Fe treatment (iron with no chelator) in July was moderately stimulatory, but less than treatments containing EDTA alone. The response from the

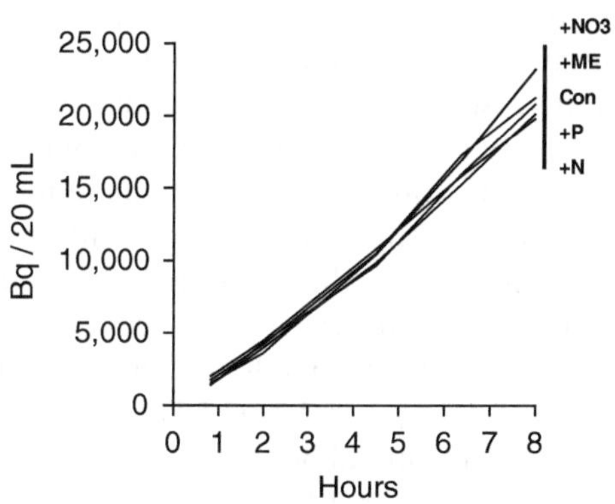

Fig. 4 Photosynthetic response (^{14}C trapped on a filter from a 20 ml aliquot) of epilimnetic Crater Lake phytoplankton to various nutrient additions (see text) over an 8 h incubation in July, 1987. Each line represents the mean of two replicates

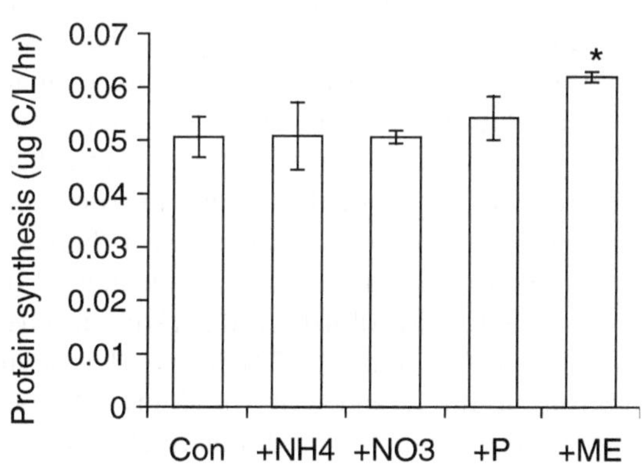

Fig. 6 Rates of protein synthesis in an August 1987 nutrient addition experiment. (* indicates significantly different from the control (Con) at $p < 0.05$; modified from Groeger & Tietjen, 1993)

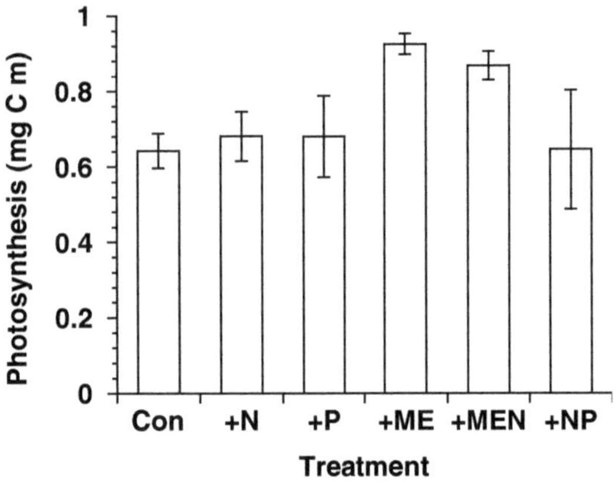

Fig. 7 Photosynthetic response of epilimnetic Crater Lake phytoplankton in 1997 after 2 days of exposure to various nutrient additions. Each treatment represents the mean of two replicates, except for the control where three replicates were used. Error bars represent ±1 standard deviation

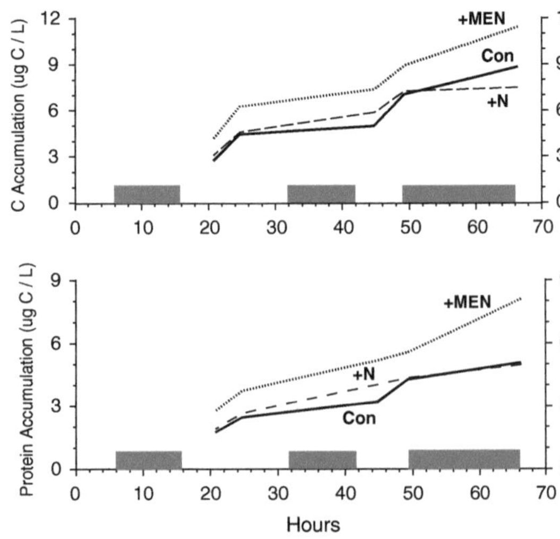

Fig. 8 Carbon accumulated through photosynthesis (upper panel) and protein carbon accumulated (lower panel) in Crater Lake epilimnetic phytoplankton in 1997 under different nutrient treatments. Each line represents the mean of two replicates. Shaded bars represent periods of dark

spring samples clearly suggested that the chemical composition of this water is very different than the Crater Lake water column.

Discussion

Summer phytoplankton in the Crater Lake epilimnion were primarily limited by a trace metal(s) or its availability. With one exception, there was no response in the long-term assays in treatments including only nitrogen or phosphorus additions, but there was commonly a significant positive response to metals-EDTA additions, or EDTA only. Due to extremely low concentrations, nitrogen quickly became limiting to the phytoplankton when the trace metals were made available. Phosphorus and silica were not limiting.

Many lakes in the western U.S. have been characterized as particularly low in trace elements, and the Oregon Cascades, with their granitic and volcanic bedrock, have been singled out as very dilute (Bradford et al., 1968; Landers et al., 1987). Goldman (1972) found the phytoplankton of three lakes of the Sierra Nevada range in California reacted positively to trace metal additions, with Lake Tahoe phytoplankton responding to individual additions of zinc, manganese, and iron. Stoddard (1987) found that trace metals and phosphorus were co-limiting in another Sierra Nevada lake. Trace metals are extremely low in Crater Lake, with concentrations of filtered and unfiltered iron in surface waters ranging from approximately 2–17 nmol l^{-1} and 36–88 nmol l^{-1}, respectively, with concentrations decreasing rapidly below 75 m (Collier et al., 1990). Because of this distribution with depth, Collier et al. (1990) concluded that atmospheric deposition, instead of spring inputs or snowmelt, are the major source of trace metals to the lake.

The role of organic chelators is critical in determining trace metal availability for phytoplankton growth (Huntsman & Sunda, 1980). For example, Rue & Bruland (1997) found that >99.9% of the dissolved iron in an equatorial region of the Pacific was chelated with organic ligands. Therefore, the extremely low DOC within Crater Lake may influence the actual availability of trace metals to phytoplankton. When the organic chelator EDTA was used without a specific metal or mixture of metals, such as in the September 1987 long-term laboratory assay and the *Selenastrum* bioassays, there was a significant positive response, very similar in magnitude to the treatments with both EDTA and metals. In some studies, a positive algal

Table 4 Response of *Selenastrum capricornutum* biomass (mg dry wt L^{-1}) at day 14 to unamended (control, no nutrients added) and various nutrient additions (see Table 1 and text) from water samples collected from various depths at station 13 in Crater L. or influent springs (*indicates significantly greater than the control, $p < 0.05$)

	Con	+NO_3	+P	+E	+ENO_3	+EP	+NO3P	+ENO3P	+Fe	+ME	+ME–Fe
June											
10 m	0.66	0.89	0.13	1.58*	6.38*	1.49*	2.27*	27.20*			
60 m	0.67	2.00*	0.31	1.56*	7.46*	1.71*	0.21	31.06*			
120 m	0.92	2.10*	3.23*	5.10*	6.67*	6.42*	10.18*	26.21*			
200 m	0.55	0.29	0.38	1.10*	8.40*	1.22*	6.33*	28.86*			
300 m	0.15	1.12*	0.65*	1.52*	8.17*	1.38*	1.13*	31.35*			
400 m	0.48	0.35	0.17	1.76*	7.85*	1.68*	0.12	26.73*			
550 m	0.14	0.25	0.19	1.75*	7.92*	2.15*	3.80*	30.09*			
Sp.20	0.40	7.61*	3.46*	1.31*	24.72*	3.29*	0.16	33.34*			
Sp.42	0.15	4.99*	0.16	8.15*	19.80*	8.61*	7.76*	37.14*			
July											
10 m	0.07	0.09	0.06	3.39*	10.93*	4.45*	0.18	28.78*	0.21	2.20*	3.73*
	0.09	0.16	0.16	4.36*	19.87*	2.59*	0.31	31.10*	0.48*	3.23*	3.57*
60 m	0.12	0.12	0.31	3.17*	13.05*	3.55*	0.13	30.15*	1.26*	4.38*	3.45*
	0.26	0.16	0.17	4.94*	7.96*	6.87*	0.43	28.18*	0.22	5.70*	5.97*
120 m	0.19	1.83*	0.15	5.23*	5.94*	6.50*	0.77*	29.01*	3.35*	5.16*	5.27*
	0.19	0.09	0.11	4.75*	7.13*	2.27*	0.10	22.59*	0.12	2.87*	4.08*

Pairs of rows in July refers to filtered (above), and filtered and autoclaved water samples (bottom row). In June all sample water was simply filtered

growth response to a chelator alone has been attributed to the dissolution of hydrous iron oxide particles providing a source of available iron (Huntsman & Sunda, 1980).

Iron is the most likely limiting trace metal in Crater Lake, similar to certain oceanic areas, because of its high cellular demand relative to other trace metals (Bruland et al., 1991). Near-surface concentrations of manganese, copper, molybdenum, and nickel in Crater Lake (Collier et al., 1990) suggest that these elements are in greater supply, relative to cellular demand, than iron is. Further research is needed to determine which trace metal or metals is limiting phytoplankton growth in this ecosystem, and what role DOC plays in regulating water column primary production.

The very low concentration of nitrogen was also an important factor in maintaining very low phytoplankton biomass in the Crater Lake epilimnion. Paerl (1982) predicted that lakes with small drainage areas relative to their volume and with a long water residence time, such as Crater Lake, would tend to be nitrogen limited. The lakes in the mountainous regions of the western U.S. tend to have very low nitrogen concentrations (Landers et al., 1987), and nitrogen is always extremely low in the upper 200 m of the Crater Lake water column (Larson et al., 1996a). Nitrogen inputs into Crater Lake from external sources are dominated by atmospheric inputs (Dymond et al., 1996; Nelson et al., 1996), though 85% of nitrogen input into the euphotic zone comes from upwelling of water from the deep lake (Dymond et al., 1996). Therefore, during the period of stratification, nitrogen supply, and possibly other critical nutrient loading, should be at its lowest in the epilimnion. McIntire et al. (1996) found that both chlorophyll *a* concentrations and chlorophyll *a* to cellular biovolume decreased in near surface phytoplankton throughout the stratified period relative to the winter and spring, possibly suggesting an increasing severity of nutrient deficiency. Planktonic N_2–fixing cyanobacteria are rare or not present in the Crater Lake water column (McIntire et al., 1996), though nitrogen fixation is important in the littoral periphyton (Loeb & Reuter, 1981). The low availability of trace metals, particularly iron, may be primarily responsible for the apparent lack of nitrogen fixation within the Crater Lake phytoplankton (Wurtsbaugh & Horne, 1983; Wurtsbaugh et al., 1985).

The 1997 in situ long-term bioassay (Fig. 3) was the sole instance in which a positive response, a significant increase in chlorophyll *a* concentration, to nitrogen alone was seen in the epilimnion. In the two short-term parallel experiments (Experiments 4 & 5) the +NH_4^+ treatments were not stimulatory relative to controls, though treatments with EDTA-metals and EDTA-metals plus NH_4^+ were. Three explanations, not mutually exclusive of each other, are possible to account for these different responses. First, the increase in chlorophyll *a* may solely represent the synthesis of a nitrogen-rich molecule, and nitrogen is not actually stimulating cellular growth. Paerl (1982) found that additions of nitrogen to some phytoplankton communities caused chlorophyll *a* to respond more dramatically than cell counts or ATP content relative to a control. Therefore, while chlorophyll *a* is normally a good indicator of algal biomass and response, its dynamics may not always be linked tightly to reproduction. The response to the trace metal additions or increase in metal availability caused a different response in which the phytoplankton increased their photosynthetic capabilities and increased growth as indicated by protein synthesis.

Second, different groups or species within the phytoplankton community may be responding differently to the various treatments; i.e., one species or group responded to the +NH_4^+ treatment by synthesizing chlorophyll *a*, and another species or group responded to the +ME treatment by increasing their capacity for a higher photosynthetic rate and growth rate. In Crater Lake in August of 1981, Lane & Goldman (1984) found that photosynthesis in near-surface nannoplankton (3–8 μm size fraction) was stimulated by +NH_4^+, +NH_4^+ and +P, and trace metals, though the ultraplankton fraction (0.45–3 μm) was only stimulated by trace metals, and netplankton fraction (8–100 μm) was not stimulated by any of the treatments. Cavender-Bares et al. (1999) fractionated the phytoplankton community in the equatorial Pacific into five distinct groups using flow cytometry during an iron enrichment experiment. They found that all five groups were stimulated by the iron, though due to differential response by grazers, biomass of some groups increased greatly while another group's biomass actually decreased.

Third, it may be that the phytoplankton community or some part of it is responding to trace metals additions more rapidly than the NH_4^+ additions. In these experiments in 1987 and 1997, the response to trace metals were observable within 2–3 days or sooner. This may be the weakest possibility, because it is clear that many phytoplankton are capable of very quick response (seconds to hours) to nitrogen additions (e.g., Groeger & Kimmel, 1988).

There are less data to determine nutrient limitation below the epilimnion, though in 1997 only combination treatments were successful in stimulating the deep phytoplankton community. McIntire et al. (1996) found that the phytoplankton community below the epilimnion was much more diverse than the epilimnion during stratification in Crater Lake, and they suggested that the region from a depth of 40 to 100 m may be a zone of high nutrient regeneration and metabolic activity due to the high zooplankton biomass, including *Daphnia pulicaria* and *Bosmina longirostris* (Larson et al., 1996b), which are absent from the epilimnion. The higher algal species diversity, and increased grazing and biologically mediated nutrient turnover may allow for a phytoplankton community highly adapted to the available nutrients and nutrient supply in these deeper waters.

The short-term assays, for the most part, did not indicate a physiological stress from nutrient deficiency in the epilimnion. In both years NH_4^+ did not stimulate the uptake of CO_2in the dark, and in 1987 neither the addition of NH_4^+ nor NO_3^- had an effect on photosynthesis or photosynthetic carbon metabolism. Photosynthetic carbon metabolism of the control treatments (Fig. 5) was not reflective of a physiological stress (Groeger & Kimmel, 1989), and may have been indicative of high growth with ≈33% of photosynthate found in protein, 48% in polysacchardie, and <10% in lipids. For example, Groeger & Kimmel (1988) and Groeger & Tietjen (1993) found high rates of lipid synthesis in N-limited phytoplankton, and the rates of lipid synthesis were dramatically reduced when NH_4^+ was added. In most of the short-term experiments in which a +ME treatment was used in Crater Lake, protein synthesis was stimulated. An increase in protein synthesis is the clearest indication of an increase in growth rate

(Perin et al., 2002), and indicates a rapid response to increased trace metal availability.

The argument has been made that in ultraoligotrophic oceans (Laws et al., 1984: Sheldon, 1984) where biomass is extremely low, phytoplankton growth rates are near maximal and are maintained at high levels by very rapid and efficient turnover of nutrients. There are a number of interesting similarities between an open-ocean water column and that of Crater Lake, including nitrogen and trace metal limitation, extremely low DOC, nutrient loading to the surface waters being dominated by atmospheric inputs and upwelling, minimal drainage basin influence, remarkable water clarity, and bacterioplankton groups that are common in the ocean and Crater but not found before in other lakes (Urbach et al., 2001). Dymond et al. (1996) found that that nitrogen must be recycled between 10 and 30 times within the euphotic zone to support the phytoplankton primary production found there, which is also characteristic of oceanic water columns where the food web is integral to rapid nutrient regeneration. This indicates that macrograzers and the microbial loop are critically important in maintaining rapid nutrient cycling to support the productivity of the Crater Lake pelagial ecosystem. Similarly, trace metals would apparently also have to be recycled rapidly. Rue and Bruland (1997) found that dissolved iron turned over approximately every 0.5 days in the upper 100 m of the ocean. In the epilimnion, where macrozooplankton and fish are rare, recycling is probably less efficient and may be responsible for lower phytoplankton diversity (McIntire et al., 1996) and more intense nutrient limitation.

The interaction between factors influencing "bottom-up" control, or nutrient loading into the euphotic zone, and "top-down" effects, or the food web structure including fish, macrozooplankton and microbial loop components, on Crater Lake phytoplankton may vary significantly between years. The first would be the annual pattern of stratification and mixing within the lake. These patterns are critical in the distribution of nitrogen within the lake (Dymond et al., 1996), and may be as equally important in trace metal supply and availability. For example, during cold years with heavy snowfall, Lake Tahoe tends to mix completely, resulting in much higher nitrogen loading into the euphotic zone and higher productivity (reviewed in Goldman, 1998). Interannual variation between *Daphnia pulicaria* and the other macrozooplankton and kokanee salmon (Buktenica & Larson, 1996; Larson et al., 1996b) indicates strong "top-down" influence on food web structure. These cyclic variations within the food web may have significant effects on the year-to-year variation on cycling and distribution of limiting nutrients.

Acknowledgements I would like to thank Gary Larson for his financial support through the National Park Service and the U.S. Geologic Survey, and for his patience and encouragement. Scott Girdner, Jeff Milder, Ashley Gibson, and Mark Buktenica all provided invaluable assistance in the field. I would like to thank Mary Debacon for conducting the *Selenastrum* bioassays and kindly providing me with the data. I would like to thank Robert Hoffman, Jr. and Gary Larson for providing the water column temperature and chlorophyll *a* data for July and August, 1997.

References

Bradford, G. R., F. L. Bair & V. Hunsaker, 1968. Trace and major element content of 170 High Sierra lakes in California. Limnology and Oceanography 13: 526–530.

Bruland, K. W., J. R. Donat & D. A. Hutchins, 1991. Interactive influences of bioactive trace metals on biological production in oceanic waters. Limnology and Oceanography 36: 1555–1577.

Buktenica, M. W. & G. L. Larson, 1996. Ecology of kokanee salmon and rainbow trout in Crater Lake, Oregon. Lake and Reservoir Management 12: 298–310.

Burnison, B. K., 1980. Modified dimethyl sulfoxide (DMSO) for chlorophyll analysis of phytoplankton. Canadian Journal of Fisheries and Aquatic Sciences 37: 729–733.

Cavender-Bares, K. K., E. L. Mann, S. W. Chisholm, M. E. Ondrusek & R. B. Bidigare, 1999. Differential response of equatorial Pacific phytoplankton to iron fertilization. Limnology and Oceanography 44: 237–246.

Collier, R., J. Dymond, J. McManus & J. Lupton, 1990. Chemical and physical properties of the water column at Crater Lake, Oregon. In Drake, E. G. Larson, J. Dymond & R. Collier (eds), Crater Lake, an Ecosystem Study. Pacific Division of the American Association for the Advancement of Science, 69–79.

Dymond, J., R. Collier, J. McManus & G. L. Larson, 1996. Unbalanced particle flux budgets in Crater Lake, Oregon: Implications for edge effects and sediment focusing in lakes. Limnology and Oceanography 41: 732–743.

Goldman, C. R., 1972. The role of minor nutrients in limiting the productivity of aquatic ecosystems. In Likens, G. E. (ed.), Nutrients and Eutrophication: The Limiting-Nutrient Controversy. Special Symposium, American Society of Limnology and Oceanography Vol. 1: 21–33.

Goldman, C. R., 1998. Four decades of change in two subalpine lakes. Verhandlungen Internationale Vereinigung für Theoretische und Angewandte Limnologie 27: 7–26.

Groeger, A. W. & B. L. Kimmel, 1988. Photosynthetic carbon metabolism in a nitrogen-limited reservoir. Canadian Journal of Fisheries and Aquatic Sciences 45: 720–730.

Groeger, A. W. & B. L. Kimmel, 1989. Relationship between photosynthetic and respiratory carbon metabolism in freshwater phytoplankton. Hydrobiologia 173: 107–117.

Groeger, A. W. & T. E. Tietjen, 1993. Physiological responses of nutrient-limited phytoplankton to nutrient addition. Verhandlungen Internationale Vereinigung für Theoretische und Angewandte Limnologie 25: 370–372.

Huntsman, S. A. & W. G. Sunda, 1980. The role of trace metals in regulating phytoplankton growth. In Morris I. (ed) The physiological ecology of phytoplankton. University of California Press, Berkeley, 285–328.

Hutchinson, G. E., 1957. Treatise on limnology, 1. Wiley & Sons, New York.

Kalff, J., 2002. Limnology: Inland Water Ecosystems. Prentice Hall, Upper Saddle River, NJ.

Kirk, J. T. O., 1983. Light and Photosynthesis in Aquatic Ecosystems. Cambridge University Press, Cambridge.

Landers, D. H., J. M. Eilers, D. F. Braake, W. S. Overton, P. E. Kellar, M. E. Silverstein, R. D. Sconbrod, R. E. Crowe, R. A. Linthurst, J. M. Omernik, S. A. Teague, & E. P. Meier, 1987. Characteristics of Lakes in the Western United States, Vol. 1. EPA/600/3–86/054a, U.S. Environmental Protection Agency, Washington, D.C.

Lane, J. L. & C. R. Goldman, 1984. Size-fractionation of natural phytoplankton communities in nutrient bioassay studies. Hydrobiologia 118: 219–223.

Larson, G. L., 1988. Crater Lake Limnological Studies 1987 Annual Report.

Larson, G. L., 1998. Crater Lake Limnological Studies 1997 Annual Report. Technical Report NPS/CCSOOSU/NRTR-98/13, National Park Service, Seattle, WA.

Larson, G. L., C. D. McIntire, M. Hurley & M. W. Buktenica, 1996a. Temperature, water chemistry, and optical properties of Crater Lake. Lake and Reservoir Management 12: 230–247.

Larson, G. L., C. D. McIntire, R. E. Truitt, & M. W. Buktenica, 1996b. Zooplankton assemblages in Crater Lake, Oregon, USA. Lake and Reservoir Management 12: 281–297.

Laws, E. A. & many others, 1984. High phytoplankton growth and production rates in oligotrophic Hawaiian coastal waters. Limnology and Oceanography 29: 1161–1169.

Li, W. K., H. E. Glover & I. Morris, 1980. Physiology of carbon photoassimilation by *Oscillatoria thiebautii* in the Caribbean Sea. Limnology and Oceanography 25: 447–456.

Loeb, S. L. & J. E. Reuter, 1981. The epilithic periphytopn community: A five lake comparative study of community productivity, nitrogen metabolism and depth-distribution of standing crop. Verhandlungen Internationale Vereinigung für Theoretische und Angewandte Limnologie 21: 346–352.

McIntire, C. D., G. L. Larson, R. E. Truitt & M. K. DeBacon, 1996. Taxonomic structure and productivity of phytoplankton assemblages in Crater Lake, Oregon. Lake and Reservoir Management 12: 259–280.

Miller, W. E., J. C. Greene & T. Shiroyama, 1978. The *Selenastrum capricornutum* Prinz algal assay bottle test: Experimental design, application, and data interpretation protocol. EPA-600/9–78–018, U.S. Environmental Protection Agency, Corvallis, Oregon.

Nelson, P. O., J. F. Riley & G. L. Larson, 1996. Chemical solute mass balance for Crater Lake, Oregon. Lake and Reservoir Management 12: 248–258.

Paerl, H. W., 1982. Factors limiting productivity of freshwater ecosystems. In Marshall K. C. (ed.), Advances in Microbial Ecology. Plenum Press, New York, 75–110.

Perin, S., D. R. S. Lean, F. R. Pick & A. Mazumder, 2002. S Photosynthetic carbon allocation: Effects of planktivorous fish and nutrient enrichment. Aquatic Sciences 64: 217–238.

Rue, E. L. & K. W. Bruland, 1997. The role of organic complexation on ambient iron chemistry in the equatorial Pacific Ocean and the response of a mesoscale iron addition experiment. Limnology and Oceanography 42: 901–910.

Sheldon, R. W., 1984. Phytoplankton growth rates in the tropical ocean. Limnology and Oceanography 29: 1342–1346.

Stoddard, J. L., 1987. Micronutrient and phosphorus limitation of phytoplankton abundance in Gem Lake, Sierra Nevada, California. Hydrobiologia 154: 103–111.

Urbach, E., K. L. Vergin, L. Young, A. Morse, G. L. Larson & S. J. Giovannoni, 2001. Unusual bacterioplankton community structure in ultra-oligotrophic Crater Lake. Limnology and Oceanography 46: 557–572.

Wurtsbaugh, W. A. & A. J. Horne, 1983. Iron in eutrophic Clear Lake, California: its importance for algal nitrogen fixation and growth. Canadian Journal of Fisheries and Aquatic Sciences 40: 1419–1429.

Wurtsbaugh, W. A., W. F. Vincent, R. Alfaro Tapia, C. L. Vincent & P. J. Richerson, 1985. Nutrient limitation of algal growth and nitrogen fixation in a tropical alpine lake, Lake Titicaca (Peru/Bolivia). Freshwater Biology 15: 185–195.

Yentsch, C. M., C. S. Yentsch & L. R. Strube, 1977. Variations in ammonium ehhancement, an indication of nitrogen deficiency in New England coastal phytoplankton populations. Journal of Marine Research 35: 539–555.

Hydrobiologia (2007) 574:217–233
DOI 10.1007/s10750-006-0354-2

Distribution and abundance of zooplankton populations in Crater Lake, Oregon

Gary L. Larson · C. David McIntire · Mark W. Buktenica · Scott F. Girdner · Robert E. Truitt

© Springer Science+Business Media B.V. 2007

Abstract The zooplankton assemblages in Crater Lake exhibited consistency in species richness and general taxonomic composition, but varied in density and biomass during the period between 1988 and 2000. Collectively, the assemblages included 2 cladoceran taxa and 10 rotifer taxa (excluding rare taxa). Vertical habitat partitioning of the water column to a depth of 200 m was observed for most species with similar food habits and/or feeding mechanisms. No congeneric replacement was observed. The dominant species in the assemblages were variable, switching primarily between periods of dominance of *Polyarthra-Keratella cochlearis* and *Daphnia.* The unexpected occurrence and dominance of *Asplanchna* in 1991 and 1992 resulted in a major change in this typical temporal shift between *Polyarthra-K. cochlearis* and *Daphnia.* Following a collapse of the zooplankton biomass in 1993 that was probably caused by predation from *Asplanchna*, *Kellicottia* dominated the zooplankton assemblage biomass between 1994 and 1997. The decline in biomass of *Kellicottia* by 1998 coincided with a dramatic increase in *Daphnia* biomass. When *Daphnia* biomass declined by 2000, *Keratella* biomass increased again. Thus, by 1998 the assemblage returned to the typical shift between *Keratella-Polyarthra* and *Daphnia.* Although these observations provided considerable insight about the interannual variability of the zooplankton assemblages in Crater Lake, little was discovered about mechanisms behind the variability. When abundant, kokanee salmon may have played an important role in the disappearance of *Daphnia* in 1990 and 2000 either through predation, inducing diapause, or both.

Keywords Zooplankton · Rotifer · Cladoceran · Caldera lake · Crater Lake

Guest editors: Gary L. Larson, Robert Collier, and Mark W. Buktenica
Long-term Limnological Research and Monitoring at Crater Lake, Oregon

Electronic Supplementary Material is available to aurhorised users in the online version of this article at http://dx.doi.org/10.1007/s10750-006-0354-2

G. L. Larson (✉)
U. S. Geological Survey, Forest and Rangeland Ecosystem Science Center, 3200 SW Jefferson Way, Corvallis, OR 97331, USA
e-mail: gary_l._larson@usgs.gov

C. D. McIntire
Department of Botany and Plant Pathology, Oregon State University, Corvallis, OR 97331, USA

M. W. Buktenica · S. F. Girdner
Crater Lake National Park, Crater Lake, OR 97604, USA

R. E. Truitt
College of Forestry, Oregon State University, Corvallis, OR 97331, USA

Introduction

Zooplankton are important to the dynamics of lakes by influencing water clarity through grazing activities, nutrient cycling, and linking lower trophic levels with higher levels in the pelagic food web of lakes (Carpenter et al., 1985). In theory, increases in zooplankton biomass in response to reduced predation from higher trophic levels, e.g., fishes, reduce those of lower trophic levels, e.g., phytoplankton (Leibold & Tessier, 1997). This is a simplistic view because vertical habitat partitioning of the water column often modifies the ways species interact with their resources, competitive interactions, and predators (Leibold & Tessier, 1997). Potential for vertical habitat partitioning of the water column increases with lake clarity (Larson et al., 1996a). Crater Lake offers an opportunity to observe habitat partitioning during summer thermal stratification owing to the extreme clarity of the lake. The average Secchi disk clarity of the lake in August is about 30 m and the depth of 1% of the incident solar radiation exceeds 80 m (Larson et al., 1996a).

Historical records of the distribution and abundance of zooplankton populations in Crater Lake, Oregon, were limited in scope and duration. Briefly, Evermann (1897) found that the dominant species near the lake surface and in the littoral zone was *Daphnia pulex* var. *pulicaria*. Kemmerer et al. (1924) reported the presence of two species of crustaceans and three species of rotifers in samples obtained with a 20-mesh silk net in 1913. In this case, *Asplanchna* and *D. pulex* were the dominant taxa, the latter species with a maximum abundance at depths between 40 and 80 m below the lake surface. No zooplankton specimens were collected in the upper 30 m of the water column and below a depth of 200 m, but swarms of *Daphnia* were observed around the shallows of Wizard Island, the larger of two islands in the lake. Although the studies by Evermann (1897) and Kemmerer et al. (1924) identified *D. pulex* as a dominant crustacean in the water column, more recent taxonomic interpretations indicated that this taxon was, in fact, *Daphnia pulicaria* for reasons reviewed by Karnaugh (1988). Brode (1938) and Hasler (1938) found that *Daphnia* reached its maximum densities in the water column at depths between 38 and 53 m and between 50 and 122 m, respectively. However, *Daphnia* was not found in any samples obtained in 1940, either by net tows near Wizard Island or by vertical tows in the water column between the water surface and a depth of 100 m (Hasler & Farner, 1942). More recent studies by Hoffman (1969) and Malick (1971) indicated that *D. pulicaria* and *Bosmina longirostris* were the dominant crustaceans in the water column of the lake and that populations of these species exhibited annual changes in relative abundance. In 1967 the relative abundance of *Bosmina* and *Daphnia* was 98% and 2%, respectively, whereas the corresponding values in 1969 were 2% (*Bosmina*) and 98% (*Daphnia*).

The first comprehensive investigation of limnetic zooplankton populations in Crater Lake extended from July 1985 to September 1987 (Karnaugh, 1988, 1990). This research also examined alternative equipment and strategies for sampling zooplankton in the water column, and served as a pilot study for a long-term sampling program that was initiated in 1988. The dominant rotifer in the water column was *K. cochlearis*, whereas the most abundant cladoceran was *B. longirostris*.

An analysis of the density, biomass, and taxonomic structure of zooplankton assemblages in Crater Lake for the period from 1985 through 1990 was reported by Larson et al. (1996b). Results of this study revealed the presence of two cladocerans (*B. longirostris* and *D. pulicaria*) and 11 species of rotifers. In general, rotifers were more numerous than cladocerans. Most zooplankton taxa were distributed from the lake surface to a depth of about 200 m during periods without thermal stratification (winter and spring). The distribution of zooplankton species was stratified in the water column to a depth of 200 m during thermal stratification. Zooplankton density was very low in the epilimnion, whereas the highest densities were found between 80 and 120 m. Diel vertical migrations were essentially absent. *Daphnia* was cyclical in abundance and, when present, densities of *Bosmina* and rotifers were relatively low. Moreover, there was a conspicuous absence of copepods in the water column, although a more recent study (Warncke,

1999) indicated that copepods sometimes were present in low densities in the near-benthic habitats of the littoral zone.

The contents of this paper compares the results of the 1985–1990 study with data obtained between 1991 and 2000, and integrates the analysis to include data collected from 1988 to 2000, a period during which sampling methods were the same and all zooplankton samples were processed by the same technical assistant (R. Truitt). Objectives of the integrated analysis were to describe long-term temporal patterns in: (1) the distribution and abundance of zooplankton populations; (2) the taxonomic composition and biomass of the zooplankton assemblages; and (3) the relationships between the abundance of kokanee salmon and the taxonomic structure, abundance, and biomass of zooplankton assemblages.

Methods

Study site

Crater Lake is located in Crater Lake National Park in south-central Oregon at a latitude of 42°56′ N and longitude of 122°06′ W. The lake covers the floor of the caldera formed after the eruption and subsequent collapse of Mount Mazama, approximately 7700 years ago. The lake is the deepest lake in the United States with a maximum depth of 594 m, a mean depth of 325 m, and a surface area of 53.2 km^2. The shoreline around the lake extends for 31 km, and the diameter varies between 8 and 10 km. There is no surface outlet stream and, thus, water is lost from the system by seepage and evaporation. The caldera functions like a large rain gauge.

The surface elevation of the lake varies about 0.5 m annually, but exhibits long-term interannual declines of elevation. The surface elevation of the lake dropped by about 4 m between 1910 and 1931 and then returned to the benchmark elevation of 1882 m by the early 1950s. The lake exhibited long-term water level fluctuations of about ± 1 m from the benchmark level through the mid-1980s and then dropped about 3 m by 1994. The lake level returned to near the benchmark level in 2000.

Annual nutrient inputs into the lake are determined by: (1) runoff from the caldera wall, a catchment area of about 15 km^2 with about 40 permanent and ephemeral springs and small inlet streams; (2) the quantity of atmospheric deposition; and (3) seepage from underwater hydrothermal springs (Collier et al., 1993; Nelson et al., 1996). Crater Lake is classified as an ultra-oligotrophic lake with low nutrient concentrations and a high Secchi disk transparency, usually between 28 and 33 m from June to September (Larson & Buktenica, 1998). Additional information about the geological, chemical, and physical properties of Crater Lake is available in publications by Byrne (1965), Phillips (1968), Bacon & Lanphere (1990), Barber & Nelson (1990), Nathenson & Thompson (1990), McManus et al. (1992), Bacon et al. (2002), Larson et al. (1996a), and Nathenson et al. (2007).

Sampling

Zooplankton samples were obtained by vertical tows of a net with a diameter of 0.5 m and mesh size of 64 μm. The net was built by Research Nets Inc. (Bothell, Washington), and was equipped with a closing apparatus that allowed sampling of selected segments of the water column. All samples were obtained at Station 13, a location approximately 3 km south of Cleetwood Cove (see Larson et al., 1996a, Fig. 1). The zooplankton assemblage was sampled on 51 dates from June 1988 through September 2000. With the exceptions of samples obtained in January 1990 and 1997, April 1989, and October 1988, all other samples were obtained in the months of June, July, August, and September. Consequently, patterns in the fauna were well documented for the summer and early fall, but much less data were available for the winter and spring months. After sample processing, the entire data set consisted of 51 sets of samples, each of which consisted of organisms collected from eight different depth segments of the water column: 0–20 m, 20–40 m, 40–60 m, 60–80 m, 80–100 m, 100–120 m, 120–160 m, and 160–200 m. There

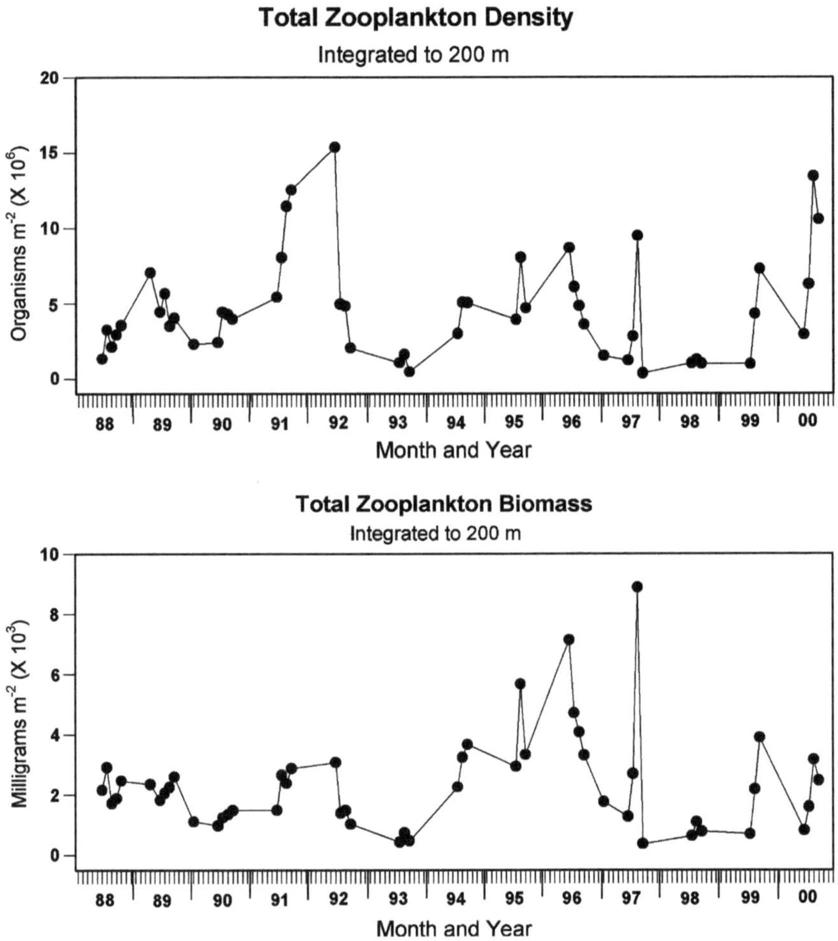

Fig. 1 Total zooplankton density and biomass integrated to 200 m, 1988–2000

were only two missing samples from the data (July 1994: 40–60 m and 160–200 m), so the total number of samples processed in the laboratory was 406. Zooplankton specimens were not found in two of the processed samples. Consequently, these samples were deleted for the statistical analysis, and the total number of samples represented in the data matrix was 404.

Laboratory methods

All zooplankton samples were preserved in the field with a 4% solution of sucrose–formaldehyde. In the laboratory, a Folsom Plankton Splitter was used to dilute each sample to a suitable density for a quantitative analysis. A sub-sample of the final concentration of organisms was transferred to a 50-ml settling chamber (Hydro-bios Kiel) and allowed to settle undisturbed for 24 h. Subsequently, each sample was mounted and counted on an inverted microscope at a magnification of 40×. The raw counts were converted to estimates of the number of organisms per cubic meter by a formula that adjusted the data for the dilution factor, the volume of lake water filtered by the net, and a factor expressing the efficiency of the net (see Larson et al., 1996a, for details). The biomass of each species in a sample was determined by multiplying the density of organisms by a factor expressing the average weight of an individual organism (Table 1). The conversion factors were determined by weighing sets of organisms on a Cahn Model 4100 electrobalance after the filtration procedure described by Larson et al. (1996a).

Table 1 List of zooplankton species included in the classification analysis of the zooplankton assemblage of Crater Lake

Group	Species	Acronym	Factor	Total Density
Crustaceans	*Bosmina longirostris*	BOLO	1.47	753,766.30
	Daphnia pulicaria	DAPU	9.93	100,763.94
Rotifers	*Keratella cochlearis macracantha*	KECO	0.11	6,133,036.64
	Keratella quadrata dispensa	KEQU	0.77	71,472.44
	Filinia terminalis	FITE	0.40	1,004,045.52
	Kellicottia longispina	KELO	0.84	3,167,867.85
	Polyarthra dolichoptera	PODO	0.39	2,005,157.55
	Synchaeta oblonga	SYOB	0.20	1,647,899.62
	Philodina cf. acuticornis	PHAC	0.35	348,586.08
	Conochilus unicornis	COUN	0.05	66,240.16
	Collotheca pelagica	COPE	0.05	20,256.32
	Asplanchna priodonta	ASPL	0.89	198,873.45

Acronyms are used in other tables and graphs to designate taxonomic entities. The Factor is a value, expressed as μg/individual, that is used to convert density to biomass. Total density is the summation of the density values over all samples

Data analysis

For the statistical analysis of the 404 zooplankton samples, the number of species in the data matrices was reduced to 12: 2 crustaceans and 10 rotifers (Table 1). These species accounted for 99.9% of the total density of zooplankton sampled from the water column of the lake. The other taxa were extremely rare and occurred infrequently in the samples.

To examine seasonal and interannual changes in the zooplankton assemblages, independent of spatial distribution, the density and biomass of each taxon under 1 m^2 of water surface was integrated to a depth of 200 m on each sampling date. Consequently, the data were expressed as the total number or weight of organisms in the upper 200 m of the water column. The corresponding data matrix was used to describe temporal patterns of species richness and dominance (McIntire & Overton, 1971). The seasonal and interannual variations in the species composition and biomass of the zooplankton assemblages were analyzed by CLUSB4, a clustering algorithm that provided a divisive, non-hierarchical, minimum variance partitioning of the data (McIntire, 1973). After an interpretable cluster structure was obtained by this analysis, the results were used to help interpret an ordination diagram produced by Detrended Correspondence Analysis (DCA) (Hill & Gauch, 1980). Covariance among zooplankton species and between zooplankton density and chlorophyll concentration and phytoplankton cell size was determined by correlation analyses (Snedecor & Cochran, 1967).

Results

Zooplankton density and biomass

Plots of total zooplankton density and biomass, integrated to a depth 200 m, revealed some lack of correspondence that was related to temporal changes in species composition and the average size of the dominant organisms (Fig. 1). Covariance between total density and total biomass, integrated to a depth of 200 m, was significant but still relatively low ($r = 0.6$, $P < 0.01$). The correlation coefficient expressing the covariance between zooplankton density and biomass matched up by depth interval was 0.75 ($P < 0.01$). Highest total densities were recorded in 1991, 1992, and 2000, whereas the corresponding period of highest total zooplankton biomass was 1995–1997.

Keratella cochlearis had the highest total density of all zooplankton taxa collected between 1988 and 2000 (Table 1), although it was primarily abundant between 1988 and 1992. Five rotifer taxa (*K. cochlearis*, *Filinia terminalis*, *Kellicottia longispina*, *Polyarthra dolichoptera*, and *Synchaeta oblonga*) exhibited higher total density values than either of the two crustaceans (*B. longirostris* and *D. pulicaria*). However, the crustaceans were larger organisms than the rotifers (see size factor in Table 1), and therefore, they made a proportionally greater contribution to the total zooplankton biomass when they were present in the fauna.

Because of the dominance of rotifers in the water column of Crater Lake, the temporal pattern of rotifer total density, integrated to 200 m, was similar to the pattern presented for total zooplankton density (Figs. 1, 2). In contrast, crustaceans had an intermediate level of density from 1988 through 1992, a relatively high density in 1994 and 1995, and a relatively low density from 1996 to 2000 (Fig. 2). Total rotifer biomass began to increase in 1995 and reached maximum values in 1996 and 1997 (Fig. 3). Total crustacean biomass was highest in 1988 and 1989, and was relatively low from 1990 to 1993 and in 1996, 1998, and 2000 (Fig. 3).

The number of species in the samples collected between 1988 and 2000 typically ranged from 8 to 10 (see Electronic Supplementary Material). Three samples had 11 species, whereas the July 1993 sample had 7 species and the July 1994 sample had 6 species. The September 1997 sample was lowest in species richness with only 4 species. Species dominance typically ranged from about 0.2 to 0.6 (see Electronic Supplementary Material). However, dominance was highest (>0.8) in 1996 and 1997.

Temporal abundance

Bosmina longirostris had the highest total density of the two dominant crustaceans (Table 1), so its temporal pattern was similar to the pattern for the total crustacean density (Figs. 2 and 4). *D. pulicaria*, the largest in body size but less numerous of the two dominant crustaceans, was abundant in 1988 and 1989, and again in 1997, 1998, and 1999, periods of time when the density of *B. longirostris* was relatively low (Fig. 4).

Fig. 2 Total rotifer and crustacean densities integrated to 200 m, 1988–2000

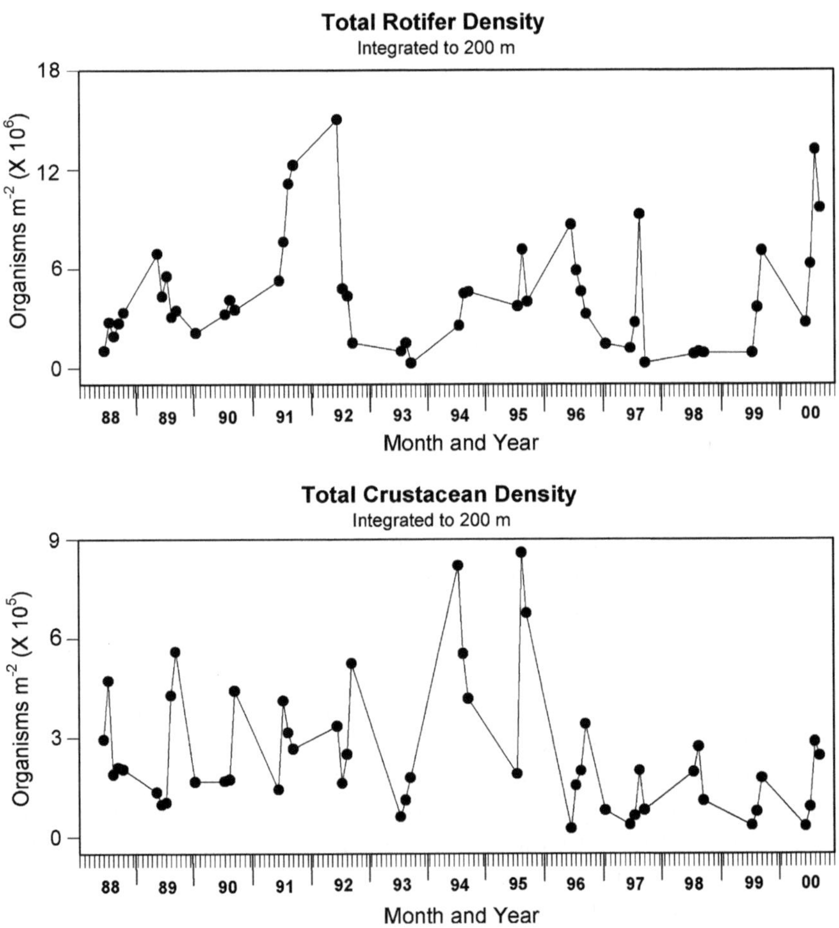

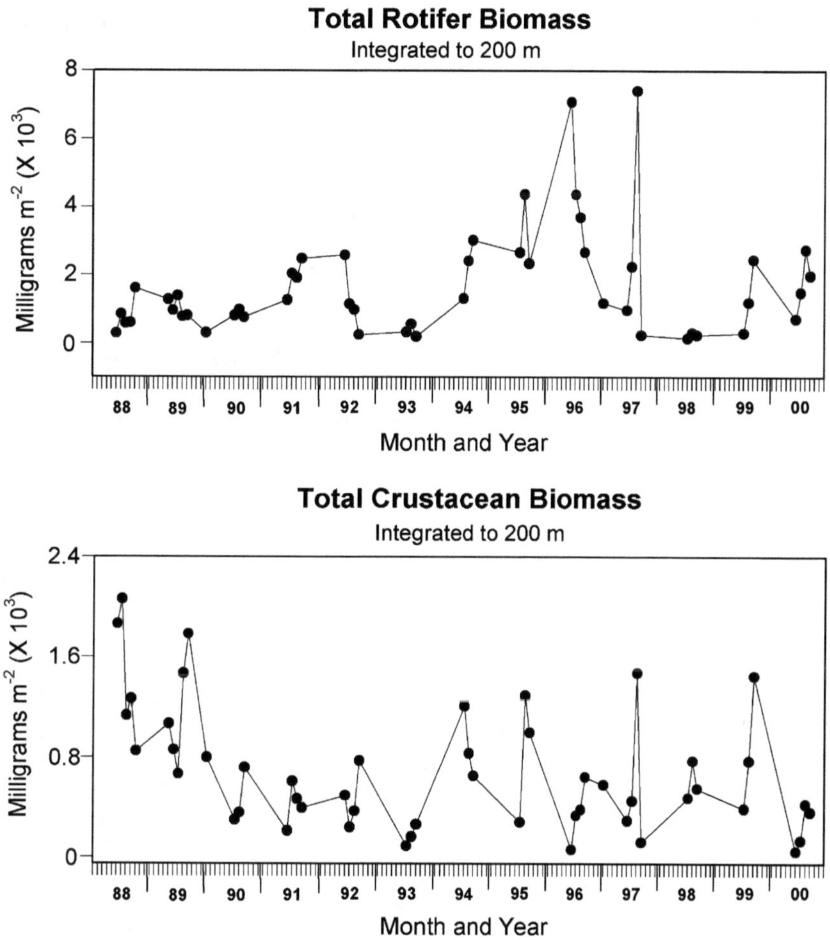

Fig. 3 Total rotifer and crustacean biomasses integrated to 200 m, 1988–2000

Rotifers also exhibited different periods of high abundance (Figs. 5, 6). *Synchaeta oblonga* had relatively high densities early in the study (1988–1992), whereas *K. longispina* increased in density in 1994 and reached maximum densities in 1996 and 1997. Rotifer species reaching their maximum abundance near the end of the study included *Conochilus unicornis* (1998), *Filinia terminalis* (1999), and *Philodina acuticornis* (1999). *Keratella cochlearis* and *Asplanchna priodonta* had a similar density maximum in 1991 and 1992, but *K. cochlearis* also was abundant again in 2000. *Polyarthra dolichoptera* had two periods of relatively high density, one in 1995 and another in 2000.

Vertical distribution

Spatial distributions of 10 dominant zooplankton species in the water column of Crater Lake are summarized in Figs. 7, 8. These values represent the summation of all density values for each species over the entire data set, in this case 404 samples.

The two crustaceans, *B. longirostris* and *D. pulicaria*, were most abundant at depths between 40 and 80 m, and between 60 and 100 m, respectively (Fig. 7). Although these two species exhibited some overlap in the water column, *Bosmina* was usually less abundant during the years when *Daphnia* reached its maximum abundance (Fig. 4).

Vertical distributions of eight dominant species of rotifers in the water column were variable (Fig. 8). Maximum abundance of *P. dolichoptera* was at a depth range of 20 to 80 m, the shallowest vertical distribution of the eight species. *Asplanchna priodonta* reached its maximum density at a depth interval between 40 and 60 m and gradually decreased in density between depths of 60 and 200 m. *Conochilus unicornis* exhibited the

Fig. 4 Densities of *Daphnia* and *Bosmina* integrated to 200 m, 1988–2000

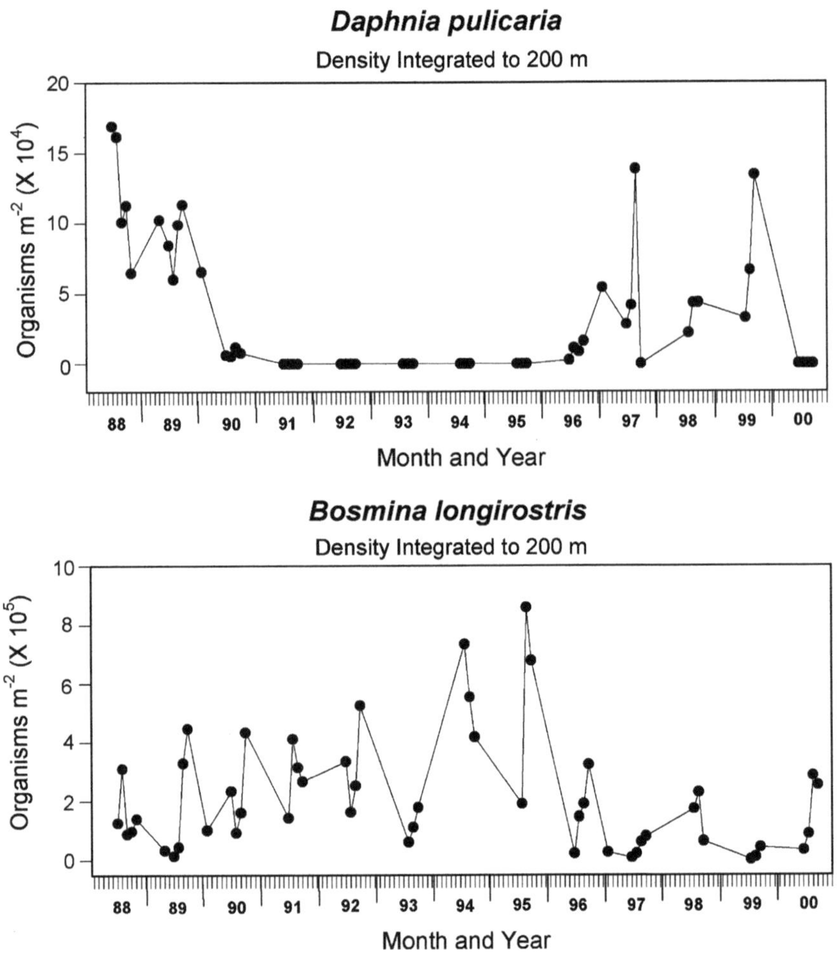

deepest vertical distribution, and was usually found in samples from a depth range between 120 m and 200 m. *Keratella cochlearis*, *F. terminalis*, *K. longispina*, and *S. oblonga* had intermediate distributions in the water column, with maxima between 60 and 120 m. The relationship between *P. acuticornis* and water depth was unclear, as it was collected at times throughout the water column between depths of 20 m and 200 m, as were *Kellicottia* in 1996 and 1997 and *Asplanchna* in 1991 and 1992 (data not shown).

Temporal patterns of assemblage composition

Temporal changes of the zooplankton assemblages in the samples from 1988 to 2000 based on biomass are illustrated in ordination space in Fig. 9. Locations of the samples correspond to the configuration of zooplankton species in the species ordination (Fig. 10). As a point of clarification, samples located in a particular region of the sample ordination (Fig. 9) would contain high biomasses of the species in that portion of the species ordination (Fig. 10).

Based on the cluster analysis (8 clusters), the sample ordination was subdivided into five regions (Fig. 9). Some regions included only one cluster, whereas other regions included up to three clusters (Table 2). Region A was dominated by *Kellicottia* from August 1994 to September 1997. Region B was dominated by *Daphnia* in 1988–1989 and 1998–1999. Region C was dominated by *Polyarthra*, *K. cochlearis*, *Bosmina*, and *Syncheata* in 1990 (except September) through July of 1991, July and August of 1993, and in 2000. Region D was dominated by *K. cochlearis*, *Asplanchna*, and *Bosmina* from August 1991

Fig. 5 Densities of *Keratella cochlearis*, *Kellicottia longispina*, *Polyarthra dolichoptera*, and *Syncheata oblongata* integrated to 200 m, 1988–2000

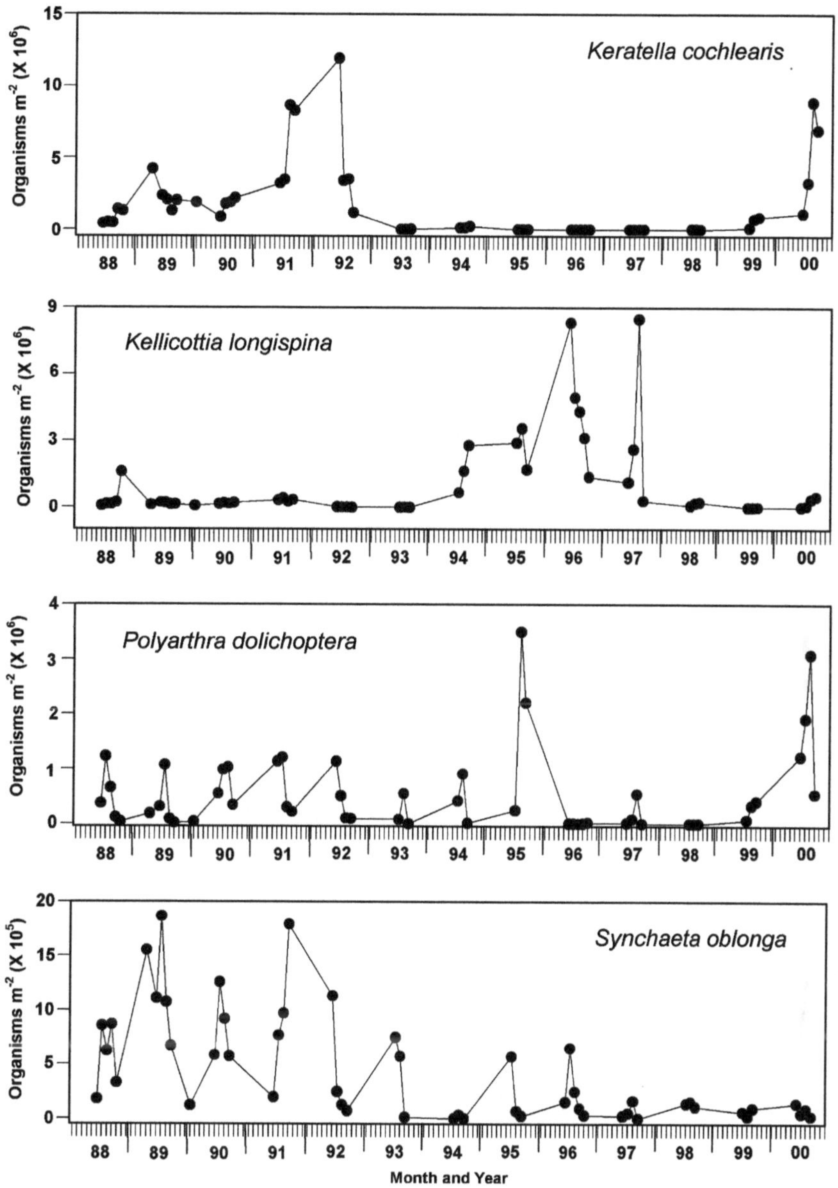

through August 1992, and Region E by *Bosmina* in September 1990, 1992, 1993, and July 1994.

Discussion

Characteristics of the zooplankton assemblage during the period from 1991 and 2000 that were not evident in the data obtained between 1988 and 1990 were: (1) *Keratella quadrata* was rare in the samples; (2) *Kellicottia* was exceptionally high in biomass and dominated the assemblages throughout the water column to 200 m during thermal stratification from 1994 to 1997; (3) *Asplanchna* was present and in high biomass in 1991 and 1992; (4) the vertical distribution of *Philodina* was not restricted to the lower stratum of the 200 m water column during thermal stratification; and (5) the assemblages exhibited a greater range of maximum and minimum biomasses. Similarities between the two sample periods were that: (1) the assemblages collectively included two crustacean taxa and 10 rotifer species (excluding rare taxa); (2) no copepod taxa where collected

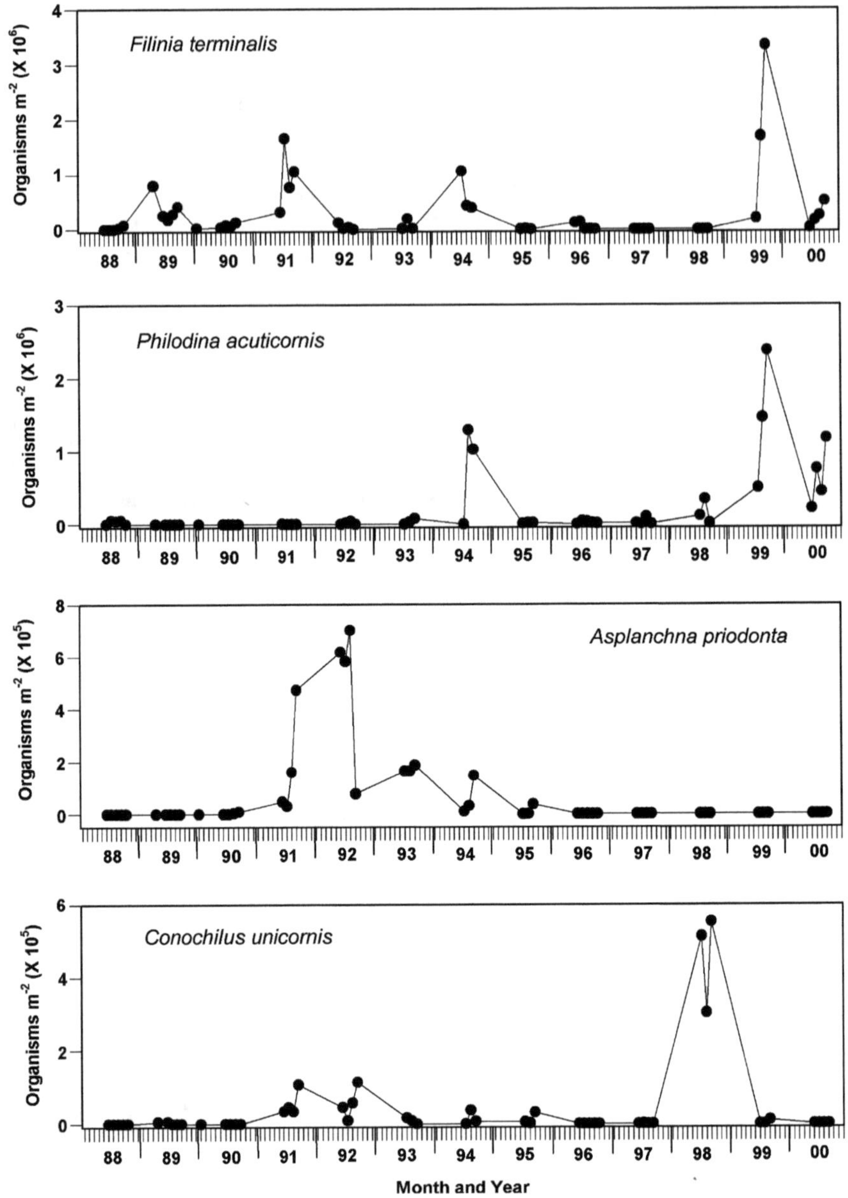

Fig. 6 Densities of *Filinia terminalis*, *Philodina acuticornis*, *Asplanchna priodonta*, and *Conochilus unicornis* integrated to 200 m, 1988–2000

from the pelagic zone; (3) with the exceptions of *Kellicottia* and *Philodina*, the zooplankton taxa that either fed on similar food items or that had the similar feeding modes and mechanisms exhibited vertical segregation in the water column; (4) there was no seasonal congeneric succession; (5) *Daphnia* returned to the assemblage (after a 6-year hiatus) and then disappeared; (6) there were virtually no zooplankton taxa in the epilimnion; and (7) the zooplankton taxa were distributed throughout the water column in winter and spring from near the lake surface to a depth of about 200 m.

Zooplankton abundance and the DCA ordination of zooplankton biomass for 1988 to 2000 indicated that the species compositions of the assemblages changed through time. Although some of this variability was undoubtedly stochastic (Pauli, 1990), other sources of variation were likely related to resource competition, interference, and predation. When *Daphnia* was in high biomass, *Bosmina* biomass was low. This

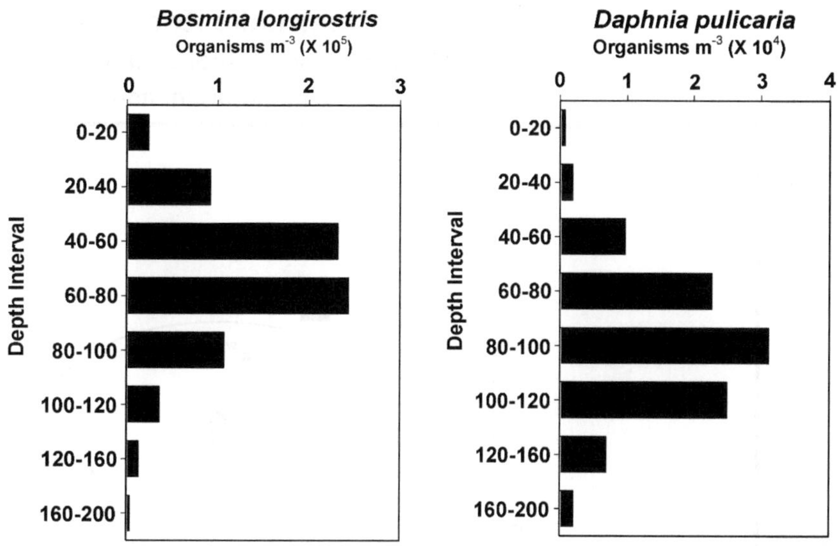

Fig. 7 Average vertical distributions for *Bosmina* and *Daphnia* in the water column to a depth of 200 m

relationship was likely due to exploitative resource competition by *Daphnia* because both species graze on phytoplankton of similar size (DeMott, 1989). However, *Bosmina* was not eliminated from the assemblages during periods of high *Daphnia* biomass because *Bosmina* was able to prey on large-bodied flagellates that are not effectively consumed by *Daphnia* (DeMott,

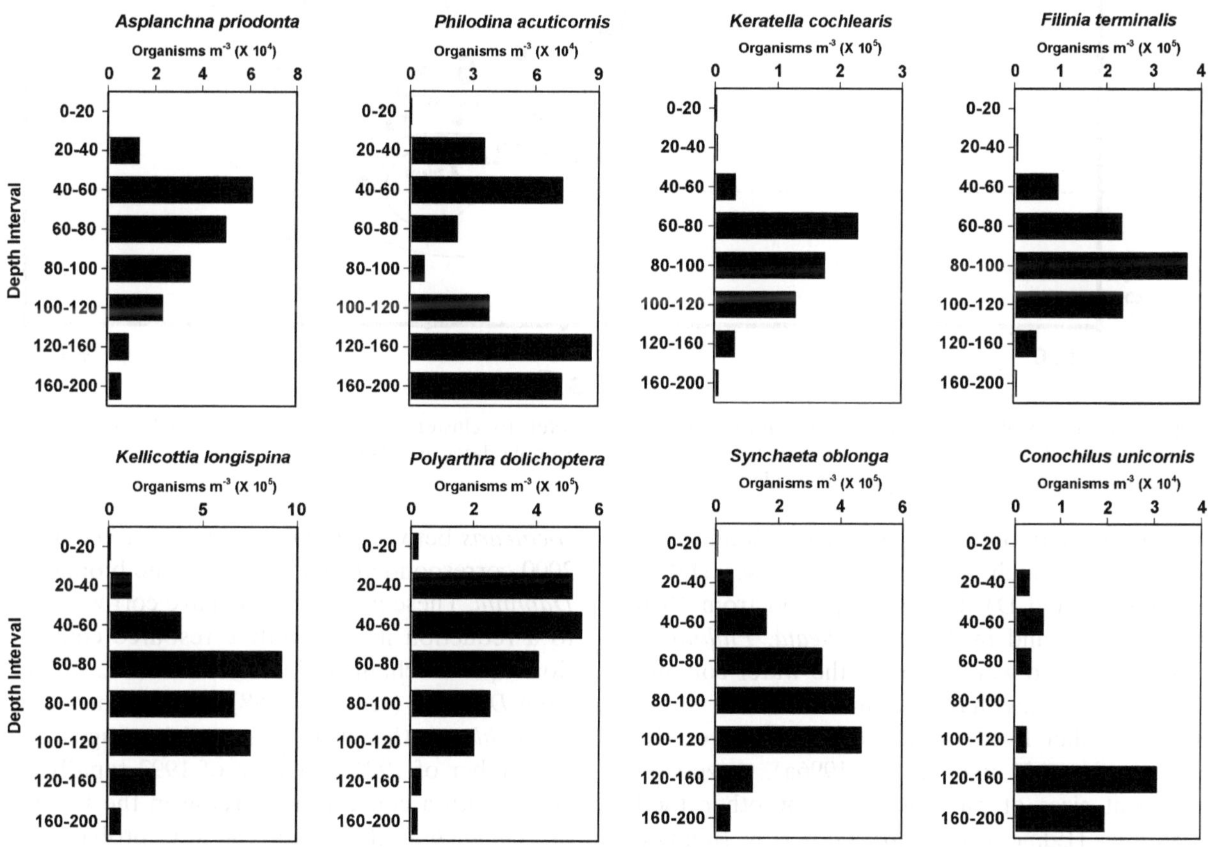

Fig. 8 Average vertical distributions of eight rotifer taxa in the water column to a depth of 200 m

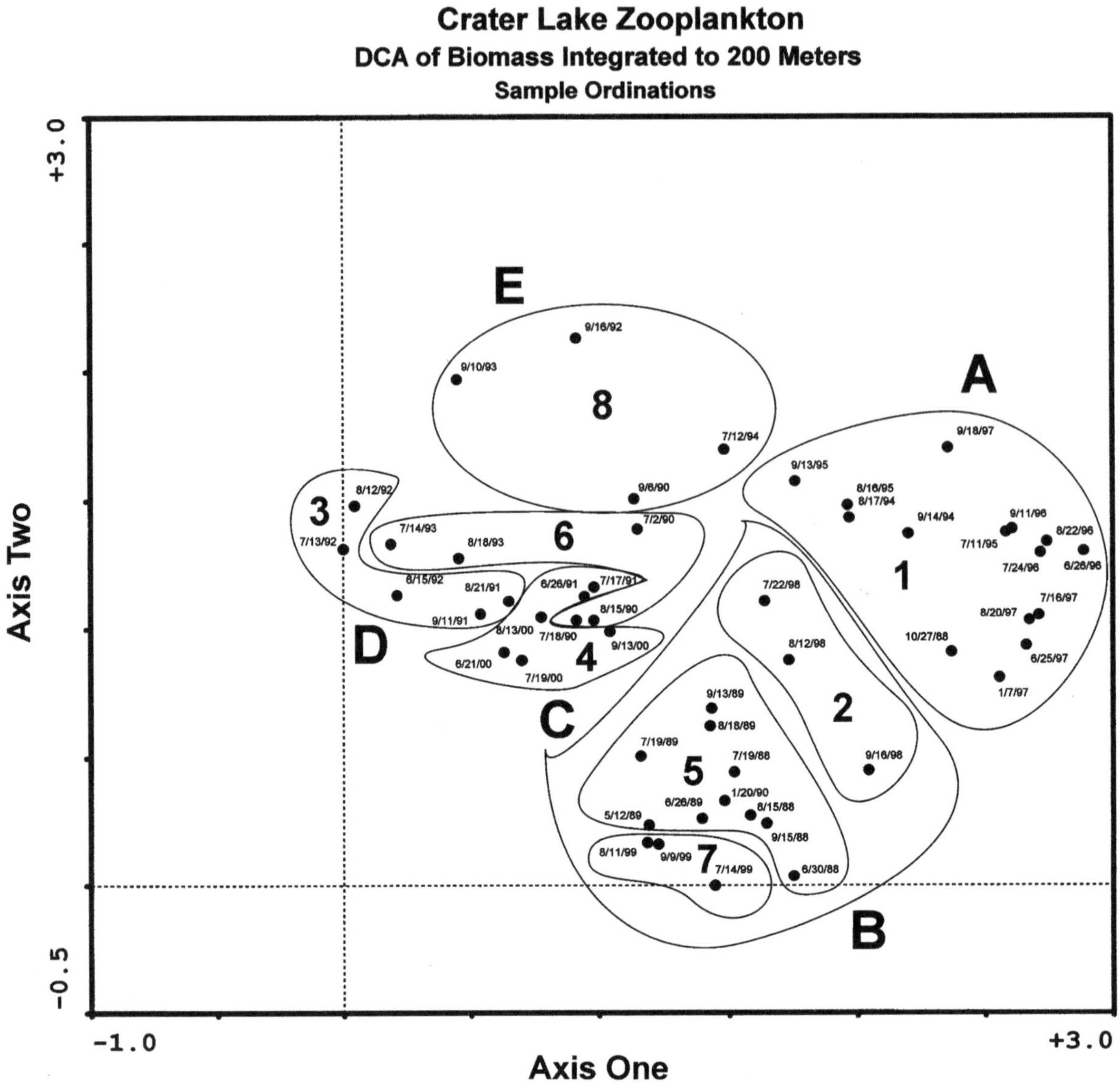

Fig. 9 Ordination of the integrated zooplankton samples (biomass) to a depth of 200 m. Letters refer to regions of the ordination (A–E), whereas numbers inside the regions refer to clusters (1–8) based on the CLUSB analysis. Sample dates are also shown

1989). Furthermore, when *K. cochlearis* and *Polyarthra* were high in relative biomasses (DCA Regions C and D) during the period from 1990 through1993 and in 2000, *Syncheata*, *Philodina*, and *Filinia* also were found in the water column. Coexistence may have occurred because these species either lived in different portions of the water column (Larson et al., 1996a), grazed on different sizes of phytoplankton, or other food resources (Edmondson, 1961; Gilbert & Bogdan, 1984; Arndt, 1993; Walz, 1997). The increase of *K. cochlearis* between 1990 and June of 1992 and in 2000 corresponded to a decline in the biomass of *Daphnia*. These increases may have corresponded to a reduction in exploitative resource competition and mechanical interference (i.e., injury) from *Daphnia* (Gilbert, 1988). High biomasses of *K. cochlearis* that were observed in August and September of 1991 and June of 1992 rapidly declined with a concurrent increase in the biomass of *Asplanchna*. This decrease was probably related to predation by *Asplanchna* on *Keratella*,

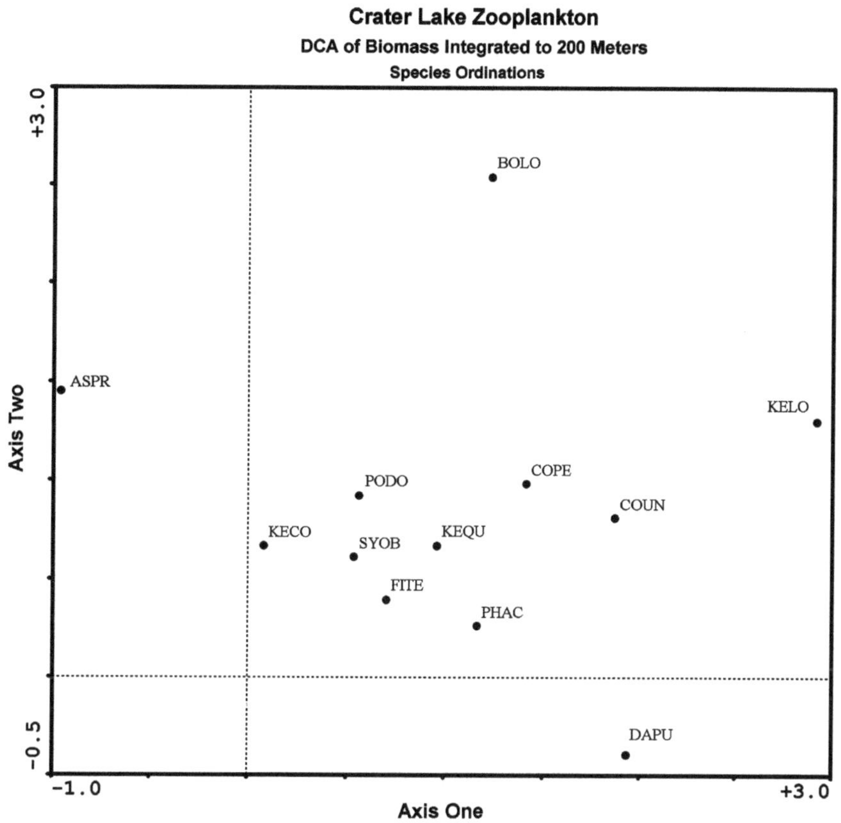

Fig. 10 Ordination of the zooplankton species based on biomass integrated to a depth of 200 m

which is one of its primary prey (Gilbert, 1980). Zooplankton biomass was extremely low in 1993, which was probably caused from predation by *Asplanchna*. Zooplankton then increased and was dominated by *Kellicottia* between 1994 and 1997. During 1996 and 1997 the zooplankton assemblages attained the highest biomasses and species dominance for the 1988–2000 sampling period. The decline of *Kellicottia* biomass by 1998 was likely caused exploitative resource competition from *Daphnia* because (1) both species graze on phytoplankton of similar size and (2) *Daphnia* is a much more efficient grazer than *Kellicottia* (DeMott, 1989). *Daphnia* apparently had an impact on the entire zooplankton assemblage because the density and biomass of the total zooplankton assemblages were low from September 1997 through July 1999.

It is unclear why *Daphnia* suddenly appeared in the zooplankton assemblage in 1987 and 1996. However, the rapid decrease in biomass of *Daphnia* after 1989 and 1999 may have been

Table 2 Relative biomasses of dominant zooplankton in five regions of the DCA sample ordination, 1988–2000

Region	Cluster	N[1]	Dominant species (≤0.05 relative biomass)
A	1	15	KELO (O.71, BOLO (0.12), DAPU (0.08), PODO (0.05)
B	2	3	DAPU (0.44), BOLO (0.28), KELO (0.16), PHAC (0.06)
	5	5	DAPU (0.52), BOLO (0.11), KECO (0.09), SYOB (0.08), PODO (0.07)
	7	3	DAPU (0.40), FITE (0.28), PHAC (0.22), PODO (0.05)
C	4	6	PODO (0.24), KECO (0.23), BOLO (0.13), FITE (0.09), PHAC (0.08)
	6	5	PODO (0.24), BOLO (0.22), SYOB (0.19), ASPR (0.11), KECO (0.08), KELO (0.08)
D	3	5	KECO (0.33), ASPR (0.23), BOLO (0.19), PODO (0.07), SYOB (0.07), FITE (0.06)
E	8	4	BOLO (0.56), ASPR (0.11), KELO (0.09), KECO (0.07), FITE (0.06), PODO (0.05)

Species acronyms are listed in Table 1. (N[1] = sample size)

related to predation by kokanee salmon (Goldman et al., 1979), induced diapause (Slusarczyk, 2001), or both. Infrequent occurrence of *Daphnia* has been documented in past studies at Crater Lake (Brode, 1938; Hasler, 1938; Hasler & Farner, 1942; Hoffman, 1969; Malick, 1971), but its presence and absence in relationship to the abundance of fish remains unclear.

Kokanee reproduction may have been affected by lake level because the highest abundances only occurred during the two periods when the level of the lake was near the benchmark level of 1882 m. A major spawning habitat for the fish is thought to be the nearshore gravel deposits at the bases of numerous avalanche shoots. These gravel areas were submerged at or near the benchmark water level, but were drained when the lake level fell about 2 m below the benchmark water level.

An interesting observation during this study was the sudden occurrence of *Asplanchna* and its apparent influence on the temporal characteristics of the zooplankton assemblage. *Asplanchna* is a large-bodied and voracious predator that preys on rotifers, especially *Keratella* and large-bodied flagellates (Gilbert, 1980). Also, *Asplanchna* apparently is an inefficient grazer on phytoplankton in the size range typically grazed on by *Daphnia, Keratella*, and *Kellicottia* (Edmondson, 1961; Gilbert & Bogdan, 1984). Much of what is known about the ecology of this genus has come from laboratory experiments because its impact on zooplankton assemblages in the field has often been masked by complex food-web interactions (Snell et al., 2001). However in one field study, a large increase in the density of *Asplanchna* coincided with the decline of the *K. cochlearis* population, which was followed by a large increase in the density of *Kellicottia* (Hofmann, 1983). These results were similar to those observed in Crater Lake in 1994, when *Kellicottia* increased in density after *Asplanchna* apparently decimated the rotifer assemblages from 1990–1993. During that time, especially in 1992 (data not shown), *Asplanchna* was distributed throughout the water column to a depth of 200 m. The overlap in the vertical distribution of *Asplanchna* with other rotifer species probably resulted in predation by *Asplanchna* on *Polyarthra, Syncheata, Philodina, Filinia*, and *Bosmina*. *Asplanchna* may also have negatively impacted *Bosmina* by preying on large-bodied flagellates, one of its primary prey. It is unlikely *Kellicottia* would have been preyed upon effectively by *Asplanchna* owing to its relatively large size and lengthy spines.

In 1913, Kemmerer et al. (1924) found that the August and September zooplankton assemblages in Crater Lake were dominated by *Daphnia* (probably *pulicaria*) and *Asplanchna. Bosmina, Kellicottia*, and *Keratella* (probably *cochlearis*) were collectively low in relative biomass at 3.6% in August and 1.9% in September. The simultaneous dominance of *Daphnia* and *Asplanchna* was not observed in the samples collected from 1985 to 2000. Our study indicated that a high relative biomass of *Daphnia* was not associated with a high biomass of *Kellicottia*, and high biomasses of *Asplanchna* and *Daphnia* were not associated with a high biomass of *Keratella*. One explanation for the unusual dual species assemblage in 1913 would be that *Asplanchna*, a voracious predator, decimated the rotifer assemblage (Gilbert, 1980) and the *Bosmina* population (Matveeva, 1989) by direct predation and indirectly by the consumption of large-bodied flagellates that are an important food resource for *Bosmina* (DeMott, 1989). Since *Daphnia* was immune to predation by *Asplanchna* and the fact that *Daphnia* would have grazed phytoplankton that are inefficiently eaten by *Asplanchna* (Gilbert, 1980), the two species appeared to co-exist, at least temporarily. The reason that *Daphnia* was present at that time remains unresolved, but was consistent with the present work when *Daphnia* suddenly appeared in the assemblages in 1987 and 1996.

Random events were likely the primary forces causing sudden occurrences of *Asplanchna* and *Daphnia*, whereas resource competition, predation (*Asplanchna* and kokanee salmon), and mechanical interference (*Daphnia* on *K. cochlearis*) were forces that changed the structure of the assemblage during the period from 1988 to 2000. For example, predation by *Asplanchna* resulted in an ecological release for *Kellicottia*, which was in low biomass prior to the occurrence of *Asplanchna*. These results suggest that *Kellicottia* would not have become abundant at that time without predatory pressure on the zooplankton assemblage. Furthermore, the occurrence of *Daphnia* in the late 1990s corresponded to a rapid decline in the biomass of *Kellicottia*. Of interest is whether *Kellicottia* would have declined without the occurrence of *Daphnia*. We speculate that the typical assemblage in Crater Lake, i.e., *K. cochlearis, Polyarthra*, and *Bosmina*, would have eventually regained dominance as these species regained their competitive advantages through habitat segregation of the water column.

Daphnia was abundant when the lake level was high in the late 1980s and early 1990s. This apparent correspondence, however, is doubtful because *Daphnia* was abundant in 1937 (Hasler, 1938) when the lake level was about 3.5 m below the benchmark level. Furthermore, declining primary production and chlorophyll concentrations during the 1990s (McIntire et al., 2007) did not correspond well with the observed changes in the species structure of the zooplankton assemblages, in that *Daphnia* and *K. cochlearis* were abundant at the beginning and the end of this period of observation. Furthermore, there were no changes in water quality (Larson et al., 2007), deep water circulation (Crawford & Collier, 2007), and phytoplankton densities, abundances, and species composition (McIntire et al., 2007) to account for the changes in species dominance or temporal changes in the species structure of the assemblage. Thus, no catastrophic shifts (Scheffer et al., 2001) in the lake system were identified to account for the changes in species dominance or assemblage structure. Crater Lake appeared to be resilient to the environmental changes that occurred during the study (In Sensu Scheffer et al., 2001).

There was a conspicuous absence of copepoda in the pelagic zooplankton samples. Copepoda and other crustacean taxa not collected in the pelagic samples were found in the fish diets, however (Buktenica et al., This Volume). These taxa may be restricted to the near shore areas of the lake. For example, samples of a deep-water moss (*Drepanocladus aduncus*) collected from 30 to about 120 m contained many of these taxa (Robert Truitt, personal communication).

In summary, results of this study suggest that there were five primary species in the zooplankton assemblages of Crater Lake: *D. pulicaria*, *B. longirostris*, *K. cochlearis*, *P. dolichoptera*, and *K. longispina*. Biomasses of *Daphnia* and the other species were inversely related. Similarly, when the biomass of *Kellicottia* was high the biomass of *Keratella* was low and vice versa. *Bosmina*, *K. cochlearis*, *Keratella quadrata*, *Philodina*, *Synchaeta*, *Polyarthra*, *Conochilus*, and *Filinia* appeared able to co-exist in various combinations as indicated by their closely positioned locations in species ordination space. Segregation within the water column by these species and different feeding modes and mechanisms undoubtedly contributed to these observations (see Larson et al., 1996b). Similar results were reported by Karnaugh (1988) during the initial studies of the Crater Lake zooplankton from 1985 to 1987, with the exception of *Kellicottia* from August 1994 to September 1997. During that period, *Kellicottia* was high in biomass and distributed from a depth of 20 m to 200 m below the lake surface, a depth zone encompassing the metalimnion and the upper portion of the hypolimnion. Furthermore, the zooplankton assemblages in Crater Lake exhibited temporal consistency in species richness and general composition, but varied in species composition, density, and biomass.

Although no congeneric replacement was observed, the dominant species in the assemblages were variable, switching primarily between periods of dominance by *Polyarthra-K. cochlearis* and *Daphnia*. The sudden occurrence of *Asplanchna* in 1991 resulted in a major change in the typical temporal shift between *Polyarthra-K. Cochlearis* and *Daphnia*, as well as a large decline in the total zooplankton biomass in 1993. *Kellicottia* dominated the zooplankton biomass for the period between 1994 and 1997, a period when the total zooplankton biomass was at the maximum observed during this study. The eventual decline in biomass of *Kellicottia* by 1998 coincided with a dramatic increase in *Daphnia* biomass. When *Daphnia* biomass declined by 2000, *Keratella* biomass increased again. Thus, by 1998 the assemblage returned to the typical shift between *Keratella-Polyarthra* and *Daphnia*.

These observations provide considerable insight about the interannual variability of the zooplankton assemblages in Crater Lake, but little is known about mechanisms behind the variability. Kokanee salmon may have played an important role, however, in the disappearance of *Daphnia* in 1990 and 2000, either through predation or perhaps by inducing diapause.

Acknowledgements We thank James Larson, Shirley Clark, Jon Jarvis, Robert Benton, Al Hendricks, Chuck Lundy, Mac Brock, Dennis Fenn, Gerald McCrea, Robert Collier, Ray Herrmann, and Stanford Loeb for their support of the long-term limnological project at Crater Lake. Many thanks are also extended to the numerous park employees who assisted with the project. Funding was provided by the National Park Service and the USGS Forest and Rangeland Ecosystem Science Center.

References

Arndt, H., 1993. Rotifers as predators on components of the microbial web (bacteria, heterotrophic flagellates, ciliates)—a review. Hydrobiologia 255/256: 231–246.

Bacon, C. R. & M. A. Lanphere, 1990. The geological setting of Crater Lake, Oregon. In Drake, E. T., G. L. Larson, J. Dymond & R. Collier (eds), Crater Lake: An Ecosystem Study. Pacific Division, American Association for the Advancement of Science, San Francisco, CA, 19–27.

Bacon, C. R., J. V. Gardner, L. A. Mayer, M. W. Buktenica, P. Dartnell, D. W. Ramsey & J. E. Robinson, 2002. Morphology, volcanism, and mass wasting in Crater Lake, Oregon. Geological Society of America Bulletin 114: 675–692.

Barber, J. H. Jr. & C. H. Nelson, 1990. Sedimentary history of Crater Lake caldera, Oregon. 1990. In Drake, E. T., G. L. Larson, J. Dymond & R. Collier (eds), Crater Lake: An Ecosystem Study. Pacific Division, American Association for the Advancement of Science, San Francisco, CA, 29–39.

Brode, J. S., 1938. The denizens of Crater Lake. Northwest Science 12: 50–57.

Buktenica, M. W., S. F. Girdner, G. L. Larson & C. D. McIntire, 2007. Variability of kokanee and rainbow trout food habits, distribution, and population dynamics, in an ultraoligotropic lake with no manipulative management. Hydrobiologia 574: 235–264.

Byrne, J. V., 1965. Morphology of Crater Lake, Oregon. Limnology and Oceanography 10: 462–465.

Carpenter, S. R., J. F. Kitichell & J. R. Hodgson, 1985. Cascading trophic interactions and lake productivity. Bioscience 35: 634–639.

Collier, R., J. Dymond & J. McManus, 1993. Studies of hydrothermal processes. In Larson, G., C. D McIntire & R. W. Jacobs (eds), Crater Lake limnological studies, Final report. Technical Report NPS/PNRO-SU/NRTR–93/03: 205–213.

Crawford, G. & R. Collier, 2007. Long-term observations of deepwater renewal in Crater Lake, Oregon. Hydrobiologia 574: 47–68.

DeMott, W. R., 1989. The role of competition in zooplankton succession. In Sommer, U. (ed.), Plankton ecology: succession in plankton communities. Springer-Verlag, New York, 195–252.

Edmondson, W. T., 1961. Reproductive rate of planktonic rotifers as related to food and temperature in nature. Ecological Monographs 35: 61–111.

Evermann, B. W., 1897. United States Fish Commission investigations at Crater Lake. Mazama 1: 230–238.

Gilbert, J. J., 1980. Feeding in the rotifer *Asplanchna*: Behavior, cannibalism, selectivity, prey defenses, and impact on rotifer communities. In Kerfoot, W. C. (ed.), Evolution and Ecology of Zooplankton Communities. University Press of New England, 158–172.

Gilbert, J. J., 1988. Susceptibilities of ten rotifer species to interference from Daphnia Pulex. Ecology 69: 1826–1838.

Gilbert, J. J. & K. G. Bogdan, 1984. Rotifer grazing: in situ studies on selectivity and rates. In Meyers, D. G. & J. R. Strickler (eds), Trophic Interactions within Aquatic Ecosystems. American Association for the Advancement of Science. Westview Press, Inc., Boulder Colorado, 97–133.

Goldman, C. R., M. D. Morgan, S. T. Threlkeld & N. Angeli, 1979. A population dynamics analysis of the cladoceran disappearance from Lake Tahoe, California – Nevada. Limnology and Oceanography 24: 289–297.

Hasler, A. D., 1938. Fish biology and limnology of Crater Lake. Journal of Wildlife Management 2: 94–103.

Hasler, A. D. & D. S. Farner, 1942. Fisheries investigations in Crater Lake. Journal of Wildlife Management 6: 319–327.

Hill, M. O. & H. G. Gauch, 1980. Detrended correspondence analysis, an improved ordination technique. Vegetation 42: 47–58.

Hoffman, F. O., 1969. The horizontal distribution and vertical migrations of the limnetic zooplankton in Crater Lake, Oregon. MS thesis. Department of Fisheries and Wildlife, Oregon State University. 60 pp.

Hofmann, W., 1983. Interactions between *Asplanchna* and *Keratella cochlearis* in the PluBsee (north Germany). Hydrobiologia 104: 363–365.

Karnaugh, E. N., 1988. Structure, abundance, distribution of pelagic zooplankton in a deep, oligotrophic caldera lake. MS Thesis. Department of Fisheries and Wildlife, Oregon State University, 167 pp.

Karnaugh, E., 1990. Sampling strategy and a preliminary description of the pelagic zooplankton community in Crater Lake. In Drake, E. T., G. L. Larson, J. Dymond & R. Collier (eds), Crater Lake: An Ecosystem Study. Pacific Division, American Association for the Advancement of Science, San Francisco, 177–183.

Kemmerer, G., J. F. Bovard & W. R. Boorman, 1924. Northwest lakes of the United States: biological and chemical studies with reference to possibilities in production of fish. Bulletin of the United States Bureau of Fisheries 39: 51–140.

Larson, G. L. & M. W. Buktenica, 1998. Variability of Secchi disk readings in an exceptionally clear and deep caldera lake. Archiv fuer Hydrobiologie 141: 377–388.

Larson, G. L., C. D. McIntire, M. Hurley & M. W. Buktenica, 1996a. Temperature, water chemistry, and optical properties of Crater Lake. Journal of Lake and Reservoir Management 12: 230–247.

Larson, G. L., C. D. McIntire, R. E. Truitt, M. W. Buktenica & E. Karnaugh-Thomas, 1996b. Zooplankton assemblages in Crater Lake, Oregon, USA. Journal of Lake and Reservoir Management 12: 281–297.

Larson, G. L., R. L. Hoffman, C. D. McIntire, M. W. Buktenica & S. F. Girdner, 2007. Thermal, chemical, and optical properties of Crater Lake, Oregon. Hydrobiologia 574: 69–84.

Leibold, M. A. & A. J. Tessier, 1997. Habitat partitioning by zooplankton and the structure of lake ecosystems. In Streit, B., T. Stadler & C. M. Lively (eds), Evolutionary Ecology of Freshwater Animals. Birkhauser, Basel, 3–30.

Malick, J. G., 1971. Population dynamics of selected zooplankton in three oligotrophic Oregon lakes. MS thesis, Department of Fisheries and Wildlife, Oregon State University, 112 pp.

Matveeva, L. K., 1989. Interrelations of rotifers with predatory and herbivorous cladocera: a review of Russian works. Hydrobiologia 186/187: 69–73.

McIntire, C. D., 1973. Diatom associations in Yaquina Estuary, Oregon: a multivariate analysis. The American Naturalist 129: 97–121.

McIntire, C. D. & Overton, W. S. 1971. Distribution patterns in assemblages of attached diatoms from Yaquina Estuary, Oregon. Ecology 52: 758–777.

McIntire, C. D., G. L. Larson & R. E. Truitt, 2007. Seasonal and interannual variability in the taxonomic composition and production dynamics of phytoplankton assemblages in Crater Lake, Oregon. Hydrobiologia 574: 179–204.

McManus, J., R. Collier, C-T. A. Chen & J. Dymond, 1992. Physical properties of Crater Lake, OR: a method for the determination of a conductivity and temperature dependent expression for salinity. Limnology and Oceanography 37: 41–53.

Nathenson, M. & J. M. Thompson, 1990. Chemistry of Crater Lake, Oregon, and nearby springs in relation to weathering. In Drake, E. T., G. L. Larson, J. Dymond & R. Collier (eds), Crater Lake: an ecosystem study. Pacific Division, American Association for the Advancement of Science, San Francisco, 115–126.

Nathenson, M., C. R. Bacon & D. W. Ramsey, 2007. Subaqueous geology and a filling model for Crater Lake, Oregon. Hydrobiologia. 574: 13–27.

Nelson, P. O., J. F. Reilly & G. L. Larson, 1996. Chemical solute mass balance of Crater Lake, Oregon. Journal of Lake and Reservoir Management 12: 248–258.

Pauli, H-R., 1990. Seasonal succession of rotifers in large lakes. In Tilzer, M. & C. Serruya (eds), Large lakes: ecological structure and function. Springer-Verlag, New York, 459–474, 691 pp.

Phillips, K. N., 1968. Hydrology of Crater Lake, East Lake, and Davis Lake, Oregon. U.S. Geological Survey Water Supply Paper 1859-E, 60 pp.

Scheffer, M., S. Carpenter, J. A. Foley, C. Folke & B. Walker, 2001. Catastrophic shifts in ecosystems. Nature 413: 591–596.

Slusarczyk, M., 2001. Food threshold for diapause in *Daphnia* under the threat of fish predation. Ecology 82: 1089–1096.

Snedecor, G. W. & W. G. Cochran, 1967. Statistical Methods, 6th edn. Iowa State University Press, Ames, Iowa, USA, 593 pp.

Snell, T. W., B. J. Dingmann & M. Serra, 2001. Density-dependent regulation of natural and laboratory rotifer populations. Hydrobiologia 446/447: 39–44.

Walz, N., 1997. Rotifer life history strategies and evolution in freshwater plankton communities. In Streit, B., T. Stadler & C. M. Lively (eds), Evolutionary ecology of freshwater animals: concepts and case studies. Birkhauser Verlag, Basel, Switzerland, 119–149.

Warncke, W. M. Jr., 1999. The species composition, density, and distribution of the littoral zooplankton assemblage in Crater Lake, Oregon. MS Thesis, Department of Fisheries and Wildlife, Oregon State University, 102 pp.

Hydrobiologia (2007) 574:235–264
DOI 10.1007/s10750-006-0355-1

Variability of kokanee and rainbow trout food habits, distribution, and population dynamics, in an ultraoligotrophic lake with no manipulative management

Mark W. Buktenica · Scott F. Girdner · Gary L. Larson · C. David McIntire

© Springer Science+Business Media B.V. 2007

Abstract Crater Lake is a unique environment to evaluate the ecology of introduced kokanee and rainbow trout because of its otherwise pristine state, low productivity, absence of manipulative management, and lack of lotic systems for fish spawning. Between 1986 and 2004, kokanee displayed a great deal of variation in population demographics with a pattern that reoccurred in about 10 years. We believe that the reoccurring pattern resulted from density dependent growth, and associated changes in reproduction and abundance, driven by prey resource limitation that resulted from low lake productivity exacerbated by prey consumption when kokanee were abundant. Kokanee fed primarily on small-bodied prey from the mid-water column; whereas rainbow trout fed on large-bodied prey from the benthos and lake surface. Cladoceran zooplankton abundance may be regulated by kokanee. And kokanee growth and reproductive success may be influenced by the availability of *Daphnia pulicaria*, which was absent in zooplankton samples collected annually from 1990 to 1995, and after 1999. Distribution and diel migration of kokanee varied over the duration of the study and appeared to be most closely associated with prey availability, maximization of bioenergetic efficiency, and fish density. Rainbow trout were less abundant than were kokanee and exhibited less variation in population demographics, distribution, and food habits. There is some evidence that the population dynamics of rainbow trout were in-part related to the availability of kokanee as prey.

Keywords *Oncorhynchus nerka* · *Oncorhynchus mykiss* · Food habits · Distribution · Population dynamics · Ultraoligotrophic

Guest Editors: Gary L. Larson, Robert Collier, and Mark W. Buktenica
Long-term Limnological Research and Monitoring at Crater Lake, Oregon.

M. W. Buktenica (✉) · S. F. Girdner
U.S. National Park Service, Crater Lake National Park, PO Box 7, Crater Lake, OR 97604, USA
e-mail: mark_buktenica@nps.gov

G. L. Larson
U.S. Geological Survey, Forest and Rangeland Ecosystem Science Center, 777 NW 9th Street, Suite 400, Corvallis, OR 97330, USA

C. D. McIntire
Department of Botany and Plant Pathology, Oregon State University, Corvallis, OR 97331, USA

Introduction

The ecology of kokanee (*Oncorhynchus nerka)* and rainbow trout (*O. mykiss*) has been described for lakes covering a broad range of productivity. Most studies of kokanee occurred in mesotrophic and oligotrophic lakes and reservoirs. Although some lakes were classified as ultraoligotrophic, Crater Lake's trophic state has been classified near the low end of the trophic-continuum as characterized by low chlorophyll concentration, primary production (McIntire et al., 2007), nutrient concentration (Larson et al., 1996), and high Secchi disk clarity (Larson & Buktenica, 1998). In addition, evaluating the ecology of fish in Crater Lake was not confounded by water withdrawals, fish stocking, a diverse fish assemblage, competition with *Mysis relicta*, or angler harvest, as has been common in other studies (e.g., Hanzel, 1984; Rieman & Bowler, 1980; Martinez & Wiltzius, 1995).

Although Crater Lake was naturally fishless, during the period from 1888 to 1941 five species of salmonids (cutthroat trout, *O. clarki;* brown trout, *Salmo trutta*; rainbow and steelhead trout, *O. mykiss*; coho salmon, *O. kisutch*; and kokanee or landlocked sockeye salmon, *O. nerka*), totaling nearly two million fish, were introduced into the lake on an irregular schedule (Buktenica, 1989). No fish were stocked into the lake after 1941. Only naturally reproducing populations of rainbow trout and kokanee persist in the lake. In absence of spawning streams, both species spawn in the littoral zone of the lake.

In 1982, the National Park Service initiated a 10-year limnological study of Crater Lake in response to concerns that the lake was experiencing human-related change (Larson, 1996). Little was known at that time about the ecology of fish in the lake, or the role of fish in the lake food web, because fish studies of the lake in the late 19th and 20th Centuries were limited in scope and duration. In view of this lack of information, the objective of this work, early on, was to describe the food habits, distribution, and select population dynamics of kokanee and rainbow trout in Crater Lake (Buktenica & Larson, 1990). At the end of the 10-year limnological study, Crater Lake was judged to be in pristine condition except for the consequences of introduced fish (Larson et al., 1993). The monitoring of fish populations has continued as part of the Crater Lake Long-term Limnological Monitoring Program, resulting in the development of a data set covering a 19-year period of time, 1986–2004. The objectives of this report are to evaluate the ecology of fish in Crater Lake relative to the entire monitoring record and the literature, and to discuss the possible mechanisms responsible for the observed food habits, distribution, and population dynamics of rainbow trout and kokanee in this unique lake environment.

Materials and methods

Study area

Crater Lake covers the floor and partially fills the Mount Mazama Caldera that formed 7,700 year B.P (Bacon et al., 2002). There is no surface outlet, and surface inlets are limited to about 40 small ephemeral and permanent streams draining the steep caldera walls. The lake has a maximum depth of 594 m (1949 ft) and a mean depth of 350 m (1148 ft) relative to a surface elevation of 1883 m (6178 ft) (Bacon et al., 2002). Due to the lake's relatively young age and precipitously steep caldera walls, there are few wave-cut beaches and shallow water habitat is limited relative to lake volume (18.7 km^3). Unlike other Cascade Mountain Lakes, Crater Lake rarely is capped by snow and ice in winter. The lake is exceptionally clear as characterized by June to August Secchi disk depth measurements that typically range between 28 and 33 m (Larson & Buktenica, 1998), and the depth of 1% incident solar radiation (PAR) occurs between 80 and 100 m (Larson et al., 1996). Free nitrate is near the limits of detection between 0 and 200 m in depth and increases slightly to approximately 0.02 mg/l in the deep basin (Larson et al., 1996). Deep light penetration and nutrient distribution account for deep chlorophyll, primary production (McIntire et al., 2007) and zooplankton (Larson et al., 2007b) maxima. The limnetic zooplankton community is relatively simple, consisting of two cladocerans and ten rotifers (Larson et al., 2007b). Although pelagic zooplankton samples are conspicuously lacking copepods, several taxa were collected in low abundance near shore and in the diet of kokanee. Aquatic macrophytes are rare between the depths of 0–30 m, and are dominated by moss (*Drepanocladus aduncus*) between 30 and 150 m (McIntire et al., 1994). The water column is well oxygenated and cold year-round, but develops a shallow thermocline (5–15 m in depth) by August each year (Larson et al., 1996). No private watercraft are permitted on the lake and recreational access is limited to a steep 1.6 km foot trail and a short section of shoreline, from June through October. Therefore, angler harvest is negligible. Because of limited seasonal access to the lake (mid-June through September), and low fish density, little is known about fish reproduction and early life history. With no suitable stream inlets, fish spawning is limited to the lakeshore. Kokanee were collected in spawning condition at the base of an avalanche chute in January 1988 and 1997, when the lake was accessed by helicopter. Rainbow trout were collected in spawning condition from April through September, and rarely observed spawning in pockets of gravel along the shore.

Food habits

Fish were collected using horizontal gill nets set overnight perpendicular from the lakeshore to the 20 m depth contour. Most nets were set between June and September (147) although several were set in January (4), April (4), May (6), and October (2). Floating and sinking multifilament nets measuring 38 m × 3 m were used. Mesh sizes ranged from 19 to 51 mm square-mesh in five 7.6 m panels. Whole stomachs were removed from fish in the field and placed in 90% ethanol in individual containers. In the laboratory, the contents of each stomach were flushed (from the esophagus to the pyloric sphincter) and stored in 70% ethanol.

Stomach samples were sorted and identified with the aid of a binocular-dissecting microscope (6× to 50×). Intact individuals or intact and readily identifiable body parts were identified to the lowest possible taxonomic group. Head capsules of midge larvae, thorax or genitalia of pupae or adult caddis flies, and thorax or head capsules of hymenopterans were sufficient for identification of these groups. When possible, prey were identified to genus or species to determine mean weights and to classify the taxa into functional groups. The taxa were grouped into 31 taxonomic categories. Hemiptera and Coleoptera were each partitioned into two groups (aquatic or terrestrial) based on habitat origin, whereas the Chironomid and *Heterotrissocladius* categories were divided into two groups (larvae or pupae) determined by life history stage.

Zooplankton in each stomach sample was diluted to 25–100 ml, depending on abundance, and two 1 ml sub-samples were removed for enumeration. Cladocerans were enumerated when either whole specimens were intact or individual eyespots were intact. Samples were counted in a rectangular counting chamber with longitudinal divisions under a binocular dissecting microscope (40×). Counts of sub-samples were multiplied by the corresponding dilution factor to obtain the total zooplankton count for each stomach sample.

Weight of each taxon was based on 1–100 of the least-digested individuals. Weights of taxa that were not available were estimated based on specimens of equivalent size preserved in ethanol in collections of Drs. R. Wisseman and N. Anderson (Oregon State University, Department of Entomology). Stomach and collection samples were dried to a constant weight at 60°C and then placed in a desiccator to cool to room temperature. Dry weights were measured on a Mettler H16 Balance (accuracy 0.05 mg) and a Cahn 4100 Electrobalance (accuracy 0.005 mg). Weights were not corrected either for partial digestion or for effects of preservation in ethanol.

Data analysis

Prey of kokanee and rainbow trout were analyzed relative to the percent composition of the diet by vertical distribution (lake surface, water column, or benthos) (by weight), the percent composition by aquatic or terrestrial origin (by weight), and the mean weight of food items. Chironomid larvae were assumed to have been of benthic origin, with the exception of *Heterotrissocladius* sp. larvae that were found in pelagic zooplankton tows (Buktenica & Larson, 1996) and were assumed to have been consumed from the water column. All percent composition values were calculated for each stomach sample, and averages were calculated for the fish groups of interest (e.g., by species and year).

A ubiquity index expressed the degree to which each food taxon was evenly distributed among the specimens under consideration. For this purpose, the niche breadth measure B was used:

$$B_i = \exp\left[-\sum_{j=1}^{s} q_j \log_e q_j\right],$$

where

$$q_j = p_{ij} \Big/ \sum_{j=1}^{s} p_{ij},$$

p_{ij} was the proportional abundance of the i-th food taxon in the j-th stomach, and s was the total number of stomachs in the sample (Whiting & McIntire, 1985). For our purposes, B can vary from 1, when an item was present in just one stomach, to s, when the item was equally common in all s stomachs. B values were rescaled and presented as a percentage of the total number of sample (s) (Table 1).

Relationships between the food groups and species and size class of fish were investigated by canonical correspondence analysis, an ordination procedure designed to detect patterns in species abundance data (food item taxa in this case) that can be explained best by a set of companion (environmental) variables (Jongman et al., 1987).

Table 1 List of food categories found in the stomachs of kokanee salmon and rainbow trout from 1986 through 1997, food group acronyms, and the corresponding niche breadth values

Kokanee Salmon fish food category	Acronym	Ubiquity index	Rainbow Trout fish food category	Ubiquity index
Heterotrissocladius—pupae	HETER	21.91	Tricoptera	35.62
Amphipoda	AMPH	18.36	Hymenoptera	28.24
Heterotrissocladius—larvae	HETEL	17.34	Chironomid pupae	22.22
Homoptera	HOMO	16.60	Gastropoda	20.62
Daphnia	DAPH	14.81	Coleoptera—terrestrial	18.53
Chironomid pupae	CHIRP	14.08	Diptera	17.65
Diptera	DIPT	12.84	Coleoptera—aquatic	15.03
Hymenoptera	HYME	11.38	Hemiptera—terrestrial	11.15
Aracnida	ARAC	10.50	Aracnida	9.05
Tricoptera	TRIC	10.36	Homoptera	9.03
Coleoptera—terrestrial	COLET	8.91	Chironomid larvae	8.56
Bosmina	BOSM	8.34	Amphipoda	6.22
Chironomid larvae	CHIRL	7.63	Ephemeroptera	6.13
Copepoda	COPE	6.62	Lepidoptera	3.38
Hemiptera—terrestrial	HEMIT	5.03	Neuroptera	2.58
Thysanoptera	THYS	4.26	Orthoptera	2.03
Coleoptera—aquatic	COLEA	3.93	Hemiptera—aquatic	1.73
Gastropoda	GAST	2.76	Odonata	1.70
Psocoptera	PSOC	2.25	Thysanoptera	1.65
Lepidoptera	LEPI	1.61	Heterorissocladius—pupae	1.57
Ostracoda	OSTR	1.25	Decapoda	1.26
Plecoptera	PLEC	1.22	Plecoptera	0.85
Chydoridae	CHYD	1.20	Salmoniformes	0.79
Ephemeroptera	EPHE	0.93	Heterotrissocladius—larvae	0.68
Neuroptera	NEUR	0.89	Psocoptera	0.48
Hemiptera—aquatic	HEMIA	0.60	Chydoridae	0.46
Orthoptera	ORTH	0.28	Caludata	0.41
Odonata	ODON	–	Daphnia	0.27
Decapoda	DECA	–	Copepoda	0.26
Salmoniformes	SALM	–	Ostracoda	0.26
Claudata	CLAU	–	Bosmina	–
Sample size		867	*Sample size*	387

The ubiquity index is expressed as a percentage of the sample size

For these analyses, species of fish (kokanee or rainbow trout) and size class were treated as nominal variables, and the food categories were ordered along axes that maximized correspondence to various sets of the nominal variables. All ordinations were performed by the computer program CANOCO (Ter Braak, 1987).

Distribution

To assess fish abundance, distribution, and diel migration patterns, annual hydroacoustic surveys were conducted in the month of August or September starting in 1996. Daytime surveys were conducted each year, while night surveys were conducted in 1996 and 1999–2004. In 1996, the hydroacoustic system consisted of a BioSonics DT4000 digital sonar and 420-kHz single-beam digital transducer with a 6 ° nominal beam angle. For the 1997–2004 surveys, a BioSonics DT6000 digital sonar with a 200-kHz 6 ° split-beam digital transducer was used. The transducer was mounted on a 1.3 m tow-fin and towed 1 m deep off the port stern of an 11 m research vessel. The BioSonics DT series is a computer controlled digital sonar where all data are stored directly to a computer hard drive in raw-digital format. Data were sampled at a rate of 2 pings/s while the transducer was towed at 2.5 m/s. We used a 0.4 ms pulse width, target detection threshold of –66 dB, and 40 Log Range varied threshold type. The lake was divided into near shore and offshore zones for acoustic sampling and analysis. The depth used to separate the zones was 100 m in 1996, but was increased to 150 m in 1997–2004 because of improved detection capabilities of the split beam transducer. The offshore zone was divided into 10 equally spaced transects running NE/SW with a total length of 69.3 km (43 mi). The near shore sampling involved a zigzag pattern (73 legs) between shore and the 150 m depth contour (100 m in 1996), totaling 56 km (34.8 mi). The single beam data from 1996 was visually echo counted to estimate fish density using the duration-in-beam (DIB) method for determining sampling volume (Dawson, 1972). Echo counts of fish in the 1997–2004 surveys were made by the BioSonics DT Visual Analyzer Target Strength Analysis software following visual confirmation of each target by a technician. Sampling volume in the 1997–2004 surveys was determined directly through the split-beam phase angle measurement in the BioSonics software. The water column was divided vertically into five segments (1996: 2–20, 20–40, 40–60, 60–80, 80–100) or eight segments (1997–2004: 2–20, 20–40, 40–60, 60–80, 80–100, 100–120, 120–140, 140–150) for target strength and density analysis. Because the sampling area of an acoustic cone increases with depth, simple averaging of all depth values associated with fish targets cannot be used to determine mean fish depth. Consequently, mean fish depth was calculated using a weighting factor for each fish target that adjusted for the diameter of the sampling cone at that target's depth. Average equivalent fish length was calculated from mean target strength based on the equation of Love (1971). Target strength measurements have inherently higher variance than direct length or weight measurements due to the effects of target attitude (Levy, 1991). Therefore, conversions of target strength to fish length should be treated as approximations only.

Population dynamics

Fulton's condition factor was used to assess condition of fish (Ricker, 1975; Anderson & Gutreuter, 1983). This condition factor (K) is defined as

$$K = (W/L^3)X,$$

where W equals whole weight (including stomach contents) in gram, L equals length in mm, and X is an arbitrary scaling constant so that the small decimal values can be more easily comprehended. Comparisons of condition factors were initially limited to comparisons of fish of the same species within 50 mm size classes (Buktenica & Larson, 1996) because body shape, and therefore condition factor, can change with length. In this report we evaluate condition by age class because we did not find large differences between 50 mm size classes (Buktenica & Larson, 1996).

Catch per unit effort (CPUE) was used to evaluate relative abundance of fish. CPUE represented the number of fish captured per m^2 of gill net, per set, multiplied by a scaling factor of

100, and averaged for only June and July. These 2 months were used because catch rates of kokanee dropped off rapidly in August and September (Buktenica & Larson, 1996). Fork length and total length of fish captured were measured to the nearest millimeter on a 0.6 m measuring board. Whole fish weights were determined to the nearest gram on a top-loading temperature-compensated spring dial scale. To age fish, scales were collected from just below the posterior margin of the dorsal fin and above the lateral line. Six scales from each fish were cleaned and mounted between two microscope slides. Mounted scales were viewed through a microfiche projector or compound microscope to determine the age of each fish. Length-frequency analysis and modal-progression analysis were used for age validation (Jearld, 1983). Our confidence in aging kokanee from scales was not high, and other methods for determining age, e.g., processing of fish otoliths were too resource intensive for the long-term monitoring program. Interpreting scales was difficult because of variation between scales, even when taken from the same location on the same fish, annuli development was not consistent from year to year, and re-absorption of scales was common during some periods. Rainbow trout scales varied less in size and shape, and abrasion and absorption of scales generally occurred in age IV and older fish. Scales were not collected from rainbow trout after 1998. After 1998 trout age for estimates of condition and growth were assigned based on length (age 1 and 2, <250 mm; age 3 through 6, >250 mm).

Growth of kokanee was evaluated by percent change in mean weight between age 1 and age 2 in consecutive years, by weight at age, and by length at age. Weight at age was used when comparing growth to fish density, zooplankton density, lake level, and indices of lake productivity because change in weight is usually greater than change in length over short periods of time and may provide greater precision (Busacker et al., 1990). Total length at age was used to compare study results to the literature. Growth of kokanee was evaluated relative to zooplankton density (#/m^3) and biomass (mg/m^3) integrated to 200 m and collected with a closing vertical net (Larson et al., 2007b), lake level (feet above sea level) (Redmond et al., 2007), primary productivity (mg Carbon/m^2/h) and chlorophyll *a* (mg/m^3) integrated to 200 m (McIntire et al., 2007), and average descending and ascending Secchi transparency (m) with a 20 cm limnological disk (Larson et al., 2007a). Values represent mean summer conditions (July, August, and September) except for lake level (October 1) and fish density (September).

Density of kokanee was estimated from acoustic surveys proportionalized by age class based on aging from scales. Density of kokanee greater than age 2 was used for regressions of fish weight verses density (Table 4) because there was little evidence that these older age classes segregated food or habitat resources. Density at each age class was used to compare study results to the literature (fish length verses density, Fig. 12).

Results

Food habits

Kokanee fed on a large number of small-bodied prey from the mid-water column (Table 2). Five food groups were of primary importance in terms of occurrence (Table 1) and relative abundance (Fig. 1): Chironomids (particularly *Heterotrissocladius* sp.), zooplankton (primarily *Daphnia pulicaria*), Amphipods, Homopterans, and Trichopterans. *Heterotrissocladius* is a deep-water midge whose larvae were captured in pelagic zooplankton tows, and abyssal benthic samples collected with a one-person submarine (Buktenica, 2001). The relative abundance of these five food groups in kokanee stomach contents varied from year to year. Although *Daphnia* were not the dominant food group for kokanee, when present in the lake, they occurred in the diets of kokanee (Fig. 2a,b), and kokanee fed to a greater extent in the mid-water column (Table 2). When *Daphnia* decreased in occurrence in the diet from 1990 through 1995, *Bosmina*, *Chydorus*, and Copepoda (Cyclopoida and Harpacticoida) increased in occurrence (Fig. 2b), diversity of taxa taken increased (Fig. 2f), and the proportion of prey from benthic and terrestrial origins increased (Table 2). In general, pelagic *Heterotr-*

Table 2 Sample size, average number of food items per stomach, mean weight of food items, percent composition by vertical distribution in the water column (by weight) and percent composition by origin (by weight) of food items for kokanee salmon and rainbow trout from Crater Lake, Oregon, 1986–1997

Category	Sample size (*N*)	Average number of food items per stomach, Count	Mean weight of food items (mg)	Percent composition by vertical distribution			Percent composition by origin	
				Surface	Column	Benthos	Aquatic	Terrestrial
Kokanee	867	1072	1.72	22	55	23	80	20
Rainbow[a]	383	180	11.06	37	13	50	66	34
Kokanee—fork length								
100–120 mm	562	678	1.82	30	46	24	72	28
221–320 mm	301	1777	1.52	7	73	20	95	5
321–420 mm	4	3470	1.71	0	45	55	100	0
Rainbow—fork length								
100–120 mm	114	158	5.97	31	11	58	72	28
221–320 mm	130	246	11.39	45	15	40	57	43
321–420 mm	117	176	14.07	33	12	55	71	29
>420 mm	26	141	17.12	37	24	39	69	31
Kokanee								
1986	97	657	1.70	7	74	19	97	3
1987	155	2500	1.18	3	83	14	98	2
1988	11	4053	0.06	0	99	1	100	0
1989	14	2545	0.55	4	82	14	96	4
1990	16	211	1.43	26	13	61	82	18
1991	69	582	1.95	11	60	29	91	9
1992	78	497	3.31	14	46	40	87	13
1993	103	348	1.42	53	34	13	51	49
1994	94	603	1.89	41	31	28	65	35
1995	150	754	1.39	35	45	20	66	34
1996	54	1823	2.81	25	60	15	76	24
1997	26	454	1.33	0	44	56	100	0
Rainbow								
1986	55	191	10.34	41	12	47	67	33
1987	61	328	9.25	32	14	54	71	29
1988	1	34	36.00	0	0	100	100	0
1989	17	249	6.11	20	2	78	80	20
1990	26	155	8.48	34	26	40	73	27
1991	108	166	14.65	34	13	53	70	30
1992	53	213	7.63	37	16	47	63	37
1993	30	129	11.96	43	21	36	59	41
1994	22	64	10.83	65	3	32	36	64
1995	8	67	6.19	8	12	80	92	8
1996	5	40	31.79	79	8	13	29	71
1997	1	140	4.34	4	49	47	95	5

[a] Stomachs containing kokanee salmon and salamanders *were not included*

issocladius occurred in kokanee stomachs to a greater extent when *Daphnia* was present; other chironomids, amphipods, and homopterans were more prevalent when *Daphnia* was absent. The deep-water midge *Heterotrissocladius* was singularly the most important food resource for kokanee in terms of percent composition of the diet by weight. In contrast to kokanee, rainbow trout fed on fewer, larger organisms from the lake bottom and lake surface (Table 2). The five food groups of primary importance to rainbow trout in terms of occurrence (Table 1) and relative abundance (Fig. 1) were Coleoptera, Trichoptera, Hymenoptera, Gastropoda, and Chironomids.

An ordination of food groups relative to species of fish separated food items by their relative association with kokanee and rainbow trout (Fig. 3). *Bosmina* was found only in the stomachs of

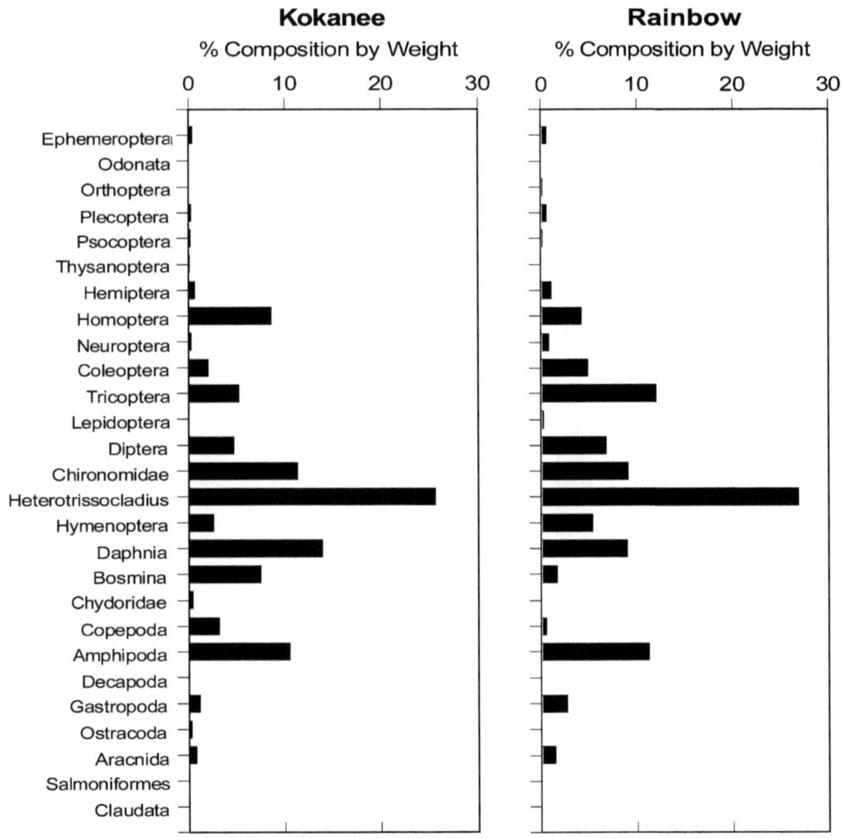

Fig. 1 Percent composition by weight of prey in kokanee and rainbow trout stomach samples from Crater Lake, Oregon, 1986–1997

kokanee, whereas Odonata, Decapoda, Salmoniformes, and Claudata were confined to the stomachs of rainbow trout. Copepoda, *Heterotrissocladius* larvae and pupae, Plecoptera, Ostracoda, *Daphnia*, and to a lesser degree Thysanoptera, Homoptera, and Amphipoda were strongly associated with the diets of kokanee, whereas aquatic Hemiptera, Orthoptera, Gastropoda, Lepidoptera, and Ephemeroptera were taken more often by rainbow trout. Taxa that occurred in the diets of both fish species, but were not strongly associated with either, included Aracnida, Chironomid larvae and pupae, Psocoptera, Neuroptera, Diptera, and terrestrial Hemiptera.

In general, rainbow trout diets varied less between years and within size groups than did kokanee diets. However, as rainbow trout increased in size, mean weight of food items increased from 5.97 to 17.12 mg (Table 2). The opposite pattern was seen for kokanee, as the larger size classes took a greater number of smaller sized prey. Ordinations of the food groups for kokanee relative to size class revealed a greater diversity of the food assemblages in the stomachs of the smaller fish (fork length of 100–220 mm), and some association between the larger fish (fork length > 220 mm) and Ephemeroptera, larval *Heterotrissocladius*, Trichoptera, and *Daphnia* (Fig. 4). The abundance of *Daphnia* in the diet of larger kokanee may be explained in part by the fact that spawning kokanee captured in January 1988 and 1997 contained stomachs full of *Daphnia*. Stomach walls of spawning kokanee lacked musculature, and were thin and translucent. Similar ordinations for rainbow trout illustrated the association between size class 4 (fork length > 420 mm) and Claudata and Salmoniformes (Fig. 5). In the latter case, small kokanee were found in the stomachs of a few large rainbow trout.

Distribution

Using hydroacoustics, fish were detected deeper in the water column during the day than at night (Fig. 6). In 1999, the difference between patterns

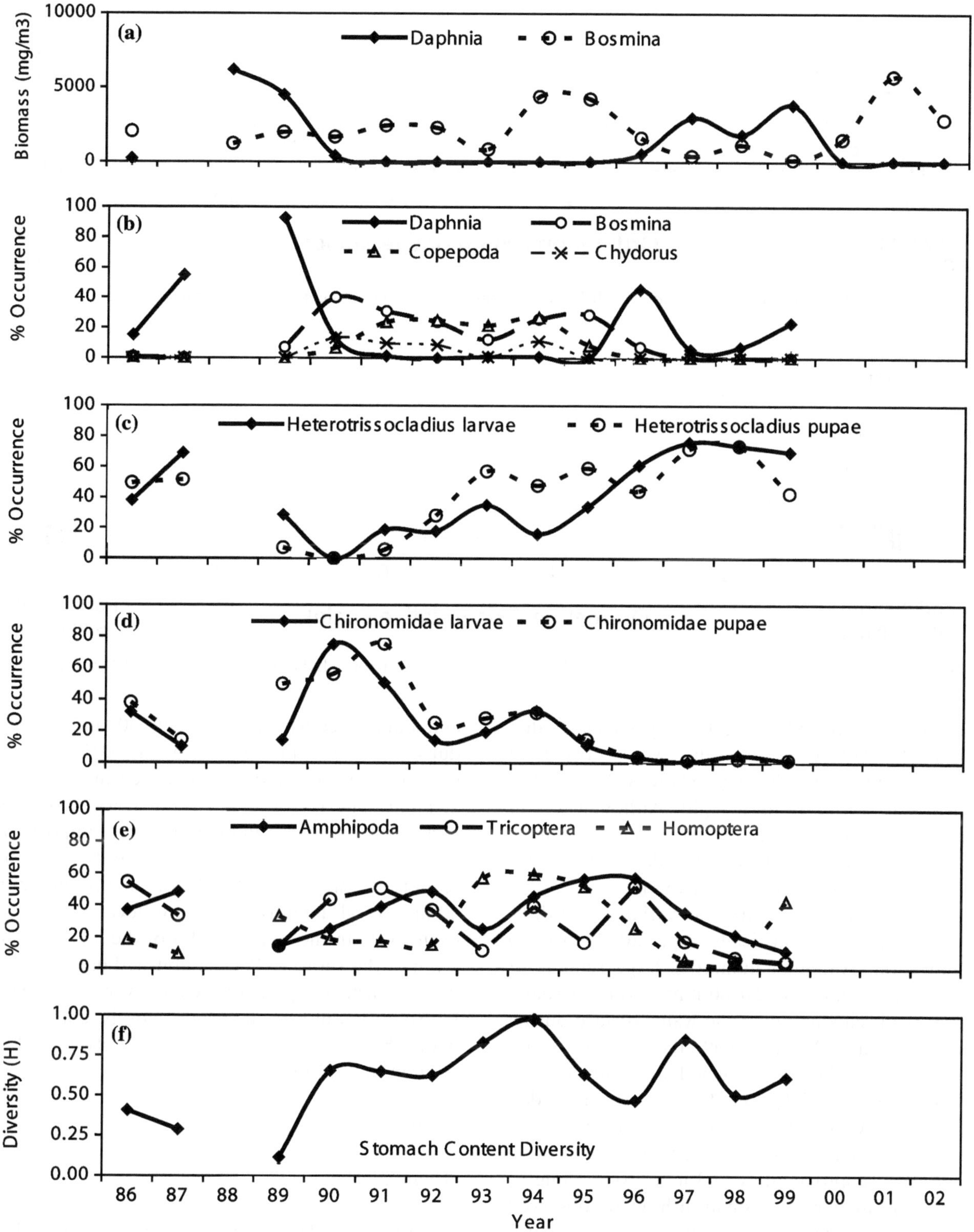

Fig. 2 Biomass of *Daphnia* and *Bosmina* (**a**) in Crater Lake (mg/m^3), percent occurrence of 11 food groups in kokanee stomachs (**b**, **c**, **d**, **e**), and mean kokanee stomach content diversity (**f**) from Crater Lake, Oregon, 1986–2002

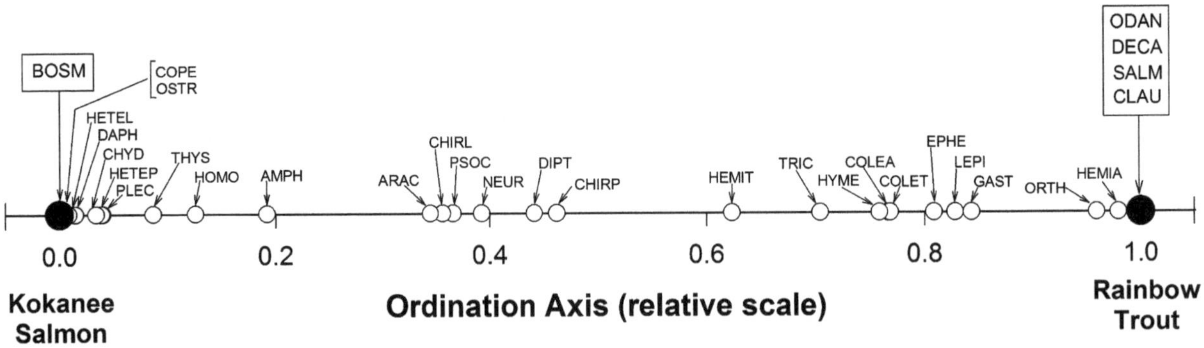

Fig. 3 Ordination of fish prey taxa relative to fish species from Crater Lake, Oregon, 1986–1997 (acronyms are defined in Table 1)

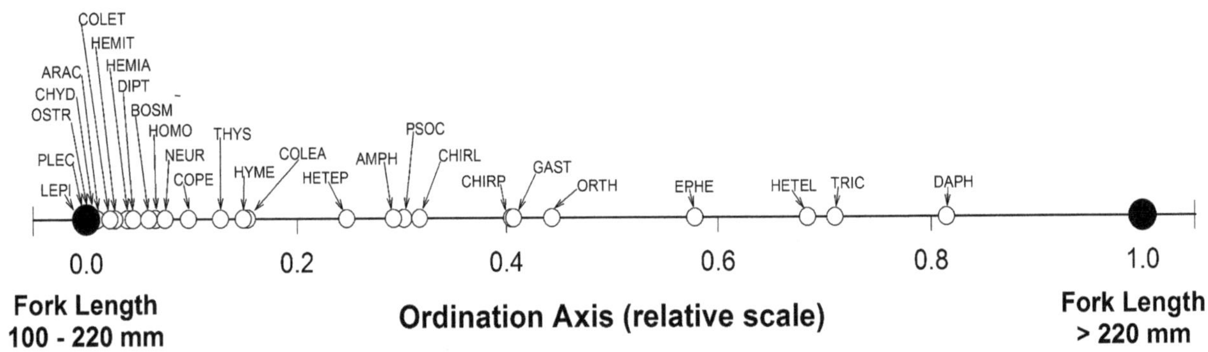

Fig. 4 Ordination of fish prey taxa relative to kokanee size class from Crater Lake, Oregon, 1986–1997 (acronyms are defined in Table 1)

during the day and at night was particularly distinct. The 96% of observed fish were between 40 and 100 m during the day, whereas 98% of the fish were above 20 m at night. Although few in number, some fish were observed below 145 m in all surveys sampling below that depth. The maximum depth of detection was 153 m.

The distribution of fry was variable and difficult to evaluate acoustically, especially between 0 and 20 m. Although numerous large schools of fry (approximately 50–80 mm) were visually observed schooling near the water surface during the day in 1998 and 1999, the schools would scatter as the boat approached, avoiding acoustic detection. At night small fish were recorded acoustically near the surface in some years, as indicated by the group of targets in 1999 around –54 dB (approximately 63 mm total length) (Fig. 6). In contrast, small fish were observed acoustically near the surface, at greater abundance during the day in 2002 than at night. Small fish were also observed at relatively high abundance acoustically between 40 and 120 m during the day in 2000 through 2004, but at greatly reduced abundance near the water surface at night. The fry observed acoustically near the surface in 2002 were unique in several important ways. The fry in 2002 were slightly smaller (centered around –58 dB or approximately 40 mm total length) and were located almost exclusively in the southeast 1/4 of the lake within 1 km of the shore (NPS, unpublished data). Most of these fish occupied a very narrow depth range between 12 and 16 m, which closely coincided with the temperature interface between the epilimnion and the metalimnion.

Population dynamics

Abundance

Kokanee abundance varied dramatically over the study period displaying two peaks in abundance separated by about 10 years (Fig. 7). Kokanee

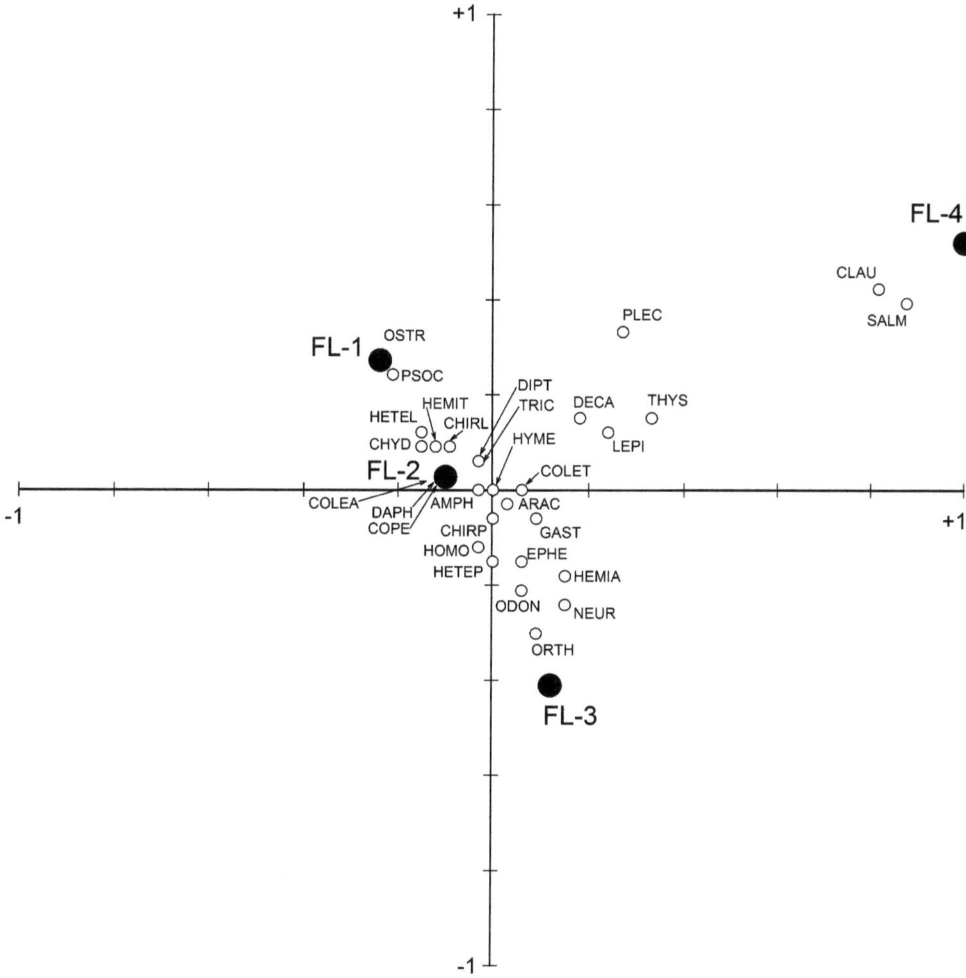

Fig. 5 Ordination of fish prey taxa relative to rainbow trout size class from Crater Lake, Oregon, 1986–1997. FL-1 = 100–220 mm, FL-2 = 221–320 mm, FL-3 = 321–420 mm, FL-4 > 420 mm (acronyms are defined in Table 1)

Catch Per Unit Effort (CPUE) increased from 1986 to 1990, and was low and variable from 1992 to 1998 (Fig. 7). CPUE again increased considerably in 1999 and 2000, decreased rapidly through 2003, and increased in 2004. Rainbow trout CPUE was low and variable throughout the study (Fig. 7). Whole lake population estimates using acoustics ranged from a low of 8,400 (1.6 fish/ha) in 1998 to a high of 633,000 (119.0 fish/ha) in 2000 (Fig. 7; Table 3). The largest population increase measured over a 1-year period occurred between 1998 and 1999, a 25-fold increase. Population estimates were higher in the offshore zone than in the nearshore zone of the lake, except during periods of extremely low fish abundance (Fig. 7; Table 3). However, fish density was consistently higher in the near shore zone (Table 3).

Catch per Unit Effort, as a measure of relative abundance was a poor substitute for acoustic derived population and density estimates early in the study. CPUE had large standard deviations, and high inherent sampling bias. Although CPUE followed the general trends demonstrated in acoustic data, when both were collected, the timing and magnitude of CPUE was not consistent, and on several occasions CPUE indicated a directional trend change not indicated in the acoustic population estimate (e.g., 2000 and 2004). Based on these limitations and our field observations in 1990, we believe the first peak in CPUE

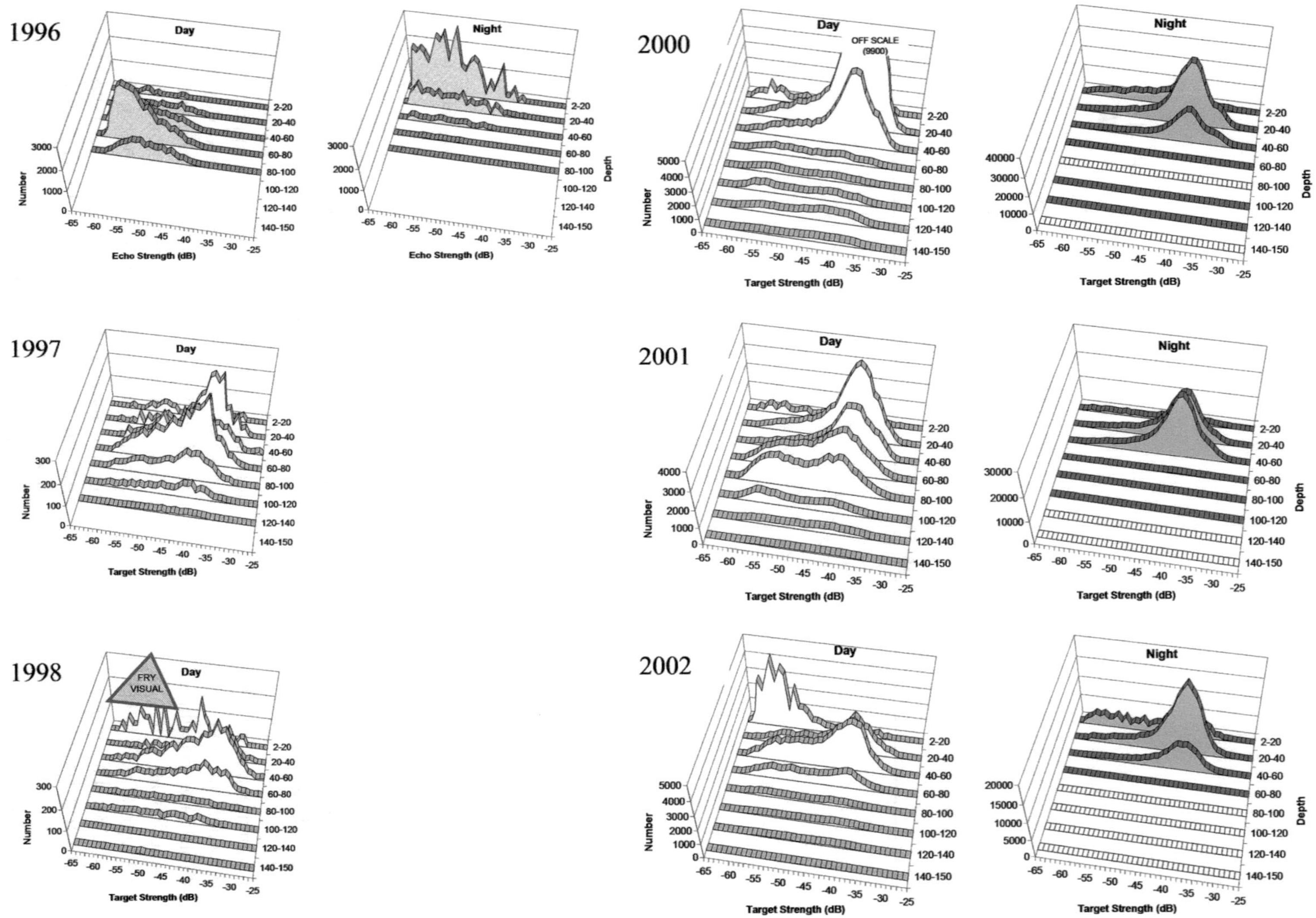

Fig. 6 Number of fish by depth stratum and target strength from acoustic surveys in Crater Lake, Oregon, 1996–2004. Below 100 m was not sampled in 1996. Target strength of 1996 is based on a single beam transducer and is not directly comparable to the 1997–2004 survey which used a split beam transducer

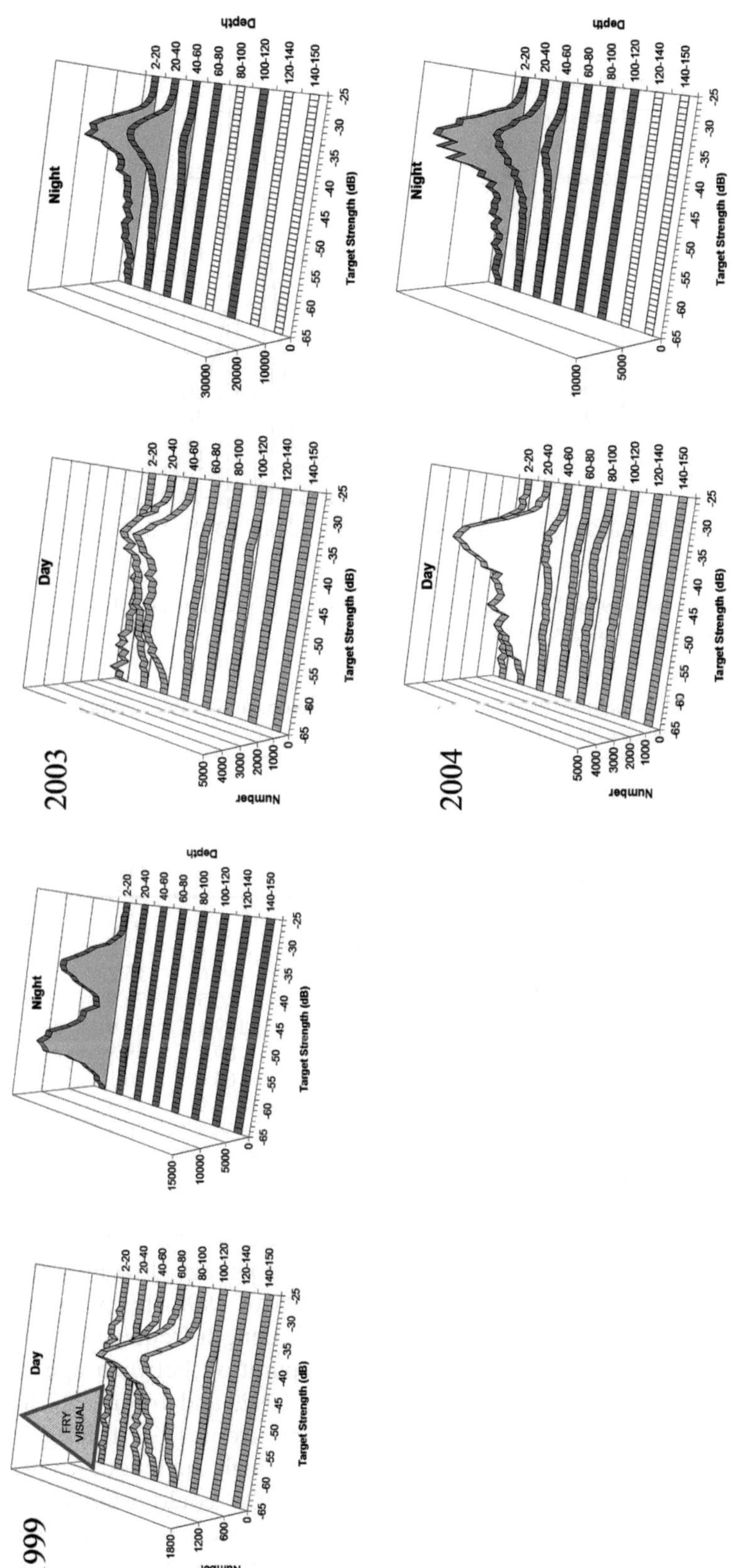

Fig. 6 continued

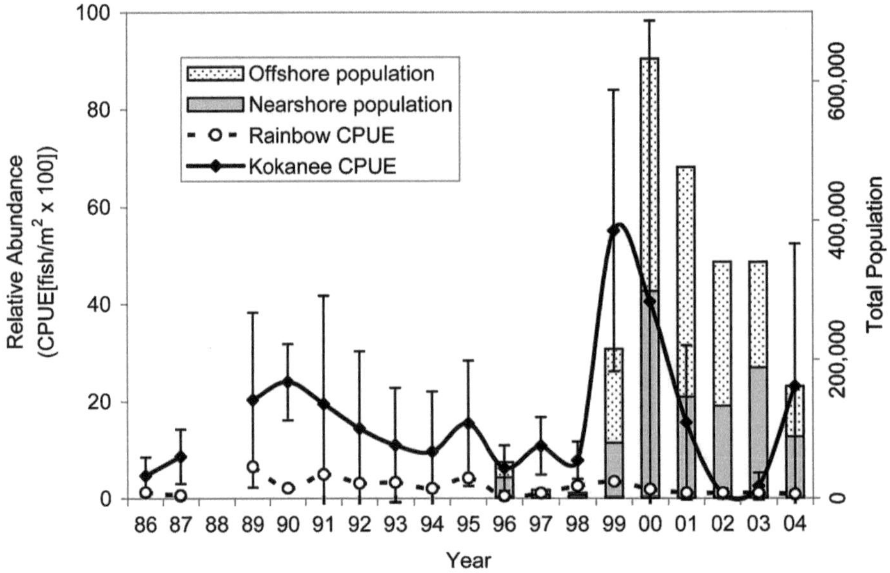

Fig. 7 Whole lake population estimates (1996–2004) and relative abundance (Catch Per Unit Effort, CPUE) (1986–2004) of kokanee and rainbow trout from Crater Lake, Oregon

Table 3 Whole lake fish population estimates, fish density, mean fish depth, and water temperature at mean fish depth during annual acoustic surveys of Crater Lake, Oregon, 1996–2004

Year	Total population			Density (fish/ha)			Mean depth (m)		Water temp. (°C) at mean depth	
	Offshore	Near shore	Total	Offshore	Near shore	Total	Day	Night	Day	Night
1996	21,820	30,895	52,751	4.8	34.4	9.7	63.3	21.9	4.8	9.0
1997	6,784[a]	5,867[a]	12,651[a]	1.6[a]	5.0[a]	2.4[a]	71.8	–	4.5	–
1998	3,220[a]	5,163[a]	8,382[a]	0.8[a]	4.4[a]	1.6[a]	55.2	–	4.7	–
1999	135,483	79,938	215,421	32.6	68.6	40.5	74.0	9.5	4.3	12.9
2000	334,501	298,489	632,990	80.6	256.2	119.1	58.9	35.9	4.9	6.0
2001	331,032	145,962	476,994	79.7	125.3	89.7	69.8	42.7	4.7	6.4
2002	207,311	132,969	340,280	49.9	114.1	64.0	56.1	34.7	5.2	7.0
2003	152,257	187,707	339,964	36.7	161.1	63.9	51.3	27.1	5.0	6.9
2004	72,439	88,709	161,148	17.4	76.1	30.3	50.3	28.3	5.6	6.9

[a] Values in 1997 and 1998 were based on daytime surveys

underrepresented the magnitude of kokanee population growth in 1990 as it did in 2000.

Age and length frequency

Kokanee age and length frequencies varied considerably over the study period (Fig. 8). In 1986 and 1987, a single year class of kokanee predominated (age 2 and 3, respectively). In 1988, fish sampling was limited to January, and only kokanee in spawning condition with reabsorbed scales were collected. After 1989, the population age structure became more diverse. From 1989 to 1995 a strong age 1 component occurred, and from 1995 to 1998 the proportion of age 1 fish declined. Length frequency histograms displayed some separation between age classes from 1986 through 1992, and from 1997 to 2000, and little separation from 1993 through 1996, and after 2000. Although the length frequency modes from 1995 to 1998 could represent one cohort, we believe there were several age classes compressed together due to the preceding period of low condition and growth (see below).

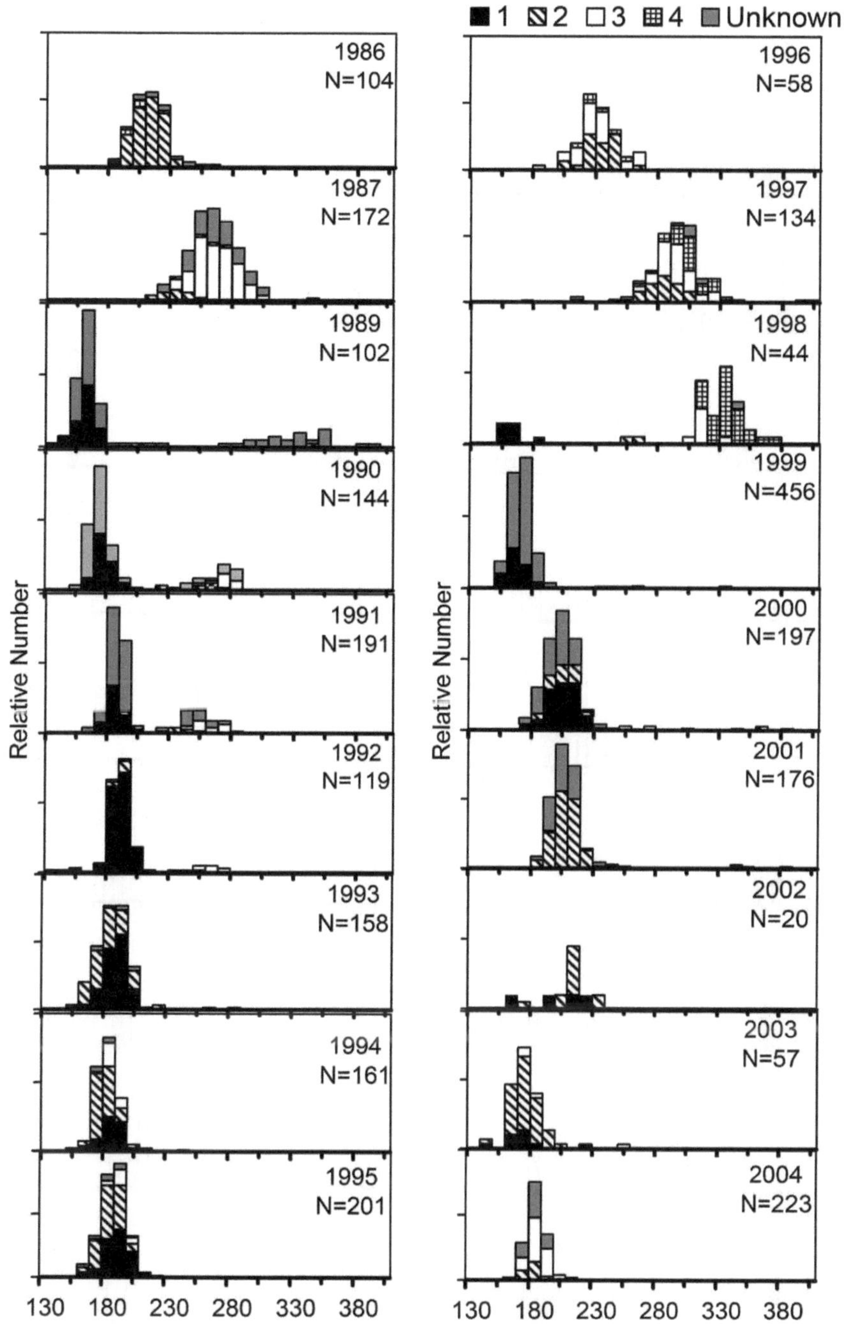

Fig. 8 Kokanee age and length frequencies in Crater Lake, Oregon, 1986–2004

Rainbow trout exhibited a more complex population structure than did kokanee (Fig. 9). Although sample sizes were often small, the composition of the rainbow trout population shifted toward larger size classes and older-aged fish between 1986 and 1991, back toward smaller size and younger fish through 1998, followed by increasing length through 2004.

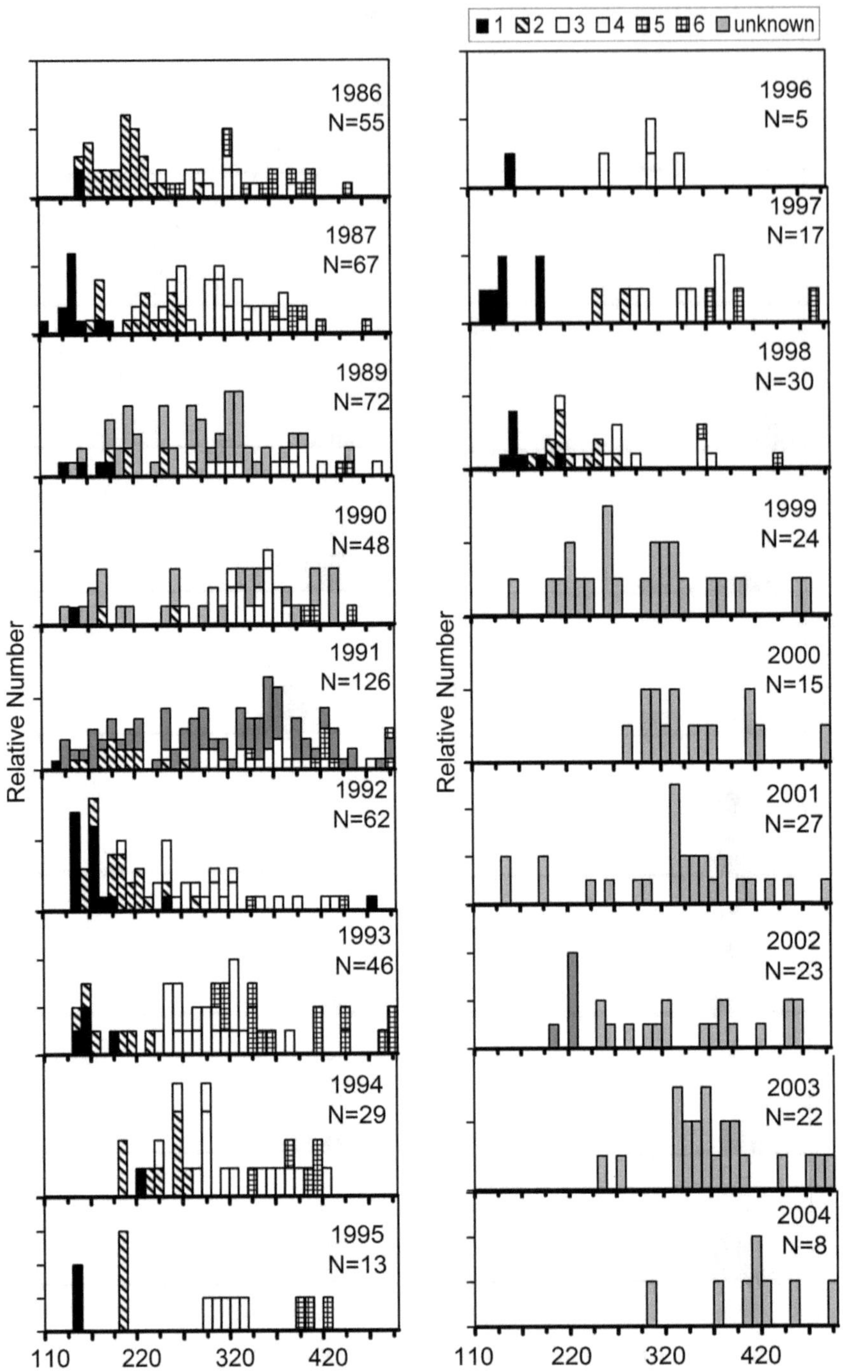

Fig. 9 Rainbow trout age and length frequencies in Crater Lake, Oregon, 1986–2004

Length, weight, condition, growth, and maturity

Kokanee length, weight, condition, growth and maturity also displayed a pattern that repeated in about 10 years (Fig. 10a–e). In general, values for older fish (age 2–4) increased through the late 1980s, declined to low levels between 1993 and 1995, increased again through 1998, and declined thereafter (Fig. 10a–c). Age 1 kokanee displayed less variation in average length and weight while

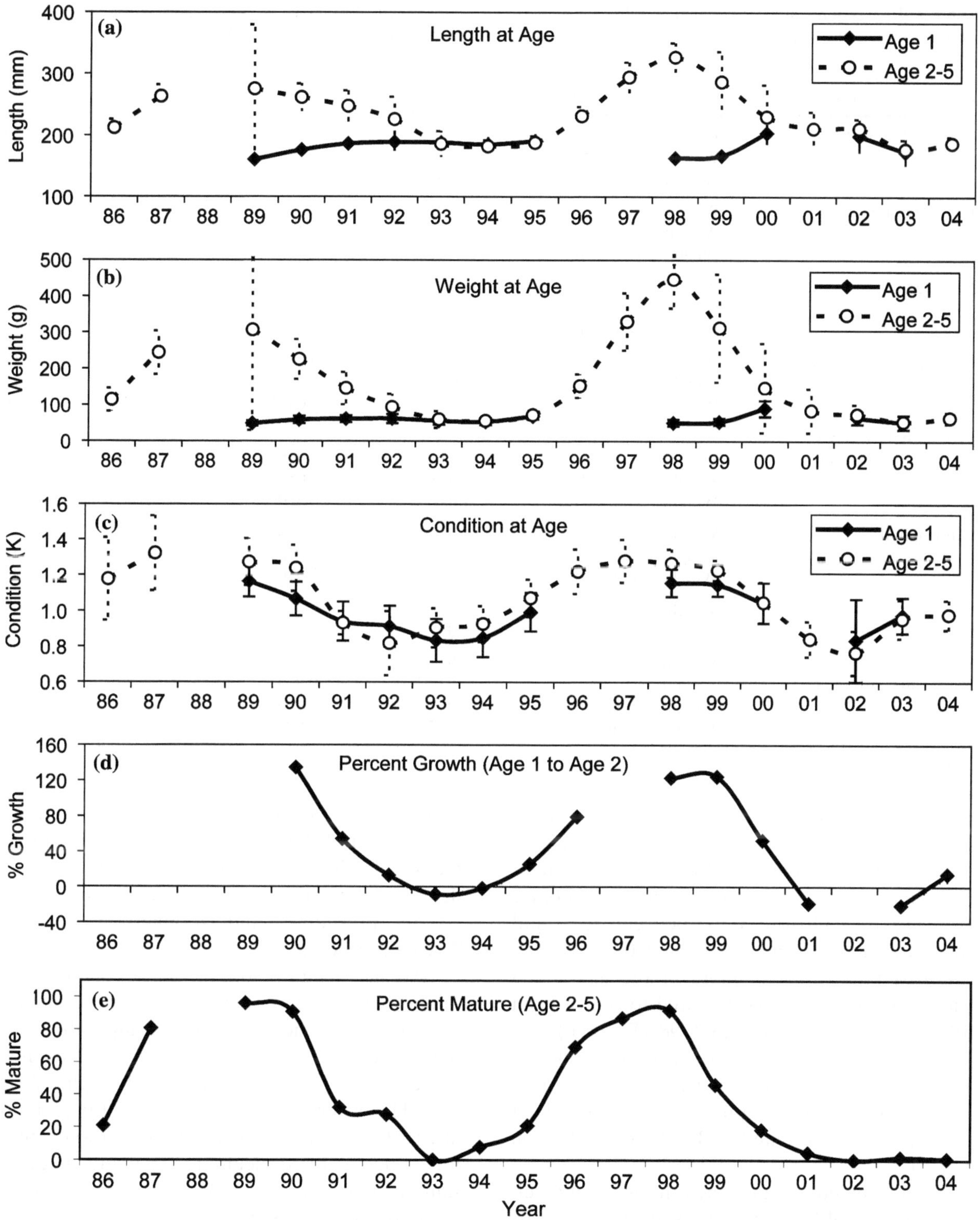

Fig. 10 Mean length, weight, and condition at age, and percent growth and maturity of kokanee in Crater Lake, Oregon, 1986–2004

Table 4 Regression results of kokanee weight (g, age 2–5) on fish density (fish/ha, age 2–5), *Daphnia* density (#/m^3), lake level (feet), primary productivity (mg C/m^2/h), chlorophyll (mg/m^2), Secchi depth (m), and total zooplankton biomass (mg/m^3)

Independent variable	Dependent variable	R^2	P-value	Coefficient	N
Log_e (fish den+1)	Weight	0.70	0.005*	–83.6	9
	Weight 1 year later	0.85	0.001*	–92.0	8
	Weight 2 years later	0.36	0.156	–58.1	7
Log_e (daphnia den+1)	Weight	0.77	< 0.001*	40.6	18
	Weight 1 year later	0.64	< 0.001*	36.0	17
	Weight 2 years later	0.16	0.119	18.1	16
Lake level	Weight	0.36	0.008*	23.6	18
	Weight 1 year later	0.22	0.058	19.0	17
	Weight 2 years later	0.001	0.884	1.7	16
Change in lake level	Weight	0.11	0.186	21.5	18
	Weight 1 year later	0.36	0.011*	39.1	17
	Weight 2 years later	0.26	0.043*	32.4	16
Productivity Indices:					
Primary productivity	Weight	< 0.001	0.968	0.1	12
	Weight 1 year later	0.36	0.052	–7.6	9
	Weight 2 years later	0.09	0.520	–3.0	7
Chlorophyll	Weight	< 0.001	0.936	0.09	18
	Weight 1 year later	< 0.001	0.970	–0.05	16
	Weight 2 years later	< 0.001	0.970	0.05	15
Secchi	Weight	0.04	0.426	7.3	18
	Weight 1 year later	0.13	0.142	13.5	17
	Weight 2 years later	0.04	0.438	7.7	16
Zooplankton biomass	Weight	0.002	0.894	< 0.01	14
	Weight 1 year later	0.16	0.159	< 0.01	14
	Weight 2 years later	0.57	0.003*	0.02	13

* indicates $p < 0.05$

condition closely tracked that of age 2 and older kokanee (Fig. 10a–c). Percent growth between age 1 and age 2 kokanee followed the general pattern of the other variables for age 2 and older fish (Fig. 10d). Growth of fish between age 1 and age 2 was negative in 1993, 1994, 2001, and 2003. Approximately 70% or more of the age 2 and older kokanee examined matured to spawn from 1987 to 1990, and 1996 to 1998 (Fig. 10e). No kokanee captured in 1993 and 2002 were maturing to spawn, and fewer then 2% were maturing in 2003 and 2004. Length at age and length at maturity were presented by Buktenica & Larson (1996).

We evaluated kokanee growth, or weight, relative to several lake characteristics that also varied during the study including fish density, *Daphnia* density (Larson et al., 2007b), lake level (Redmond et al., 2007), and several indices of productivity (Larson & Hoffman, 2002; McIntire et al., 2007) (Table 4). The strongest simple predictors of fish growth were fish density, *Daphnia* density, lake level, change of lake level 1 and 2 years preceding, and zooplankton biomass 2 years preceding.

Average length, weight, and condition of rainbow trout may have displayed a similar pattern as kokanee with decreasing values for older fish through the early 1990's, but then values for trout increased through 2004 (Fig. 11), whereas kokanee values dropped off after 1998. Length, weight, and condition trends for trout were difficult to evaluate because sample sizes were small and corresponding standard deviations were high. We were not able to obtain reliable estimates of growth and percent mature for rainbow trout because of small sample sizes.

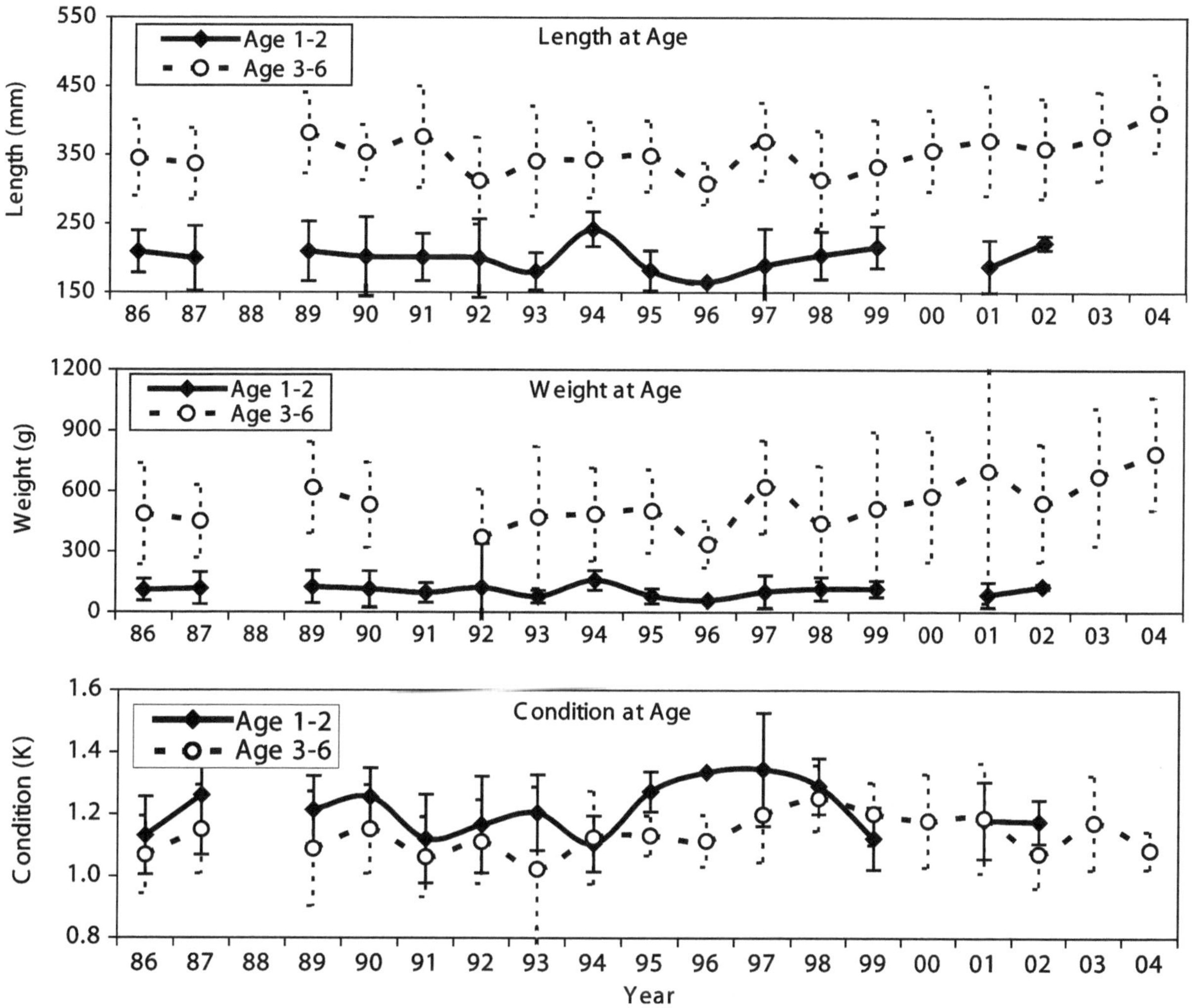

Fig. 11 Mean length, weight, and condition at age of rainbow trout in Crater Lake, Oregon, 1986–2004

Discussion

Food habits

Although early work on fish food habits in Crater Lake was limited in scope and duration, results of that research fit the pattern of food resource partitioning between kokanee and trout reported here. For example, the importance of *Daphnia* in the diet of kokanee, varied with its abundance in the lake (Hubbard, 1933; Brode, 1935, 1938; Hasler & Farner, 1942; Patten & Thompson, 1957), and chironomids, amphipods, and small-bodied trichopterans were important components in the diet (Brode, 1937; Patten & Thompson, 1957). Patten & Thompson (1957) also noted that kokanee preferred smaller taxa generally taken in the water column, whereas rainbow trout fed actively on Trichoptera, Hymenoptera, Gastropoda, Coleoptera, and Diptera, at the surface and in benthic habitats.

Prey segregation between kokanee and rainbow trout was not an unexpected result owing to differences in morphology and behavior. Kokanee are small-bodied pelagic schooling salmonids. They are efficient endurance swimmers because they have a forked tail and a fusiform body. In addition they have small mouths, large eyes, wide visual fields, and long and numerous closely spaced gill rakers (Scott & Crossman, 1973; Buktenica & Larson, 1990). Because of these morphological characteristics and swimming abilities, kokanee are well adapted to feed on zooplankton and other small-bodied invertebrates in offshore areas of lakes (Carlander, 1969; Bond, 1979). In contrast, rainbow trout are not efficient swimmers, relative to kokanee, because they are more laterally compressed and possess a marginate tail fin. Rainbow trout have large mouths, small eyes, reduced visual fields, and possess short and widely spaced gill rakers (Scott & Crossman, 1973; Buktenica & Larson, 1990). These characteristics are better suited to feeding on large-bodied prey, including kokanee, in the near shore zone of lakes.

The food habits of rainbow trout in Crater Lake were consistent with what has been documented in the literature. Rainbow trout typically feed on benthic and terrestrial insects, snails, worms, amphibians, and other fish species, and prey size increases with body size (Scott & Crossman, 1973; Beacham & McDonald, 1982). The diets of kokanee in Crater Lake were generally consistent with the literature with several notable exceptions. Kokanee and juvenile sockeye salmon feed heavily on crustacean zooplankton in other lakes and reservoirs (Norhtcote & Lorz, 1966; Cordone et al., 1971; Hoag, 1972; McDonald, 1973; Dobble & Eggers, 1977; Beacham & McDonald, 1982; Beauchamp et al., 1995), and *Daphnia* is often preferred (Finnell and Read, 1969; Collins, 1971; Cordone et al., 1971; Rieman & Bowler, 1980). When *Daphnia* are not abundant, other large Crustaceans and Copepods increase in the diet (Richards et al., 1975), as can chironomid larvae and pupae (Northcote & Lorz, 1966; Collins, 1971; Richards et al., 1975). In contrast to our results, different size classes (Beacham & McDonald, 1982; Beauchamp et al., 1995; Baldwin et al., 2000) and age classes (Rieman & Bowler, 1980; Leathe & Grahm, 1981; Thiede et al., 2002) of kokanee, excluding fry, have been found to have similar summer-season feeding habits, while feeding habits during other seasons may vary. Beauchamp et al. (1995) found that adult kokanee and sockeye in Lake Ozette, WA, fed on *Daphnia*, insects, and copepods from late summer to early spring, while juveniles fed on *Daphnia* year round. It is surprising that larger kokanee in Crater Lake fed on smaller prey than smaller kokanee. A possible explanation for this result may be that forage swimming speed increases with fish size. Therefore, larger kokanee could encounter more zooplankton especially at low abundance (Dobble & Eggers, 1977). In contrast, Zaret (1980) suggested that as fish get larger relative to their prey, their foraging and growth efficiencies should decline, particularly when shifting from large-bodied zooplankton to small-bodied zooplankton. Alternately, the smaller size class of kokanee in Crater Lake foraged to a greater extent at the lake surface, on larger prey (Table 2), for reasons unrelated to forage efficiency.

Kokanee in Crater Lake fed more heavily on aquatic and terrestrial insects than previously reported in the literature. We believe that this result was related to Crater Lake's low productivity, and that kokanee of all size classes will feed on large cladocerans when abundant. As cladoceran abundance decreases, from a decrease in productivity or increase in predation, size class differentiation will occur as a result of competition, and that eventually all size classes may switch to small zooplankton, chironomids and other small-bodied aquatic and terrestrial insects.

Distribution

Kokanee and rainbow trout exhibited different patterns of spatial distribution in the lake. Both species were found unusually deep, kokanee in the offshore and near shore zones of the lake, and rainbow trout near shore (Buktenica & Larson, 1990). Schooling behavior and diel migration of kokanee were variable during the study period and varied with fish density and prey availability. Additionally, this study indicates that kokanee fry distribution may vary as population size and perhaps food availability fluctuate. The patterns of diel migration relative to fish abundance and lake conditions over a 7-year period provide insights, explored below, into the mechanisms affecting kokanee distribution patterns in this ultraoligotrophic lake.

Mean daytime depth of kokanee in Crater Lake (50–74 m) was deep compared to the distribution of kokanee in many lakes and reservoirs in the west (5–50 m Beauchamp et al., 1997; 10–18 m Finnell & Reed, 1969; 10–30 m Luecke & Wurtsbaugh, 1993; 10–30 m Stockwell & Johnson, 1999) but similar to clear, deep, sockeye salmon lakes in British Columbia and Alaska (40–80 m) (Narver, 1970; Levy, 1991; Schmidt et al., 1994). The maximum depth reported here (153 m) is the maximum depth of *O. nerka* reported for freshwater. Kokanee and juvenile sockeye salmon are noted to occur deeper and migrate over greater vertical distances in clear lakes (Levy, 1990; Rieman & Myers, 1992). Crater Lake is one of the clearest lakes in the world, with an average Secchi disk reading of about 30 m during thermal stratification (Larson & Buktenica, 1998) and a maximum descending reading of 43.3 m. Depth of 1% incident light (PAR) 2007 is 80–100 m (Hargreaves et al., 2007; Larson et al., 1996). The light level at 150 m in Crater Lake was greater than the 0.1 lx demonstrated necessary for successful *O. nerka* feeding (Ali, 1959). The exceptional clarity of Crater Lake in conjunction with stable summer thermal stratification and a deep nutrient pool, account for deep distribution of phytoplankton (McIntire et al., 2007), zooplankton (Larson et al., 2007b), and consequently kokanee. Highest cladoceran densities were observed between 80 and 120 m, while the upper 20 m of the water column were virtually devoid of zooplankton.

In late summer, 1996–2004, fish were found in the cooler hypolimnion during the day and shallower in the warmer epilimnion and metalimnion at night. Similar patterns were observed in Crater Lake during previous acoustic surveys (Marino, 1987; Buktenica, 1989) and have been observed for *O. nerka* in other lakes and reservoirs in the Western United States and Canada (Narver, 1970; Eggers, 1978; Clark & Levy, 1988; Levy, 1990; Beauchamp, 1994; Beauchamp et al., 1997). Several hypotheses have been proposed to explain diel vertical migration behavior in kokanee and sockeye salmon, including maximization of food consumption (Wurtsbaugh & Neverman, 1988), predator avoidance (Eggers, 1978; Clark & Levy, 1988), and the maximization of bioenergetic efficiency (Brett, 1971). Diel vertical migration behavior is likely a combination of these factors (Clark & Levy, 1988; Johnston, 1990; Levy, 1990; Bevelhimer & Adams, 1993; Stockwell & Johnson, 1999), where populations respond differently to the unique combination of constraints imposed by each lake (Beauchamp et al., 1997). Our results agreed most closely with the maximization of bioenergetic efficiency hypothesis described by Brett (1971) and Beauchamp et al. (1997). When *Daphnia* were abundant and kokanee growth was high in Crater Lake in 1999, kokanee may have been able to consume larger rations of food in deeper hypolimnetic waters during the day and increase food assimilation in the warm epilimnetic waters at night (Fig. 6; Table 3). Conversely, when *Daphnia* abundance and fish growth declined after 1999, growth efficiency was maximized by remaining at lower temperatures day and night.

Compared with older kokanee (age 1–4), limited information exists on the movement of fry in lakes (Levy, 1991). Several authors suggested it was more advantageous for kokanee fry to remain in surface waters than to migrate (Stockwell & Johnson, 1999; Johnston, 1990). Levy (1987) suggested that by remaining in the epilimnion during the day, the visually-feeding fry could obtain higher growth rates as a consequence of increased foraging opportunities. In contrast, kokanee and sockeye fry have been observed in deep waters during the day in other lakes with diel migration to shallower waters at night (Clark & Levy, 1988; Levy, 1991). Factors other than bioenergetic efficiency, such as balancing predation risk with energy returns, were suggested to affect fry behavior in these lakes.

We observed both distribution patterns for fry in Crater Lake. When adult kokanee growth and condition were high in 1998 and 1999, kokanee fry remained near the surface during the day (visually observed) and remained above 20 m at night (Fig. 6). When adult kokanee growth and condition dropped after 1999, most fry were not observed at the surface during the day but were observed mostly below 40 m (Fig. 6). The movement of fry to deeper water may suggest a shift away from growth maximization and toward maximizing bioenergetic efficiency when exposed to high fish density and perhaps limited food resources. It is not known to what extent fry compete with older aged kokanee or what food resources fry utilized in Crater Lake particularly in the upper 20 m as this area had very low abundance of large zooplankton. The influence of predation pressure on kokanee fry during this period is difficult to assess and cannot be discounted. Although catch per unit effort (CPUE) of rainbow trout did not fluctuate greatly during the study (Fig. 7), average rainbow length and weight increased between 1998 and 2004. Larger rainbow trout may have resulted in increased predation pressure on kokanee.

The relative importance of factors driving kokanee and sockeye diel migration and distribution have been suggested to vary between lakes and between seasons within a year (Clark & Levy, 1988; Bevelhimer & Adams, 1993; Beauchamp et al., 1997; Steinhart & Wurtsbaugh, 1999; Stockwell & Johnson, 1999). The change in kokanee fry distribution patterns in Crater Lake during the study period suggests that the relative importance of factors influencing fry diel migration may also vary on a longer time scale (2–7 year) in lakes with large fluctuations in fish density and (or) food resources.

The fry observed during the day in 2002 between 2 and 20 m (Fig. 6) behaved differently than those in previous years and may have been rainbow trout. The fry near the surface in 2002 were unique in that fish target strength indicated smaller size (40 mm total length versus 63 mm in 1999), they were located almost exclusively in the south-east 1/4 of the lake within 1 km of shore, and they occupied a very narrow depth range (12–16 m), which closely coincided with the temperature interface between the epilimnion and the metalimnion. The limited horizontal and vertical distributions of these small fish were unlike those of kokanee fry in 1999 and 2000. In those years the abundant kokanee fry were distributed around the entire lake in both offshore and near shore areas, and had a wider distribution vertically. In addition, kokanee spawning success was likely low 2001–2002 because of very low condition and lack of maturing fish (Fig. 10). Conversely, trout size and condition suggested that older age rainbow trout were in favorable spawning condition in 2001.

Population dynamics

Kokanee displayed a great deal of variation in abundance, length, weight, condition, growth, maturity, and age and length frequencies, all of which exhibited a pattern that repeated in about 10 years. At the beginning of this study (1986), the kokanee population was dominated by a single age class, in low abundance and high body condition. The percentage of maturing kokanee was high, abundance increased through 1990, and the age structure of the population became more diverse. As abundance increased, growth, condition, and the percentage of maturing adults decreased. The population declined after 1990, growth became negative, there was poor separation in length frequency between age classes, and few kokanee matured. By 1996, growth and condition improved a higher proportion of kokanee matured, and the population increased rapidly to a maximum for the study period in 2000. It is possible that rainbow trout abundance tracked kokanee abundance, but sample sizes were low and variable, and this pattern could be an artifact of the collection method. However, the proportion of older and larger rainbow trout increased during both periods of high kokanee abundance.

Oncorhynchus nerka abundance and age-class dominance is often variable (Lewis, 1971; Goodlad et al., 1974; Hanzel, 1984; Rieman & Myers, 1992; Thiede et al., 2002), may exhibit a cyclic pattern (Goodlad et al., 1974; Levy & Wood, 1992; Ricker, 1997; Myers et al., 1998), and is the result of their life history strategy and complex interactions among environmental variables and stochastic events, species interactions (e.g., predation and competition), harvest, disease, and prey limitation. Sockeye often exhibit strong 4-year cycles, as observed in Fraser River populations (Levy & Wood, 1992; Ricker, 1997). Sockeye cyclic abundance usually occurs in populations with a single age-at-maturity, resulting in one dominant and three subdominant "subpopulations." Myers et al. (1998) concluded that stochastic environmental events can set the cyclic pattern in motion, and that delayed density dependent mortality may help to perpetuate the pattern and become increasingly important as the population approaches or exceeds carrying capacity. Gerking (1994), in his review of fish feeding ecology, notes that oligotrophic lakes in general are more likely to exhibit wide swings in fish population abundance than in eutrophic lakes. Population responses to environmental changes should vary with increasing fish density, exhibiting even greater fluctuations in ultraoligotrophic lakes and reservoirs (Rieman & Myers, 1992).

Kokanee in Crater Lake did not have a single age-at-maturity, and their pattern of abundance was not a strong 4-year cycle, however it appears that kokanee in Crater Lake exhibited density dependent and perhaps delayed density dependent growth (Table 4) (where a high abundance of kokanee 1 year inhibited growth in subsequent years) and associated reductions in spawning and recruitment, resulting in large-scale changes in abundance and biomass. Other processes that could account for large-scale changes in growth and abundance include changes in the physical or chemical environment, nutrients leading to changes in biological productivity, competition, predation, angler harvest, and fish health. There were no apparent large-scale changes in physical or chemical limnological variables (Larson et al., 2007a; Crawford & Collier, 2007) that could readily account for changes in fish abundance observed during this study, with the possible exception of lake level (Redmond et al., 2007). Lake level may influence habitat availability for fish, and may be an indicator of allochthonous inputs to the lake. At high lake levels, wave cut platforms are inundated around the caldera, perhaps providing more foraging, spawning, and rearing habitat for fish, and influencing reproduction, growth, and survival. Growth of kokanee was positively associated with lake level, and change in lake level, one and 2 years preceding (Table 4).

McIntire et al. (2007) observed a decline in chlorophyll during a period of lowering lake surface elevation through 1996 and suggested this may have been associated with a decrease in allocthonous nutrients. However, a subsequent rise in lake level between 1995 and 2000 was not associated with a concurrent increase in chlorophyll. McIntire et al. (2007) suggested that primary producer biomass may not have responded to the latter increase in lake level, or allochthonous nutrients, because of grazing by zooplankton. Caution should be used in making direct comparisons between fish population dynamics and allochthonous inputs because allochthonous inputs may be small relative to climate-driven upwelling (Dymond et al., 1996), and biological responses may exhibit significant time delays and are increasingly complex and dynamic as they move up trophic levels.

Although competition, predation, angler harvest, and fish health may affect kokanee population dynamics, we believe the relative importance of these factors was low relative to density dependence. Although competition between trout and salmon undoubtedly occurred, we believe the observed prey and habitat resource partitioning by kokanee and rainbow trout was driven more by morphological and life history differences than from competition. Trout predation on kokanee was probably low because trout abundance was low, trout were primarily found near shore, and kokanee were not a large component of trout stomach samples. However, predation could be depensatory (increasing with decreasing number of kokanee) and therefore a destabilizing force. Angler harvest was very low, because there is no recreational boat access and shoreline access is limited to one steep, 1.6 km foot trail that is snow-free for approximately 5 months of the year. Although a comprehensive fish health survey has not been conducted, several dying kokanee captured on the lake surface in 2001 were evaluated by the Oregon Department of Fish and Wildlife (ODFW) Fish Disease Laboratory at Oregon State University, Corvallis. These fish were emaciated, and exhibited a loss of body fat and muscle, indicating that the fish quit eating or their food disappeared. However, no indication of pathogens commonly associated with fish loss was found (ODFW fish exam report number CB01-190).

Density dependent growth of sockeye (Burgner, 1964; Johnson, 1965; Goodlad et al., 1974) and kokanee populations (Rieman & Myers, 1992; Martinez & Witzius, 1995) has been well documented in many lakes in the United States and Canada, and is commonly thought to result from intense size-selective predation and corresponding reductions in number and size of zooplankton prey (Johnson, 1965; Goodlad et al., 1974; Kyle et al., 1988; Rieman & Myers, 1992). The decline in kokanee growth with increasing density in Crater Lake is similar to the continuous exponential decline shown for a series of oligotrophic lakes in Idaho (Fig. 12, after Rieman & Myers, 1992). Rieman & Myers (1992) hypothesized that lake productivity should mediate density dependent growth in kokanee, or conversely in less productive lakes kokanee should experience a steeper decline in growth with increasing density. Based on their results, we expected that values for Crater Lake would fall below the values for Idaho lakes with Secchi transparency > 7 m because of Crater Lake's higher transparency and lower productivity. However, values for Crater Lake kokanee generally plotted within the range of observed values for all Idaho lakes, and the slopes of the lines were similar. Crater Lake kokanee attained longer total length at each size class. The Crater Lake and Idaho data sets are not directly comparable and several factors may account for the deviation from predicted values, including: long-term data on Crater Lake vs. short-term data on 10 Idaho lakes; a simple fish community in Crater Lake relative to the Idaho Lakes; different techniques for estimating fish density (acoustics vs. trawl) and for collecting fish (gill nets vs. trawl); stock of kokanee; and estimators of age (scales vs. length frequency mode) and length (mean summer vs. adjusted to September). Rieman & Myers estimated the Idaho trawl derived density estimates were 75% efficient relative to limited acoustic data. The Idaho trawl (smallest mesh 6 mm, stretch-measure) sampled fish less then 100 mm total length where the Crater Lake gill nets (smallest mesh 19 mm, square-measure) did not effectively collect fish smaller then 150 mm total length. Nonetheless results from the two studies are strikingly similar.

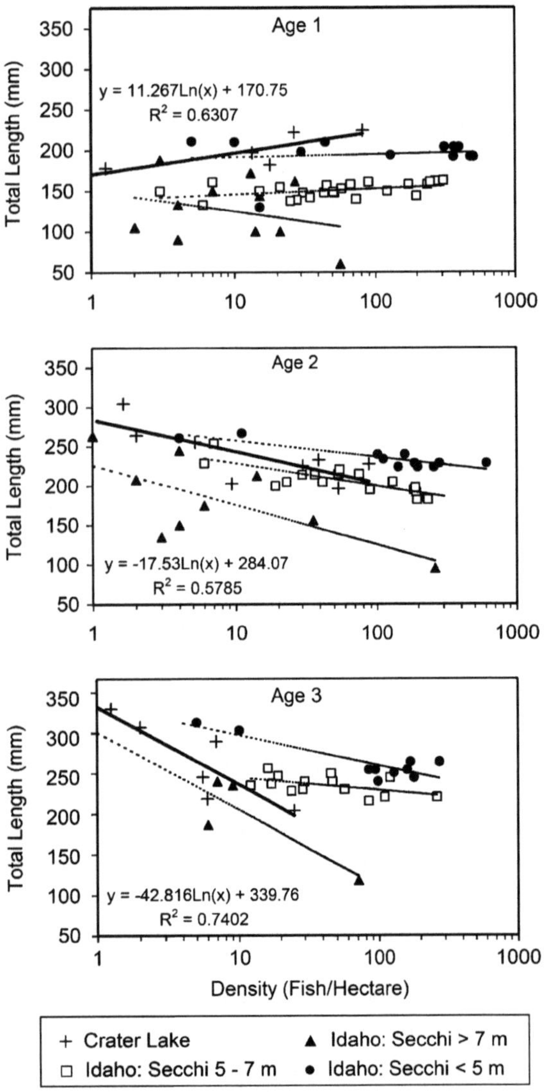

Fig. 12 Relationships between total length at age and density of kokanee in Crater Lake and 10 Idaho lakes and reservoirs (after Rieman & Myers, 1992). Regression equation is for Crater Lake data

Rieman & Myers (1992) found that density dependent growth was less important for age 1 fish at low density than for older fish (Fig. 12), and suggested that a higher density may be necessary to reduce growth of age 1 kokanee in oligotrophic lakes of moderate productivity. We did not document density dependent growth for age 1 kokanee in Crater Lake. However, this may have resulted because our gill nets did not effectively sample kokanee below 150 mm total length.

A reduction in salmonid fecundity (Collins, 1971; Thiede et al., 2002) egg size (Taylor, 1980), and egg to recruitment survival (Thiede et al., 2002) may be associated with decreasing adult length and condition. Thiede et al. (2002) noted a decrease in size, growth, fecundity, and egg-to-recruitment survival, with increasing kokanee abundance in Lake Billy Chinook, OR. They concluded that egg-to-recruitment survival was the most significant factor determining kokanee abundance in Lake Billy Chinook.

Richards et al. (1975) recorded a decrease in the abundance of spawning kokanee that corresponded to the disappearance of *Daphnia* in Lake Tahoe; later, Morgan et al. (1978) concluded that *Daphnia* were necessary for growth and recruitment of Lake Tahoe kokanee. Zooplankton density and size of *Daphnia* were the most important factors in influencing kokanee fry survival in Lake Pend Oreille (Paragamian & Bowles, 1995). Although *Daphnia* were not the dominant food resource for kokanee in Crater Lake, it appears that growth, spawning and recruitment of kokanee may be somewhat dependent on *Daphnia* availability. Maturation, condition, and growth of kokanee were greater when *Daphnia* was abundant, and kokanee captured during spawning fed almost exclusively on *Daphnia*. Alternately, *Heterotrissocladius* abundance may be important to kokanee growth, condition, and maturation, or an independent variable (e.g., nutrient availability) may influence *Daphnia* and *Heterotrissocladius* abundance and kokanee spawning success. Little is known about *Heterotrissocladius* life history, distribution, patterns of abundance, and response to predation by kokanee.

We believe that kokanee in Crater Lake are food-limited because of the lake's low productivity, exacerbated by prey consumption at high fish abundance. At high density, relative to lake productivity, sockeye (Ricker, 1937; Goodlad et al., 1974; Kyle et al., 1988), kokanee (Martinez & Witzius, 1995; Larson et al., 2002), and other salmonids in mountain lakes (Liss et al., 1998; McNaught et al., 1999; Donald et al., 2001) may greatly reduce zooplankton abundance and shift dominant taxa from large-bodied crustaceans to small-bodied crustaceans or rotifers. It is unclear to what extent kokanee regulate zooplankton abundance and community structure in Crater Lake. However, kokanee increased in size, condition, and abundance (1986–1990) when *Daphnia* were present in the lake (1987–1989). At high abundance, kokanee may have drastically reduced *Daphnia* directly through predation and perhaps through predation induced diapause (Slusarczyk, 2001). Continued grazing by kokanee appeared to prohibit *Daphnia's* re-appearance until the kokanee population crashed (1996–1998). *Daphnia* returned to the lake (1996–1999), the kokanee population rebounded (1998–2000), and *Daphnia* subsequently disappeared again (2000–2002). Alternatively, *Daphnia* and kokanee may be responding to an increase in nutrient loading, not detected by the monitoring program but implied by changes in lake surface elevation, or changes in habitat associated with lake level.

Rainbow trout abundance was low, and population demographics were variable in part because sample sizes were small. Differences in the magnitude of change in population demographics reported here between the two fish species was in part related to the general life history strategies of kokanee and rainbow trout. In contrast to kokanee, rainbow trout are relatively long-lived, mature at an older age, and spawn several times in subsequent years. This life history strategy accounts for a more diverse and stable population structure. In addition, the frequency of larger trout in the population and maximum body size may be related to the availability of kokanee as a food resource.

Widespread global introductions of non-native salmonids to lakes and streams, and rapid loss of pristine ecosystems, have led to an increased interest in the impacts of trout and salmon in aquatic environments. Fish introductions can result in system changes at the individual, population, community, and ecosystem levels (Simon & Townsend, 2003). Changes at the Individual level may include changes of within lake prey distribution and behavior (Macan, 1966a, b; Larson & Hoffman, 2002; Tyler et al., 1998). Population changes may include altered prey abundance including local extinction (Reimers, 1979), reduced mean weights of prey, and reduced biomass of benthic prey populations (Post & Cucin, 1984). Direct and indirect interactions among populations may induce trophic cascades (Carpenter et al., 1985), and result in changing pathways and magnitudes of movement of energy and nutrients within the system (Simon & Townsend,

2003). Kokanee and rainbow trout in Crater Lake clearly altered food web dynamics from its prior fishless condition. Salmon and trout fed heavily on pelagic and benthic invertebrate grazers, respectively. A reduction of grazers is expected to result in an increase in abundance and production of algae (Simon & Townsend, 2003). Trout and salmon alter nutrient flux and transport nutrients within and between the pelagic and littoral zones of the lake. Trout introduced into fishless alpine lakes in Canada and the Sierra Nevada Mountains enhanced phosphorus recycling from the littoral to the pelagic zones, resulting in an increase in algal production (Leavitt et al., 1994; Schindler et al., 2001). The magnitude of phosphorus regeneration in the Sierra Nevada lakes was approximately equal to that introduced by atmospheric deposition (Schindler et al., 2001). This transfer of nutrients to the water column can enhance abundance and production of algae already enhanced by the reduction of invertebrate grazers. However, reducing grazers alone may not release primary producers in nutrient limited systems (Parker et al., 2001, in Simon & Townsend, 2003). In addition to the nutrients transferred by fish from the benthos to the water column in Crater Lake (23% diet by weight for kokanee, 50% diet by weight for rainbow), fish also took food items from the lake surface (22% diet by weight for kokanee, 37% diet by weight for rainbow), presumably transferring these nutrients to the water column as well.

The impacts of invasive species are expected to be high in ecosystems like Crater Lake, where vertebrate predators were previously absent, feeding behavior linked previously unlinked ecosystem components (e.g., littoral and pelagic zones), predation impacted native taxa with strong food web relationships and trophic cascade implications (e.g., littoral and pelagic grazers), and physiological, behavioral, or demographic traits lead to high biomass and resistance to natural disturbance (Simon & Townsend, 2003). Crater Lake was judged to be in pristine condition at the end of the 10-year lake study in 1993 with the exception of the consequences of introduced fish (Larson et al., 1993). Although the food habits, distribution, and select population dynamics of the two fish species have been evaluated, the impacts of fish on the system have not been fully evaluated. For this reason, long-term monitoring is needed to evaluate relationships between fish abundance and such variables as plankton density and biomass; primary productivity; nutrient cycling; water clarity; chironomid, other invertebrate, and amphibian density and biomass; and associated littoral processes. And special studies should be designed to evaluate the principal environmental and biological controls of food web components.

Acknowledgments We thank current Crater Lake National Park managers Mac Brock and Charles Lundy for their continued support and contributions to the Crater Lake Long-Term Limnological Monitoring Program, and the field and lab crews over the years for tireless and dedicated work, in particular Ashly Gibson, Scott Stonum, Steve Brady, and Abigail Buktenica. Robert Wisseman identified and enumerated fish stomach contents. Comments by Patrick J. Martinez, Bruce E. Rieman, and two anonymous reviewers are gratefully acknowledged.

References

Ali, M. A., 1959. The ocular structure, retinomotor and photobehavioral responses of juvenile pacific salmon. Canadian Journal of Zoology 37: 965–996.

Anderson, R. O. & S. J. Gsutreuter, 1983. Length, weight, and associated structural indices. Chapter 15. In Nielson L., & D. Johnson (eds), Fisheries Techniques. American Fisheries Society, Bethesda, MD: 283–300.

Bacon, C. R., J. V. Gardner, L. A. Mayer, M. W. Buktenica, P. Dartnell, D. W. Ramsey & J. E. Robinson, 2002. Morphology, volcanism, and mass wasting in Crater Lake, Oregon. Geological Society of America Bulletin 114: 675–692.

Baldwin, C. M., D. A. Beauchamp & J. J. Van Tassell, 2000. Bioenergetic assessment of temporal food supply and consumption demand by salmonids in the Strawberry Reservoir food web. Transactions of the American Fisheries Society 129: 429–450.

Beacham, T. D. & J. G. McDonald, 1982. Some aspects of food and growth of fish species in Babine Lake, British Columbia. Canadian Technical Report of Fisheries and Aquatic Sciences 1057: iv+23 p.

Beauchamp, D. A., 1994. Spatial and temporal dynamics of piscivory: implications for food web stability and the transparency of Lake Washington. Lake and Reservoir Management 9: 151–154.

Beauchamp, D. A., M. G. LaRiviere & G. L. Thomas, 1995. Evaluation of competition and predation as limits to juvenile kokanee and sockeye salmon production in Lake Ozette, Washington. North American Journal of Fisheries Management 15: 193–207.

Beauchamp, D. A., C. Luecke, W. A. Wurtsbaugh, H. G. Gross, P. E. Budy, S. Spaulding, R. Dillenger & C. P. Gubala, 1997. Hydroacoustic assessment of abundance and diel distribution of sockeye salmon and kokanee in the Sawtooth Valley Lakes, Idaho. North American Journal of Fisheries Management 17: 253–267.

Bevelhimer, M. S. & S. M. Adams, 1993. A bioenergetics analysis of diel vertical migration by kokanee salmon, *Oncorhynchus nerka*. Canadian Journal of Fisheries and Aquatic Sciences 50: 2336–2349.

Bond, C. E., 1979. Biology of Fishes. Saunders College Publishing, Philadelphia. 514 pp.

Brett, J. R., 1971. Energetic responses of salmon to temperature. A study of some thermal relations in the physiology and freshwater ecology of sockeye salmon (*Oncorhynchus nerka*). American Zoolologist 11: 99–113.

Brode, J. S., 1935. Food habits of Crater Lake fish. Crater Lake Nature Notes 8: 11–13.

Brode, J. S., 1937. Food habits of Crater Lake Fish. Unpublished Preliminary Report. National Park Service, Crater Lake, Oregon. 7 pp.

Brode, J. S., 1938. The denizens of Crater Lake. Northwest Science 12: 50–57.

Buktenica, M. W., 1989. Ecology of kokanee salmon and rainbow trout in Crater Lake, a deep ultraoligotrophic caldera lake (Oregon). MS Thesis, Oregon State University, Corvallis. 89 pp.

Buktenica M. W. & G. L. Larson, 1990. Ecology of kokanee salmon and rainbow trout in Crater Lake. In Drake E. T., G. L. Larson J. Dymond R. & Collier (eds.), Crater Lake: An Ecosystem Study. Pacific Division of the American Association for the Advancement of Science. California Academy of Sciences, Golden Gate Park, San Francisco, California: 185–195.

Buktenica, M. W. & G. L. Larson, 1996. Ecology of kokanee salmon and rainbow trout in Crater Lake, Oregon. Lake and Reservoir Management 12: 298–310.

Buktenica, M. W., 2001. Journey to the bottom of the lake—results and personal observations from studies conducted by submarine in Crater Lake, 1988–1989. The Journal of the Shaw Historical Library 15, 2001. ISSN 0889-0277.

Burgner, R. L., 1964. Factors influencing production of sockeye salmon (*Oncorhynchus nerka*) in lake of Southwestern Alaska. Verhandlungen Internationale vereinigung fur theoretische und angewandte Limnologie 15: 504–513.

Busacker G. P., I. R. Adelman & E. M. Goolish, 1990. Growth. In Schreck C. G. & P. B. Moyle (eds.), Methods for Fish Biology. American Fisheries Society, Bethesda, Maryland: 363–387.

Carlander, K. D., 1969. Handbook of Freshwater Fishery Biology. Volume 1. Life History Data on Freshwater Fishes of the United States and Canada, Exclusive of the Perciformes. Iowa State University Press, Ames, Iowa: 752 pp.

Carpenter, S. R., J. F. Kitchell & J. R. Hodgson, 1985. Cascading trophic interactions and lake productivity. BioScience 35: 634–639.

Clark, C. W. & E. A. Levy, 1988. Diel vertical migrations by juvenile sockeye salmon and the antipredation window. American Naturalist 131: 271–290.

Collins, J. J., 1971. Introduction of kokanee salmon (*Oncorhynchus nerka*) into Lake Huron. Journal of Fisheries Research Board of Canada 28: 1857–1871.

Cordone, A. J., S. J. Nicola, P. H. Baker & T. C. Frantz, 1971. The kokanee salmon in Lake Tahoe. California Fish and Game 57: 28–43.

Crawford, G. B. & R. W. Collier, 2007. Long-term observations of deepwater renewal in Crater Lake, Oregon. Hydrobiologia 574: 47–68.

Dawson, J. J., 1972. Determination of seasonal distribution of juvenile sockeye salmon in Lake Washington by means of acoustics. MS Thesis, University of Washington.

Dobble, B. D. & D. M. Eggers, 1977. Diel Feeding Chronology, Rate of Gastric Evacuation, Daily Ration, and Prey Selectivity in Lake Washington Juvenile Sockeye Salmon (*Oncorhynchus nerka*). Contribution No. 299 from the Coniferous Forest Biome, and Contribution No. 476 from College of Fisheries, University of Washington, Seattle.

Donald, D. B., F. D. Vinebrooke, R. S. Anderson, J. Syrgiannis & M. D. Graham, 2001. Recovery of zooplankton assemblages in mountain lakes from the effects of introduced sport fish. Canadian Journal of Fisheries and Aquatic Sciences 58: 1822–1830.

Dymond, J., R. Collier & J. McManus, 1996. Unbalanced particle flux budgets in Crater Lake, Oregon: implications for edge effects and sediment focusing in lakes. Limnology and Oceanography 41: 732–743.

Eggers, K. M., 1978. Limnetic feeding behavior of juvenile sockeye salmon in Lake Washington and predator avoidance. Limnology and Oceanography 23: 1114–1125.

Finnell, L. M. & E. B. Reed, 1969. The diel vertical movements of kokanee salmon, *Oncorhynchus nerka*, in Granby Reservoir, Colorado. Transactions of the American Fisheries Society 98: 245–252.

Gerking, S., 1994. Feeding Ecology of Fish. Academic Press, San Diego, CA.

Goodlad, J. C., T. W. Gjernes & E. L. Brannon, 1974. Factors affecting sockeye salmon (*Oncorhynchus nerka*) growth in four lakes of the Fraser River system. Journal of the Fisheries Research Board of Canada 31: 871–892.

Hanzel, D. A., 1984. Measure annual trends in recruitment and migration of kokanee populations and identify major factors affecting trends. Job Completion

Report, Project No. F-33-R-18, Job No. I-b, Fisheries Division, Montana Department of Fish, Wildlife, and Parks.

Hargreaves, B. R., S. F. Girdner, M. W. Buktenica, R. W. Collier, E. Urbach & G. L. Larson, 2007. Ultraviolet radiation and bio-optics in Crater Lake, Oregon. Hydrobiologia 574: 107–140.

Hasler, A. D. & D. S. Farner, 1942. Fisheries investigations in Crater Lake, Oregon, 1937–1940. Journal of Wildlife Management 6: 319–327.

Hoag, S. H., 1972. The relationship between the summer food of juvenile sockeye salmon, Oncorhynchus nerka, and the standing stock of zooplankton in Iliamna Lake, Alaska. NOAA Fishery Bulletin 70: 355–362.

Hubbard A. C., 1933. Fact and fancy about Crater Lake fish. Unpublished report. National Park Service, Crater Lake, Oregon. 50 pp.

Jearld, A., Jr., 1983. Age determination. pp. 301–324, chapter 16. In Nielson L. & D. Johnson (eds), Fisheries Techniques. American Fisheries Society, Bethesda, MD.

Johnson, W. E., 1965. On mechanisms of self-regulation of population abundance in *Oncorhynchus nerka*. Mitteilungen Internationale Vereinigung fur theoretische und angewandte Limnologie Verhandlungen 13: 66–87.

Johnston, N. T., 1990. A comparison of the growth of vertically-migrating and nonmigrating kokanee (*Oncorhynchus nerka*) fry. Canadian Journal of Fisheries and Aquatic Sciences 47: 486–491.

Jongman, R. H. G., C. J. F. ter Braak & O. F. R. van Tongeren, 1987. Data Analysis in Community and Landscape Ecology. Centre for Agricultural Publishing and Documentation (Pudoc), Wageningen. 299 pp.

Kyle, G. B., J. P. Koenings & B. M. Barett, 1988. Density-dependent trophic level responses to an introduced run of sockeye salmon (*Oncorhynchus nerka*) at Frazer Lake, Kodiak Island, Alaska. Canadian Journal of Fisheries and Aquatic Sciences 45: 856–867.

Larson, G. L., R. L. Hoffman, B. R. Hargreaves & R. W. Collier, 2007a. Predicting Secchi disk depth from average beam attenuation in a deep, ultra-clear lake. Hydrobiologia (this issue).

Larson, G. L., C. D. McIntire, M. W. Buktenica, S. F. Girdner & R. E. Truitt, 2007b. Distribution and Abundance of Zooplankton Populations in Crater Lake, Oregon (USA). Hydrobiologia (this issue).

Larson, G. L. & R. L. Hoffman, 2002. Abundances of Northwestern Salamander larvae in Montane Lakes with and without fish, Mount Rainier National Park, Washington. Northwest Science 76(No. 1): 35–40.

Larson, G. L., R. L. Hoffman & C. D. McIntire, 2002. Persistence of an unusual pelagic zooplankton assemblage in a clear, mountain lake. Hydrobiologia 468: 163–170.

Larson, G. L. & M. W. Buktenica, 1998. Variability of Secchi disk readings in an exceptionally clear and deep caldera lake. Archiv Fuer Hydrobiologie 141: 377–388.

Larson, G. L., 1996. Development of a 10-year limnological study of Crater Lake, Crater Lake National Park, Oregon, USA. Lake and Reservoir Management 12: 221–229.

Larson, G. L., C. D. McIntire M. Hurley & M. W. Buktenica, 1996. Temperature, water chemistry, and optical properties of Crater Lake. Lake and Reservoir Management 12: 230–247.

Larson, G. L., C. D. McIntire, R. Collier & M. W. Buktenica, 1993. Long-term monitoring program. Pages 711–722. In Larson, G. L., J. Dymond & R. Collier (eds), Crater Lake, Limnological Studies final Report. Natl. Park Serv. Tech. Rep. NPS/PNROSU/NRTR-93-03.

Leathe, S. A. & P. J. Graham, 1981. Flathead Lake fish food habits study. Montana Department of Fish Wildlife and Parks Project Report. 93 pp.

Leavitt, P. R., D. E. Schindler, A. J. Paul, A. K. Hardie & D. W. Schindler, 1994. Fossil pigment records of phytoplankton in trout-stocked alpine lakes. Canadian Journal of Fisheries and Aquatic Sciences 51: 2411–2423.

Levy, D. A. & C. C. Wood, 1992. Review of proposed mechanisms for sockeye salmon population cycles in the Fraser River. Bulletin of Mathematical Biology 54: 241–261.

Levy, D. A., 1987. Review of the ecological significance of diel vertical migrations by juvenile sockeye salmon (*Oncorhynchus nerka*), pp. 44–52. In Smith, H. D., L. Margolis & C. C. Wood (eds), Sockeye Salmon (*Oncorhynchus nerka*) Population Biology and Future Management. Canadian Special Publication of Fisheries and Aquatic Sciences 96.

Levy, D. A., 1990. Sensory Mechanism and selective advantage for diel vertical migration in juvenile sockeye salmon, *Oncorhynchus nerka*. Canadian Journal of Fisheries and Aquatic Sciences 47: 1796–1802.

Levy, D. A., 1991. Acoustic analysis of diel vertical migration behavior of *Mysis relicta* and kokanee (*Oncorhynchus nerka*) within Okanagan Lake, British Columbia. Canadian Journal of Fisheries and Aquatic Sciences 48: 67–72.

Lewis, S. L., 1971. Life history and ecology of kokanee in Odell Lake. Oregon State Game Commission, Research Division, Job Progress Report. F-71-R-6. 48p.

Liss, W. J., G. L. Larson, E. A. Deimling, L. M. Ganio, R. L. Hoffman & G. A. Lomnicky, 1998. Factors influencing the distribution and abundance of diaptomid copepods in high-elevation lakes in the Pacific Northwest, U.S.A. Hydrobiologia 379: 63–75.

Love, R. H., 1971. Measurements of fish target strength: a review. Fisheries Bulletin 69: 703–715.

Luecke, C. & W. A. Wurtsbaugh, 1993. Effects of moonlight and daylight on hydroacoustic estimates of pelagic fish abundance. Transactions of the American Fisheries Society 122: 112–120.

Macan, T. T., 1966a. The influence of predation on the fauna of a moorland fish pond. Archiv fuer Hydrobiologie 61: 432–452.

Macan, T. T., 1966b. Predation by *Salmo trutta* in a moorland fish pond. Verhandlungen der Internationalen Vereinigung fur Theoretische und Angewandte Limnologie 16: 1091–1087.

Marino, D. A., 1987. Dual-beam hydroacoustic assessment of kokanee salmon spatial and temporal distribution and abundance in three Pacific Northwest lakes. Masters thesis, University of Washington.

Martinez, P. J. & W. J. Wiltzius, 1995. Some factors affecting a hatchery-sustained kokanee population in a fluctuating Colorado reservoir. North American Journal of Fisheries Management 15: 220–228.

McDonald, J., 1973. Diel vertical movements and feeding habits of underyearling sockeye salmon (*Oncorhynchus nerka*), at Babine Lake, B.C. Fisheries Research Board of Canada Technical Report No. 378. 19 pp.

McIntire, C. D., G. L. Larson & R. E. Truitt, 2007. Seasonal and interannual variability in the taxonomic composition and production dynamics of phytoplankton assemblages in Crater Lake, Oregon. Hydrobiologia 574: 179–204.

McIntire, C. D., H. K. Phinney, G. L. Larson & M. W. Buktenica, 1994. Vertical distribution of a deep-water moss and associated epiphytes in Crater Lake, Oregon. Northwest Science 68: 11–21.

McNaught, A. S., D. W. Schindler, B. R. Parker, A. J. Paul, R. S. Anderson, D. B. Donald & M. Agbeti, 1999. Restoration of the food web of an alpine lake following fish stocking. Limnology and Oceanography 44: 127–136.

Morgan, M. D., S. T. Threlkeld & C. R. Goldman, 1978. Impact of the introduction of kokanee (*Oncorhynchus nerka*) and opossum shrimp (*Mysis relicta*) on an alpine lake. Journal of the Fisheries Research Board of Canada 35: 1572–1579.

Myers, R. A., G. Mertz, J. M. Bridson & M. J. Bradford, 1998. Simple dynamics underlie sockeye salmon (*Oncorhynchus nerka*) cycles. Canadian Journal of Fisheries and Aquatic Sciences 55: 2355–2364.

Narver, D. W., 1970. Diel vertical movements and feeding of underyearling sockeye salmon and the limnetic zooplankton in Babine Lake, British Columbia. Journal of the Fisheries Research Board Canada 27: 281–316.

Northcote, T. G. & H. W. Lorz, 1966. Seasonal diel changes in food of adult kokanee *(Oncorhynchus nerka)* in Nicola Lake, British Columbia. Journal of the Fisheries Research Board of Canada 23: 1259–1263.

Paragamian, V. L. & E. C. Bowles, 1995. Factors affecting survival of kokanees stocked in Lake Pend Oreille, Idaho. North American Journal of Fisheries Management 15: 208–219.

Parker, B. R., D. W. Schindler, D. B. Donald & R. S. Anderson, 2001. The effects of stocking and removal of a nonnative salmonid on the plankton of an alpine lake. Ecosystems 4: 334–345.

Patten, B. G. & R. B. Thompson, 1957. Food studies of a small sample of rainbow trout *(Salmo gairdneri)* and kokanee salmon *(Oncorhynchus nerka)* from Crater Lake, Oregon. Unpublished report. National Park Service, Crater Lake, Oregon. 14 pp.

Post, J. R. & D. Cucin, 1984. Changes in the benthic community of a small Precambrian lake following the introduction of yellow perch, *Pera flavescens*. Canadian Journal of Fisheries and Aquatic Sciences 41: 1496–1501.

Redmond, K. T., 2007. Evaporation and the hydrologic budget of Crater Lake, Oregon. Hydrobiologia 574: 29–46.

Richards, R. C., C. R. Goldman, T. C. Frantz & R. Wickwire, 1975. Where have all the *Daphnia* gone? The decline of a major cladoceran in Lake Tahoe, California-Nevada. Verhandlungen Internationale vereinigung fur theoretische und angewandte Limnologie 19: 835–842.

Ricker, W. E., 1975. Computation and Interpretation of Biological Statistics of Fish Populations. Bulletin 191. Bulletin of the Fisheries Research Board of Canada, Department of the Environment Fisheries and Marine Service, Ottawa, Canada.

Ricker, W. E., 1937. The food and the food supply of sockeye salmon (*Oncorhynchus nerka* Walbaum) in Cultus Lake, British Columbia. Journal of the Biological Board of Canada 3: 450–468.

Ricker, W. E., 1997. Cycles of abundance among Fraser River sockeye salmon (*Oncorhynchus nerka*). Canadian Journal of Fisheries and Aquatic Sciences 54: 950–968.

Rieman, B. E. & D. L. Myers, 1992. Influence of fish density and relative productivity on growth of kokanee in ten oligotrophic lakes and reservoirs in Idaho. Transactions of the American Fisheries Society 121: 178–191.

Rieman, B. E. & B. Bowler, 1980. Kokanee trophic ecology and limnology in Pend Oreille Lake. Fisheries Bulletin No. 1. Idaho Department of Fish and Game. 27 pp.

Reimers, N., 1979. A history of a stunted brook trout population in an alpine lake: a life span of 24 years. California Fish and Game 65: 106–215.

Schmidt D. C., J. P. Koenings & G. B. Kyle, 1994. Predator-induced changes in copepods vertical migration: explanations for decreased overwinter survival of sockeye salmon. In Stouder D. J., K. L. Fresh & R. J. Feller (eds.), Theory and Application in Fish Feeding Ecology. University of South Carolina Press, Charleston: 188–209.

Scott, W. B. & E. J. Crossman, 1973. Freshwater fishes of Canada. Fisheries Research Board of Canada. Bulletin 184. 966 pp.

Simon, K. S. & C. R. Townsend, 2003. Impacts of freshwater invaders at different levels of ecological organization, with emphasis on salmonids and ecosystem consequences. Freshwater Biology 48: 982–994.

Schindler, D. R., K. A. Knapp & P. R. Leavitt, 2001. Alteration of nutrient cycles and algal production resulting from fish introductions into mountain lakes. Ecosystems 4: 308–321.

Slusarczyk, M., 2001. Food threshold for diapause in *Daphnia* under the threat of fish predation. Ecology 82: 1089–1096.

Steinhart, G. B. & W. A. Wurtsbaugh, 1999. Under-ice diel vertical migrations of *Oncorhynchus nerka* and their zooplankton prey. Canadian Journal of Fisheries and Aquatic Sciences 56: 152–161.

Stockwell, J. D. & B. M. Johnson, 1999. Field evaluation of a bioenergetics-based foraging model for kokanee (*Oncorhynchus nerka*). Canadian Journal of Fisheries and Aquatic Sciences 56: 140–151.

Taylor, S. G., 1980. Marine survival of pink salmon fry from early and late spawners. Transactions of the American Fisheries Society 109: 79–82.

Ter Braak, C. J. F., 1987. CANOCO—a FORTRAN Program for Canonical Community Ordination by [partial] [detrended] [canonical] Correspondence Analysis, Principal Components Analysis and Redundancy Analysis (version 2.0). TND Institute of Applied computer Science, Wageningen, The Netherlands. 95 pp. 1987. All ordinations were performed by the computer program CANOCO.

Thiede, G. P., J. C. Kern, M. K. Weldon, A. R. Dale, S. L. Thiesfeld & M. A. Buckman, 2002. Lake Billy Chinook sockeye salmon and kokanee research study 1996–2000. Draft Project Completion Report. Pelton-Round Butte Hydroelectric Project. For Portland General Electric Company. 92 pp.

Tyler, T. J., W. J. Liss R. L. Hoffman & L. M. Ganio, 1998. Experimental analysis of trout effects on survival, growth, and habitat use of two species of ambystomatid salamanders. Journal of Herpetology 32: 345–349.

Whiting, M. C. & C. D. McInitire, 1985. An investigation of distributional patterns in the diatom flora of Netarts Bay, Oregon, by correspondence analysis. Journal of Phycology 21: 655–661.

Wurtsbaugh, W. A. & D. Neverman, 1988. Post-feeding thermotaxis and daily vertical migration in a larval fish. Nature 333: 846–848.

Zaret, T. M., 1980. Predation and Freshwater Communities. Yale University Press, New Haven, Connecticut.

Hydrobiologia (2007) 574:265–280
DOI 10.1007/s10750-006-2615-5

CRATER LAKE, OREGON

Seasonal nutrient and plankton dynamics in a physical-biological model of Crater Lake

Katja Fennel · Robert Collier · Gary Larson · Greg Crawford · Emmanuel Boss

© Springer Science+Business Media B.V. 2007

Abstract A coupled 1D physical-biological model of Crater Lake is presented. The model simulates the seasonal evolution of two functional phytoplankton groups, total chlorophyll, and zooplankton in good quantitative agreement with observations from a 10-year monitoring study. During the stratified period in summer and early fall the model displays a marked vertical structure: the phytoplankton biomass of the functional group 1, which represents diatoms and dinoflagellates, has its highest concentration in the upper 40 m; the phytoplankton biomass of group 2, which represents chlorophyta, chrysophyta, cryptomonads and cyanobacteria, has its highest concentrations between 50 and 80 m, and phytoplankton chlorophyll has its maximum at 120 m depth. A similar vertical structure is a reoccurring feature in the available data. In the model the key process allowing a vertical separation between biomass and chlorophyll is photoacclimation. Vertical light attenuation (i.e., water clarity) and the physiological ability of phytoplankton to increase their cellular chlorophyll-to-biomass ratio are ultimately determining the location of the chlorophyll maximum. The location of the particle maxima on the other hand is determined by the balance between growth and losses and occurs where growth and losses equal.The vertical particle flux simulated by our model agrees well with flux measurements from a sediment trap. This motivated us to revisit a previously published study by Dymond et al. (1996). Dymond et al. used a box model to estimate the vertical particle flux and found a discrepancy by a factor 2.5–10 between their model-derived flux and measured fluxes from a sediment trap. Their box model neglected the exchange flux of dissolved and suspended organic matter, which, as our model and available data suggests is significant for the

Guest Editors: Gary L. Larson, Robert Collier, and Mark W. Buktenica
Long-term Limnological Research and Monitoring at Crater Lake, Oregon

K. Fennel (✉)
Institute of Marine and Coastal Sciences and Department of Geological Science, Rutgers University, New Brunswick 08901, New Jersey, USA
e-mail: kfennel@marine.rutgers.edu

R. Collier
College of Oceanic and Atmospheric Sciences, Oregon State University, Corvallis 97331, Oregon, USA

G. Larson
USGS Forest and Rangeland Ecosystems Center, Forest Science Laboratory Oregon State University, Corvallis 97331, Oregon, USA

G. Crawford
Department of Oceanography, Humboldt State University, Arcata 95521, California, USA

E. Boss
School of Marine Sciences, University of Maine, Orono 04473, Maine, USA

vertical exchange of nitrogen. Adjustment of Dymond et al.'s assumptions to account for dissolved and suspended nitrogen yields a flux estimate that is consistent with sediment trap measurements and our model.

Keywords Physical-biological model · Deep chlorophyll maximum · Photoacclimation · Crater Lake

Introduction

Crater Lake is an ultra-oligotrophic, isolated caldera lake in the Cascade Mountains, Oregon. With a maximum depth of 590 m it is the deepest lake in the US and one of the clearest bodies of water on Earth. Crater Lake is surrounded by steep caldera walls, has a very small watershed and only small inputs of allochthonous matter. Most of the water entering the lake is direct precipitation onto its surface, which makes up 78% of its watershed. The remainder enters as run off from the caldera walls and as hydrothermal influx from the bottom (Collier et al. 1991). Despite its great depth, Crater Lake is relatively well-mixed. The upper portion of the lake (upper 200 m) is homogenized twice a year in early winter and late spring by free convection and wind-mixing (McManus et al. 1993). Partial ventilation of the deep lake occurs during these mixing periods when the upper water column is nearly isothermal (McManus et al., 1993) and during sporadic deep-mixing events (Crawford & Collier, 1997). Independent estimates of the lake ventilation time agree that the lake is ventilated on timescales of 1 to 5 years (Simpson, 1970; McManus et al., 1993; Crawford & Collier, 1997). Depletion of inorganic nitrogen species in the upper 200 m of the lake in summer implies nitrogen-limitation of phytoplankton growth, although bioassays suggest co-limitation by trace metals (Lane & Goldmann, 1984). Concentrations of inorganic phosphorus and silicic acid are relatively high and nearly uniform throughout the water column (Larson et al., 1996a). Vertical mixing is the most important source of new nutrients for primary production in the euphotic zone (Dymond et al., 1996). Comparisons of upward nutrient fluxes based on box-model considerations and measurements of vertical particle fluxes from sediment traps reveal large discrepancies, i.e. the particle flux falls short of balancing the estimated upwelled nitrogen (Dymond et al., 1996).

The populations of microorganisms show distinct vertical maxima during stratified conditions (McIntire et al., 1996; Larson et al., 1996b; Urbach et al., 2001). In July and August, diatoms and dinoflagellates reach their maximum biomass in the upper 20 m of the water column, isolated from below by a pronounced thermocline. Chlorophyta, chrysophyta and cryptophyta dominate below the thermocline with biomass maxima between 80 and 100 m (Fig. 1) coincident with the vertical maximum of primary production (McIntire et al., 1996). The chlorophyll concentration has a deep maximum at a depth of 120 m, separated from the phytoplankton biomass maximum by about 50 m (Fig. 2). The vertical separation of the maxima in phytoplankton biomass (in units of carbon or nitrogen) and chlorophyll is mainly due to photoacclimation (Fennel & Boss, 2003) which results in dramatic vertical changes in the chlorophyll to biomass ratio (Fig. 2) and implies that the chlorophyll concentration does not represent phytoplankton biomass.

We developed a coupled physical-biological model of Crater Lake that captures the processes thought to determine nutrient cycling and phytoplankton production at first-order in a simplified description of the food-web. We consider the simplicity as valuable since it allows us to keep the number of poorly known model parameters at a minimum, but can elucidate dependencies and regulating mechanisms not easily seen in the observations alone. The model allows us to test hypotheses about the factors determining spatial and temporal patterns in species distribution, chlorophyll concentration, and primary production in relation to physical circulation processes. Here we present results of a 1-year model simulation in comparison with measurements obtained during a 10-year monitoring study (Larson, this issue), integrating the available data in a biomass-based framework. We discuss factors leading to the observed vertical distribution of phytoplankton populations and compare model-simulated vertical fluxes with particle flux measurements

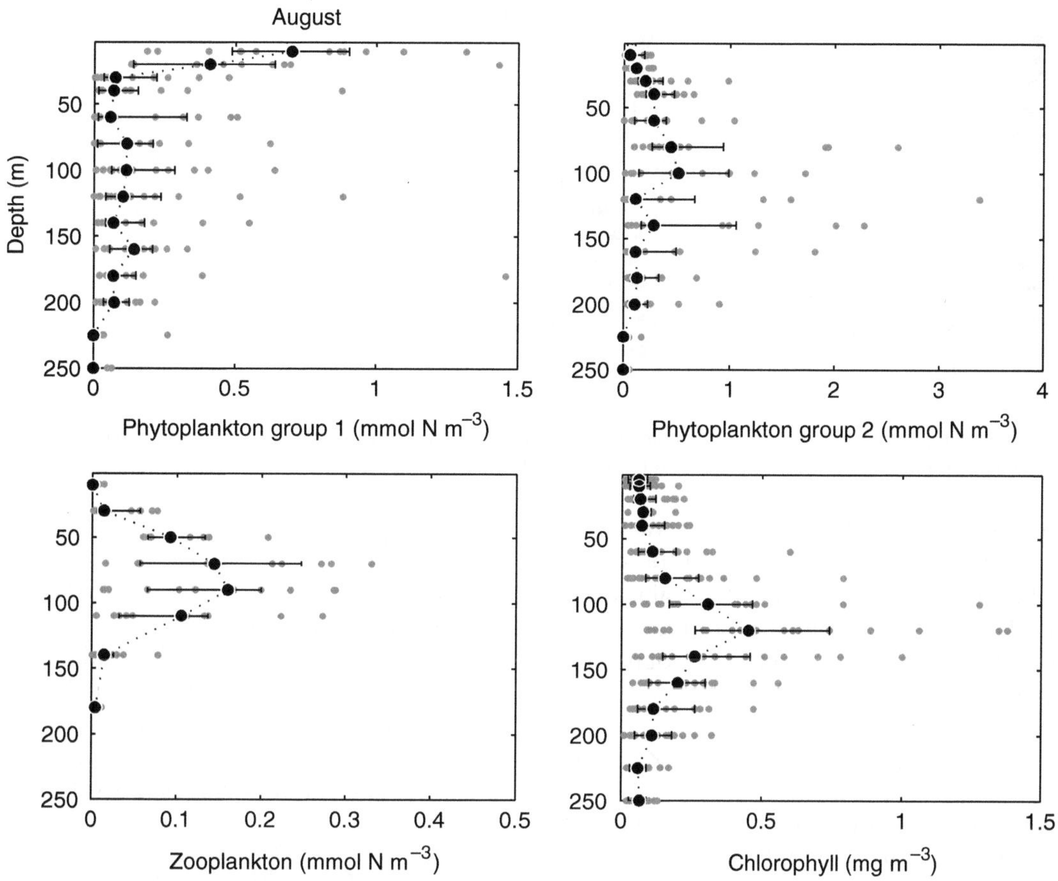

Fig. 1 Available August data from 1989 to 2001 (gray bullets) and median with 25- and 75-percentiles (black bullets). Diatoms and dinoflagellates comprise phytoplankton group 1. Chlorophyta, chrysophyta, cryptomonads and cyanobacteria comprise phytoplankton group 2

and box-model estimates (Dymond et al., 1996) reconciling a previously reported mismatch.

Materials & methods

The biological model

We formulated a relatively simple biological model containing the following 7 state variables: dissolved nitrogen N, two phytoplankton groups P_1 and P_2, one group of zooplankton Z, one detrital pool D, and the chlorophyll concentrations C_1 and C_2 of P_1 and P_2, respectively (Fig. 3). All variables are expressed in units of mmol N m^{-3} except the chlorophyll variables, which are in mg chl m^{-3}. Nitrogen was chosen as the nutrient currency since Crater Lake is considered nitrogen-limited (Lane & Goldman, 1984; Larsen et al., 1996a). Note that the nutrient variable N also includes dissolved organic nitrogen in addition to inorganic nitrogen. Technically detritus includes both, particulate and dissolved pools of organic nitrogen. However, the detritus variable D in our model is subject to vertical sinking. Hence, the dissolved organic nitrogen is more realistically treated as part of the nutrient pool N. The phytoplankton group P_1 represents diatoms and dinoflagellates, i.e. the phytoplankton dominating in the upper 20 m of the water column. P_2 comprises the remaining phytoplankton divisions (chlorophyta, chrysophyta, cryptomonads and cyanobacteria). The chlorophyll concentrations C_1 and C_2 of P_1 and P_2 are included as dynamical variables, to allow an explicit inclusion of photoacclimation.

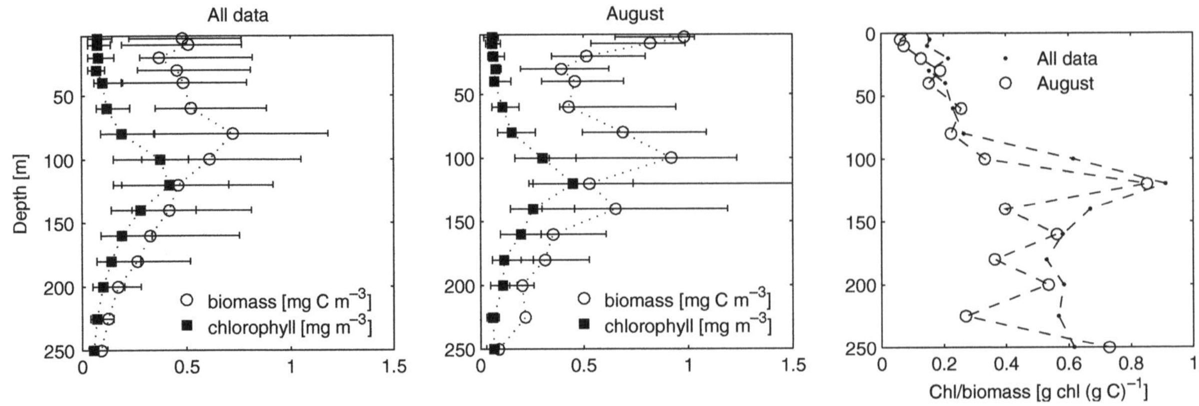

Fig. 2 Median and 25- and 75-percentiles of chlorophyll and phytoplankton biomass. All data (left panel), August data (middle panel), and chlorophyll to biomass ratio (right panel) are shown

The dynamics of the biological state variables are determined by the following set of equations. The corresponding parameter values are given in Table 1. The biological sources minus sinks (sms) of the phytoplankton groups are defined as

$$sms(P_i) = \mu_i \cdot P_i - (L_{iN} + L_{iD}) \cdot P_i - g_i \cdot Z, \quad (1)$$

with i = 1,2, where μ_i is the phytoplankton growth rate, L_{iN} and L_{iD} are phytoplankton loss rates representing fluxes due to respiration and natural mortality, and g_i is the zooplankton grazing rate.

The phytoplankton growth rates μ_i depend on the nutrient concentration N, following the Michaelis–Menten response, the photosynthetically available radiation E and the chlorophyll concentration C_i as

$$\mu_i^m = \mu_i^{\max}\left(1 - \exp\left(-\frac{\alpha_i E\, C_i}{\mu_i^{\max} P_i}\right)\right), \quad (2)$$

$$\mu_i = \mu_i^m \frac{N}{k_{iN} + N} \text{ with } i = 1,2. \quad (3)$$

Here $\mu_i^{\max}$ is the maximum phytoplankton growth rate, α_i is the chlorophyll-specific initial slope of the photosynthesis-irradiance curve, and k_{iN} is the half-saturation concentration for nutrient uptake. Note that we assume a tight coupling of nutrient uptake, growth and photosynthesis.

E represents the light that is available for photosynthesis, which is determined for a given depth as

$$E(z) = E_0 \cdot \mathrm{par} \cdot \exp\{-z(k_W + k_{Chl}\mathrm{chl}|_z)\}, \quad (4)$$

where z is the water depth, E_0 is the incoming shortwave radiation just below the lake surface, and *par* is the fraction of this radiation that is active in photosynthesis and assumed to be equal to 43%. k_W = 0.04 m^{-1} and k_{Chl} = 0.03 m^{-1} (mg chl m^{-3})$^{-1}$ are the light attenuation coefficients for water and chlorophyll, respectively. chl$|_z$ is the

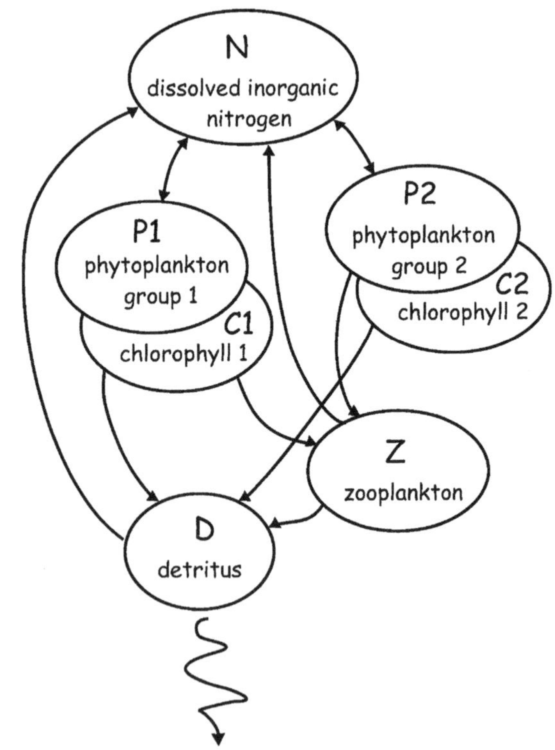

Fig. 3 Schematic of the biological model component

Table 1 Biological model parameters used in this study and range of published parameter values. N.D. denotes non-dimensional parameter

Symbol	Description	Value	Unit	Range
μ_1^{max}	maximum growth rate of P_1	1.50	d^{-1}	0.62[a]–3.0[b]
μ_2^{max}	maximum growth rate of P_2	0.85	d^{-1}	0.62[a]–3.0[b]
k_{1N}	half-saturation concentration for nutrient-uptake of P_1	0.01	mmol N m^{-3}	0.0005–3.86[c]
k_{2N}	half-saturation concentration for nutrient-uptake of P_2	0.01	mmol N m^{-3}	0.0005–3.86[c]
α_1	chlorophyll-specific initial slope of the photosynthesis-irradiance curve of P_1	0.05	mol N (g chl)$^{-1}$ (W m^{-2})$^{-1}$	0.007–0.13[d]
α_2	chlorophyll-specific initial slope of the photosynthesis-irradiance curve of P_2	0.7	mol N (g chl)$^{-1}$ (W m^{-2})$^{-1}$	0.007–0.13[d]
$\Theta^{max}{}_1$	maximum ratio of chlorophyll to biomass of P_1	0.5	g chl (mol N)$^{-1}$	maximum 5.6[e]
$\Theta^{max}{}_2$	maximum ratio of chlorophyll to biomass of P_2	6.0	g chl (mol N)$^{-1}$	maximum 5.6[e]
L_{1N}	loss rate from P_1 to N	0.13	d^{-1}	0.02[i]–0.1[g]
L_{1D}	loss rate from P_1 to D	0.06	d^{-1}	0.05–0.2[f]
L_{2N}	loss rate from P_2 to N	0.15	d^{-1}	0.02[i]–0.1[g]
L_{2D}	loss rate from P_2 to D	0.15	d^{-1}	0.05–0.2[f]
g_{max}	maximum grazing rate	0.25	d^{-1}	0.5[g]–1.0[h]
K_3	half-saturation concentration for grazing	0.1	mmol N m^{-3}	0.75[i]–1.89[j]
P_1	grazing preference for P_1	0.2	N.D.	
P_2	grazing preference for P_2	0.8	N.D.	
L_{ZN}	loss rate from Z to N	0.1	d^{-1}	0.04[g]–15.0[j]
L_{ZD}	loss rate from Z to D	0.1	d^{-1}	0.04[g]–15.0[j]
w_D	sinking rate of D	0.5	m d^{-1}	0.009[k]–10[l]
r	remineralization rate for D	0.3	d^{-1}	0.05[k]–0.18[m]

[a] Taylor, 1988
[b] Anderson et al., 1987
[c] Moloney & Field, 1991
[d] Geider et al., 1997
[e] Geider et al., 1998
[f] Taylor et al., 1991
[g] Wroblewski, 1989
[h] Fasham, 1995
[i] Palmer & Totterdell, 2001
[j] Ross et al., 1994
[k] Moskilde, 1996
[l] Fasham, 1993
[m] Jones & Henderson, 1986

mean chlorophyll concentration above the actual depth z and is determined by integrating over C_1 and C_2 as

$$\mathrm{chl}|_z = \frac{1}{z}\int_0^z (C_1 + C_2)\mathrm{d}z'. \tag{5}$$

The chlorophyll concentrations are determined following the photoacclimation model of Geider et al. (1996, 1997)

$$sms(C_i) = \rho_i \mu_i P_i - L_{iD} C_i - g_i Z \frac{C_i}{P_i}, \tag{6}$$

with

$$\rho_i = \Theta_i^{max}\left(\frac{\mu_i P_i}{\alpha_i E C_i}\right). \tag{7}$$

Θ_i^{max} is the maximum ratio of chlorophyll to biomass (in units of nitrogen) of phytoplankton group i. ρ_i represents the fraction of growth of phytoplankton i that is devoted to chlorophyll synthesis and is regulated by the ratio of achieved-to-maximum potential photosynthesis $(\mu_i P_i)/(\alpha_i E C_i)$ (Geider et al., 1997).

The zooplankton dynamics are determined by

$$sms(Z) = (g_1 + g_2)\, Z - (L_{ZN} + L_{ZD})\, Z, \tag{8}$$

$$g_i = g_{\max} \frac{p_i P_i}{k_3 + p_1 P_1 + p_2 P_2} \tag{9}$$

where i = 1,2. $g_{\max}$ is the maximum grazing rate, p_i are the grazing preferences (note that $0 < p_1, p_2 < 1$ and $p_1 + p_2 = 1$), k_3 is the half-saturation concentration of grazing, and L_{ZN} and L_{ZD} are the zooplankton excretion and mortality rates.

The sources and sinks of the detrital pool are given by

$$sms(D) = L_{1D}P_1 + L_{2D}P_2 + L_{ZD}Z - rD - w_D \frac{\partial D}{\partial z}, \tag{10}$$

where r is the remineralization rate of detrital matter and w_D is the sinking rate.

The nutrient equation follows as

$$sms(N) = -\mu_1 P_1 - \mu_2 P_2 + L_{1N}P_1 + L_{2N}P_2 + L_{ZN}Z + rD. \tag{11}$$

The physical model

The biological model is coupled to a one-dimensional physical model such that the evolution of any biological scalar X is given by

$$\frac{\partial X}{\partial t} = \frac{\partial}{\partial z}\left(k_z \frac{\partial X}{\partial z}\right) + \mathrm{sms}(X), \tag{12}$$

where t is time and z is water depth. The first term on the right-hand side is the turbulent flux of X, and the second term represents the biological sources minus sinks of X, including the sinking of particles defined in the previous section. The physical model component is an implementation of the turbulent mixing scheme developed by Large et al. (1994) to simulate the planetary boundary layer in oceanic applications. Given surface fluxes of wind stress, heat and freshwater, the model predicts the evolution of the surface boundary layer depth, the vertical eddy diffusivity profile $k_z(z)$, and the vertical profiles of temperature T and salinity S. The turbulent mixing of T, S and biological scalars in the surface boundary layer is controlled by finite eddy diffusivities which decrease below the boundary layer. The boundary layer depth represents the penetration depth of surface-generated turbulence and does not resemble a surface mixed layer a priori.

Model set up and forcing

The physical model is set up on an equidistant grid with 150 vertical levels of 4 m thickness. The model is forced with hourly values of the surface wind stress, the net shortwave radiation, the net longwave radiation, and sensible and latent heat fluxes. No evaporation or precipitation is included. Wind stress and heat fluxes were either measured directly or estimated from continuous measurements at one of the two meteorological stations located on a tower at the southwest caldera rim and on a buoy in the North Basin of the lake. The u and v components of the wind stress were determined from hourly averages of wind speed and direction. The net shortwave radiation was estimated from measurements of the downwelling shortwave radiation at the rim station and a parameterization of albedo following Payne (1972). The downwelling radiation at the rim was corrected to account for the shading effect of the caldera walls on the lake surface by a simple geometric argument. The net longwave heat flux is given by the downwelling longwave heat flux minus the longwave back radiation which was determined assuming the lake radiates like a grey body. The downwelling longwave heat flux was determined by a bulk parameterization depending on air temperature, water vapor pressure (determined from measurements of air temperature and relative humidity) and a cloudiness parameterization. The sensible heat flux was determined from the difference between the lake surface and air temperatures based on the bulk formula of Large & Pond (1982). The latent heat flux was determined from the evaporation rate following Large & Pond (1982).

Note that our model does not include the atmospheric precipitation of nitrogen and the nitrogen loss due to burial in the sediment and seepage. However, as the input from precipitation and the loss by burial and seepage balance (Dymond et al., 1996), this does not affect the overall nitrogen inventory of the lake. The initial conditions for the biological variables were obtained as follows: N was defined by interpolating the mean profile all available nitrate and ammonium data onto the model grid, while all other biological variables were set to 0.01 mmol N m^{-3}. The model was then started on January 1 and run for 3 year to allow an adjustment to the initial conditions. All the results discussed here are from the third year of the simulation.

Limnological data

Water samples were collected from the RV Neuston at station 13 (42°56′ N 122°06′ W, located at the deepest part of Crater Lake) between June 1988 and September 2001. Nitrate and ammonia were measured by automated cadmium reduction and phenate calorimetric methods, respectively, and a Technicon autoanalyzer (Larson et al., 1996a). Chlorophyll concentrations were determined by fluorometry after filtration of samples onto 0.45-μm pore-size filters and extraction with 90% acetone. Phytoplankton were preserved with Lugol's solution and concentrated by gravity settling (96 h). Cells > 1 μm were identified and counted by means of inverted microscopy. Biovolume conversion factors were determined for each taxon by geometric approximation (McIntire et al., 1996). We converted biovolume to biomass (in units of carbon) by applying the following formula: cell C = 0.142 $volume^{0.996}$ [pg C μm^{-3}] for each taxon. The conversion formula has been obtained by Rocha & Duncan (1985) by fitting 47 determinations of cell carbon and cell volume for 25 freshwater algal species.

Zooplankton samples were obtained by vertical towing of a 64-μm mesh-size tow net. Zooplankton were diluted with pre-filtered lake water, preserved with 4% formaldehyde/4% sucrose, concentrated by gravity settling (24 h), and counted by means of inverted microscopy (Larson et al., 1996b). Zooplankton weight conversion factors were estimated for each taxon from dried animals and used to convert organism densities to dry weight biomass (Larson et al., 1993). Here we converted from dry weight to biomass in units of carbon and nitrogen assuming that carbon constituted 48\% and nitrogen 10% of dry weight respectively (Andersen & Hessen 1991; Brett et al., 2000).

Results

We present model results of a simulation for 1995 in comparison with observations. The simulated annual evolution of vertical distributions of the phytoplankton groups, total chlorophyll and zooplankton are shown in Fig. 4. Quantitative comparisons of the simulated concentrations with mean observations are given in Figs. 5–7.

The simulated total phytoplankton biomass is low from January through March. The biomass of phytoplankton group 2 (comprised of chlorophyta, chrysophyta, cryptomonads and cyanobacteria) starts to increase in April with increasing solar radiation before the upper water column stratifies thermally. Phytoplankton group 1 (diatoms and dinoflagellates) does not increase until thermal stratification is established in late May/early June (Fig. 4). During the stratified period in summer and early fall a marked vertical structure persists. The phytoplankton groups 1 and 2 express pronounced vertical maxima in the upper 40 m, and between 50 and 80 m, respectively (Figs. 4, 5). The low biomass in January and the vertical distribution of P_1 and P_2 compare well with the observed biomass of the "surface" and "deep" groups (Fig. 5). In particular, the profiles of phytoplankton group 2 agree remarkably well with the observed mean profiles of the "deep group" from June through September. The simulated vertical structure of phytoplankton group 1 agrees qualitatively with mean profiles of the "shallow group," but the absolute surface concentrations are overestimated by our model in June and July.

The simulated chlorophyll profiles have a vertical maximum at about 120 m from June through August, about 50 m below the vertical maximum of phytoplankton group 2 (Figs. 4–6). The vertical

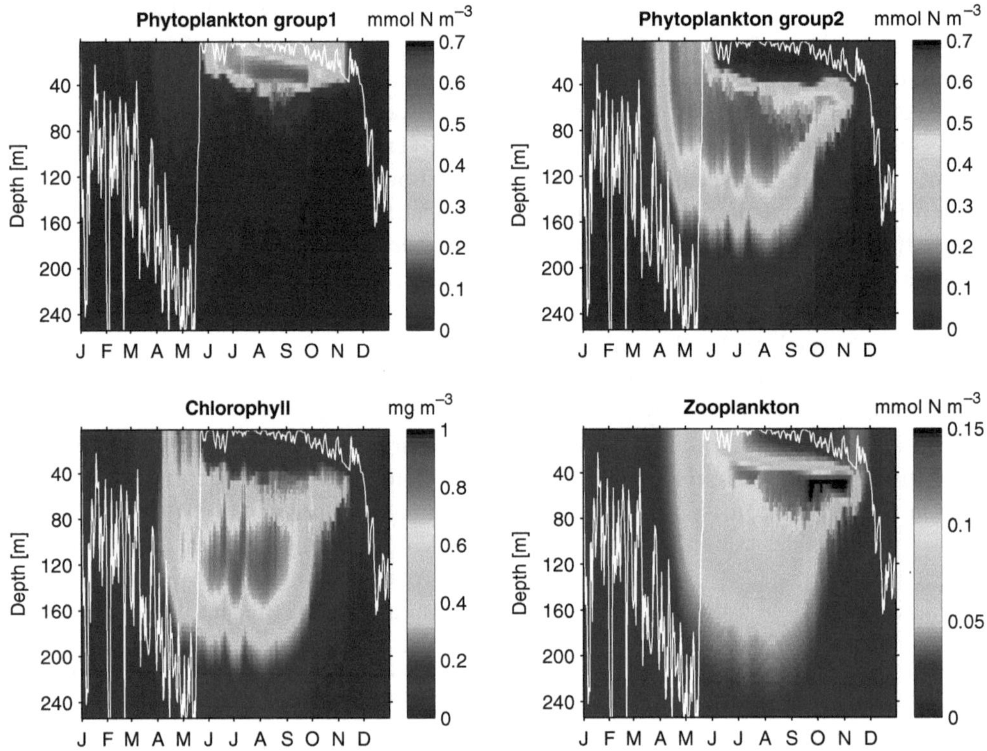

Fig. 4 Simulated evolution of phytoplankton groups 1 and 2, total chlorophyll and zooplankton concentrations in the upper 250 m of the water column. The white line represents the depth of the simulated surface boundary layer

distribution of the simulated chlorophyll concentrations, which are low at the surface and increase monotonically to the deep maximum, agrees well with mean profiles of extracted chlorophyll (Fig. 6), but the absolute chlorophyll values are overestimated by the model.

The zooplankton concentrations are highest between 40 and 80 m and increase over the course of the growing season (Fig. 4). The magnitude of zooplankton biomass compares well with observed values, but a discrepancy in the vertical structure is apparent (Fig. 7). Some zooplankton species in Crater Lake exhibit diel vertical migrations with displacements between 20 and 40 m (Larson et al. 1993, 1996b). This behavior is not easily accounted for in the model but may explain the apparent discrepancy in vertical structure between simulated and observed zooplankton.

In summary, the model captures essential features of the system. Quantitative discrepancies exist and may stem from unresolved and/or poorly quantified processes.

Discussion

We now discuss the seasonal evolution of vertical mixing and thermal stratification and its implications for the distribution of nutrients and plankton (section Physical controls of nutrient and plankton dynamics); suggest biochemical and bio-optical factors that contribute to the vertical structure of phytoplankton during the stable period in summer and early fall (section Biological factors for the vertical distribution of biomass and chlorophyll); and discuss a nitrogen budget for the watercolumn of Crater Lake (section Nitrogen budget).

Physical controls of nutrient and plankton dynamics

Vertical mixing exerts important controls on the biological system, i.e., it affects nutrient supply to the euphotic zone, phytoplankton spatial distributions and light levels received by the phytoplankton community. We briefly describe the

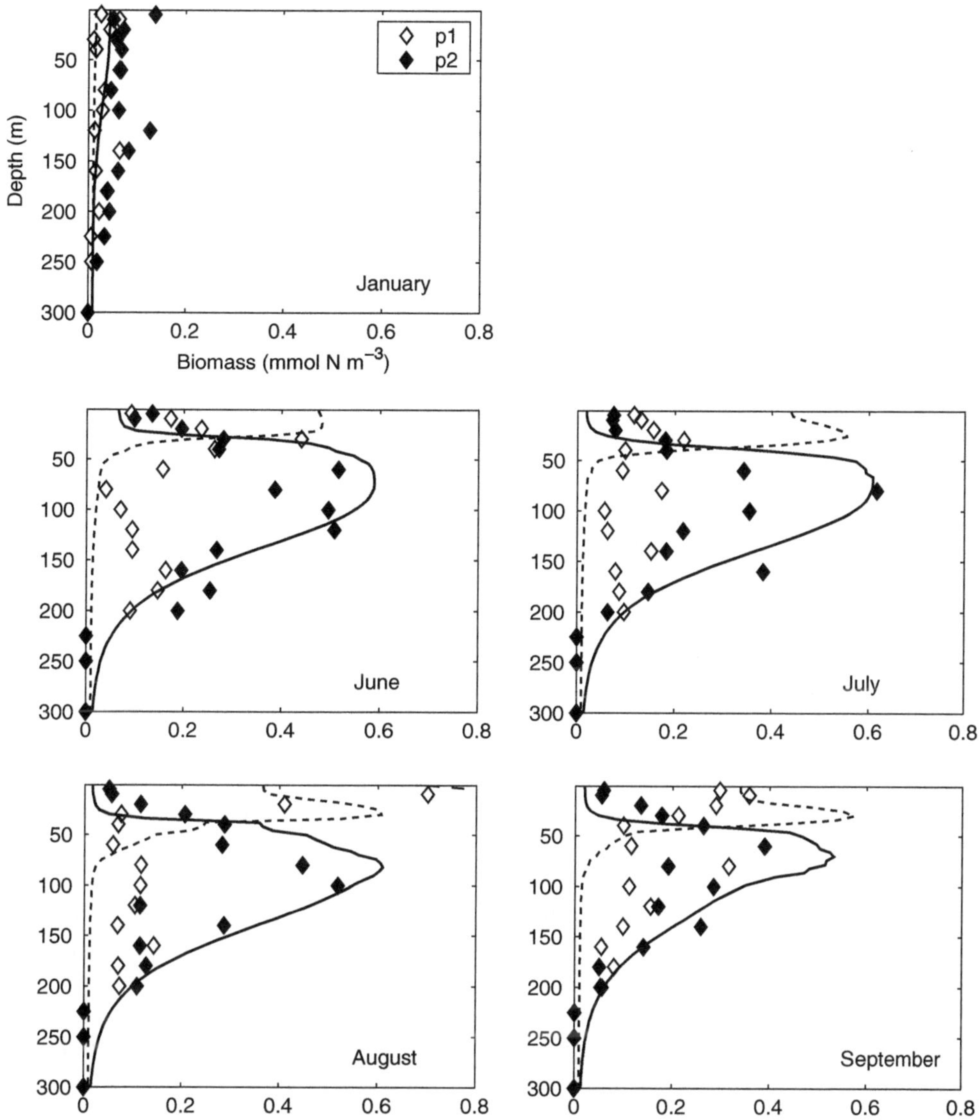

Fig. 5 Simulated phytoplankton concentrations given as monthly means (solid and dashed lines) in comparison with medians of observed phytoplankton biomass (diamonds). P1 (open diamonds and solid lines) includes diatoms and dinoflagellates. P2 (filled diamonds and dashed lines) includes chlorophyta, chrysophyta, cryptomonads and cyanobacteria

annual evolution of vertical mixing and stratification.

In winter and early spring the upper portion of Crater Lake (upper 200–250 m) is relatively well mixed. The water column is isothermal at two distinct events—once in early winter and once in early spring (McManus et al., 1993; Crawford & Collier 1997). The timing of these isothermal events and the intensity of mixing at a given time are determined by the current thermal structure and the surface wind stresses and heat fluxes. In fall decreasing solar radiation and surface cooling erode the summer thermocline and produce cool, dense surface water that is convected. Convective mixing continues until the temperature of maximum density (about 4 °C) is reached. At this point the upper part of the water column is isothermal and a direct exchange of water between the upper and deep portions of the lake is possible. Continued cooling of surface water below the temperature of maximum density leads to inverse thermal stratification. Heating in early spring

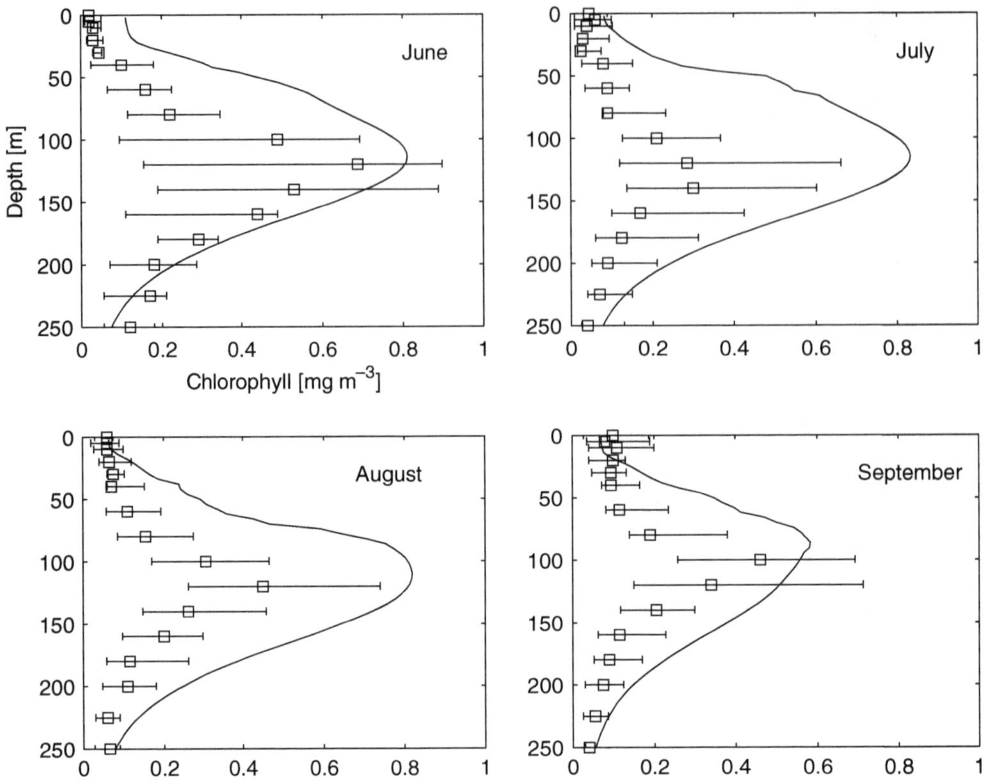

Fig. 6 Simulated chlorophyll concentrations given as monthly means (solid lines) in comparison with observed chlorophyll data plotted as median with 25- and 75-percentiles (squares)

results in convective mixing until the temperature of maximum density is reached again—the second isothermal event. Subsequent heating leads to pronounced thermal stratification in summer. This evolution of convection and inverse thermal stratification is typical for high latitude systems below 20 PSU (e.g., Fennel, 1999; Botte & Kay, 2000).

The history of winter mixing affects processes in the euphotic zone in two important ways:(i) the most significant input of inorganic nutrients from the deep lake occurs during the isothermal periods (Dymond et al., 1996; Crawford & Collier, 1997); and (ii) as microorganisms are mixed vertically the photosynthetically active radiation (PAR) received by the phytoplankton community is strongly determined by mixing depth (Sverdrup, 1953). According to Sverdrup (1953) net phytoplankton growth in spring can only occur when the depth of vertical mixing is shallower than the critical depth (the depth at which the depth-integrated phytoplankton growth equals depth-integrated respiration). In other words, net phytoplankton growth in spring can only occur in Crater Lake after the temperature of maximum density is exceeded and convective mixing ceases—a link that has been demonstrated for the spring bloom in the Baltic Sea (Fennel, 1999). Consistent with this concept phytoplankton biomass increases in our simulation only in April after the temperature of maximum density is exceeded in late March.

In summer a strong seasonal thermocline is established between 20 and 30 m depth and effectively limits vertical exchange. The vertical transport of nutrients is at a minimum and particles are not displaced vertically by advective forcings other than their own buoyancy forcing. The vertical transport of nitrate is restricted to turbulent diffusion. Since nitrate concentrations in the upper 100 m are low without a notable vertical gradient, a significant diffusive input to the euphotic zone can only occur below 100 m. Due to the stable stratification, microorganisms

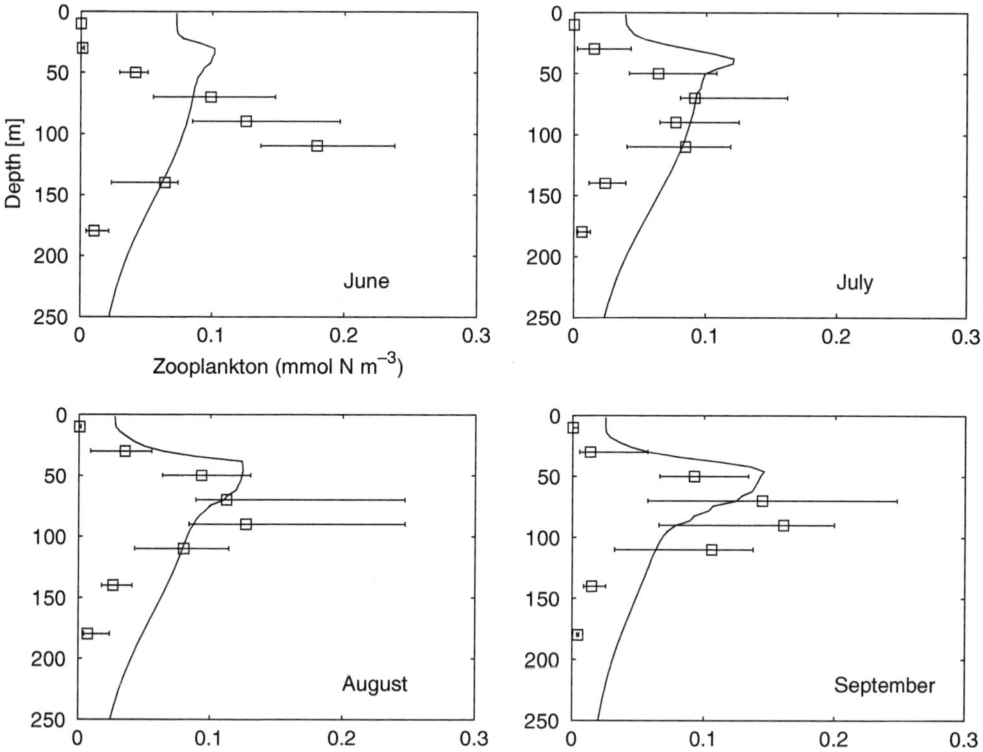

Fig. 7 Stimulated zooplankton concentrations given as monthly means (solid lines) in comparison with observed zooplankton biomass plotted as median with 25- and 75-percentiles (squares)

can maintain their vertical position allowing for the observed and simulated dominance of distinct species groups at different depths (McIntire et al., this issue: Fig. 5).

Biological factors for the vertical distribution of biomass and chlorophyll

Our model captures a separation between the vertical maxima of phytoplankton biomass and chlorophyll (by up to 50 m) and the vertical differentiation between the phytoplankton functional groups 1 and 2 in agreement with observations. General criteria that determine the vertical structure of phytoplankton biomass and chlorophyll in stable, oligotrophic environments were discussed recently in Fennel & Boss (2003). We suggested that the particle maximum and the chlorophyll maximum are generated by fundamentally different processes. Photoacclimation results in a deep chlorophyll maximum vertically separated from the biomass maximum. During stable conditions in summer and early fall, this vertical separation is pronounced in Crater Lake. The inclusion of photoacclimation in our model results in a vertical separation similar to the one observed. The location of the chlorophyll maximum is ultimately determined by light attenuation and the physiological ability of phytoplankton to increase their cellular chlorophyll-to-biomass ratio. The particle maximum on the other hand, is determined by the balance between biomass growth and losses including respiration, grazing, mortality, and divergence in sinking velocity.

Under steady-state conditions the biomass maximum is located at the depth where growth and losses equal (Riley et al., 1949; Fennel & Boss, 2003). This criterion can be derived as follows. Assuming that growth, biological losses (due to respiration, mortality and grazing), sinking and vertical mixing are the main processes determining the distribution of phytoplankton, a general 1D phytoplankton equation can be written as

$$\frac{\partial P}{\partial t} + \frac{\partial (wP)}{\partial z} = (\mu - R)P + \frac{\partial}{\partial z}\left(k_z \frac{\partial P}{\partial z}\right). \quad (13)$$

Here P represents the phytoplankton biomass, w the phytoplankton settling velocity, μ the growth rate, R the rate of biological losses including respiration, mortality and grazing, and k_z the eddy diffusion coefficient. We want to solve Eq. 13 for steady state conditions, in which case the first term can be neglected and the equation simplifies to

$$\frac{d(wP)}{dz} = (\mu - R)P + \frac{d}{dz}\left(k_z \frac{dP}{dz}\right). \quad (14)$$

The condition for the location of a vertical phytoplankton maximum $P(z_{max})$ then follows as

$$\left.\frac{\partial P}{\partial z}\right|_{z_{max}} = 0 \quad (15)$$

$$\Rightarrow \left(\mu - R - \frac{dw}{dz}\right) + k_z \frac{d^2 P(z_{max})}{dz^2} = 0. \quad (16)$$

Since the particle maximum is usually located in a region with low diffusivity we neglect the diffusive flux term in Eq. 16 and obtain

$$\mu - R - \frac{dw}{dz} = 0. \quad (17)$$

In other words, at the particle maximum the community growth rate μ equals the sum of the biological losses R and the divergence of particles due to changes in the settling velocity. This criterion can be applied to the phytoplankton groups 1 and 2 to predict the location of their vertical maxima (Fig. 8).

In our model the growth and loss term profiles of both phytoplankton groups differ mainly due to different parameter choices for the initial slope of the PI-curve α_{chl} and the maximum growth rate μ_{max} (see Table 1), although grazing and respirative losses differ as well. The higher maximum growth rate of group 1 results in an advantage at high light levels near the surface, while the larger initial slope of phytoplankton group 2 is advantageous at lower light levels deeper in the water column. Unfortunately, no data are available on the light-physiological parameters of the algae present in Crater Lake. Data from oceanic algae, however, are consistent with our parameter choices for α_{chl} and μ_{max}. Geider et al. (1997) collected photo-physiological parameters of 15 different marine algae from a wide range of environments (obtained from a total of 19 studies). We calculated the mean values of μ_{max} and α_{chl} for the 8 bacillariophyta and 11 chrysophyta included in this collection. The mean μ_{max} equals 3.6 ± 1.3 d^{-1} and 1.3 ± 0.9 d^{-1} for bacillariophyta and chrysophyta, respectively, and the mean α_{chl} equals 0.05 ± 0.36 mol N (g chl W m^{-2})$^{-1}$ and 0.054 ± 0.34 mol N (g chl W m^{-2})$^{-1}$ for bacillariophyta and chrysophyta, respectively. Consistent with our parameter choices the mean maximum growth rate is significantly higher for the bacillariophyta. There is no significant difference in α_{chl} between both divisions, due to large variations between different taxa. Moore & Chisholm (1999) have shown that different isolates of *Prochlorochoccus* express distinct photo-physiological parameters. Comparing 10 different isolates they found two clusters: a high-light adapted group that had higher maximum light-saturated growth rates (corresponding to μ_{max}) and lower growth efficiencies under sub-saturating light intensities (corresponding to α_{chl}) than the low-light adapted group. These distinct ecotypes have been found to coexist in the same water column.

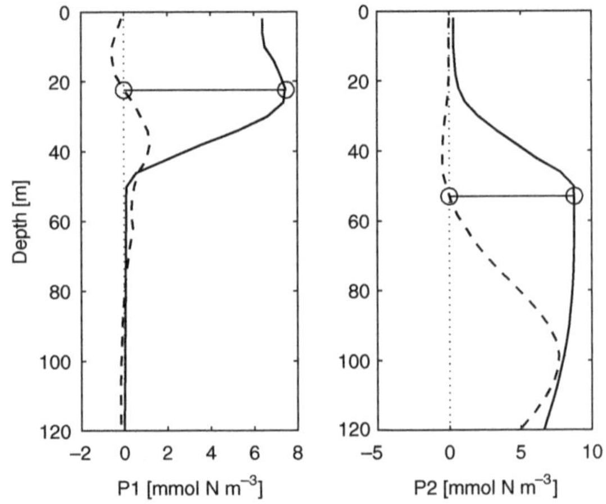

Fig. 8 Mean July biomass of phytoplankton groups 1 and 2 (solid lines) and sum of their growth, respiration, grazing, mortality and diffusion terms on an arbitrary scale (dashed lines). The vertical biomass maxima occur where sources and sinks are equal as indicated by the circles

A potentially strong selective factor in the epilimnion in Crater Lake, not considered in our model is ultra-violet (UV) radiation. UV radiation is a significant stressor in aquatic environments. In Crater Lake, UV levels are particularly high. Incident UV radiation is elevated due to the high altitude and attenuation in the water column is small because concentrations of colored dissolved organic matter are low (Boss et al., this issue). Widely observed deleterious effects of UV on planktonic microorganisms include inhibition of photosynthesis (Lorenzen, 1979; Cullen et al., 1992; Smith et al., 1992) and bacterial heterotrophic potential (Herndl et al., 1993), and damage of DNA (Karentz et al., 1991). Two physiological acclimation mechanisms that decrease algal sensitivity to UV are efficient repair of photodamage, and the synthesis of photoprotective pigments (Litchman et al., 2002). In Crater Lake, the efficiency of both of these acclimation mechanisms is likely compromised by nitrogen limitation which has been observed to significantly depress the potential to repair photodamage and, to a lesser extent, the synthesis of photoprotective pigments (Litchman et al., 2002).

Despite the nitrogen limitation in Crater Lake, microorganisms in the epilimnion contain photoprotective pigments (Lisa Eisner, personal comm.). The usefulness of these pigments in decreasing sensitivity to UV radiation is strongly linked to organism size: for picoplankters (radii < 1 μm) pigment sunscreens are not of any relevance for photoprotection, but for microplankter (cell radii > 10 μm) these sunscreens can be very effective (Garcia-Pichel, 1994). Consequently, cell size represents a selective criterion, favoring larger cells in the epilimnion. This is consistent with optical measurements of the slope of the beam attenuation spectrum which point to the dominance of larger cells near the surface than at the deeper biomass maximum (Boss et al., this issue).

Nitrogen budget

The simulated sinking flux of detritus at 200 m in our model is 92 mg N m^{-2} yr^{-1}. This value agrees well with vertical particulate nitrogen fluxes from a sediment trap at 200 m measured between 1984 and 2002 (Dymond et al., 1996; Collier, R.W., unpublished data). The sediment trap data vary between 40 and 260 mg N m^{-2} yr^{-1} with a mean of 130 ± 67 mg N m^{-2} yr^{-1}. Note that our estimate is conservative as our model does not account for atmospheric inputs of nitrogen.

This agreement between simulated and observed fluxes stands in contrast to box-model estimates of vertical particle flux by Dymond et al. (1996), who found a significant discrepancy between estimated vertical particle fluxes and the sediment trap data. The authors derived an internal nitrogen budget, dividing Crater Lake into an upper and deep box (at 200 m) and formulating the nitrogen mass balance equations for both boxes as

$$F_p = F_a + F_r + F_u - F_d - F_{sp,upper} \tag{18}$$

and

$$F_p = \frac{1}{1-\alpha}\left(F_u - F_d + F_{sp,deep}\right). \tag{19}$$

F_p is the vertical flux of nitrogen due to settling particles. F_a and F_r are nitrogen sources to the upper box due to atmospheric inputs and runoff, respectively. F_u and F_d represent the nitrogen exchange between the upper and deep box due to upward mixing of deep-lake nutrients and downward mixing of surface nitrate pools, respectively. F_{sp} is the seepage loss of nitrogen through the lake floor, split into seepage losses from the upper and deep portions, $F_{sp,upper}$ and $F_{sp,deep}$. The sediment burial term F_b is assumed to relate proportionally to the settling of particles F_p with α representing the fraction of settling particles buried in the sediments ($F_b = \alpha F_p$).

By inserting their best estimates of atmospheric, runoff and seepage inputs and losses, and a reasonable range of values for the burial fraction of settling particles into the right-hand-sides of Eqs. (18) and (19), Dymond et al. obtained model estimates for the vertical particle flux that were 2.5–10 times larger than the measured sediment trap fluxes. In their discussion the authors state that this discrepancy could be due to errors in the direct measurements, errors in their box-model assumptions, or unaccounted processes. They suggested a combination of sediment focusing and significantly higher productivity at the edges of Crater Lake to explain the mismatch. Both of these processes are hard to quantify with the available data.

An analysis of our model results suggests a different explanation. We calculated the simulated exchange of nitrogen between the upper and deep box due to winter mixing as 0.57×10^6 mol yr^{-1}. This value is significantly smaller than the flux assumed by Dymond et al. (2.0–4.0×10^6 mol yr^{-1}). The authors obtained this exchange flux of nitrogen between the upper and deep boxes assuming (i) that winter mixing replaces 2–4×10^{12} liters of water between both boxes and (ii) a nitrate inventory of 0 and 1 μM for the upper and deep boxes, respectively. They neglected the possibility of dissolved and suspended organic matter being exchanged between the upper and deep portions in addition to the exchange of nitrate. A flux of organic dissolved and suspended nitrogen could counterbalance an exchange of inorganic nitrogen. Interestingly, the simulated exchange of DIN in our model equals 1.13×10^6 mol yr^{-1}, a value closer to Dymond et al.'s estimate, while the simulated exchange of organic matter is -0.56×10^6 mol yr^{-1}, reducing the net exchange to 0.57×10^6 mol yr^{-1}.

In order to revisit the box-model calculations we estimate a "corrected" exchange flux (F_u–F_d) based on the following arguments. Dymond et al. assumed a ΔN (defined as the difference between nitrogen in the upper and deep portions of the lake) of 1 μM neglecting dissolved and suspended nitrogen. We suggest a reduction of ΔN to 0.2 μM. We obtain this value assuming that the nitrate concentrations in the upper box equals 0 in agreement with Dymond et al., but reduce the assumed nitrate concentration in the deep box from 1 to 0.6 μM. Our value is more representative of nitrate concentrations between 200 and 400 m depth—a water mass more likely to be mixed up above 200 m than water from below 400 m depth (Fig. 9). For estimates of dissolved and organic nitrogen we compare total Kjeldahl nitrogen (TKN) values between the upper and deep portions of the lake. TKN represents nitrogen in the ammonium, dissolved and particulate organic matter pools. TKN in the upper box is typically 1.4 μM and TKN in the deep box is about 1 μM (Fig. 9). With these nitrogen concentrations in the upper and deep boxes we obtain a ΔN of 0.2 μM and an exchange flux (F_u–F_d) of 0.4–0.8×10^6 mol yr^{-1}. This flux estimate compares well with our simulated flux of 0.57×10^6 mol yr^{-1}. Iterating Dymond et al.'s box-model calculation with our estimate of F_u–F_d, we obtain vertical particle fluxes of 0.69–1.38×10^6 mol yr^{-1} from Eq. (18) and 0.5–1.1×10^6 mol yr^{-1} from Eq. (19). These revised box-model estimates are consistent with the sediment trap fluxes of $0.49 \pm 0.25 \times 10^6$ mol N yr^{-1}, our model estimates, and the deep lake's oxygen budget (McManus et al., 1996).

Conclusions

We developed a simple physical-biological model of Crater Lake that captures the observed vertical structure of two distinct phytoplankton groups and chlorophyll, and predicts the biomass of phytoplankton and zooplankton in good quantitative agreement with observations from a 10-year monitoring study. Our model suggests

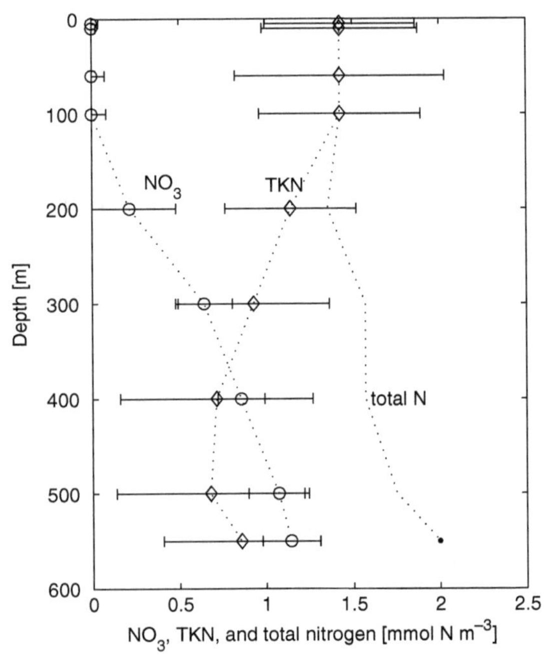

Fig. 9 Mean profiles of nitrate and Kjeldahl nitrogen (represents the sum of ammonia, and particulate and dissolved nitrogen) with standard deviation, and total nitrogen

that phytoplankton production starts in spring only after the temperature of maximum density (about 4°C) is exceeded. That is, the onset of phytoplankton production is dependent on the evolution of vertical mixing in winter. Stable stratification in summer allows a pronounced vertical structure of phytoplankton distributions to emerge. The vertical separation of the deep chlorophyll maximum (at about 120 m) and the biomass maximum (between 50 and 80 m) results from photoacclimation. In our model the biomass maximum of both functional phytoplankton groups during stable conditions in summer is located where their respective growth and loss rates are equal. The differences in the vertical position of the biomass maximum of phytoplankton groups 1 and 2 at 20 m and 50–80 m, respectively, result mainly from parametric differences in the initial slope of the PI-curve and the maximum growth rate. While no experimental data on the growth and light-response parameters are available for the specific species present in Crater Lake, parameters determined for oceanic species are consistent with our choices. Sensitivity to UV radiation is likely to be an important selective factor as well, but is not included in our model since quantitative information on its effect is lacking. The average annual vertical flux of particles in our model compares well with the average annual flux measured in sediment traps at a depth of 200 m. We suggest correcting previous model estimates of vertical exchange by Dymond et al. (1996), who found a mismatch between model-estimated and observed vertical fluxes. Our model simulation pointed to a significant downward flux of suspended and dissolved organic matter which had been neglected by Dymond et al. Inclusion of this flux reconciles the box-model estimates and the trap measurements.

Acknowledgements We wish to thank Leon Tovey for reviewing the manuscript and three anonymous reviewers for helpful comments. Funding for KF was provided by USGS.

References

Andersen, T. & D. O. Hessen, 1991. Carbon, nitrogen and phosphorus content of freshwater zooplankton. Limnology and Oceanography 36: 807–814.

Andersen, V., P. Nival & R. Harris, 1987. Modelling of a planktonic ecosystem in an enclosed water column. Journal of the Marine Biological Association of the U.K. 67: 407–430.

Boss, E., R. Collier, G. Larson, K. Fennel & W. S. Pegau, this issue. Measurements of spectral optical properties and their relation to biogeochemical variables and processes in Crater Lake.

Botte, V. & A. Kay, 2000. A numerical study of plankton population dynamics in a deep lake during the passage of the spring thermal bar. Journal of Marine Systems 26: 367–386.

Brett, M. T., D. C. Mueller-Navarra & S.-K. Park, 2000. Empirical analysis of the effect of phosphorus limitation on algal food quality for freshwater zooplankton. Limnology and Oceanography 45: 1564–1575.

Collier, R. W., J. Dymond & J. McManus, 1991. Studies of hydrothermal processes in Crater Lake, Oregon. College of Oceanography Report, 90. Oregon State University.

Crawford, G. B. & R. W. Collier, 1997. Observations of a deep-mixing event in Crater Lake, Oregon. Limnology and Oceanography 42: 299–306.

Cullen, J. J., P. J. Neale & M. P. Lesser, 1992. Biological weighting functions for the inhibition of phytoplankton photosynthesis by ultraviolet radiation. Science 258: 646–650.

Dymond, J., R. Collier & J. McManus, 1996. Unbalanced particle flux budgets in Crater Lake, Oregon: Implications for edge effects and sediment focusing in lakes. Limnology and Oceanography 41: 732–743.

Fasham M. J. R., 1993. Modelling marine biota. In Heimann M. (ed.), The Global Carbon Cycle. Springer Verlag, New York, 457–504.

Fasham M. J. R., 1995. Variations in the seasonal cycle of biological production in subarctic oceans: A model sensitivity analysis. Deep-Sea Research I 42: 1111–1149.

Fennel K., 1999. Convection and the timing of the phytoplankton spring bloom in the Western Baltic Sea. Estuarine, Coastal and Shelf Sciences 49: 113–128.

Fennel K. & E. Boss, 2003. Subsurface maxima of phytoplankton and chlorophyll: Steady state solutions from a simple model. Limnology and Oceanography 48: 1521–1534.

Garcia-Pichel F., 1994. A model for internal self-shading in planktonic organisms and its implications for the usefulness of ultraviolet sunscreens. Limnology and Oceanography 39: 1704–1717.

Geider R. J., H. L. McIntyre & T. M. Kana, 1996. A dynamic model of photoadaptation in phytoplankton. Limnology and Oceanography 41: 1–15.

Geider R. J., H. L. McIntyre & T. M. Kana, 1997. Dynamic model of phytoplankton growth and acclimation: Responses of the balanced growth rate and the chlorophyll a:carbon ratio to light, nutrient-limitation and temperature. Marine Ecology Progress Series 148: 187–200.

Geider R. J., H. L. McIntyre & T. M. Kana, 1998. A dynamic regulatory model of phytoplanktonic acclimation to light, nutrients and temperature. Limnology and Oceanography 43: 679–694.

Herndl G. J., G. Müller-Niklas & J. Frick, 1993. Major role of ultraviolet-b in controlling bacterioplankton in the surface layer of the ocean. Nature 361: 717–719.

Jones R. & E. W. Henderson, 1986. The dynamics of nutrient regeneration and simulation studies of the nutrient cycle. Journal du Conceil 43: 216–236.

Karentz D., J. E. Cleaver & D. L. Mitchell, 1991. Cell survival characteristics and molecular responses of Antarctic phytoplankton to ultraviolat-B radiation. Journal of Phycology 27: 326–341.

Lane J. L. & C. R. Goldman, 1984. Size-fractionation of natural phytoplankton communities in nutrient bioassay studies. Hydrobiologia 118: 219–223.

Large W. G. & S. Pond, 1982. Sensible and latent heat flux measurements over the ocean. Journal of Physical Oceanography 12: 464–482.

Large W. G., J. C. McWilliams & S. C. Doney, 1994. Oceanic vertical mixing: A review and a model with non-local boundary layer parameterization. Reviews of Geophysics 32: 363–403.

Larson, G. L., this issue. Overview over the Crater Lake program.

Larson, G. L., C. D. McIntire, R. E. Truitt, M. W. Buktenica & K. E. Thomas, 1993. Zooplankton assemblages in Crater Lake. US Department of the Interior. Report, NPS/PNROSU/NRTR-93/03.

Larson, G. L., C. D. McIntire, M. Hurley & M. W. Buktenica, 1996a. Temperature, water chemistry, and optical properties of Crater Lake. Lake and Reservoir Management 12: 230–247.

Larson, G. L., C. D. McIntire, R. E. Truitt, M. W. Buktenica & E. Karnaugh-Thomas, 1996b. Zooplankton assemblages in Crater Lake, Oregon, USA. Lake and Reservoir Management 12: 281–297.

Litchman, E., P. J. Neale & A. T. Banaszak, 2002. Increased sensitivity to ultraviolet radiation in nitrogen-limited dinoflagellates: Photoprotection and repair. Limnology and Oceanography 47: 86–94.

Lorenzen, C. J., 1979. Ultraviolet radiation and phytoplankton photosynthesis. Limnology and Oceanography 24: 1117–1124.

McIntire, C. D., G. L. Larson, R. E. Truitt & M. K. Debacon, 1996. Taxonomic structure and productivity of phytoplankton assemblages in Crater Lake, Oregon. Lake and Reservoir Management 12: 259–280.

McIntire, C. D., G. L. Larson & R. E. Truitt, this issue. Taxonomic composition and production dynamics of phytoplankton assemblages in Crater Lake, Oregon.

McManus, J., R. W. Collier & J. Dymond, 1993. Mixing processes in Crater Lake, Oregon. Journal of Geophysical Research 98C: 18295–18307.

McManus, J., R. Collier, J. Dymond, C. G. Wheat & G. Larson, 1996. Spatial and temporal distribution of dissolved oxygen in Crater Lake, Oregon. Limnology and Oceanography 41: 722–731.

Moloney, C. L. & J. G. Field, 1991. The size-based dynamics of plankton food webs. I. A simulation of carbon and nitrogen flows. Journal of Plankton Research 13: 1003–1038.

Moore, L. R. & S. W. Chisholm, 1999. Photophysiology of the marine cyanobacterium prochlorochoccus: Ecotypic differences among cultured isolates. Limnology and Oceanography 44: 628–638.

Moskilde, E., 1996. Topics in Non-linear Dynamics: Application to Physics, Biology and Economic Systems. World Scientific Publishing Co., London, U.K.

Palmer, J. R. & I. J. Totterdell, 2001. Production and export in a global ocean ecosystem model. Deep-Sea Research I 48: 1169–1198.

Payne, R. E., 1972. Albedo at the sea surface. Journal of Atmospheric Science 29: 959–970.

Riley, G. A., H. Stommel & D. F. Bumpus, 1949. Quantitative ecology of the plankton of the western North Atlantic. Bulletin of the Bingham Oceanographic Collection 12: 1–169.

Rocha, O. & A. Duncan, 1985. The relationship between cell carbon and cell volume in freshwater algal species used in zooplankton studies. Journal of Plankton Research 7: 279–294.

Ross, A. H., W. S. C. Gurney & M. R. Heath, 1994. A comparative study of the ecosystem in four fjords. Limnology and Oceanography 39: 318–343.

Simpson, H. J., 1970. Tritium in Crater Lake, Oregon. Journal of Geophysical Research 75: 5195–5207.

Smith, R. C., B. B. Prezelin, K. S. Baker, R. R. Bidigare, N. P. Boucher, T. Coley, D. Karentz, S. MacIntyre, H. A. Matlick, D. Menzies, M. Ondrusek, Z. Wan & K. J. Waters, 1992. Ozone depletion: Ultraviolet radiation and phytoplankton biology in Antarctic waters. Science 255: 952–959.

Sverdrup, H. U., 1953. On the conditions for the vernal blooming of phytoplankton. Journal du Conseil Permanent International pour l'exploration de la Mer 18: 287–295.

Taylor, A. H., 1988. Characteristic properties of models for the vertical distribution of phytoplankton under stratification. Ecological Modelling 40: 175–199.

Taylor, A. H., A. J. Watson, M. Ainsworth, J. E. Robertson & D. R. Turner 1991. A modelling investigation of the role of phytoplankton in the balance of carbon at the surface of the North Atlantic. Global Biogeochemical Cycles 5: 151–171.

Urbach, E., K. L. Vergin, L. Young, A. Morse, G. L. Larson & S. J. Giovannoni, 2001. Unusual bacterioplankton community structure in ultra-oligotrophic Crater Lake. Limnology and Oceanography 46: 557–572.

Wroblewski, J. S., 1989. A model of the spring bloom in the North Atlantic and its impact on ocean optics. Limnology and Oceanography 34: 1563–1571.

MIX
Papier aus verantwortungsvollen Quellen
Paper from responsible sources
FSC® C105338
FSC
www.fsc.org

If you have any concerns about our products,
you can contact us on
ProductSafety@springernature.com

In case Publisher is established outside the EU,
the EU authorized representative is:
Springer Nature Customer Service Center GmbH
Europaplatz 3, 69115 Heidelberg, Germany

Printed by Libri Plureos GmbH
in Hamburg, Germany